The Transfer and Licensing of Know-How and Intellectual Property

Understanding the Multinational Enterprise in the Modern World

The Transfer and Licensing of Know-How and Intellectual Property

Understanding the Multinational Enterprise in the Modern World

David J. Teece
University of California, Berkeley

World Scientific

NEW JERSEY · LONDON · SINGAPORE · BEIJING · SHANGHAI · HONG KONG · TAIPEI · CHENNAI

Published by

World Scientific Publishing Co. Pte. Ltd.
5 Toh Tuck Link, Singapore 596224
USA office: 27 Warren Street, Suite 401-402, Hackensack, NJ 07601
UK office: 57 Shelton Street, Covent Garden, London WC2H 9HE

Library of Congress Cataloging-in-Publication Data
Teece, David J.
The transfer and licensing of know-how and intellectual property : understanding the multinational enterprise in the modern world / by David J. Teece.
p. cm.
Includes bibliographical references.
ISBN-13 978-981-256-849-6
ISBN-10 981-256-849-2
1. License agreements--United States. 2. Technology transfer--Law and legislation--United States. 3. Technological innovations--Law and legislation--United States. 4. License agreements. 5. Technology transfer--Law and legislation. 6. Technological innovations--Law and legislation. 7. Technology transfer.
KF3145.T44 2007
346.7304'8--dc 22

2006049156

British Library Cataloguing-in-Publication Data
A catalogue record for this book is available from the British Library.

Printed in Singapore by World Scientific Printers

Contents

Introduction

This volume contains a selection of previously published articles on technological innovation, technology transfer, and the multinational enterprise. Different business models to effectuate technology commercialization are analyzed, along with licensing and the technology transfer process itself. Both management and public policy issues are addressed.

The papers in Part I focus on international technology transfer, which was the focus of my dissertation research at the University of Pennsylvania. It was at the University of Pennsylvania in the 1970s that I developed my interests in the transfer and licensing of know-how.

The first chapter in the volume, originally published in *The Economic Journal*, contains the results of my dissertation research on the resource cost of international technology transfer. This study provided the first published scientific evidence of the real costs of technology transfer. It also provided some insights into the factors that determine those costs. The study found that, in general, technology transfer costs decrease with each subsequent transfer of the same or similar technology, particularly in the chemical industry. The age of the technology was most important in the cost of transferring technology machinery. This paper also evidences that, even back in the mid-1970s, I had begun to recognize and analyze the role of "capabilities" and the effort, time and cost of conveying these capabilities across international borders.

Chapter 2, from *Management Science*, was also research done as part of my doctoral dissertation. This paper investigates the tradeoff between reduced time and higher cost that a multinational corporation faces when transferring technology internationally. I estimated that the elasticity of cost with respect to time was greater than one, and was significantly greater for technology that had not been previously commercialized, for projects that were large, and for projects carried out by larger firms.

Chapter 3 contains my first attempt to elucidate the distinctive issues associated with exchanging knowledge (as compared to tangible assets), and to analyze how these characteristics of knowledge affect the choice of organizational arrangements for international technology transfer. This chapter highlights the advantages that internal transfer within a multinational corporation has over arms-length transfer (between independent firms) in overcoming these problems. It also explains some of the practices used by multinational firms to control the transfer of technology, and advocates against restrictions on the flow of technology.

Chapter 4 in this volume, written for the *Journal of Technology Transfer*, extends this analysis, emphasizing how the nature of knowledge compounds the difficulties of technology transfer. It also explains how a multinational enterprise (MNE) overcomes some of these difficulties. This is the first chapter in this volume to mention and challenge the Hymer hypothesis — that the MNE is a tool for the extraction of monopoly rents and, by implication, the amplification of poverty. The chapter also relates my work on international technology transfer to Edwin Mansfield's earlier work on technological innovation more generally, and highlights the significant influence that Mansfield had on my understanding of research and development and innovation.

The series of articles in Part II discuss licensing issues.

"Profiting from Technological Innovation" (Chapter 5 in this volume) has become a highly cited article in technology management, and was selected by the editors of *Research Policy* as one of the ten best papers that the journal had published for the period 1971–1991. It is perhaps the most cited article ever published in *Research Policy*. This article analyzes the optimal strategy of a firm commercializing an invention, and seeks to explain why innovating firms often fail to capture the economic returns from their inventions. Drawing on my understanding of the nature of knowledge and functioning of the markets for know-how, I introduce a new framework based on three basic concepts — the appropriability regime, complementary assets, and the dominant design paradigm — and a set of alternative organizational modes — or business models — available to entrepreneurs to commercialize their inventions. Applying this framework, this chapter argues that in order to capture value from its innovation when the appropriability regime is weak (i.e., when imitation is easy), an innovating firm must often integrate into the complementary assets as soon as a dominant design is becoming established. If imitation is easy, however, there is no guarantee of success even when the innovating firm follows this strategy.

Chapter 6, written with Peter C. Grindley and first published in the *California Management Review*, investigates intellectual property (IP) management and licensing strategies in the semiconductor and electronics industries. The article discusses how royalty-free cross-licensing evolved as a response to regulatory and judicial constraints; but, as the regulatory environment changed, global competition sharpened, and the value of intellectual capital increased, thus IP owners began to enforce their intellectual property rights more aggressively. This chapter illustrates how preferred licensing strategies change with the importance of intellectual capital, and provides lessons for managing intellectual capital.

Chapter 7, written with Deepak Somaya, analyzes the strategies for commercializing innovation in "multi-invention" contexts — that is, where innovation requires the combination of a very large number of inventions. This article builds on the framework developed in Chapter 5 by adding a set of "patent strategies" that support the entrepreneur's choice of organizational mode. It revisits the analysis of IP management in semiconductors and electronics. In Chapter 7, we examine the patent strategies most appropriate to support the integrated commercialization modes chosen by AT&T, IBM, and Texas Instruments. We also examine the patent strategies used to support non-integrated licensing or "component" modes chosen by ARM and Kentron Technologies.

Chapter 8, co-authored with Edward F. Sherry and published in *Research Policy*, discusses methods for calculating "reasonable royalty" rates when a patent is found to be valid and infringed. The article draws the distinction between the value of innovation and the value of intellectual property, and explains how the value of intellectual property changes over the different stages of the life of a patent. The value of patents proven valid and infringed through litigation should have higher royalty rates attached to them as compared to royalties negotiated in licensing agreements for the equivalent but untested patents. The article argues that the standards used to calculate "reasonable royalty" rates in litigation, and thereby to set intellectual property damage awards, have a huge impact on the incentives not only to litigate patent disputes to trial, but also to invest in innovation in the first place.

Chapter 9, also written with Edward F. Sherry and first published in the *Minnesota Law Review*, looks at patent licensing issues that arise in the context of setting design and compatibility standards for technological products. There has been debate over whether, when the patented technology of a participant in a standard-setting organization (SSO) overlaps with a voluntary standard, that firm has an obligation to license its patents to other participants, and on what terms. We carefully examine the standard-setting process and the roles played by the different participants in these SSOs, and argue that there are very limited situations in which a patent holder can exercise "hold up" by "manipulating" standards. We discuss the problems with mandatory search, disclosure and compulsory licensing rules that SSOs contemplate to try to prevent hold up. We conclude that the most important requirement is that the rules of SSOs be clearly stated.

Chapter 10, written with Michaelyn Corbett and Mohan Rao, outlines licensing issues in trademarks. We discuss the economic principles behind the granting of a trademark, and relate these to the different approaches for valuing them — namely the cost approach, the market approach, and the income approach.

Part III focuses on the organization of research and development.

Chapter 11 attempts to move the economic debate on technological innovation beyond its preoccupation with market structure (or market power) to explore how different organizational arrangements inside and amongst firms affect the rate and direction of innovation. I develop a framework based on considering the depth of hierarchy, the scope of product market activities, the degree of vertical integration, the strength of organizational culture, and the number of external linkages. I use this framework to analyze how different organizational types — or governance modes — suit different forms of innovation.

In the *Economica* article reproduced in Chapter 12, Edwin Mansfield, Anthony Romeo and I examine the amount and reasons for overseas R&D investment by US-based firms three decades ago. We find a significant relationship between the percentage of R&D carried out overseas and the percentage of sales derived overseas. However, the relationship between sales and R&D is stronger when the sales are made by foreign subsidiaries than through direct exports. Meanwhile, we found that those R&D activities carried out overseas had large minimum efficient scale and the firm had access to inputs at substantially lower cost than in the US. In addition, those activities were focused more on development than research, on product and process improvements rather than developing new products and processes, and on relatively short-term, technically "safe" work. The data is not available to know just how the situation differs today.

The article reproduced as Chapter 13 assesses the importance of inter-firm and inter-organizational linkages for the development of and commercialization of new technology. These include linkages between firms in an industry and universities, the users of its products, the suppliers of its inputs, and other firms engaged in similar activities. The chapter compares the merits of market-based cooperation (or contractual arrangements) and vertical integration for establishing and managing these linkages.

Chapter 14, written with my Berkeley colleague David C. Mowery, examines the decrease in in-house R&D by US firms, and the corresponding increase in firm participation in alliances and research consortia. The chapter considers three broad forms of R&D collaboration that have developed over the past 20 years or so — international strategic alliances, pre-commercial research consortia, and university–industry research collaborations. The chapter examines the comparative advantages of each of these arrangements for R&D, as well as the limitations of each as a substitute for in-house R&D.

In Chapter 15, in an article co-authored with Henry W. Chesbrough for the *Harvard Business Review*, I counter the sometimes popular arguments for organizing innovation through "virtual" structures. We argue that virtual corporations, which do not have control over all the important aspects of technology development, are likely to lose out in races to develop systemic innovations. We show several examples of firms that have failed to capture the economic returns to an innovation when they have contracted out the important aspects of technology development. Meanwhile, we illustrate how Motorola in wireless communication, and the successful firms in chemicals, steel and railroads during the late 19th and early 20th century, were companies that made major internal investments to shape markets, rather than relying on others to do the R&D for them.

Chapter 16, written with John M. de Figueiredo, examines the hazards to which downstream, non-integrated firms in leading-edge industries are exposed when they procure important inputs from vertically-integrated competitors. The paper outlines the main strategic hazards in such procurement, and examines the conditions under which a constellation of transaction-specific and relational safeguards — rather than (upstream) integration by the downstream firm or (downstream) divestiture by its supplier — can be used to protect the downstream firm against these hazards.

Part IV contains a collection of papers focused on the multinational enterprise.

Chapter 17, which originally appeared in a 1981 book titled *Technology Transfer and Economic Development*, examines the role of the MNE in stimulating technology transfer to and R&D activities in developing countries. Drawing on the "Markets & Hierarchies" (or "transaction-cost economics") approach, I demonstrate how the MNE can attenuate some of the problems of transferring technology through an international market transaction between distinct firms. This chapter discusses the types of technology transfer and R&D activities that are conducted by MNEs, and examines the role of the MNE in affecting technological choice in the host country.

Chapter 18 discusses whether the MNE has efficiency-enhancing or market power (i.e., efficiency-reducing) effects. Although the theory identifies certain instances in which the organizational structure of the MNE may be used to support anticompetitive behavior, in general it suggests that the MNE structure is efficiency-enhancing. Moreover, I argue in this chapter that domestic antitrust regulation should be adequate to deal with those instances where the MNE structure has anticompetitive consequences.

Chapter 19, published in the *American Economic Review*, analyzes the development of economic theory on the MNE from an arbitrageur of capital, through an arrangement for achieving monopoly power (as argued by Hymer), to an efficient organizational structure for overcoming failures in the market for know-how. This paper argues that in order to understand the MNE, it is necessary to compare it against a feasible alternative, rather than against an unrealistic benchmark of a product market with perfect competition. Using this framework, it illustrates that the MNE's anticompetitive properties are overblown. The paper also analyzes the desirability of host countries imposing controls on MNEs, and MNEs requiring investment safeguards from the host country.

Chapter 20 contrasts the transaction-cost analysis of the MNE with the "internalization school", which emphasizes the advantages and costs associated with internalizing economic activity within the MNE. The distinctive feature of the transaction-cost approach as against the internalization school is the emphasis on a comparative institutional analysis of the MNE against feasible alternatives. This chapter highlights the tradeoffs between the governance costs associated with transactions in MNEs and the transaction costs associated with international market transactions. The chapter also uses the transactions cost approach to analyze the relationship between the MNE and the host country with respect to foreign direct investment, as well as the implications of nationalization on that relationship.

Chapter 21 extends the theory on the MNE to incorporate the recent developments in the field of strategic management, particularly the development of the "dynamic capabilities" approach. The chapter reflects on Penrose's contribution to the resource-based theory of the firm, in the context of the capabilities theory, which holds that the source of a firm's competitive advantage derives from distinctive and non-imitable capabilities. It argues that the distinctive feature of the MNE is the ability to leverage these non-imitable, firm-specific assets across international borders.

In addition to the co-authors already identified, there are numerous colleagues, students, and friends who have helped with preparing this volume. They include: Mie Augier, Sebastien Belanger, Frances Darnley, Doug Kidder,

Patricia Lonergan, Patricia Murphy, Edward F. Sherry, Anita Stephens, and Simon Wakeman. I am deeply thankful for their assistance.

David J. Teece
Thomas W. Tusher Chair in Global Business
Director, Institute of Management, Innovation and Organization
Haas School of Business
University of California, Berkeley
Berkeley, CA 94720-1930
USA

Part I

INTERNATIONAL TECHNOLOGY TRANSFER

The Economic Journal, 87 (*June 1977*), 242–261
Printed in Great Britain

TECHNOLOGY TRANSFER BY MULTINATIONAL FIRMS: THE RESOURCE COST OF TRANSFERRING TECHNOLOGICAL KNOW-HOW*

I. INTRODUCTION

The essence of modern economic growth is the increase in the stock of useful knowledge and the extension of its application. Since the origins of technical and social innovations have never been confined to the borders of any one nation, the economic growth of all countries depends to some degree on the successful application of a transnational stock of knowledge (Kuznets, 1966). In other words, the economic growth of every nation is inextricably linked to the successful international transfer of technology. Nevertheless, economists have been remarkably slow in addressing themselves to the economics of international technology transfer. The result is that "at both the analytic and factual level very little is known about the international transfer of knowhow" (Reynolds, 1966).

This paper addresses itself to this need. The starting-point is Arrow's suggestion that the cost of communication, or information transfer, is a fundamental factor influencing the world-wide diffusion of technology (Arrow, 1969).[1] The purpose of the paper is to examine the level and determinants of the costs involved in transferring technology. The value of the resources which have to be utilised to accomplish the successful transfer of a given manufacturing technology is used as a measure of the cost of transfer. The resource cost concept is therefore designed to reflect the ease or difficulty of transferring technological know-how from manufacturing plants in one country to manufacturing plants in another.

II. TECHNOLOGY TRANSFER AND THE PRODUCTION OF KNOWLEDGE

The literature on technological change recognises that it takes substantial resources to make a new process or product feasible (Mansfield, 1968). However, it is common to assume that the cost of transferring the innovation to other firms is very much less, so that the marginal costs of successive application is trivial compared to the average cost of research, development, and application. This paradigm is sometimes extended to international as well as domestic technology transfer (Rodriguez, 1975).[2] Buttressing this view is a common belief that

* The findings described in this paper resulted from research undertaken for my Ph.D. dissertation, "The Multinational Corporation and the Resource Cost of International Technology Transfer" (Cambridge, Mass.: Ballinger, 1976). The trenchant comments of Professor Edwin Mansfield were much appreciated during all phases of the study. My particular gratitude goes to the participating firms, without whose co-operation this paper would not have been possible. I should also like to acknowledge the financial support provided for this study by the National Science Foundation, under a grant to Professor Edwin Mansfield of the University of Pennsylvania.

[1] Arrow asks: "If one nation or class has the knowledge which enables it to achieve high productivity, why is not the other acquiring that information?...The problem turns on the differential between costs of communication within and between classes" (or nations). P. 33.

[2] "Transmission of technology between countries is assumed costless. Thus, it is possible for the

technology is nothing but a set of blueprints that is usable at nominal cost to all. Nevertheless, it has been pointed out that generally "only the broad outlines of technical knowledge are codified by non-personal means of intellectual communication, or communication by teaching outside the production process itself" (Berrill, 1964). The cost of transfer, which can be defined to include both transmission and absorption costs, may therefore be considerable when the technology is complex and the recipient firm does not have the capabilities to absorb the technology. The available evidence is unfortunately very sketchy. From the case studies of Mueller and Peck, Arrow inferred that transfer costs must be high (Arrow, 1962). From the Hall and Johnson study of the transfer of aerospace technology from the United States to Japan, it is not clear that this is true (Hall and Johnson, 1970). Robinson believes that economists' views on transfer costs are exaggerated (Robinson, 1973) while Mansfield and Freeman take the opposite view (Freeman, 1965; Mansfield, 1973). The lack of compelling evidence is apparent, and the appeals for further research (Mansfield, 1974; UNCTAD, 1970) seem to be well founded.

III. THE SAMPLE

The domain of this study is the transfer of the capability to manufacture a product or process from firms in one country to firms in another. Consequently the transfers can be considered as horizontal,[1] and in the design phase.[2] Data on 26 fairly recent international technology transfer projects were obtained. The proprietary nature of much of the data meant that sampling costs were high, which in turn severely limited the size of the sample that could be collected. All 26 transfers were conducted by firms which were multinational in the scope of their manufacturing activity, although they varied considerably in sales value (10–20 billion U.S. dollars) and R & D expenditures (1·2–12·5% of sales value). All had headquarters in the United States. The transferees were on the average much smaller and less research-intensive. In 12 instances they were wholly owned subsidiaries of the transferor, in 8 instances the transferor and transferee were joint ventures partners, in 4 instances transfers were to wholly independent private enterprises, and the remaining 3 were to government enterprises. Table 1 shows that 17 of the projects fall into a broad category which will be labelled "chemicals and petroleum refining." The remaining 9 projects fall into a category which will be labelled "machinery".[3] Table 1 also indicates the wide geographical dispersion of the transferees.

---

country which owns the technology to operate a plant in a foreign country without any transfer of factors." P. 122.

[1] Horizontal transfer refers to the transfer of technical information from one project to another. It can be distinguished from vertical transfer, which refers to the transfer of technical information within the various stages of a particular innovation process, e.g. from the basic research stage to the applied research stage.

[2] For the distinctions between materials transfer, design transfer, and capacity transfer, see Hayami and Ruttan (1971).

[3] Chemicals and petroleum refining thus embrace ISIC categories (United Nations, 1968), 351, 353, and 356, while "machinery" embraces categories 381, 382, and 383.

Table 1

Twenty-six Technology Transfer Projects: 3 Digit ISIC Category and Transferee Location

Location	"Chemicals and Petroleum Refining"			"Machinery"			Total
	351: industrial chemicals	353: petroleum refineries	356: plastic products	381: fabricated metal products machinery and equipment	382: machinery except electrical	383: electrical machinery, appliances, and supplies	
Canada	1	1	0	1	0	0	3
Northern and Western Europe	3	1	0	0	4	1	9
Australia	0	0	1	0	0	0	1
Japan	3	0	0	0	1	0	4
Eastern Europe	2	0	0	0	0	0	2
Latin America	3	0	0	0	0	1	4
Asia (excluding Japan)	0	1	0	0	0	1	2
Africa	1	0	0	0	0	0	1
Total	13	3	1	1	5	3	26

IV. DEFINITION OF TECHNOLOGY TRANSFER COSTS

An economic definition of transfer cost is developed below. The emphasis is on the resources which must be utilised to transfer technological know-how. Of course royalty costs or rents must be incurred merely to secure access to the technology, but these costs are not the focus of attention of this paper.[1] In order to appreciate the import of the definition that will be presented, a distinction must first be made between two basic forms in which technology can be transferred. The first form embraces physical items such as tooling, equipment, and blue prints. Technology can be embodied in these objects. The second form of technology is the information that must be acquired if the physical equipment or "hardware" is to be utilised effectively. This information relates to methods of organisation and operation, quality control, and various other manufacturing procedures. The effective conveyance of such "peripheral" support constitutes the crux of the process of technology transfer, and it typically generates the associated information flows. It is towards discovery of the cost of transfer of this "unembodied"[2] knowledge that the attention is directed.

Technology transfer costs are therefore defined as the costs of transmitting and absorbing all of the relevant unembodied knowledge. The costs of performing the various activities which have to be conducted to ensure the transfer of the necessary technological know-how will represent the cost of technology transfer.[3] Clearly, a great many skills from other industries (e.g. design engineering) will be needed for plant design, plant construction, and equipment installation. However, not all of these skills will have to be transferred to ensure the success of the project. As defined, the costs of transfer clearly do not include all of the costs of establishing a plant abroad and bringing it on stream.

The definition of transfer costs presented at the conceptual level can be translated into operational measures by considering the nature of a given project activity. At the operational level the subset of project costs identified as transfer costs fall into four groups. The first group is the cost of pre-engineering technological exchanges. During these exchanges the basic characteristics of the technology are revealed to the transferee, and the necessary theoretical insights are conveyed. The second group of costs included are the engineering costs associated with transferring the process design and the associated process engineering[4] in the case of process innovations, or the product design and production engineering[5] in the case of product innovations. If the technology has already

[1] Many observers equate the cost of technology with royalty fees (Mason, 1973; Gillette, 1973). Royalty costs are considered in the dissertation from which this paper was taken.

[2] Unembodied knowledge is the term used here to denote knowledge not embodied in capital goods, blueprints, and technical specifications, etc.

[3] All of the relevant costs are included, irrespective of which entity initially or eventually incurs them.

[4] Process engineering for continuous flow technology involves the compilation of flow diagrams, heat balances, control instrumentation, etc. It can be distinguished from detailed engineering which involves the translation and elaboration of the process engineering into a manufacturing facility.

[5] Production engineering for a specified item can be divided into two phases: production design and process planning. Production design is the modification of the functional design in order to reduce manufacturing costs. (Functional product design is the design of a product to fulfil certain specifications and requirements.) Given the design, process planning for manufacture must be carried out to specify,

been commercialised,[1] transmission may simply involve transferring existing drawings and specifications with the minimum of modification. However, the process of absorption may be more difficult, requiring the utilisation of considerable consulting or advisory resources. "Engineering" costs not falling into the specified categories[2] are excluded from transfer costs. The excluded engineering costs are essentially the plant or detailed engineering costs, net of advisory or consulting costs. This residual is assumed to correspond with routine drafting costs. Routine drafting is generally performed by technicians under the supervision of engineers. Drafting skills do not have to be transferred for the viability of the project to be assured. Accordingly, drafting is not considered to represent a transfer activity.[3]

The third group of costs are those of R & D personnel (salaries and expenses) during all phases of the transfer project. These are not the R & D costs associated with developing the underlying process or product innovations. Rather, they are the R & D costs associated with solving unexpected problems and adapting or modifying the technology. For instance, research scientists may be utilised during the transfer if new and unusual technical problems are encountered[4] with the production inputs. These R & D costs are generally small or non-existent for international transfers falling into the "design transfer" category.

The fourth group of costs are the pre-start-up training costs and the "excess manufacturing costs". The latter represent the learning and debugging costs incurred during the start-up phase, and before the plant achieves the design performance specifications. It is quite possible that no marketable output will be produced during the initial phases of the start-up. Nevertheless, normal labour, materials, utilities, and depreciation costs will be incurred, together with the costs of the extra supervisory personnel that will inevitably be required to

---

in careful detail, the processes required and their sequence. The production design first sets the minimum possible costs that can be achieved through the specification of materials, tolerances, basic configurations, methods of joining parts, etc. Process planning then attempts to achieve that minimum through the specification of processes and their sequence to meet the exacting requirements of the design specifications. The accepted end-point for production design is manifested by the drawing release. Process planning takes over from this point and develops the broad plan of manufacture of the part or product. A distinction can also be drawn between process planning and the layout of the physical facilities. Some process planning will take place during the layout phases of the design of a production system, Process plans can be regarded as inputs to the development of a layout. (*McGraw-Hill Encyclopedia*. 1960.)

[1] An innovation is said to have been commercialised if it has already been applied in a facility of economic size which is essentially non-experimental in nature. Thus pilot plant or prototype application is not considered to represent commercialisation.

[2] These categories are (*a*) process or design engineering costs and related consultation for process innovations or (*b*) production engineering expenses for product innovations; and (*c*) costs of engineering supervision and consultation (salaries plus travel and living) for the plant engineering.

[3] Drafting costs can be considered an implementation cost rather than a transfer cost, the implication being that if the host country does not have these skills, the viability and cost of the project is unlikely to be affected. The advisory and consulting costs, on the other hand, represent transfer costs since these activities are necessary if the technology is to be adjusted to the local circumstance and requirements. Clearly, if an existing plant was to be duplicated in its own environment, consulting costs could be expected to go to zero, whereas routine drafting would still have to be performed.

[4] Referring to process technologies, it is possible that differences in feedstocks amongst various locations may create problems that only research scientists can effectively handle. Similarly, changes in atmospheric conditions or water supply could have unexpected consequences for some highly complex processes.

assist in the start-up. The operating losses incurred during initial production are very often a close approximation to excess manufacturing costs.[1]

V. TRANSFER COSTS: DATA AND HYPOTHESES

1. *The Level of Transfer Costs*

The above definition was used to calculate the transfer costs for 26 projects. The results are presented in Table 2. The costs are given in absolute dollars, and then normalised by total project costs.[2] For the sample as a whole, transfer costs average 19 % of total project costs. Clearly, the data do not support the notion that technology is a stock of blueprints usable at nominal cost to all. Nevertheless, there is considerable variation in the sample data, with transfer costs ranging from 2 % to 59 % of total project costs. The number of factors influencing transfer costs is undoubtedly very great,[3] but some factors are likely to have a more pervasive influence than others. The discussion to follow is restricted to hypotheses for which statistical testing is feasible, given the available data. Two groupings of testable hypotheses can be identified: characteristics of the technology/transferor, and characteristics of the transferee/host country.

2. *Technology/Transferor Characteristics*

A critical factor in the transfer of technology is the extent to which the technology is completely understood by the transferor. The number of manufacturing start-ups[4] or applications which the transferor has already conducted with a specific technology can be used as an index of this knowledge.[5] An increase in the number of applications is likely to lower transfer costs since with each start-up additional knowledge about the technology is acquired. Since no two manufacturing start-ups are identical, each start-up provides the firm with the opportunity to observe the effects of different operating parameters and differences in equipment design. Each application can be regarded as a new experiment which

[1] An important consideration is the extent to which excess manufacturing costs correctly reflect technology transfer costs rather than the costs of discovering and overcoming the idiosyncrasies of a new plant. One way to confront this issue is to consider the level of excess manufacturing costs when an absolutely identical plant is constructed in a location adjacent to an existing plant. Further, assume the second plant embodies the same technology as the first plant, and the labour force from the first is transferred to the second for the purpose of performing the manufacturing start-up. The assumption is that under these circumstances excess manufacturing costs in the second plant will be zero, or very nearly so. The correctness of this assumption was corroborated by a subsample of project managers subsequently questioned about this matter. The postulated circumstance would be identical to shutting down the first plant and then starting it up again. Some excess manufacturing costs might be incurred during the initial hours of operation if the plant embodies flow process technology. (For the projects in the sample the average duration of the manufacturing start-up was 8·2 months.) However, these costs are unlikely to be of sufficient magnitude to challenge the validity of classifying excess manufacturing costs as a component of technology transfer costs.

[2] Total project costs are measured according to the inside boundary limits definition commonly employed by project accountants. Installations outside the plant perimeter are thereby excluded.

[3] For a broader view of the spectrum of hypotheses, see the author's Ph.D. dissertation.

[4] Manufacturing start-ups are synonymous with the number of applications of the technology. If a new plant is built for each application, it would also by synonymous with the number of plants that are built which utilise the technology.

[5] Corporations engaged in technology transfer ventures not grounded on their own technology are known to have encountered massive transfer problems and costs.

yields new information and new experience.[1] Transfer will be facilitated the more fully the technology is understood. Besides these engineering economies, additional applications provide expanded opportunities for the pre-start-up, training of the labour force. Clearly, if identical or similar plants exist elsewhere,

Table 2

Sample Data on the Resource Costs of Technology Transfer: 26 International Projects

Chemicals and Petroleum Refining		Machinery	
Transfer costs: dollar amount (thousands)	Transfer costs: dollar amount / total project cost	Transfer costs: dollar amount (thousands)	Transfer costs: dollar amount / total project cost
49	18	198	26
185	8	360	32
683	11	1,006	38
137	17	5,850	45
449	8	555	10
362	7	1,530	42
643	6	33	59
75	10	968	24
780	13	270	45
2,142	6		
161	2		
586	7		
877	7		
66	4		
2,850	19		
7,425	22		
3,341	4		

then experienced operators from these plants can be used to assist the start-up in the new plant. In addition, untrained operators can be brought into existing plants for pre-start-up training.

The second variable to be considered is the age of the technology. The age of the technology is defined as the number of years since the beginning of the first commercial application of the technology[2] anywhere in the world, and the end[3] of the technology transfer programme. The age of an innovation will determine the stability of the engineering designs and the transferor's knowledge of the manufacturing procedures. The older the technology, *ceteris paribus*, then the greater have been the opportunities for interaction between the development groups and the manufacturing and operating groups within the firm. Problems

[1] The first application represents first commercialisation of the technology. This will result in the creation of a set of basic engineering drawings and specifications. Duplication and alteration of these for subsequent start-ups will involve a modest cost compared to the initial cost of constructing them.

[2] If there is more than one key innovation embodied in the technology, then the date of commercial application of the most recent key innovation is the reference date.

[3] Age is defined up to the end of the transfer programme since any knowledge about the technology acquired up to this point is potentially useful for the transfer. For the very first start-up, age will be the length of the transfer minus the development overlap.

stand a better chance of already being ironed out, and the drawings are likely to be more secure. Further, since technology is not embodied in drawings alone, there is a great deal of uncodified information – the relevant "art". This kind of knowledge is carried by the supervisors, engineers, and operators. As the age of the technology increases, more individuals in the firm have the opportunity to acquire this non-codified information, and hence are potentially available to assist in the transfer. There will, however, be some point after which greater age will begin to increase the cost of transfer. When the length of stay of corporate personnel begins to be outstripped by the age of technology, then the non-codified dimensions of design knowledge may be lost to the firm.[1]

It is necessary to distinguish the cost reductions resulting from additional start-ups from the cost reductions resulting from greater age of the technology. For continuous flow technologies, additional applications of an innovation in entirely new plants will allow experimentation with scale and with the basic parameters of the design. This will generate a greater understanding of the technology. On the other hand, greater age, given the number of applications or start-ups, generally permits experimentation only with operating parameters, the design of the plant remaining fixed throughout.

The third technology variable to be considered is the number of firms utilising the technology, or one that is "similar and competitive". This is taken to represent the degree to which the innovation and the associated manufacturing technology is already diffused throughout the industry. The greater the number of firms with the same or similar and competitive technology, then the greater the likelihood that technology is more generally available, and can therefore be acquired at lower cost.[2]

These technology variables and the attendant hypotheses begin to take on some extra significance when viewed together. Taken singly they define the technology to only a limited degree. Together, they hypothesise, *ceteris paribus*, that the most difficult and hence costly technology to transfer is characterised by very few previous applications, a short elapsed time since development, and limited diffusion. Technology displaying such characteristics can be termed "leading-edge" technology. "Leading-edge" technology is likely to be in a state of flux; the engineering drawings will be constantly altering, thus frustrating the transfer. In comparison, state-of-the-art technology is hypothesised, *ceteris paribus*, to involve lower transfer costs since the engineering drawings are more likely to be finalised and the fundamentals of the technology stand a better chance of being more fully understood.

[1] In the limit, the firm could terminate its utilisation of a particular technology, and the non-codified information associated with it could be gradually lost for ever as the technology becomes historic. Further, the drawings associated with technology that is very old may suffer from so many small alterations that the very essence of even the codified technology may become quite obscure. Since none of the technology transfer projects in the sample were historic in the above sense, the relevant range of the hypothesised age–transfer cost function involves an inverse relationship between the age of the technology and the cost of transfer.

[2] An identification problem may exist here because more firms may have applied the technology because the transfer cost is low.

3. *Transferee and Host Country Characteristics*

The technical and managerial competence of the transferee will be presented as an important determinant of the ease with which technology can be absorbed. The years of manufacturing experience of the transferee in a given 4-digit ISIC industry (United Nations, 1968) is used as an index of the extent to which managers, engineers, and operators have command over the general manufacturing skills of an industry. A firm skilled in the manufacture of a group of products is likely to have less difficulty absorbing a new innovation in that industry group than is the firm which has had no previous experience manufacturing products in a particular industry group (Rawski, 1975). Older enterprises, with their skilled manufacturing personnel, seem more likely to be able to understand and apply codified knowledge to the manufacture of a new product, or the utilisation of a new process.[1]

Another variable to be considered is the size of the transferee. Although less compelling, the reasoning behind the hypothesis that transfer costs decline with firm size is that larger firms generally have a wider spectrum of technical and managerial talent which can be called on for assistance during the transfer. A small firm may be technically and managerially quite competent yet unable to absorb new technology easily because of the extra demands placed on its scarce managerial and technical manpower. Consultants may have to be engaged by the smaller firms to perform tasks that are typically handled internally in larger firms.

A third variable considered is the R & D activity of the transferee. When unusual technical problems are unexpectedly encountered, an in-house R & D capability is likely to be of value. Oshima has argued that the R & D capability of Japanese firms facilitated the low-cost importation of foreign technology by Japanese firms (Oshima, 1973). The R & D to sales ratio of the transferee is taken as an index of its R & D capability, and an inverse relationship between this and transfer cost is postulated.

The final variable considered is designed to reflect the level of development of the host country infrastructure, which is hypothesised to be a determinant of the cost of transfer. For example, the level of skill formation in the host country will influence the amount and type of training that the labour force will require. Similarly, if the new venture is to acquire its inputs domestically, the quality of the inputs available will undoubtedly influence the level of start-up costs. There are many other considerations of similar kind which could be discussed. However, the high degree of cross-sectional collinearity between indices of development (Kuznets, 1966) makes the identification of separate effects statistically difficult. However, GNP *per capita*, a measure of productive capacity, can

[1] According to Rawski, recent experience of the People's Republic of China shows that during at least some phases of industrialisation, production experience may be a key determinant of the level and fungibility of industrial skills. Rawski notes that "with their skilled veteran workers and experienced technical persons, old industrial bases and old enterprises find it easier to tackle complicated technical problems than new enterprises and new industrial bases. With these advantages, it is the established centers which are best able to copy foreign equipment samples, to extract useful information from foreign technological publications, and to apply it to current domestic problem areas." (Rawski (1975), p. 386.)

be expected to capture some of the above considerations, and it will be used in this study as an index of economic development. A negative relationship between transfer cost and GNP/*per capita* is postulated.[1]

VI. DETERMINANTS OF THE COST OF INTERNATIONAL TECHNOLOGY TRANSFER: TESTS AND RESULTS

1. *The Model*

The basic model to be tested is

$$C_i = f(U_i, G_i, E_i, R_i, S_i, N_i, P_i, Z_i), \tag{1}$$

where C_i is the transfer cost divided by the total project cost for the *i*th transfer; U_i is the number of previous applications or start-ups that the technology of the *i*th transfer has undergone by the transferor;[2] G_i is the age of the technology in years; E_i is the number of years of manufacturing experience that the recipient of the *i*th transfer has accumulated; R_i is the ratio of research and development to sales for the recipient of the *i*th transfer, calculated for the year the transfer commenced; S_i is the volume of sales, measured in millions of dollars, of the recipient of the *i*th transfer; N_i is the number of firms identified by the transferor as having a technology that is identical or "technically similar and economically competitive" to the technology underlying the *i*th transfer; P_i is the level of GNP *per capita* of the host country (International Bank, 1973); Z_i is the random error term for the *i*th transfer. The expected derivatives are:

$$\frac{\partial C_i}{\partial U_i} < 0, \quad \frac{\partial C_i}{\partial G_i} < 0, \quad \frac{\partial C_i}{\partial E_i} < 0, \quad \frac{\partial C_i}{\partial R_i} < 0, \quad \frac{\partial C_i}{\partial S_i} < 0, \quad \frac{\partial C_i}{\partial N_i} < 0, \quad \frac{\partial C_i}{\partial P_i} < 0.$$

Since one of the best tests of any hypothesis is to look for the convergence of independent lines of evidence, the testing of this model will proceed in two phases. First, cross-section data on 26 completed projects is utilised in a linear version of the model estimated by ordinary least-squares procedures. Secondly, cost estimates provided by project managers for comparable projects are pooled to test a more specific non-linear version of the model.

2. *Statistical Tests: Phase 1*

The model to be tested is

$$C_i = \alpha_0 + \alpha_1 \bar{U}_i + \alpha_2 G_i + \alpha_3 E_i + \alpha_4 R_i + \alpha_5 S_i + \alpha_6 N_i + \alpha_7 P_i + Z_i, \tag{2}$$

where $\bar{U}_i$ is a dummy variable taking the value 1 if the transfer represents the first manufacturing start-up, and zero otherwise. $\bar{U}_i$ is used rather than U_i for empirical reasons, since the first start-up is often of critical importance. The sample was dichotomised because of the large differences between continuous flow process technology, and product technology. One category includes chemicals and petroleum refining and the other includes machinery (see Table 1).

[1] The sample did not include countries where high GNP statistics were grossly dependent on oil revenues.

[2] The number of previous manufacturing start-ups was significant in Phase 1 only when it was included as a dummy variable taking the value 1 if there had been no previous manufacturing start-ups of this technology by the transferring firm, and zero otherwise.

The results in Table 3 indicate that in chemicals and petroleum refining $\bar{U}_i$, N_i, and E_i are significant at the 0·05 level and carry the expected signs. In the machinery category the variables N_i, G_i, and E_i all carry the expected signs and are significant at the 0·05 level. N_i and E_i are thus significant in both industry

Table 3

Regression of Coefficients and t Statistics in Regression Equations to Explain C (The Cost of Transfer)

Independent variable	Chemicals and petroleum refining: Equation (1)*	Chemicals and petroleum refining: Equation (2)*	Machinery: Equation (1)*	Machinery: Equation (2)*
Constant	12·79 (6·82)	13·42 (6·98)	16·67 (8·27)	65·98 (6·60)
Novelty dummy variable $\bar{U}$†	6·73 (1·92)	6·11 (1·75)	—	1·62 (0·15)
Number of firms variable	−0·37 (−2·06)	−0·39 (−2·22)	−1·29 (−2·28)	−1·26 (−1·95)
Age of technology variable (years)	—	—	−2·43 (−3·53)	−2·35 (−2·51)
Experience of transference variable (years in 4-digit ISIC)	−0·09 (−1·66)	−0·08 (−1·42)	−0·84 (−3·37)	−0·85 (−2·95)
Size of transferee variable (thousands of dollars of sales)	—	−0·0009 (−1·18)	—	—
Number of observations	17	17	9	9
R^2	0·56	0·61	0·78	0·78
F	5·66	4·73	6·00	5·22
Significance level of F	0·01	0·02	0·04	0·12

* Omitted coefficient indicates variable dropped from the regression equation.
† Note 2, p. 251.

groupings, strongly supporting the hypothesis that transfer costs decline as the number of firms with identical or "similar and competitive" technology increases, and as the experience of the transferee increases. However, R_i and P_i were not significant in any of the equations, and although S_i carries the expected sign and approaches significance in one of the regressions it is not possible to be more than 85 % sure that the sign is correct or that the coefficient is different from zero.[1]

[1] Multicollinearity does not appear to be a serious problem in any of the equations. Correlations amongst pairs of the independent variables were never significant at the 0·05 level. The stability of the regression coefficients further suggests that multicollinearity is not a serious problem. Dummy variables were introduced to test for the effects of the organisational relationship between transferor and transferee (affiliate/non affiliate, public enterprise/private enterprise), but they were not found to be statistically significant determinant of transfer costs. Application of a forward step by step procedure did not reveal a preferred subset of variables. However, it is possible that the correct model is the simultaneous equation model $Ci = f(N_i, \ldots)$, $N_i = f(C_i, \ldots)$. To eliminate simultaneous equation bias it would be desirable to use a two-stage procedure. A predictor of N could first be obtained by regressing N_i on arguments other than C_i. This could then be used as an argument in the transfer cost regression. It was not possible to obtain a good predictor of N using the available cross-section data, so this procedure was not employed. Consistency was sacrificed for efficiency. It is therefore possible that simultaneous equation bias remains in the model. Therefore, the estimates of the parameters may not be consistent.

The results therefore generally support the hypotheses advanced earlier, but there are differences in the size of coefficients as well as in the specification of the equations between the industry groups. In particular, the novelty variable $\bar{U}_i$ is significant in chemicals and petroleum refining, but insignificant in machinery. The converse is true for the age variable G_i. The reason may be that there exists relatively less latitude for production experimentation with continuous flow process technology than with product technology. Once the plant is constructed, the extent to which the design parameters can be changed is rather minimal because of the degree of interdependence in the production system. In comparison, many product technologies allow greater design flexibility. Innumerable small changes to the technology are very often possible without massive reconstruction of the plant. It is also of interest that the coefficient of the experience variable E_i is considerably larger in machinery than in chemicals and petroleum refining. This is consistent with other findings that reveal important learning economies in fabrication and assembling (Tilton, 1971).

3. *Statistical Tests: Phase II*

The above analysis is handicapped by the small sample size and the very high costs of adding additional observations. Limited variation in exogenous variables coupled with the problem of omitted variables can imply difficulties with bias and identification. For the projects in the sample, a procedure was therefore devised to hold the missing variables constant while generating large variation in the exogenous variables. The respondent firms were asked to estimate how the total transfer costs would vary for each project if one particular exogenous variable happened to take a different value, assuming all other variables remain constant. The responses were taken into account only if the exercise generated circumstances within the bounds of an executive's experience. Given these limitations, the change specified was quite large in order to provide a robust sample. Generally the actual value of a selected variable was hypothesised first to halve and then to double. The estimated impact on transfer costs was noted. The exercise was performed for the following independent variables: the number of applications or start-ups that the technology has undergone; the age of the technology; the number of years of previous manufacturing experience possessed by the transferee in a given four-digit industry; the research and development expenditures to sales ratio for the transferee; the size (measured by sales value) of the transferee. For each variable this exercise generated at most three observations (including the actual) or transfer costs for each project. Pooling across projects produces enough observations for ordinary least-squares regression analysis.

The estimation procedure is commenced by assuming that the shape of the cost function can be represented by the following relatively simple but quite specific equation

$$C_j = V e^{\phi / X_j}. \tag{3}$$

C is the estimated transfer cost as a percentage of total project cost, X represents the value of various independent variables, j refers to the jth observation.

With this specification, the transfer cost for a project asymptotically approaches a minimum non-zero value as the value of each X increases. That is, as X goes to infinity, C goes to V. Therefore V is the minimum transfer cost with respect to the X variable. However, there is no maximum cost asymptote for the range of the data. The expression for the elasticity of transfer cost with respect to X is given by

$$\frac{-X}{C}\frac{dC}{dX} = \frac{\phi}{X}. \tag{4}$$

Thus for a specified value of X, the elasticity of transfer cost with respect to X is determined by ϕ. Hence the elasticity depends only on ϕ and X. In order to estimate the function, the log of the arguments in (3) are taken:

$$\log C_j = \log V + \frac{\phi}{X_j}. \tag{5}$$

Dummy variables are used to pool the observations across projects. Inclusion of dummy variables allows the minimum cost asymptote to vary from project to project. It is assumed that ϕ is constant across projects. These assumptions provide a pooled sample with intercepts which vary across projects.

Ordinary least-squares regressions of log C_j on the dummy variables and $1/X_j$ then proceeded for five different X variables, and for five data sets. These were: total transfers; transfers within the chemical and petroleum refining category; transfers in the machinery category; transfers of chemicals and petroleum refining technology to developed countries; and transfers of chemical and petroleum refining technology to less developed countries.[1] The Chow test (Chow, 1960) of equality between sets of coefficients in two linear regressions revealed that the separation of the sample along industry lines was valid, except for the research and development variable. However, there was no statistically valid reason for disaggregating the chemical and petroleum refining subsample according to differences in GNP *per capita* in the host countries.

The results of the estimation are contained in Table 4. The high R^2 values are partly because the large across-project variation in costs is being captured by the project dummies. The intercept term was always highly significant and the coefficients on all the dummies were significantly different from each other. All of the coefficients are significantly greater than zero at the 0·20 level and the age of the technology, the number of manufacturing start-ups, transferee size and experience achieve at least the 0·05 significance level in one or other of the subsamples. In several cases the coefficients are highly significant, providing strong statistical support for the hypotheses that have been advanced. The number of previous applications once again has a sizeable impact. Diffusion and manufacturing experience are particularly important in the machinery category.

The calculation of elasticities allows interpretation and comparisons of estimated effects. Average or point elasticities for some typical sample values of X are presented in Table 5. These estimates suggest that in the chemicals and petroleum-refining category, the second start-up could lower transfer costs by

[1] A purely arbitrary classification was used where less developed countries were defined as those with GNP/*per capita* less than $1,000.

34% over the first start-up, other variables held constant. The corresponding change for conducting a third start-up is 19%. The other elasticities can be interpreted similarly.

Table 4

Estimated values of ϕ *(obtained from regressing* $\log C_j$ on $\log V + \phi/X_j$*) together with corresponding t-statistics, sample size, degrees of freedom, and coefficient of determination* R^2

Variable	ϕ	t-statistic	Sample size	Degrees of freedom	R^2
Start-ups					
Chemicals and petroleum refining	0·46	4·23	45	25	0·92
Machinery	0·19	1·76	20	10	0·91
Age					
Chemicals and petroleum refining	0·04	1·29	47	30	0·89
Machinery	0·41	2·19	21	13	0·94
Experience					
Chemicals and petroleum refining	0·007	0·85	52	33	0·78
Machinery	0·57	6·08	23	14	0·91
Size					
Chemicals and petroleum refining	0·008	1·17	54	35	0·88
Machinery	0·081	5·18	17	10	0·99
R & D sales					
Total sample	0·06	1·58	59	30	0·90

VII. DIFFERENCES BETWEEN INTERNATIONAL AND DOMESTIC TECHNOLOGY TRANSFER

Although this is primarily a study of international technology transfer, it is apparent that many of the characteristics of international technology transfer are also characteristic of the technology transfer that occurs within national borders, but there are differences. For instance, distance and communication costs very often differentiate international from domestic transfers. Although the communications revolution of the twentieth century has enormously reduced the barriers imposed by distance,[1] the costs of international communication are often significant.[2] Language differences can also add to communication costs, especially if the translation of engineering drawings is required. The experience of Polyspinners Ltd at Mogilev in the Soviet Union (Jones, 1973) is ample testimony to the extra costs that can be encountered.[3] International differences in units of measurements and engineering standards can compound the problems encountered (Meursinge, 1971). Additional sources of difficulty are rooted in

[1] Facsimile equipment exists which can be used to transmit messages and drawings across the Atlantic instantaneously.

[2] One of the participating companies indicated that travel, telegraph, freight, and insurance added about 10% to the total cost of a project established in New Zealand.

[3] The project manager estimated that documentation alone cost £500,000, and the translation a similar amount.

the cultural and attitudinal differences between nations, as well as differences in the level of economic development and the attendant socioeconomic structure.

It is of interest to know the magnitude and determinants of the "international component" of the transfer cost. Unfortunately, foreign and domestic transfers

Table 5

Arc or Point Elasticity of Transfer Costs With Respect to Number of Start-ups, Age of Technology, Experience, Size and R & D/Sales of Transferee

Independent variable	Chemicals and petroleum refining	Machinery
	Arc elasticity	
Number of start-ups		
1–2	0·34	0·14
2–3	0·19	0·08
3–4	0·13	0·05
9–10	0·05	0·02
14–15	0·03	0·01
Age of technology (years)	Point elasticity	
1	0·04	0·41
2	0·02	0·20
3	0·01	0·14
10	0·00	0·04
20	0·00	0·02
Experience of transferee (years)	Point elasticity	
1	0·007	0·57
2	0·003	0·28
3	0·002	0·19
10	0·001	0·06
20	0·000	0·03
Size of transferee (millions of sales dollars)	Point elasticity	
1·0	0·008	0·081
10	0·001	0·008
20	0·000	0·004
100	0·000	0·001
1000	0·000	0·000
R & D/Sales of transferee (%)	Total sample point elasticity	
1	0·06	
2	0·03	
3	0·02	
4	0·01	
5	0·01	
6	0·01	

are rarely identical in scope or in timing, and so it is not possible to gather comparative data on implemented projects at home and abroad. It was therefore found necessary to rely on estimates provided by the firms involved in international transfers. For the projects in the sample, project managers were asked to estimate the dollar amount by which transfer costs would be different if the international transfers in the sample had occurred domestically, holding firm and technology characteristics constant. The procedure was designed to highlight the effects of country characteristics such as differences in language, differences in engineering and measurement standards, differences in economic infrastructure and business environment, and geographical distance from the transferor. The international component of the transfer cost for the projects in the sample could be obtained by subtracting the estimated transfer cost from the

Table 6

International Component of Transfer Cost*

Chemicals and petroleum refining		Machinery	
Dollar amount (thousands)	As % of actual transfer cost	Dollar amount (thousands)	As % of actual transfer cost
3·03	6·07	35·55	17·88
0·00	0·00	−399·37	−110·93
−12·81	−1·87	50·06	4·93
43·90	31·00	830·70	14·20
0·00	0·00	−4·59	−0·02
5·17	1·42	226·80	14·82
132·75	20·63	0·67	1·99
0·00	0·00	−134·40	13·87
342·00	43·84	34·98	12·95
0·00	—		
0·00	—		
0·00	0·00		
−10·77	−6·66		
−50·16	−8·52		
0·00	—		
637·32	72·60		
−1·33	−1·99		
1,723·81	60·48		
1,370·25	18·45		
524·25	15·69		

* Amount of actual transfer costs attributable to the fact that transfer was international rather than domestic. (Accordingly negative values indicate that firms estimated that transfer costs would be higher had the transfer been domestic.) In general, these numbers were derived from taking the weighted average of estimated changes in the various identifiable components of transfer costs.

actual transfer cost. The data, contained in Table 6, reveal that the difference in cost is not always positive. This indicates that in at least some of the cases, the international transfer of an innovation was estimated to cost less than a comparable domestic transfer. This may seem paradoxical at first, given that international technology transfer generally augments the transfer activities that have to be performed.[1] An analysis of the determinants of the international component of transfer costs may yield an explanation.

Several hypotheses are presented and tested. The first is that the difference is

[1] The source of the apparent paradox may be differences in labour costs. Nevertheless, the identification of the transfer for which international transfer costs less than domestic transfer is an issue of importance.

large and positive when the technology has not been previously commercialised. National boundaries are often surrogates for cultural and language barriers, differences in methods and standards of measurement, and distance from the home country. During first commercialisation of a product or process, there are generally enormous information flows across the development-manufacturing interface. The hypothesis is that placing a national boundary at this interface can complicate matters considerably, and escalate the costs enormously. The second hypothesis is that transfers to government enterprises in centrally planned economies will involve higher transfer costs. Transferors can expect numerous delays and large documentation requirements (Jones, 1973). The third hypothesis is that the less the diffusion of the technology, measured as before by the number of firms utilising the innovation, the greater the positive differential associated with international technology transfer. The fourth hypothesis is that whereas, in general, low levels of economic development are likely to add to transfer costs because of inadequacies in the economic infrastructure, this may be more than offset, in some circumstances, by low labour costs. Labour costs can have a substantial impact on excess manufacturing costs, especially in relatively labour intensive industries. Since machinery manufacture is relatively labour intensive, the hypothesis is that the GNP *per capita* in the host country is positively related to the transfer cost differential in this classification, but is negatively associated with the differential in the chemicals and petroleum refining category.

To test these hypotheses it is assumed that

$$D_i = \alpha_0 + \alpha_1 d_i + \alpha_2 \bar{U}_i + \alpha_3 N_i + \alpha_4 P_i + Z_i,$$

where D_i is the "international component" as a percentage of actual transfer cost for the ith transfer. d_i is a dummy variable which takes the value 1 if the recipient of the ith transfer is a government enterprise in a centrally planned economy, and zero otherwise. The other variables carry the same definitions as previously. The expected derivations are:

$$\frac{\partial D_i}{\partial d_i} > 0, \quad \frac{\partial D_i}{\partial \bar{U}_i} > 0, \quad \frac{\partial D_i}{\partial N_i} < 0;$$

$\partial D_i/\partial P_i \gtrless 0$ according to the industry category (the partial is postulated positive for the machinery category, and negative otherwise). Least-squares estimates of the α's were obtained, the results being:

Chemicals and petroleum:

$$D_i = 0{\cdot}285 + 3{\cdot}84 d_i + 4{\cdot}46 \bar{U}_i \quad (n = 17,\ r^2 = 0{\cdot}71).$$
$$(0{\cdot}91) \quad (5{\cdot}01) \quad (4{\cdot}89)$$

Machinery:[1]

$$D_i = -8{\cdot}59 \quad -1{\cdot}39 N_i + 0{\cdot}005 P_i \quad (n = 9,\ r^2 = 0{\cdot}94).$$
$$(-1{\cdot}96) \quad (-5{\cdot}98) \quad (3{\cdot}90)$$

[1] O_i was omitted from the machinery regression since none of the actual transfers in this category were to government enterprises in centrally planned economies.

The hypotheses are to some extent borne out by the data, but the small sample size must counsel caution in the interpretation of these results.[1] In chemicals and petroleum, the results indicate that transfers to government enterprises, and transfers before first commercialisation, involve substantial extra costs. Furthermore, both N_i and P_i are significant in the machinery category, despite the small number of observations, yet they are insignificant in chemicals and petroleum refining, where there are more than twice as many degrees of freedom. Apparently, the level of host country development and the degree of diffusion of an innovation have no bearing on the international–domestic transfer cost differential in the chemicals and petroleum grouping. This calls for an explanation. The diffusion variable N_i is taken to indicate the degree to which the requisite skills are generally available. The statistical results suggest that the relevant skills for highly capital intensive industries, such as chemicals and petroleum refining, are more easily transferred internationally than are the requisite skills in the machinery category.[2] Furthermore, P_i was not significant in chemicals and petroleum refining, suggesting that costs of transfer are independent of the level of economic development in this category. This is consistent with speculation that international transfer is no more difficult than domestic transfer when the underlying technology is highly capital intensive. The perceived reluctance of multinational firms to adapt technology to suit the capital–labour endowments of less developed countries could well be rooted in the desire to avoid escalating transfer costs to unacceptable levels.

VIII. CONCLUSION

The resources required to transfer technology internationally are considerable. Accordingly, it is quite inappropriate to regard existing technology as something that can be made available to all at zero social cost. Furthermore, transfer costs vary considerably, especially according to the number of previous applications of the innovation, and how well the innovation is understood by the parties involved. It is equally inappropriate, therefore, to make sweeping generalisations about the process of technology transfer and the costs involved. For instance, technology transfer in chemicals and petroleum refining displayed relatively low transfer costs, presumably because it is possible to embody sophisticated process technology in capital equipment, which in turn facilitates the transfer process.

The analysis of the determinants of technology transfer costs provided some interesting findings with development implications. The success of the more experienced enterprises, indicated by lower transfer costs, points towards economic models which emphasise the accumulation of skills, rather than fixed

[1] If the second observation on D_i in the machinery category is excluded, and the regression results recomputed, the estimates of the coefficients exhibit considerable instability and the "goodness of fit" deteriorates. The estimated equation is

$$D_i = \underset{(1\cdot14)}{-4\cdot96} \underset{(2\cdot40)}{-0\cdot66N_i} + \underset{(1\cdot94)}{0\cdot003P_i} \quad (n = 8,\ r^2 = 0\cdot45).$$

These estimates are nevertheless significant at the 0·05 level for a one-tail test.

[2] This is consonant with the views expressed by several project managers in the chemical industry. It was asserted that technology could be transferred with equal facility to almost anywhere in the world, including less developed countries, assuming host government interference is held constant.

assets or capital, in facilitating the technology transfer process. This seems consonant with the findings of several economic historians (Rosenberg, 1970; Rawski, 1975).

The results also provide some managerial implications for the multinational firm. Consider the costs associated with separating production from development (Arditti, 1968). The results indicate that the international transfer of technology is most likely to be viable when production runs are long enough to allow second sourcing. The especially high cost of transfer before first application favours the development location, at least for production of initial units. However, transfer costs will be lowered once the first production run has been commenced, and international transfer then becomes more likely, a finding consistent with the product cycle model (Vernon, 1966). However, inter-industry differences are important, and the costs involved in separating first production from development did not prove to be an insurmountable transfer barrier for an important subset of the sample projects.

A second implication is that since transfer costs decline with each application of a given innovation, technology transfer is a decreasing cost activity. This can be advanced as an explanation for the specialisation often exhibited by engineering firms in the design and installation of particular turnkey plants,[1] a characteristic particularly noteworthy of the petrochemical industry.

A third set of managerial implications relate to the criteria which might be used for the selection of a joint venture or licensing partner to utilise the innovating firms' technology abroad. While the manufacturing experience, size, and R & D to sales ratio of the transferee were identified as statistically significant determinants of transfer costs for the sample, there was also evidence to suggest that, *ceteris paribus*, any firm moderately matured in these dimensions is a good candidate to absorb the technology at the minimum possible transfer cost. It is not clear, therefore, that super giant firms have any advantage in this respect over moderately sized firms. Nor is it clear that highly research intensive firms have more than a slight cost advantage in absorbing technology over firms with a minimal commitment to research and development activity. However, manufacturing experience is important, especially for transferring machinery technology. In addition, there is evidence that transfers to governments in centrally planned economies involve substantial extra costs, perhaps because of high documentation requirements, or differences in language and managerial procedures.

Technology transfer by multinational firms is clearly a complex matter. Collection and analysis of proprietary data has provided some helpful insights. Few issues have been settled although many have been raised. Further analytic research and more extensive data collection is required if our understanding of international technology transfer is to be improved.

D. J. TEECE

Stanford University, California

Date of receipt of final typescript: November 1976

[1] Turnkey plants generally embody state-of-the-art technology.

References

Arrow, K. (1969). "Classificatory Notes on the Production and Transmission of Technological Knowledge." *Amercian Economic Review; Papers and Proceedings*, vol. 52 (May), pp. 29–35.

—— (1962). Comment in Universities-National Bureau Committee for Economic Research. *The Rate and Direction of Inventive Activity*. Princeton: Princeton University Press.

Arditti, F. (1968). "On the Separation of Production from the Developer." *Journal of Business*, vol. 41 (July), pp. 317–28.

Baranson, J. (1967): *Manufacturing Problems in India: The Cummings Diesel Experience*. Syracuse, N.Y.: Syracuse University Press.

Berrill, K. (ed.) (1964). *Economic Development with Special Reference to East Asia*. New York: St Martins Press.

Chow, G. C. (1960). "Tests of Equality between Sets of Coefficients in Two Linear Regressions." *Econometrica*, vol. 28 (July), pp. 591–605.

Freeman, C. (1965). "Research and Development in Electronic Capital Goods." *National Institute Economic Review*, no. 34, vol. 34 (November), pp. 1–70.

Gillette, R. (1973). "Latin America: Is Imported Technology Too Expensive?" *Science*, vol. 191 (6 July), pp. 4–44.

Hall, G. R. and Johnson, R. E. (1970). "Transfers of United States Aerospace Technology to Japan." In *The Technology Factor in International Trade* (ed. R. Vernon). N.Y.: National Bureau of Economic Research.

Hayami, Y. and Ruttan, V. (1971). *Agricultural Development and International Perspective*. Baltimore: Johns Hopkins.

International Bank for Reconstruction and Development (1973). *World Bank Atlas*. Washington, D.C.: I.B.R.D.

Jones, D. (1973). "The 'Extra Costs' in Europe's Biggest Synthetic Fiber Complex at Mogilev, U.S.S.R." *Worldwide Projects and Installations*, vol. 7 (May/June), pp. 30–5.

Kuznets, S. (1966). *Modern Economic Growth: Rate, Structure, Spread*. New Haven: Yale University Press.

Mansfield, E. (1974). "Technology and Technical Change." In *Economic Analysis and the Multinational Enterprise* (ed. J. Dunning). London: Allen and Unwin.

—— (1968). *The Economics of Technological Change*. New York: Norton.

—— (1973). "Discussion of the Paper by Professor Griliches." In *Science and Technology in Economic Growth* (ed. B. R. Williams). New York: John Wiley.

Mansfield, E., Rapoport, J., Schnee, J., Wagner, S. and Hamburger, M. (1971). *Research and Innovation In the Modern Corporation*. New York: W. W. Norton.

Mason, R. Hal (1973). "The Multinational Firm and the Cost of Technology to Developing Countries." *California Management Review*, vol. 15 (Summer), pp. 5–13.

Meursinge, J. (1971). "Practical Experience in the Transfer of Technology." *Technology and Culture*, vol. 12 (July), pp. 469–70.

McGraw-Hill Encyclopedia of Science and Technology (1960). Vols. 4, 10, pp. 639–44. New York: McGraw-Hill.

Oshima, K. (1973). "Research and Development and Economic Growth in Japan." In *Science and Technology in Economic Growth* (ed. B. R. Williams). New York: John Wiley.

Rawski, T. (1975). "Problems of Technology and Absorption in Chinese Industry." *American Economic Review*, vol. 65 (May), pp. 363–88.

Reynolds, L. (1966). Discussion. *American Economic Review*, vol. 56 (May), pp. 112–14.

Robinson, E. A. G. (1973). "Discussion of the Paper by Professor Hsia." In *Science and Technology in Economic Growth* (ed. B. R. Williams). New York: John Wiley.

Rodriguez, C. A. (1975). "Trade in Technical Knowledge and the National Advantage." *Journal of Political Economy*, vol. 93 (February), pp. 121–35.

Rosenberg, N. (1970). "Economic Development and the Transfer of Technology: Some Historical Perspectives." *Technology and Culture*, vol. 11 (October), pp. 550–75.

Teece, D. (1976). *The Multinational Corporation and the Resource Cost of International Technology Transfer*. Cambridge: Ballinger.

—— (1977). "Time–Cost Tradeoffs: Elasticity Estimates and Determinants for International Technology Transfer Projects." *Management Science*, vol. 23 (April).

Tilton, J. (1971). *International Diffusion of Technology: The Case of Semiconductors*. Washington, D.C.: Brookings Institution.

United Nations (1968). *International Standard Industrial Classification of all Economic Activities*. United Nations Statistical Papers, Series M., Number 4. New York: United Nations.

UNCTAD (1970). "The Transfer of Technology." *Journal of World Trade Law*, vol. 4 (September/October), pp. 692–718.

Vernon, R. (1966). "International Investment and International Trade in the Product Cycle. *Quarterly Journal of Economics*, vol. 80 (May), pp. 190–207.

Reprinted by permission, MANAGEMENT SCIENCE, Vol. 23, No. 8, April, 1977, pp. 830–837.

TIME-COST TRADEOFFS: ELASTICITY ESTIMATES AND DETERMINANTS FOR INTERNATIONAL TECHNOLOGY TRANSFER PROJECTS*

DAVID TEECE†

This paper postulates and tests for a time-cost tradeoff during the establishment of manufacturing plants abroad based on U.S. technology. Data on twenty international projects are used to estimate negatively sloped tradeoff functions for which time-cost elasticities are subsequently calculated. The determinants of these elasticities are then analyzed and shown to bear some similarities with the determinants of time-cost tradeoffs in technological innovation. The elasticity measurements were higher for projects where the technology had not been previously commercialized, for large-scale projects, and for projects carried out by the larger firms.

1. Introduction

This paper shows that a time-cost tradeoff confronts the multinational firm in the scheduling of the design, construction, and startup of a manufacturing project abroad based on U.S. technology. In other words, the total cost of a project is revealed to increase if the time span between project commencement and project completion is reduced. Furthermore, the elasticity of this tradeoff can be explained by reference to characteristics of the technology being transferred and to characteristics of the participating firms.

Although the establishment of foreign plants is by no means an everyday activity for most multinational firms, foreign direct investment and international technology transfer is of some consequence for many U.S. and foreign firms. Therefore, the time-cost tradeoff identified is of practical importance.[1] The concept also has analytic interest, since the identification of the tradeoff under a number of different guises strengthens the generality of the principle that "buying" time in various kinds of investment decisions will require additional resources. Furthermore, an understanding of the concept is enhanced if comparisons can be made of the various kinds of time-cost tradeoff.

In the following sections reasons will be advanced for the existence of a time-cost tradeoff during the establishment of a manufacturing plant abroad. Using data from a sample of twenty such projects, the parameters of a tradeoff function will then be estimated. Finally, hypotheses will be advanced and tested to explain variation in the elasticity of the tradeoff function for these same twenty projects.

2. Foundations of the Time-Cost Tradeoff

While the existence of a time-cost tradeoff for research and development has been recognized [5] and demonstrated [3], the applicability of this discovery to nonresearch activity has at the same time been questioned [5].[2] Foreign direct investment, not obviously a research related activity, nevertheless seems to be rooted in the exploita-

* Accepted by Burton V. Dean; received December 8, 1975. This paper has been with the author 1 month, for 1 revision.

† Stanford University.

[1] The importance of the concept has been recognized in a number of different applications, and critical path analysis has been used to operationalize the time-cost tradeoff for the purpose of assisting in the efficient scheduling of complex development and construction projects. See, for example, [4, Chapter 7] and [2].

[2] Scherer contends that the time-cost tradeoff in research and development has "few close parallels in nonresearch activities."

tion of technological know-how [1]. Furthermore, the application of U.S. technology abroad is often replete with technological uncertainty, just as is the development of new technology itself [6]. The utilization of a technological innovation in a new context is likely to require, among other things, adjustment of some of the basic design parameters. For example, differences in the market size between home and abroad will induce scale adaptations to the plant; and differences in materials inputs, operator skills and engineering standards will frequently necessitate design changes in the process and/or the product. The implementation of design changes will produce uncertain responses in the quality and cost of the final product. Although the uncertainties generated are undoubtedly modest compared to those encountered during the original product or process development, they are still important. When uncertainty precludes immediate identification of the best design, it may be desirable to "hedge" by supporting several different designs [5]. By incurring higher project costs, hedging can reduce the project time relative to a procedure which explores different designs sequentially.

Besides "hedging" activities there are a number of other procedures which can be used to reduce project time, but they can all be expected to increase project costs. As additional engineers are brought on to the project to speed the design, diminishing returns can generally be expected. The concomitant increases in job segmentation will eventually augment coordination costs. Attempts to reduce project time by speeding equipment procurement can also be expected to increase project costs. Lead-times on major items of equipment can be reduced in a number of ways. For instance, the multinational firm can by-pass the equipment bidding procedure and the attendant delays by negotiating cost plus contracts with equipment suppliers. The disabilities and costs associated with this kind of contracting have been set out adequately elsewhere [7]. An alternative procedure is to solicit bids before the plant and equipment designs have been finalized. This may save several weeks, but firms generally run the risk of incurring penalty fees if the design specifications are subsequently modified. A number of procedures are also available to reduce manufacturing startup time, which commonly accounts for about 20% of total project time. For instance, the number and duration of pre-startup training programs can be increased. A more radical and costly procedure to facilitate a quick and smooth startup would involve the importation of large numbers of trained operators from established plants to assist during the startup period. Of course, if the new plant is the first of its kind, there may be little advantage to be gained from such costly procedures.

All of these various considerations provide the foundation for postulating a time-cost tradeoff that within some range has a negative slope and is convex to the origin. If the existence of some fixed costs is also postulated, then increasing project time need not always lower expected costs. The postulated tradeoff function is therefore

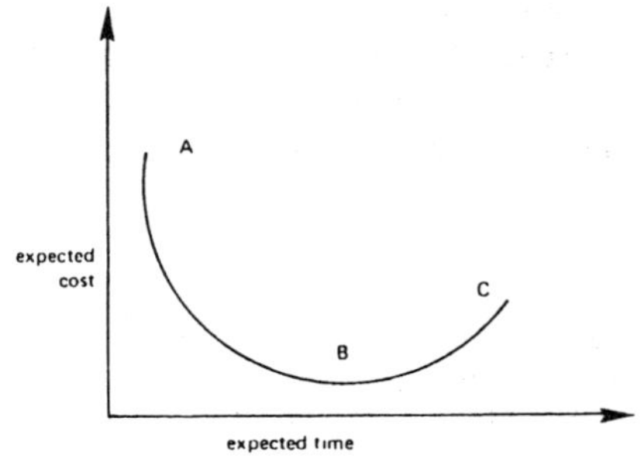

FIGURE 1

U-shaped (see Figure 1). Clearly the firm will not wish to operate to the right of B under any sort of sensible conditions, and so the range of the tradeoff function that is of greatest interest is AB.

3. Estimation of the Time-Cost Tradeoff

The relevant tradeoff function is hypothesized to be negatively sloped and convex to the origin. $\partial C/\partial T < 0$, $\partial^2 C/\partial T^2 > 0$, where C is the expected project cost, and T is the expected project time. In order to test this hypothesis, data were obtained for a sample of twenty international projects. All of the projects embodied U.S. technology, and fifteen of the twenty were in chemicals or petroleum refining. The remainder were in the machinery industry. The projects varied considerably in size and in geographical dispersion.[3]

Project managers were asked to estimate the percentage change in the actual cost of the project that would result from expected changes in the actual time.[4] The actual project cost and time were used as reference points. The project managers were also asked to estimate the minimum possible time in which the project could be completed. Five observations on time and cost were obtained for each project, including the actual time and cost.[5]

In order to estimate the function, the assumption is made that the downward sloping section of the time-cost tradeoff can be represented by:

$$C = V \exp\{\phi/((t/\alpha) - 1)\} \tag{1}$$

where C is the expected cost of the projects, t is the expected time, and V, α, and ϕ are parameters that vary from project to project. Figure 2 shows the nature of this function. It is convex and has time and cost asymptotes.

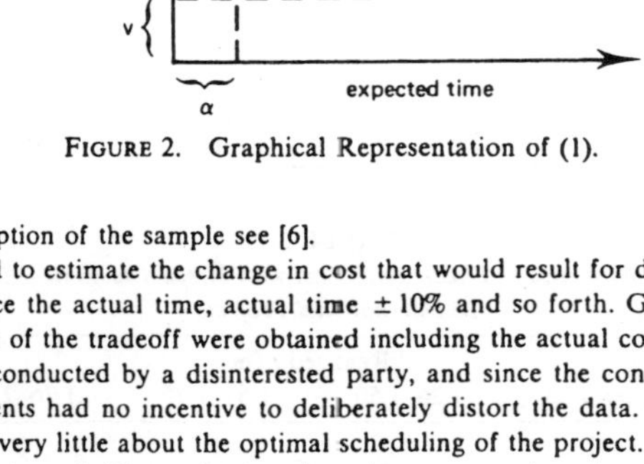

FIGURE 2. Graphical Representation of (1).

[3] For a detailed description of the sample see [6].

[4] Managers were asked to estimate the change in cost that would result for different time spans, such as half the actual time, twice the actual time, actual time ± 10% and so forth. Generally, five points on the negatively sloped portion of the tradeoff were obtained including the actual cost-time configuration. Since this was an experiment conducted by a disinterested party, and since the confidentiality of the data was guaranteed, the respondents had no incentive to deliberately distort the data. Furthermore, the time-cost tradeoff by itself implies very little about the optimal scheduling of the project. To discover this, a revenue function must also be estimated. The optimal project time occurs when the marginal revenue from time shaving equals the marginal cost from time shaving. Accordingly, the data supplied could not possibly be used to make inferences about the performance records of the respondent managers. These factors attenuate whatever opportunistic proclivities managers might possibly entertain.

[5] Although it was decided to estimate only the negatively sloped portion of the time-cost tradeoff, it is of interest to note that for 13 of the projects in the sample, costs would have increased if the expected time were doubled. Several respondents pointed out that inept management could quite easily create situations where it might be realized ex post that a project had proceeded on the positively sloped portion of the tradeoff.

Since C approaches V as t becomes larger, V can be considered the minimum expected cost of the project. Since t approaches α as C becomes larger, α can be considered the minimum expected time to complete the innovation. The elasticity of cost with respect to time, $(-dC/dt)\cdot(t/C)$, is equal to $\phi(t/\alpha)/[(t/\alpha)-1]^2$. Thus, for a given value of t/α, the elasticity of cost with respect to time is determined by ϕ. A logarithmic transformation of (1) together with the addition of an error term yields:

$$\ln C_i = \ln V_i + \phi[(t_i/\alpha) - 1]^{-1} + Z_i. \tag{2}$$

Since for each project direct estimates of α have been obtained from the respondents, estimates of V_i and ϕ could be obtained by regressing $\ln C_i$ on $1/[(t_i/\alpha) - 1]$. The error term Z_i is assumed to be distributed with mean zero and constant variance. The results are summarized in Table 1. In general, it can be said that the goodness of fit is acceptable, although in each case the number of observations is very small.

Using the estimated value of ϕ, the elasticity of cost with respect to time was calculated for a given value of t_i/α. (t_i was set at its actual value.) The results, given in Table 2, show that in fifteen out of twenty cases, strategies aimed at shaving the

TABLE 1

Estimates of V, α, and ϕ: 20 International Projects

Project	V (Thousands of Dollars)	α (Months)	ϕ	r^2
1	260	9	0.024	0.61
2	1,998	20	0.068	0.69
3	3,964	14	0.065	0.99
4	796	11	0.146	0.99
5	578	32	0.174	0.90
6	1,808	28	0.070	0.98
7	9,228	24	0.089	0.55
8	3,197	15	0.030	0.95
9	111	3	0.279	0.96
10	459	10	0.072	0.94
11	1,615	21	0.007	0.82
12	11,395	30	0.119	0.96
13	29,971	61	0.028	0.98
14	2,470	20	0.115	0.95
15	654	12	0.053	0.94
16	3,901	22	0.122	0.91
17	12,100	27	0.560	0.79
18	4,745	36	0.185	0.78
19	10,872	36	0.021	0.97
20	620	17	0.041	0.97

TABLE 2

Estimates of Elasticity of Cost with Respect to Time:
20 International Projects

Point Elasticity	Realized t/α 1.00–1.25	1.26–1.50	1.51–1.75	1.76–2.00	Total
0–0.50	1	0	1	0	2
0.51–1.00	2	1	0	0	3
1.01–1.50	1	1	0	1	3
1.51–2.00	4	0	0	0	3
2.01–2.50	3	1	0	0	3
2.51–3.00	1	1	0	0	2
3.01–3.50	2	1	0	0	3
Over 3.50	1	0	0	0	1
Total	13	5	1	1	20

actual time by one percent would raise costs by more than one percent. By comparison, Mansfield [3] discovered that for innovation, a one percent shaving in project time could be obtained for a cost increase of less than one percent for almost three quarters of the innovation projects examined. Mansfield's results are contained in Table 3. The time cost tradeoff function for international transfer of manufacturing technology is apparently much more elastic than the time cost tradeoff for innovation.

TABLE 3
Number of Innovations with Indicated Values of Elasticity of Cost with Respect to Time, at Actually Realized Values of t/α, for 29 Innovations

Value of	Actually Realized Value of t/α			
Elasticity*	1.00–1.50	1.51–2.00	Over 2.00	Total
	(Numbers of Innovations)			
0	0	0	1	1
0.01–0.49	1	2	7	10
0.50–0.99	2	3	0	5
1.00–2.00	1	3	2	6
Over 2.00	5	1	1	7
Total	9	9	11	29

* Arc elasticities were computed between the point at the actually realized value of t and the point at the next lower value of t given by the respondent. The difference in t/α is often quite large.
Source: Mansfield [3, p. 140].

4. Determinants of the Elasticity of Cost with Respect to Time

If seems likely that the elasticity is related in some way to the structure of the network of tasks involved in the technology transfer project. Ideally, resources should be provisionally allocated in a project so that all tasks on the critical path are at the same marginal cost level [4]. In order to reduce total project time, resources are allocated to the critical path tasks. If in this process the critical path shifts widely, the elasticity of the time-cost tradeoff is apt to be high. If, on the other hand, the critical path involves much the same tasks throughout the process, the elasticity of the time-cost tradeoff is apt to be lower. The elasticity is also likely to be higher the higher the proportion of critical path task expenditures are to total project expenditures.

Unfortunately, data on these underlying considerations are not readily available, and it is therefore difficult to test the above hypotheses directly. In the analysis of the determinants of the time-cost elasticity which follows, attention is confined to variables for which data are readily available. Nevertheless, it is recognized that some of the variables selected may simply be surrogates for more fundamental considerations such as the ones outlined above.

The first hypothesis to be advanced is that the elasticity will be lower the longer the duration of the preliminary planning stage, hereafter stage A,[6] relative to the other stages. Stage A usually can be telescoped, if the need arises, since it utilizes resources entirely at the firm's own command. The converse of this hypothesis is that the engineering, construction, installation, and startup can be telescoped, but only at a relatively greater expense. This is because these activities generally involve contractual relations with other firms in which the authority of the transferor is diminished.

The second hypothesis is that the elasticity will be lower if the technology to be embodied in the new facility has been applied previously. If there has been at least

[6] For a detailed elaboration of the sequence of activities in transferring technology, see [6].

one previous application, then attention can be directed away from problems of the technology per se to problems of the transfer; that is, because of the uncertainty involved, a strategy to speed up the project by maximizing overlap will present colossal problems if the technology has not been applied previously.

The third hypothesis is that the elasticity will also be determined by the size of the primary transfer agent. The primary transfer agent is defined as the enterprise with the controlling equity in the new venture. Thus, the transferor will be the primary agent if the transfer is to a subsidiary, but the transferee will be the primary agent if the transfer is to an independent enterprise. The primary agent is generally the entity which will hire the engineering contractors and authorize the bidding on key pieces of equipment. It will have the responsibility for expediting the project. It seems reasonable to hypothesize that insofar as speeding up a project requires a certain flexibility of approach, a larger organization may be handicapped by inertia and a more complex decision making procedure. Even reaching agreement on how to proceed will take longer if more people and a longer chain of command are involved.

The fourth hypothesis is that the elasticity may be influenced by total project costs. On a priori grounds it seems reasonable to hypothesize that because large projects require more coordination and integration of different tasks, they may be more costly to speed up than smaller projects.

Finally, it is hypothesized that the actually realized value of the elasticity is also a function of whether or not the foreign market can be satisfactorily supplied by exports in the interim. If trade barriers are not prohibitive, and if exporting has traditionally been used to supply a particular foreign market, then the marginal profits from bringing the new plant on stream more quickly are not likely to be enormous. If, on the other hand, prohibitive tariffs or import controls prohibit sourcing from abroad, then the returns from time shaving are likely to be enhanced and the elasticity measure higher since the project will be pushed a little faster.[7]

To test these hypotheses it was assumed that

$$\epsilon_i = \alpha_0 + \alpha_1 A_i + \alpha_2 U_i + \alpha_3 S_i + \alpha_4 C_i + \alpha_5 X_i + Z_i,$$

where ϵ_i is the estimated elasticity of cost with respect to time for the ith project calculated at the actual time taken for the transfer; A_i is the percentage of total time allocated to stage A of the ith project; U_i is a dummy variable that takes the value 1 if the ith project represents the first application of the technology, and 0 otherwise; S_i is the sales volume (in millions of dollars) of the primary transfer agent for the ith project; C_i is the total cost of the ith project (measured in thousands of dollars); X_i is a dummy variable which takes the value 1 if the foreign market which the new plant will supply was previously supplied by export by the transferor. (If it was not supplied in this manner $X_i = 0$.) Z_i is a random error term with mean zero and constant variance. The expected derivatives are

$$\partial\epsilon/\partial A_i < 0, \quad \partial\epsilon/\partial U_i > 0, \quad \partial\epsilon/\partial S_i < 0, \quad \partial\epsilon/\partial C_i > 0, \quad \partial\epsilon/\partial X_i < 0.$$

Ordinary least square estimates of the α's yielded:

$$\epsilon_i = \underset{(10.11)}{2.20} - \underset{(2.02)}{1.82 A_i} + \underset{(2.81)}{1.20 U_i} + \underset{(2.15)}{0.00014 S_i} + \underset{(1.48)}{0.00011 C_i} - \underset{(2.39)}{0.83 X_i}$$

$$r^2 = 0.71, \quad n = 20.$$

[7] A profit maximizing firm will choose a location on the tradeoff function such that the marginal benefit and the marginal cost of time shaving are equalized. The actually realized elasticity measure will therefore be sensitive to both the parameters of the time-cost tradeoff, as well as the parameters of the benefit function. The four previous hypotheses represent parameters entering the cost function, whereas the existence of a viable export option will be a parameter entering the benefit function.

All of the variables are significant and take the expected signs. Most interestingly, U_i is highly significant and has a sizable impact. If the technology has not been previously commercialized, the elasticity measure is increased by 1.20. Technological uncertainty is clearly an important determinant of the elasticity estimates. X_i is also highly significant, suggesting that the incentive to shave time is reduced if the market can be simultaneously supplied from the parent plant or from foreign subsidiaries. The coefficient on S_i indicates that a hundred million dollar increase in the size of the primary transfer agent increases the elasticity by 0.14. A million dollar increase in project size increases the elasticity by a similar amount. The large coefficient on A_i indicates considerable potential for shaving time when the preliminary planning stage has been protracted. By juxtaposing these results against those derived from analysis of the time-cost tradeoff in technological innovation [3], it becomes apparent that variables indicating how well a technology is understood[8] are particularly significant determinants of the time-cost elasticity. Measures of project size and firm size are likewise significant in both studies.

5. Conclusion

The examination and estimation of time-cost tradeoff for the establishment of foreign manufacturing facilities have been instructive for several reasons. First, the existence of a negatively sloped function has once more been demonstrated. This is of some importance since the existence of time-cost tradeoffs has at times been treated with considerable skepticism.[9] Second, the elasticity estimates were generally greater than one, indicating that time shaving would involve rather high incremental costs. Third, it was observed that the elasticity estimates were highest for projects where the technology had not been previously commercialized, for projects that were large, and for projects carried out by the larger firms.

Several important implications follow from the analysis. First, when the transferred technology involves a change in the state of the art, the extra costs of speeding a project would seem to be considerable. In fact, the sensitivity of cost with respect to time indicates the extreme importance of scheduling the project precisely. Alternatively, if accurate scheduling is difficult to achieve, the necessity of commercializing the technology at home before its transfer abroad is suggested. Secondly, smaller firms implementing smaller projects seem to possess more versatility than the larger firms implementing the larger projects. Both of these implications rest comfortably with a view of technological change which recognizes the deficiencies as well as the strengths of the larger firm as agents of technological change, and which also recognizes the high cost of innovation and technology transfer when an alteration of the state of the art is involved.[10]

[8] The variables referred to are the extent of the "state of the art" advance in the study of innovation [3], and secondly, a dummy variable used to indicate whether or not the technology has been previously applied.

[9] The reference is to military men who have argued that although hastening a project will undoubtedly increase costs per time period, the reduction in total project time will offset the higher rates of expenditure (see [5]).

[10] This paper is based on the author's Ph.D. dissertation, "The Multinational Corporation and the Resource Cost of International Technology Transfer," University of Pennsylvania, 1975. The research was supported by a grant from the National Science Foundation to Professor Edwin Mansfield of the University of Pennsylvania, whose trenchant comments were much appreciated during all phases of the study. The author's particular gratitude goes to the participating firms, without whose cooperation this paper would not have been possible.

References

1. Caves, R., "International Corporations: The Industrial Economics of Foreign Investment," *Economica*, Vol. 38 (Feb. 1971), pp. 1–27.
2. Malcolm, D., "Integrated Research and Development Management Systems," in *Operations Research in Research and Development*, Burton Dean, ed., John Wiley, New York, 1963.
3. Mansfield, E., Rapoport, J., Schnee, Wagner, S. and Hamburger, M., *Research and Innovation in the Modern Corporation*, W. W. Norton, New York, 1971.
4. Moder, J. and Phillips, C., *Project Management with CPM and PERT*, Reinhold, New York, 1964.
5. Scherer, F. M., "Government Research and Development Programs," in *Measuring Benefits of Government Expenditures*, R. Dorfman, ed., The Brookings Institution, Washington, D. C., 1965.
6. Teece, David, *The Multinational Corporation and the Resource Cost of International Technology Transfer*, Ballinger, Cambridge, 1977.
7. Williamson, O. E., *Markets and Hierarchies: Analysis and Antitrust Implications*, Free Press, New York, 1975.

ANNALS, *AAPSS*, 458, November 1981

The Market for Know-How and the Efficient International Transfer of Technology

By DAVID J. TEECE

ABSTRACT: This article explores the nature of international technology transfer and the operation of the market for know-how. It begins by examining the relationship between codification and transfer costs and then analyzes various imperfections in the market for know-how. The special properties of know-how are shown to confound various aspects of the exchange process when arms-length contracting is involved. The internalization of the exchange process within multinational firms serves to bypass many of these difficulties, and explains why the multinational firm is of such importance. Several forms of regulation of technology imports and exports are examined. It is discovered that the process is insufficiently well understood to permit the design of effective regulation that, moreover, appears unlikely to eliminate inefficiency. An efficiency focus is maintained throughout since I feel no qualification to pontificate on complex and confused distributional issues.

David J. Teece is associate professor of business economics at the Graduate School of Business, Stanford University. He has a bachelor's and master's degree from the University of Canterbury, Christchurch, New Zealand, and a doctorate in economics from the University of Pennsylvania, which he received in 1975. Professor Teece specializes in the fields of industrial organization and the economics of technological change, and has published numerous scholarly articles and monographs on the multinational firm, the organization of the petroleum industry, the behavior of the Organization of Petroleum Exporting Countries (OPEC), and the relationship between the internal organization and performance of large enterprises.

NOTE: The financial support of the National Science Foundation is gratefully acknowledged, together with the valuable comments from Max Boisot, Almarin Phillips, and Oliver Williamson.

ECONOMIC prosperity rests upon knowledge and its useful application. International, interregional, and interpersonal differences in levels of living can be explained, at least in part, by differences in the production techniques employed. Throughout history, advances in knowledge have not been uniformly distributed across nations and peoples, but have been concentrated in particular nations at particular times. According to Kuznets,

> . . . the increase in the stock of useful knowledge and the extension of its application are of the essence of modern economic growth. . . . No matter where these technological and social innovations emerge . . . the economic growth of any given nation depends on their adoption. In that sense, whatever the national affiliation of resources used, any single nation's economic growth has its base somewhere outside its boundaries—with the single exception of the pioneering nations.[1]

The rate at which technology is diffused worldwide depends heavily on the resource costs of transfer—both transmittal and absorption costs—and on the magnitude of the economic rents obtained by the seller. The resource costs of transfer depend on the characteristics of the transmitter, the receiver, the technology being transferred, and the institutional mode chosen for transfer.[2] These are matters explored in the following section. The rents obtained are a function of the working of the market for know-how, a matter explored in a subsequent section. The last two sections explore regulatory issues with respect to this market from the perspective first of less-developed country (LDC) importers and from the perspective of the United States as a net exporter of know-how. What emerges is an understanding of the technology transfer process, the role of the multinationals, and the difficulties and occasional contradictions associated with regulation. In no sense can the market for know-how and the transfer process be said to operate in an ideal fashion. However, internalization of the process appears to offer considerable efficiencies, and "codes of conduct" are likely to confound the very objectives of importers, while export controls can be expected to yield only limited benefits, and then only under special conditions.

CODIFICATION AND TRANSFER COSTS

The fact that different individuals, organizations, or nations possess different types of knowledge and experience creates opportunities for communication and mutually profitable transfer. Yet, paradoxically, such transfer as does take place among individuals and organizations can only do so on the basis of similarities in the knowledge and experience each possess. A shared context appears necessary for the formulation of meaningful messages. Transmittal and receiving costs are lower the greater the similarities in the experience of the transmitting unit and the receiving unit; for the greater these similarities, the easier it is to transfer technology in codified form, such as

1. S. Kuznets, *Modern Economic Growth: Rate, Structure, Spread.* (New Haven: Yale University Press, 1966).

2. The concept and measurement of the resource cost of transfer can be found in David Teece, *The Multinational Corporation and the Resource Cost of International Technology Transfer* (Cambridge: Ballinger, 1976), and in "Technology Transfer by Multinational Firms: The Resource Cost of International Technology Transfer," *Economic Journal* (June 1977).

blueprints, formulas, or computer languages.

Furthermore, there appears to be a simple but powerful relationship between codification[3] of knowledge and the costs of its transfer. Simply stated, the more a given item of knowledge or experience has been codified, the more economically it can be transferred. This is a purely technical property that depends on the ready availability of channels of communication suitable for the transmission of well-codified information—for example, printing, radio, telegraph, and data networks. Whether information so transferred will be considered meaningful by those who receive it will depend on whether they are familiar with the code selected as well as the different contexts in which it is used.[4]

Uncodified or tacit knowledge, on the other hand, is slow and costly to transmit. Ambiguities abound and can be overcome only when communications take place in face-to-face situations. Errors of interpretation can be corrected by a prompt use of personal feedback. Consider the apprenticeship system as an example. First, a master craftsman can cope with only a limited number of pupils at a time; second, his teaching has to be dispensed mostly through examples rather than by precept—he cannot easily put the intangible elements of his skill into words; third, the examples he offers will be initially confusing and ambiguous for his pupils so that learning has to take place through extensive and time-consuming repetition, and mastery will occur gradually on the basis of "feel"; finally, the pupil's eventual mastery of a craft or skill will remain idiosyncratic and will never be a carbon copy of his master's. It is the scope provided for the development of a personal style that defines a craft as something that goes beyond the routine and hence programmable application of a skill.

The transmission of codified knowledge, on the other hand, does not necessarily require face-to-face contact and can often be carried out largely by impersonal means, such as when one computer "talks" to another, or when a technical manual is passed from one individual to another. Messages are better structured and less ambiguous if they can be transferred in codified form. Take for example Paul Samuelson's introductory textbook for students of economics. Year after year, thousands of students all over the globe are introduced to Samuelson's economic thinking without being introduced to Samuelson himself. The knowledge acquired will be elementary and standardized, an idiosyncratic approach at this level being considered by many as a symptom of error rather than of style. Moreover

3. Codification—the transformation of experience and information into symbolic form—is an exercise in abstraction that often economizes on bounded rationality. Instead of having to respond to a hopelessly extensive and varied range of phenomena, the mind can respond instead to a much more restricted set of information. At least two obstacles stand in the way of effective codification. First, abstracting from experience can be accomplished in an almost infinite number of ways. Ask a group of painters to depict a given object and each will select different facets or features for emphasis. Furthermore, the choice of what to codify and how to codify it is often personal. Second, to structure and codify experience one way can make it difficult, subsequently, to do so in an alternative way. The conceptual channels through which experience is made to flow appear to run deep and resist rerouting.

4. These ideas are developed further in C. E. Shannon and W. Weaver, *The Mathematical Theory of Communication* (Chicago: University of Illinois Press, 1949). I am grateful to Max Boisot for drawing them to my attention.

the student can pick up the sage's book or put it down according to caprice; he can scan it, refer to it, reflect upon it, or forget it. This freedom to allocate one's attention or not to the message source is much more restricted where learning requires interpersonal contact.

With respect to the international transfer of technology, the costs of transfer are very much a function of the degree to which know-how can be codified and understood in that form by the recipient. Typically, only the broad outline of technical knowledge can be codified by nonpersonal means of intellectual communication or communication by teaching outside the production process itself. Accordingly, the transfer of technology generally requires the transfer of skilled personnel, even when the cultural and infrastructural differences are not great. History has illustrated this time and time again. For instance, the transfer of technological skills between the United States and Britain at the end of the nineteenth century was dependent upon the transfer of skilled personnel. One also observes that the diffusion of crafts from one country to another depends on the migration of groups of craftsmen, such as when the Huguenots were driven from France by the repeal of the Edict of Nantes under Louis XIV.

The costs of transfer so far examined are simply the resource costs of transfer—the costs of the labor and capital that must be employed to effect transfer. An empirical investigation of these issues based upon a sample of 26 international transfers indicated that the resource cost of international transfer is nontrivial.[5] Transfer costs ranged from 2.25 percent to 59 percent of total project costs with a mean of 19.16 percent. They declined with each subsequent application of the technology and were typically lower the greater the amount of related manufacturing experience possessed by the transferee. Experience with transfer and experience with the technology appear to be key considerations with respect to the ease with which technology can be transferred abroad. In order to understand these costs, as well as other aspects of the transfer process, it will be necessary to examine the market for know-how. In so doing, the focus is on private transactions between firms of different national origins.

5. See David Teece, "Technology Transfer by Multinational Firms: The Resource Cost of International Technology Transfer," *Economic Journal* (June 1977).

CHARACTERISTICS OF THE MARKET FOR KNOW-HOW

The differential distribution of know-how and expertise among the world's enterprises means that mutually advantageous opportunities for the trading of know-how commonly exist. However, these opportunities will be realized only if the institutional framework exists to provide the appropriate linkage mechanisms and governance structures to identify trading opportunities and to surround and protect the associated know-how transfers. Unfortunately, unassisted markets are seriously faulted as institutional devices for facilitating trading in many kinds of technological and managerial know-how.

The imperfections in the market for know-how for the most part can be traced to the nature of the commodity in question. Know-how has some of the characteristics of a public good, since it can often be used in another enterprise without its value being substantially impaired. Furthermore the marginal cost of employing know-how abroad is likely to be much less than its average cost of production and transfer. Accordingly the international transfer of proprietary know-how is likely to be profitable if organizational modes can be discovered to conduct and protect the transfer at low cost.

An examination of the properties of markets for know-how readily leads to the identification of several transactional difficulties. These difficulties can be summarized in terms of recognition, disclosure, and team organization. Consider a team that has accumulated know-how that can potentially find application in foreign markets. If there are firms abroad that can apply this know-how with profit, then according to traditional microeconomic theory, trading will ensue until the gains from trade are exhausted. Or, as Calabresi has put it, "if one assumes rationality, no transactions costs, and no legal impediments to bargaining, all misallocations of resources would be fully cured in the market by bargains."[6] However, one generally cannot expect this happy result in the market for proprietary know-how. Not only are there high costs associated with obtaining the requisite information, but there are also organizational and strategic impediments associated with using the market to effect transfer.

Consider the information requirements associated with using markets. In order to carry out a market transaction, it is necessary to discover potential trading partners and acceptable terms of trade. It is also necessary to conduct negotiations leading up to the bargain, to draw up the contract, to undertake the inspection needed to make sure that the terms of the contract are being observed, and so on. As Kirzner has explained,

> for an exchange transaction to be completed it is not sufficient merely that the conditions for exchange which prospectively will be mutually beneficial be present; it is necessary also that each participant be aware of his opportunity to gain through the exchange. . . . It is usually assumed . . . that where such scope is present, exchange will in fact occur. . . . In fact, of course, exchange may fail to occur because knowledge is imperfect, in spite of the presence of the conditions for mutually profitable exchange.[7]

The transactional difficulties identified by Kirzner are especially compelling when the commodity in question is proprietary information. One reason is that protecting the ownership of technological know-how often requires the suppression of information on exchange possibilities. By its very nature, industrial R&D requires that the activities and outcomes of the R&D establishment be disguised or concealed.

Even where the possessor of the technology recognizes the opportunity and has the capability to absorb know-how, markets may break down. This is because of the problems of disclosing value to buyers in a way that is convincing and that does not destroy the basis for exchange. Due to informational asymmetries, the less informed party must be wary of opportunistic representations by the seller. Moreover, if there is sufficient disclosure to assure the buyer that the information possesses great value, the "fundamental paradox" of information arises: "its value for the purchases is not known until he has the

6. G. Calabresi, "Transactions Costs, Resource Allocation, and Liability Rules: A Comment," *Journal of Law and Economics*, (April 1968).

7. I. Kirzner, *Competition and Entrepreneurship* (Chicago: University of Chicago Press, 1962), p. 215.

information, but then he has in effect acquired it without cost."[8]

Appropriability issues emerge not only at the negotiating state but also at all subsequent stages of the transfer. Indeed, as discussed elsewhere in this issue, Magee has built a theory of multinational enterprise around the issue of appropriability, hypothesizing that

> multinational corporations are specialists in the production of information that is less efficient to transmit through markets than within firms.[9]

However, the transactional difficulties in the market for know-how that provide an incentive for firms to internalize technology transfer go beyond issues of recognition and appropriability. Thus suppose that recognition is no problem, that buyers concede value and are prepared to pay for information in the seller's possession, and that enforceable use restrictions soften subsequent appropriability problems. Even if these assumptions are satisfied, there is still the problem of actually transferring the technology.

In some cases the transfer of a formula or a chemical compound, the blueprints for a special device, or a special mathematical algorithm may be all that is needed to effect the transfer. However, more is frequently needed. As mentioned earlier, know-how cannot always be codified, since it often has an important tacit dimension. Individuals may know more than they are able to articulate.[10] When knowledge has a high tacit component, it is extremely difficult to transfer without intimate personal contact, demonstration, and involvement. Indeed, in the absence of intimate human contact, technology transfer is sometimes impossible. In a slightly different context Polanyi has observed, "It is pathetic to watch the endless efforts—equipped with microscopy and chemistry, with mathematics and electronics—to reproduce a single violin of the kind the half literate Stradivarius turned out as a matter of routine more than 200 years ago."[11]

In short, the transfer of knowledge may be impossible in the absence of the transfer of people. Furthermore, it will often not suffice just to transfer individuals. While a single individual may sometimes hold the key to much organizational knowledge, team support is often needed, since the organization's total capabilities must be brought to bear upon the transfer problem. In some instances the transfer can be effected through a one-time contract providing for a consulting team to assist in the startup. Such contracts may be highly incomplete and may give rise to dissatisfaction during execution. This dissatisfaction may be an unavoidable—which is to say, an irremediable—result. Plainly, foreign investment would be a costly response to the need for a one-time international exchange. In the absence of a superior organizational alternative, one-time, incomplete contracting for a consulting team is likely to prevail.

Reliance on repeated contracting is less clearly warranted, however, where a succession of transfers is contemplated, or when two-way communication is needed to promote the recognition and disclosure of opportunities for information

8. K. J. Arrow, *Essays in the Theory of Risk Bearing* (Chicago: Chicago University Press, 1971).

9. See Stephen Magee, "Information and Multinational Corporation: An Appropriability Theory of Direct Foreign Investment," in *The New International Economic Order*, ed. Jagdish Bhagwati (Cambridge, MA: MIT Press, 1977), p. 318.

10. See Michael Polanyi, *Personal Knowledge: Towards a Post Critical Philosophy* (Chicago: University of Chicago Press, 1958).

11. Polanyi.

transfer as well as the actual transfer itself. In these circumstances a more cooperative arrangement for joining the parties would enjoy a greater comparative institutional advantage. Specifically, intrafirm transfer to a foreign subsidiary, which avoids the need for repeated negotiations and attenuates the hazards of opportunism, has advantages over autonomous trading. Better disclosure, easier agreement, better governance, and more effective team organization and reconfiguration all result. Here lies the incentive for internalizing technology transfer within the multinational firm.

The preceding discussion has emphasized that an important attribute of the multinational firm is that it is an organizational mode capable of internally transferring know-how among its various business units in a relatively efficient and effective fashion. Given the opportunities that apparently exist for international trade in know-how, and given the transactional difficulties associated with relying on markets, one should expect to find multinational enterprises (MNEs) frequently selecting internal channels for technology transfer. However, when problems of recognition, disclosure, and team transfer are not severe, one should expect that market processes will be utilized, in which case the licensing of know-how among nonaffiliated enterprises will be observed.

Recognition, disclosure, and team transfer problems will be modest, it would seem, when the following exist: (1) the know-how at issue is not recent in origin so that knowledge of its existence has diffused widely; (2) the know-how at issue has been commercialized several times so that its important parameters and performance in different situations are well understood, thereby reducing the need for start-up assistance; and (3) the receiving enterprise has a high level of technological sophistication. Some evidence supportive of these propositions has recently been presented. Mansfield, employing a sample of 23 multinationals, discovered that foreign subsidiaries were the principal channel of transfer during the first five years after commercialization.[12] For the second five-year period after commercialization, licensing turned out to be more important. Larger firms also tended to rely more on internal transfer than did smaller firms, although this might not reflect relative efficiency considerations but rather the sunk costs larger firms have already made in foreign subsidiaries.

One implication for a potential technology purchaser is that it will have to look among the smaller firms in the industry, and at firms in different industries, to find willing technology suppliers. This does not result in an easy search process. It is made more difficult by the fact that few firms actively market their know-how. Thus the apparent size and nature of the market is likely to be a function of the search costs buyers are willing to incur.

Another implication is that because the marginal cost of successive applications of a technology is less than the average cost of production and transfer, and because know-how is often unique—implying that trading relations are

12. See Edwin Mansfield, "Statement to the Senate Commerce Committee Concerning International Technology Transfer and Overseas Research and Development," Hearings before the Subcommittee on International Finance of the Committee on Banking, Housing, and Urban Affairs of the Committee on Commerce, Science, and Transportation, United States Senate, Ninety-fifth Congress, Second Session, Part 7: Oversight on U.S. High Technology Exports (Washington, DC: Government Printing Office, May 1978).

characterized by small numbers—there is often a high degree of indeterminacy with respect to price. Killing's field research confirmed that "neither buyer nor seller of technology seems to have a clear idea of the value of the commodity in which they are trading," fueling speculation that "royalty rates may simply be a function of negotiating skills of the parties involved."[13] This is because the market for know-how commonly displays aspects of bilateral monopoly, at least at the level of the individual transaction. So in many important cases there is likely to be a wide range of indeterminacy.

The existence of elements of bilateral monopoly has led some countries to advocate regulation of the market for know-how. Indeed, some Third World countries, as well as the antitrust authorities in some developed countries, have already imposed various regulatory regimes.

> By 1974, over 20 countries had enacted specific legislation to control and direct foreign capital and technology. Their actions and regulations focused on lowering the royalties paid for foreign technology, forcing local participation in management and ownership, and in increasing the government capability to screen and direct foreign activities—the major focus of the governments was initially to limit the kind of restrictive clauses allowed in contracts for technology transfer with foreign firms.[14]

Governmental and intergovernmental intervention in the market for know-how appears to be growing in significance. In the following sections, several dimensions of this phenomenon are explored in more detail.

CODES OF CONDUCT AND THE REGULATION OF TECHNOLOGY IMPORTS

Since the United Nations Conference on Trade and Development (UNCTAD) IV decided to set up an intergovernmental group of experts to prepare a draft of an international code of conduct on the transfer of technology, discussion has intensified on matters associated with the transfer and development of technology, particularly on topics of concern of developing countries.[15] A number of draft codes have emerged in which representatives from less-developed countries have argued that technology is part of the universal heritage of mankind and that all countries have right of access to technology in order to improve the standards of living of their peoples. Such contentions obviously involve fundamental challenges to the world's industrial property system. They also fail to recognize the constitutional restraints in countries such as the United States that pre-

13. Peter Killing, "Technology Acquisition: License Agreement or Joint Venture," *Columbia Journal of World Business* (Fall 1980).

14. See Harvey Wallender, "Developing Country Orientations Towards Foreign Technology in the Eighties: Implications for New Negotiation Approaches," *Columbia Journal of World Business* (summer 1980): 21-22.

15. The movement toward an international code on the transfer of technology is but a reflection of larger, exceedingly complex political problems that have been engendered by an international society undergoing profound changes. Demands for a new international economic order, international regulation of transnational enterprises, and the like form the backdrop of UNCTAD's activities in the technology transfer area. These broader demands raise the possibility that the work now being carried on by UNCTAD in moving toward a code of conduct for the transfer of technology will be subsumed by the development of a more comprehensive code of conduct for transnational enterprises by the U.N. Commission on Transnational Corporations.

vent the government from confiscating private property.

The stated objective of the UNCTAD code is "to encourage the transfer of technology transactions, particularly those involving developing countries, under conditions where bargaining positions of the parties to the transaction are balanced in such a way so to avoid abuses of a stronger position and thereby to achieve mutually satisfactory agreement." One of the principal mechanisms by which this is to be achieved is through the elimination of "restrictive business practices."[16] A long litany of these is typically advanced, including tying or packaging, use restrictions, exclusive dealing, and territorial restrictions. An examination of recent legislation on the transfer of technology, particularly in Latin America and Yugoslavia, shows that many of these ideas have been uncritically accepted into national law.[17]

It is not possible to attempt a comprehensive review of restrictive business practices in this article. However, I submit that insufficient analysis has been given to the efficiency-enhancing attributes of many practices surrounding the generation and transfer of technology. Many restrictive clauses in licensing and know-how agreements are designed to protect the transaction and the underlying know-how; in their absence less technology might be transferred, to the mutual detriment of all, or technology might be transferred less efficiently. In the space that follows, two "restrictive business practices" —use restrictions and tying—are analyzed in order to illustrate that "restrictive business practices" can be in fact procompetitive and may serve to promote economic efficiency.

Use restrictions

The interesting question associated with use restrictions is whether they are anticompetitive, designed merely to extract monopoly rents, or whether they are efficiency instruments, the removal of which might leave both parties worse off. Since know-how is the principal resource upon which the value of many private enterprise firms is based, firms facing market competition are not going to sell it carte blanche to a firm that might use it to compete with their own products, for to do so would reduce the value of the firm. Thus reasonable limitations on use are commonly necessary to provide adequate incentives for transfers to occur and for those transfers to operate efficiently. This is especially true when the transferor and the transferee are competitors or potential competitors.

When know-how is transferred by a market transaction (contract) the buyer does not acquire the asset to the exclusion of use by the seller in the same sense as occurs when a physical item is bought and sold. The seller of know-how retains the knowledge even after it has been transferred to a buyer. Furthermore, technology is constantly evolving. Indeed, static technology is generally obsolete technology. Accordingly, a buyer of intangible know-how typically needs ongoing, future cooperation from the seller to obtain the full benefit of the know-how purchased, since all of the

16. See UNCTAD, "Draft International Code of Conduct on the Transfer of Technology," TD/CODE/TOT/20.

17. See UNCTAD, "Selected Legislation, Policies and Practices on the Transfer of Technology," TD/B/C.6/48.

learning and experience of the developer of the know-how cannot be captured in the codified descriptions, drawings, and data that are amenable to physical transfer.

Limitations on the use of technological know-how are often needed to provide adequate incentives for the buyer and the seller to effect a continuous transfer of the knowledge in question. If the seller is limited in his use of the know-how, the buyer can rely more confidently on the seller's full disclosure and cooperation in the buyer's use of the know-how. Where the seller contemplates some use of the know-how himself, limitations on the buyer's use of the know-how in competition with the seller are necessary to provide the seller with the incentive to transfer this know-how and to share fully in his mental perceptions, understandings, working experience, and expertise.

A partial analogue to these principles is when business enterprises are sold. These transactions have traditionally included ancillary limitations on the economic activities of the seller after the business is sold. Such limitations bring about economically efficient transfers of ongoing businesses by ensuring that the buyer acquires exclusively the enterprise—or part of the enterprise—he is contracting to purchase, including its intangible goodwill. Similarly, in the sale of a business the seller is often retained as a consultant for the purpose of ensuring that the intangible knowledge that comes from the seller's experiences in conducting the business is fully transferred in the transaction. Without contractual or other limitations on the seller's use of the assets being transferred, and without the seller's continued cooperation, a buyer would not pay the full economic value of those assets. As a result, the efficient transfer of the assets would be inhibited.

Use limitations are particularly beneficial when two or more uses exist for the products that can be derived from know-how and when some of the uses are for some reason foreclosed to the developer of the know-how. In this instance, transfer of the know-how to a buyer having access to one or more of these otherwise foreclosed uses may be beneficial to both parties, since economies of scope will be generated. The seller of the know-how requires adequate incentives to transfer his knowledge, however. The seller will not transfer the know-how to a buyer for the otherwise foreclosed uses if, in doing so, he is likely to lose more in the uses that are available to him with no transfer than he gains through the expanded uses made possible by transfers. The availability of limitations on the buyer's use of the know-how provides possible means to prevent such losses.

Use limitations are also beneficial in providing incentives for the contracting parties to share complementary know-how in order to reach a new market that neither acting independently could efficiently serve. If each of the parties has one or more of the technology elements critical for a particular new use, if neither of the parties has all of the critical technology elements for that use, and if through sharing of the complementary technologies for the new use one or both of the parties could enter markets that neither party could serve without sharing, then use limitations are necessary to effect the bilateral technology transfers. Without use limitations, one or both of the parties may lack the incentive to share, since the losses that might occur in an existing market through sharing could exceed the gains derived from reaching the new market.

Tying and packaging

In a tying arrangement, the seller requires the buyer to purchase a second product as a condition of sale of the first, such as when a petrochemical firm licenses its process technology to another firm on the condition that it purchase certain inputs on a continuous basis, or when an automobile company agrees to build a facility abroad so long as it is able to select equipment and designs for the whole facility and not just for part of it.

In the context of the international transfer of technology, there are often very genuine managerial and technical reasons for tying the sale of products. For instance, coordinated design and construction might allow important systems engineering functions to be carried out more efficiently. Furthermore, processing facilities may require raw materials and components that meet certain narrow technological standards, and tying may be necessary to ensure that the requisite amount of quality control is exercised. These problems are likely to be especially severe when the technological distance between the transferor and transferee is great.

It is only under rather special circumstances that tying will enable a monopolist to expand the amount of monopoly profit that would be obtained in the absence of tying. One such circumstance is if tying can be used as a method of price discrimination.

Accordingly, blanket prohibitions against tying and packaging are likely to be costly to the country imposing the prohibitions. Technology suppliers may have good reasons for wanting to supply know-how and other products and services in a package. Certainly some striking examples exist of problems that have arisen when adequate packaging and systems design have not been performed. Consider the Soviet Union's experience in constructing and starting up its Kama River truck plant, as related by Lee Iacocca, then with Ford Motor Company:

> Well, one example of acquiring technology in its unbundled state is the Kama River truck plant in Russia. After first attempting to get a foreign company to build the plant (we were approached but decided against it) the Russians decided to do it themselves and to parcel out contracts to foreign firms for various parts of the project. That was in 1971. As of December 1976, the project was almost two years behind schedule. By year's end, only about 5,000 trucks were expected to roll off the line, instead of the 150,000 vehicles and 100,000 diesel engines and transmissions originally scheduled for annual production. According to published reports, only four of nine projected furnaces in the iron foundry were operating and those only at half capacity. What's more, 35 percent of the castings were being rejected as unserviceable. There were bottlenecks on the assembly line, and because the components and designs were bought from different suppliers all over the world, replacement parts were not interchangeable.
>
> Now compare that with Ford's recent investment in Spain. It took us just three years to the day to build a complex that includes an assembly plant, a stamping and body plant and an engine plant on a manufacturing site 2 1/2 miles long and half a mile wide, with 55 acres under roof. The first Fiesta, our new minicar, was driven off the assembly line last August, well ahead of schedule. To get from farmland to an annual capacity of 250,000 cars and 400,000 engines in

three years, we drew on the experience of our personnel and our technological resources from all over the world—experience and resources that couldn't be bought and that we probably wouldn't even know how to sell.[18]

REGULATION OF TECHNOLOGY EXPORTS

Pressures for restricting trade in technological know-how have also come from technology exporters. The reasons advanced for controls are almost the complete opposite of those advanced by the LDCs. In the United States concern is often expressed in industry and government that the United States is either selling its technology for far less than its economic value, or allowing it to be stolen through industrial espionage, principally to other developed countries, or simply transferring it abroad too soon. According to J. Fred Bucy, the president of Texas Instruments:

> Today our toughest competition is coming from foreign companies whose ability to compete with us rests in part on their acquisitions of U.S. technology. ... The time has come to stop selling our latest technologies, which are the most valuable things we've got.[19]

Labor groups in the United States go further and argue that not only is the know-how underpriced, but that one consequence of the export of technology is the export of jobs.[20] According to one labor leader:

> I recognize that technology will flow across national lines no matter what we do. But certainly we do not have to cut our own throats with aid, trade, tax and tariff policies that actively encourage and promote the export of American jobs and technology, without regard for the impact on either those who give or those who receive.[21]

Before proceeding further, it will be helpful to outline the available evidence with respect to these considerations. Unfortunately, only very sketchy data are available. Conclusive evidence on the net impacts of foreign investment and technology transfer on U.S. jobs and welfare does not exist. The available evidence suggests that the impact is likely to vary from one instance to another. Baranson has presented case studies that suggest that U.S.-based firms, driven by competitive necessity, are transferring their newest technology abroad more frequently than in the past.[22] To investigate this issue further, Mansfield and Romeo obtained information concerning the age of the technology transferred abroad in a sample of 65 transfers taken from 31 U.S.-based

18. See Lee Iacocca, "Multinational Investment and Global Purpose," speech delivered before the Swiss-American Chamber of Commerce," Zurich, June 17, 1977. Reprinted in *Vital Speeches*, 15 Sept. 1977.

19. See "Those Worrisome Technology Exports," *Fortune*, 22 May 1978, p. 106.

20. An example commonly cited is that of Piper aircraft. Until a few years ago, Brazil was the leading purchaser of light aircraft manufactured in the United States. However, the Brazilian government levied prohibitive taxes on the import of American-produced light aircraft and it invited an American manufacturer, Piper, to bring in U.S. technology and produce with Brazilian workers. As a result, hundreds of U.S. citizens who were directly employed in light aircraft production became unemployed, some permanently. Now Brazil is selling light aircraft to other Latin American countries and is also planning to export planes to the United States in competition with American producers.

21. William Winpisinger, "The Case Against Exporting U.S. Technology," *Research Management* (March 1978): 21.

22. Jack Baranson, "Technology Exports Can Hurt," *Foreign Policy*, 25 (Winter 1976-77).

TABLE 1
MEAN AND STANDARD DEVIATION OF NUMBER OF YEARS BETWEEN TECHNOLOGY'S TRANSFER OVERSEAS AND ITS INITIAL INTRODUCTION IN THE UNITED STATES, FOR 65 TECHNOLOGIES

CHANNEL OF TECHNOLOGY TRANSFER	MEAN (YEARS)	STANDARD DEVIATION (YEARS)	NUMBER OF CASES
Overseas subsidiary in developed country	5.8	5.5	27
Overseas subsidiary in developing country	9.8	8.4	12
Licensing or joint venture	13.1	13.4	26

SOURCE: Edwin Mansfield and Anthony Romeo, "Technology Transfer to Overseas Subsidiaries by U.S.-Based Firms," Research Paper, University of Pennsylvania, 1979.

multinationals.[23] As shown in Table 1, they found that the mean age of the technologies transferred to overseas subsidiaries in developed countries was about 6 years, which was significantly less than the mean age of technologies transferred to overseas subsidiaries in developing countries—about 10 years. Table 1 also suggests that the mean age of the technologies transferred through licenses, joint ventures, and channels other than subsidiaries is commonly higher than the mean age of the technologies transferred to subsidiaries, indicating that firms tend to transfer their newest technology overseas through wholly owned subsidiaries rather than via licenses or joint venture, but the latter channels become more important as the technology becomes older.

Another concern of countries that generate new technology is that the transfer of technology to overseas subsidiaries will hasten the time when foreign producers have access to this technology. Some evidence has recently become available on the speed with which technology "leaks out" and the extent to which international transfer actually hastens its "leaking out." The evidence, which is based on a sample of 26 technologies transferred abroad, indicated that the mean lag between the transfer and the time when foreign firms had access to the technology was about four years.[24] In over half the cases, the technology transfer was estimated to have had no effect at all on how quickly foreign competitors had access to the technology. On the other hand, in about one-fourth of the cases, it was estimated to have hastened their access to the technology by at least three years.

Technology transfer hastened the spread of process technologies to a greater degree than it did the spread of product technologies. According to the study, the most frequent channel by which the technology "leaked out" was reverse engineering.[25] That is, foreign com-

23. See Edwin Mansfield and Anthony Romeo, "Technology Transfer to Overseas Subsidiaries by U.S.-Based Firms," Research Paper, University of Pennsylvania, 1979.

24. Ibid.

25. Reverse engineering is very common in the semiconductor industry. it involves

petitors took apart and analyzed the new or modified product to gain insights into the relevant technology. Clearly, this evidence gives only a very sketchy impression of the level and nature of the returns from international technology transfer, and the role that technology exports are having on the U.S. competition position. However, there is little evidence that the technological lead of the United States in various industries is about to disappear as a result of the technology transfer activities of American firms. Indeed, there is some evidence, admittedly of a conjectural nature, that the international transfer of technology stimulated R&D activities by multinational firms.[26]

From a public policy perspective the interesting question is whether the United States could increase its economic welfare through restrictions on technology exports. It is a well-known theorem of international trade that if a country has monopoly (monopsony) power in world markets, then imposing a tax (tariff) on exports (imports) will serve to improve welfare in the absence of retaliation. This, of course, assumes that such a policy can be effectively administered.

The economic intuition behind this theorem is fairly apparent. By transferring technology abroad, American firms increase the likelihood of foreign competition in the future. While firms face incentives to consider this when setting prices at which technology is transferred, each firm will evaluate the future effects on themselves, not on the rest of the economy. The company that exports the technology is not usually the one that loses out. It receives payment of some kind. The victim is likely to be another American company, one that prior to the technology transfer enjoyed a competitive advantage over the foreign company. Fujitsu, for example, has used the technology it got from Amdahl to compete with IBM. Therefore, in strictly nationalist terms, private firms will have a tendency to set the price of technology too low and to transfer too much technology abroad. Where several U.S. firms have similar technology that does not exist abroad, their competition will tend to lower the price of technology transfers. The United States could prevent this by reducing competition and by establishing monopoly prices through control of such transfers. For instance, an export tax would serve to restrict exports, thereby driving up the price and enabling the United States economy to capture monopoly rents from the export of know-how. A similar result could be obtained by enabling

---

stripping down a competitor's chip to recreate an outright copy, to figure out how a chip works in order to design a functionally equivalent emulator chip, or merely to determine whether a new chip contains any new ideas that might be adaptable to other products. Creating a copy is surprisingly simple: the necessary tools include a microscope, acid to etch away the circuits layer by layer, and a camera to record the successive steps; $50,000 of equipment will suffice. Reverse engineering enables a rival to obtain the same advantages as could be obtained by pirating the masks—the negatives that are used to lay down the circuit elements on silicon wafers—used in manufacturing the product. Intel Corporation of California has accused the Soviet Union of copying one of its 4K memory chips and Japan's Toshiba Corporation of making a "dead ringer" of another. See *Business Week*, 21 April 1980, p. 182.

26. See Edwin Mansfield, Anthony Romeo, and Samuel Wagner, "Foreign Trade and U.S. Research and Development," *Review of Economics and Statistics*, 1979.

domestic industry to cartelize foreign markets.[27]

There is, in fact, a long history of government attempts to limit the export of technology and trade secrets. A prime example is England during the Industrial Revolution. There are serious disadvantages in limiting technology transfers, however. One problem is that while levels of restriction that are optimal on nationalist grounds can be determined in theoretical models, there is little reason to be confident that government policies will approach such optima in practice. Domestic firms seem able to circumvent restrictions on the export of know-how,[28] while foreign firms can engage in "reverse engineering of products and designs" to circumvent many controls.

An alternative approach to technology controls might involve placing more emphasis on technical data and critical manufacturing equipment and less on commodities.[29] However, it is enormously difficult to control the export of technical data, since it can move in many informal ways that are often difficult to detect.[30] Clearly the transfer of highly visible turnkey plants is more readily controlled than are surreptitious, casual conversations. Furthermore, the effectiveness of controls depends on the degree of monopoly power possessed by the United States. In most instances where controls are applicable, the United States does not have a clear superiority vis-a-vis other Western countries. The effectiveness of controls therefore depends upon cooperation with other suppliers and potential suppliers.[31]

CONCLUSION

In the foregoing discussion, the arms-length market for know-how has been shown to be exposed to a number of hazards and inefficiencies, many of which can be overcome by internalizing the process within the multinational firm. Despite the shortcomings identified, it was not apparent that regulation by either technology importers or exporters could substantially improve the efficiency with which this market operates; indeed, for the instances examined it appeared that the impairment of efficiency through regulation was the more likely outcome. Yet the strongest argument against controls on the transfer of

27. One difference is that with a cartel as compared with a tax, the industry would capture a larger portion of the economic rents, as there would be no revenues accruing to the government.

28. In 1980, allegations of export control violations in the United States numbered 350, up from 200 in 1979. *Business Week*, 27 April 1981, p. 131.

29. On the other hand, some authorities suggest that Soviet spies might do better acquiring consumer products in large department stores. Buying consumer and industrial products such as toys, appliances, and industrial tools in many cases may be more useful than technical data because of the delays in Defense Department procurement of new chips and the rapidity with which new chips become incorporated into consumer products.

30. According to one source, the KGB has 30 agents in California's Silicon Valley, plus others in Phoenix and Dallas, charged with obtaining data on microeconomics technology. *Business Week*, 27 April 1980, p. 128.

31. The Coordinating Committee on Export Controls (COCOM), an organization consisting of all NATO members plus Iceland and Japan, is the forum usually chosen to attempt the necessary coordination. However, the members have no legal obligation to participate in COCOM or to abide by its recommendations.

technology is the same as the argument for liberal trade policies in general. Many kinds of economic restrictions can be used to bring gains to some at the expense of others. But almost everyone is likely to end up worse off if they all succeed. This holds just as true for nations within the world economy as for individuals and groups within a national economy. The basic case for liberal policies is not that they always maximize short-run gains, but that they serve enlightened and longer-run interests in avoiding a world riddled with restrictions.

Technology and Technology Transfer: Mansfieldian Inspirations and Subsequent Developments

David J. Teece

ABSTRACT. This paper discusses the foundational work and ideas of Edwin Mansfield to the economics of technological change and innovation, and introduces some of the recent work in the field. I argue that much of the recent work on patenting, technology strategy and the economics of knowledge has roots to the early Mansfield contributions, and that he should be recognized as a pioneer for these recent developments.

Key words: economics of innovation, knowledge, intangible assets, R&D management

JEL Classification: O32, O34, L10

1. Introduction

At least since Joseph Schumpeter, scholars have struggled to understand the nature and the dynamics of the economics of technical change. Edwin Mansfield was born into that struggle and was for many decades a true pioneer in the study of the economics of technological change. His early books including 'The Economics of Technological Change' (1968), and 'Technological Change: An Introduction to a Vital Area of Modern Economics' (1971) summarize his early insights and display his passionate desire to wake up the field of economics to a critical area of research. He undoubtedly was the leader in the study of the nature of industrial research in America, certainly during the period of his active scholarship, and arguably to this day.

Although both classical economists and 'modern' economists such as Solow, Nelson, David, Rosenberg, and Kuznets had recognized the importance of innovation and understood it's key role in economic growth and wealth creation, it was not until Mansfield that anyone had performed serious empirical studies of industrial research. Mansfield provided leading insights into issues such as the role of academic and basic research in increasing innovation and productivity, the diffusion of technological innovations, the private and social returns to innovation, and the role of patents and the patent system. With great wisdom, Mansfield chose areas of study that have emerged as being critically important to managers and policy makers.

However, Ed Mansfield showed considerable frustration with modern economics and the work of economic theorists. Indeed, by the 1970s Ed openly displayed almost a disdain for modern economic theory because of the field's infatuation with static analysis, and its abject failure to embrace the study of technology and technological change.

As one of Ed Mansfield's students, I must first acknowledge my huge dept to him personally and intellectually. As a graduate student at Penn in the early 1970s, I was fortunate to end up in his Ph.D. class on the economics of technological change. He opened my eyes to a set of issues for which I had no previous exposure. Because I had a background in international trade and finance and economic development, he encouraged me to study technology transfer. No one at that time, including Ed, knew much about the topic. We learned together with Ed sending me into the field to collect data and absorb what I could from corporate R&D managers, from licensing executives, and from the experiences of the international departments of the Fortune 500. Some of my findings, along

Director, Institute of Management
Innovation and Organization Professor
Haas School of Business
University of California
Berkeley, U.S.A.

Journal of Technology Transfer, 30 1/2, 17–33, 2005

with my reflections on those findings, are discussed in Section 3.

Besides developing a substantive understanding of technology transfer, I learned quite a lot methodologically from Ed. He was a well-recognized statistician with a good nose for data. He was comfortable working with small samples. He let the data, not theory, lead him to answers. In fact, much of my work and methodological approaches can be seen as combining Mansfield's insights and approaches with other traditions, in particular transaction cost economics, and evolutionary and behavioral theory.

In the rest of this paper I shall describe in more detail the intellectual influence of Ed Mansfield on my work on the economics of technological change and technology transfer. I shall track some of the recent developments with respect to these early ideas and mention how recent work builds on the early Mansfield studies. In doing so I hope to demonstrate that his influence was substantial, and that his legacy in the field deserves more recognition. If Schumpeter founded the study of the economics of innovation,[1] then Mansfield was the first to give it empirical meaning at the micro level.

2. Mansfield's vision and early work

One of many lessons that I learned from Mansfield—and he in turn was undoubtedly shaped by his early years at Carnegie Mellon University (which in the late 1950s and early 1960s when Mansfield was there had scholars such as Herb Simon, Dick Cyert, Jim March, Franco Modigliani and Bill Cooper, among others)—was the importance of interdisciplinary research. As a young graduate student, I wanted to believe that the hard problems of the world were solvable. I came to realize with Ed's help that this would require a multidisciplinary approach. Mansfield always made the case for interdisciplinary research. In his later years he wrote:

> "[the economics of technological change] remains an area where there is particular need for people who are comfortable working in, and drawing on, a variety of disciplines. Very few problems of any consequence can be solved within the confines of a single discipline. It continues to require persons who have a lively interest in both basic and applied work, and who are able to use each to enrich the other. It is still an area needing people who like to work on ill-defined problems where little is known and nothing is tidy, but where the rewards for even a partial solution are very high. Those with such attributes should be encouraged to enter this field because the opportunities continue to be enormous. While a lot more is known now than 40 years ago, the truth is that economists have only scratched the surface' (Mansfield, 1995, p. xxi)."

This was the mantra Mansfield had been advancing to his students for over 20 years. It was good advice, although risky for a young economist to follow. Ed was keenly aware how little was known about innovation and industrial research. Mansfield, like March and Simon and the Carnegie School, was ahead of his time, substantively and methodologically. Half a century later David Kreps would write: 'I am increasingly convinced that economists should—and will—have to change large pieces of the paradigm that has kept us relatively monolithic for the past 50 years.- We'll increasingly look like and work with our colleagues in the other.. social sciences' (2004). Were David Kreps a Mansfield student, he would have realized this much earlier.

Thinking outside the box of conventional economics was particularly necessary when it came to issues of the economics of technical change. For one thing, neoclassical economics can not address issues of change other than comparative statics (Machlup, 1967) because even adjustments to equilibrium are outside the domain of neoclassical economics. As a result, neoclassical theory can not really deal with issues of innovation. Ed recognized this, but few others did.[2]

Mansfield's methodological response was always to start first with observation (influenced, perhaps, by the 'problem driven' research that was present at Carnegie). He encouraged me—as well as his other students and colleagues—to collect data in the field. This was extremely wise. Late in life he reflected on this method, in the introduction to the two volumes of collected papers of his:

> "In general, my approach has been to try to get a reasonably solid empirical footing before attempting to model complex phenomena about which very little is known; to keep the theoretical apparatus as simple, transparent and robust as possible;

to collect data directly from firms (and other economic units) carefully tailored to shed light on the problem at hand (rather than to try to adapt readily available general-purpose data, which often is hazardous), and to check the results as thoroughly as possible with technologists, executives, government officials and others who are close to whatever phenomenon is being studied'. (Mansfield, 1995, p. ix)."

'It was', Mansfield continued, 'a privilege and a great pleasure to have contributed to the formation and growth of this young field, which is now a major and vibrant sector of economics' (1995, p. ix). Similarly, I must say it was a privilege and a great pleasure to study under Mansfield, and help advance understanding of technological change and technology transfer. I only wish many more scholars had followed Ed's lead. The field would be further ahead had they done so.

My own work in technology transfer (and technological change in general) took Mansfield's advice to heart: it was interdisciplinary in the sense that it endeavored to reach out to other disciplines (although there was not much at the time to reach out to); and it tried to be methodologically rigorous. I will first summarize this work and then link it to recent developments in the economics and management of knowledge (including industrial knowledge). In particular, I shall focus on issues relating to the nature of knowledge and the importance of intellectual capital and intellectual property.

3. The economics of (international) technology transfer

In the early 1970s, literature on (international) technology transfer was basically non-existent. Indeed, to the extent that there was a literature, the focus was on the challenges of transferring know-how from the laboratory into practice. Indeed, there was almost no conceptual apparatus available to help one think through the issues.

The doctoral thesis I wrote under Ed Mansfield was an early effort to understand technology transfer. It truly involved writing on a clean sheet of paper. Doing research in an area where there had been almost no scholarly exploration is a daunting task, even to an established scholar, let alone a graduate student.But as Ed explained, it was sometimes a little easier to receive recognition if you were the first into a field or a new subject matter area. This has been my research strategy and my comparative advantage ever since.

My doctoral thesis, published as a book (Teece, 1976) and as journal articles (Teece, 1977a, b) was ably guided by Ed.[3] It was the first, and I believe the only study to date directed at measuring the costs associated with the (international) transfer of industrial knowledge. The topic was important because scholars at the time really had no idea as to what was the true state of affairs. Many economic theorists treated technology transfer as though it was costless—and while good intuition might suggest that the process was somewhat costly, there were no empirical studies to settle the issue. It's not like it was a hotly debated issue—the zero transfer cost assumption was made, and just not challenged. Undoubtedly, there are instances where assuming zero transfer cost is a sufficiently good approximation. For instance, once certain scientific knowledge is published, it can sometimes be absorbed at low cost by other scientists knowledgeable in the field. But there were no studies at the time with respect to the transfer and absorption costs associated with replicating industrial knowledge in different contexts.

Mansfield's instinct was that technology was expensive to transfer; and his instinct turned out to be well founded. However, what my dissertation study did unearth was that there was a learning curve with respect to technology transfer—the more experience (as measured by number of transfers) a transferor had at replicating a particular technology, the lower the cost of transfer/replication. The data showed that industrial enterprises simply got better at the transfer process the more they worked on it—so long as the technology in question did not change very much, and the environment to which it was transferred was familiar. Put differently, if companies could 'freeze' designs and transfer technology only to familiar 'places', in familiar configurations, then replication costs would decline with each instance of replication (replication is a topic returned to in Section 6 'Replicability, imitabililty, and appropriability of knowledge').

A rather counterintuitive finding of my study was that the costs of international technology transfer were sometimes (although not generally) less than the costs of domestic transfer. This result follows naturally if either (i) skills abroad are better than skills at home/or (ii) the factor (resource) cost of offshore skills are cheaper than equivalent domestic skills (these two factors mean that absorption costs could be lower abroad).

Another implicit finding of my study—which with three decades of refection I can now appreciate much better—is that learning industrial knowledge often involves expensive lessons. Industrial knowledge cannot generally be transferred just with the transfer of blueprints or even the transfer of people. It frequently involves the actual running (i.e. operating) of industrial facilities in a quasi-experimental way before yields/performance become acceptable. 'Switching on' a plant, however, can be a very expensive operation if non-marketable (i.e. substandard) products are produced during the startup period. This can lead to the waste of large amounts of resources, and cost overruns associated with the replication of manufacturing plants.

Indeed, some of the anecdotes I remember from my field research relate to the horrific expenses that Rolm and Haas experienced in starting up chemical (industrial) processes in the U.K. Differences in materials and environmental factors often led to surprising cost overruns, particularly if a technology was transferred and embedded in a plant configuration which had not already been tested and validated close to the home R&D facility. Another way to state this is that learning how to apply and reapply industrial knowledge can be costly—and in my study tens of millions of dollars of cost overruns in a technology transfer project were not uncommon, especially if a technology not properly understood was transferred prematurely. Upon reflection, this remains an important insight.

My doctoral dissertation study actually endeavored to measure various components of transfer costs. I endeavored to measure not just from the actual costs of transfer activities, but also the costs flowing from the consequences of poorly executed transfer activities. For many years I felt awkward about my results because my methodology included project startup costs as part of transfer (replication) costs. However, upon further reflection, I'm increasingly comfortable with this definition. The results simply drive home that transfer costs can be high because of the 'knock on' effects if replication/transfer is not properly accomplished. This is a lesson worth remembering. Put differently, the failure to achieve smart transfer can have very serious cost implications.

Needless to say, these insights required scores of interviews to develop. I remain forever grateful for Ed Mansfield's mandate that I do field research. I'm also grateful to scores of unnamed executives who gave of their time without recognition or reward, and to the Penfield Traveling Fellowship in International Affairs and Lettres (at the University of Pennsylvania) which provided the financing for me to travel throughout the U.S., interview executives, and collect data. The process itself was insightful and valuable.[4]

After I had completed my doctoral dissertation, an independent of my own efforts, a literature began to emerge on the nature of knowledge. For some reason, I did not know of Polyani (1966) even by the time I had finished by Ph.D My dissertation would have displayed better conceptual underpinnings had I been a bit more aware of the concept of tacit knowledge, and Polyani's teachings. It was too early to benefit from Nelson and Winter's work, but I have subsequently learned that they were incubating similar ideas.

Indeed, post-1980 there has been a flowering of work on the nature of know-how and the problems of replication. In what follows I introduce some of the learning which has emerged in the last 25 years on the nature of innovation—and knowledge replication/transfer, some of it having been leveraged off of Mansfield's early contributions.

4. Summarizing elements of received wisdom on replication and transfer

Developing an understanding of knowledge and intangible assets, critical to the formulation of technology strategy and the management of R&D. I will endeavor to summarize some of this literature below, and where appropriate make connections to some of Ed's contributions.

Understanding the nature of knowledge and other intangible assets remains perplexing.

Know-how, whether scientific or industrial, is not a physical commodity—it is arguably not a commodity at all. Accordingly, new concepts, language and terminology have had to be developed so that one can begin to understand and grasp the fundamental nature of knowledge. Key concepts that have developed over the years and accepted into the literature are outlined below.

Ed Mansfield's early efforts to come to grips with technology transfer benefited from the field research done at Penn. He was amongst the first to note:

> "Economists sometimes assume that technology is like a sheaf of blueprints and that all one has to do is ship off the right set of papers. Unfortunately, it isn't that simple or costless. For one thing, the available evidence, both recent and for earlier periods, indicates that publications and reports are a much less effective way of transferring technology than the movement of people. To transfer 'know-how', much of which is not written down in any event, there is frequently no substitute for person-to-person training and assistance, some of which may have to go on for extensive periods of time' (Mansfield, 1975, p. 373)."

We can address these issues better now, leveraging off of the work of many who either walked in Ed's footsteps, or were fellow travelers. Important concepts that help in the understanding of innovation and technology transfer include the following:

Codified/tacit knowledge

Tacit knowledge is (as Mansfield hints) that knowledge which is difficult to write down in a way that is meaningful and readily understood.[5] It is often hard to explain to others things which one only knows intuitively (Polyani, 1966; Teece 1981). The fact that we know more than we can tell speaks to the tacit dimension of knowledge. Moreover, stand-alone codified knowledge—knowledge which can be written down such as blueprints, formulas, or computer code—need not convey much meaning. It's more akin to information than knowledge.

Consider how to sail a yacht. It can be readily written down and explained by simple mechanics. But if one simply provides 'the book' and puts the student into a sailing dinghy with a good breeze afoot, for sure the dinghy will soon be capsized. The transfer of codified knowledge is insufficient. Tacit knowledge built with just a few hours of real experience—how to hold the mainsheet, where to put ones weight, just how to 'point' as the wind shifts, etc.—is critical to establish even a modest level of proficiency.

It is now recognized that there is a simple but powerful relationship between the codification of knowledge and the cost of its transfer. Simply stated, the more a given item of knowledge or experience has been codified, the more economically at least that part of it can be transferred. This is a purely technical property that depends on the ready availability of channels of communication suitable for the transmission of well-codified information—for example, printing, radio, telegraph, and data networks. However, it has long been recognized that whether information so transferred will be considered meaningful by those who receive it will depend on whether they are familiar with the code selected as well as the different contexts in which it is used (Shannon and Weaver, 1949).

Tacit knowledge is especially slow and costly to transmit (Teece, 1976, 1977, 1981a). Ambiguities abound and can be overcome only when communications take place in face-to-face situations. Errors or interpretation can be corrected by a prompt use of personal feedback. Mansfield (1975) pointed to the differences between types of (technology) transfer in the context of the transfer of know-how.

Other scholars have built on the earlier Mansfield and Teece work on technology transfer, and have shown that knowledge does not necessarily flow easily, even from unit to another within the firm (Grant, 1996; Szulanski, 1996).

The transmission of codified knowledge, on the other hand, does not necessarily require face-to-face contact and can often be carried out largely by impersonal means, such as when one computer 'talks' to another, or when a technical manual is passed from one individual to another. Messages are better structured and less ambiguous if they can be transferred in codified form.

Observable(not-observable) in use

Much technology is available for public examination and reverse engineering can be enabled the moment the product which embodies it is sold into the market. This is simply an unavoidable consequence of engaging in commerce; reverse engineering and copying, with or without improvements, is the harsh reality that must often be faced. For example, a new CT scanner, laser printer, or microprocessor is available for conceptual imitation and reverse engineering once it has been released in the market. The technology behind new products is typically ascertainable and, absent patents, may well be immitable.[6] In the studies conducted by Mansfield and his colleagues (Mansfield *et al.*, 1982, Chapter 2) reverse engineering was the most frequent channel by which technology leaked out.

Process technology, however, is often different. You can not easily find out the manufacturing process by which something was made simply by inspecting the product. It is rare that the 'signature' of a process is ascertainable through reverse engineering. While clues about a manufacturing process may sometimes be gleaned by closely inspecting the product, much about process technology can be protected if the owners of process technology are diligent in protecting the trade secrets used in the factory. In short, absent patents, process technology is inherently more protectable than product technology.

Positive/negative knowledge

Technological innovation involves considerable uncertainty. Research efforts frequently go down what turns out to be a blind alley. It is well recognized that a discovery (positive knowledge) can focus research on promising areas of inquiry, thereby avoiding blind alleys. However, it is frequently forgotten that negative knowledge —knowledge of failures ('this approach does not work')—is also valuable, as it can help steer resource allocation into more promising avenues. For this reason, firms often find it desirable to keep their failures as well as their successes secret, even setting to one side issues of embarrassment.

The paradigmatic nature of technological innovation

One of the best modern contributions to understanding technological change comes from Dosi's analogy between technological evolution and Thomas Kuhn's view on scientific evolution.'In broad analogy with the Kuhnian definition of a 'paradigm', we shall define a 'technological paradigm' as 'model' and a 'pattern' of solution of selected technological problems, based on selected principles derived from the natural sciences and on selected material technologies' (Dosi, 1982, p. 152). Even more Kuhnian is the view that a technological paradigm is constituted by the existence of an 'exemplar' and a set of heuristics for elaborating the relevant paradigm. The broad characteristics of technological evolution begin with a pre-paradigmatic phase where product design and technology is flexible, then a paradigmatic phase follows with the emergence of a standard.[7]

Intangible assets, tangible assets, and intellectual property

Knowledge assets are simply one class of intangible assets; they differ from tangible assets in several important respects. These are summarized in Figure 1.

First, knowledge has aspects of what economists refer to as public goods—when consumption by one individual does not reduce the amount left for another. This is especially true for scientific knowledge. One engineer's use of Newton's laws does not subtract from the ability of others to use the same laws. However, the distinction erodes quickly as one moves towards

CHARACTERISTICS	KNOW-HOW/ IP	PHYSICAL COMMODITIES
1. Recognition of trading opportunities	Inherently difficult	Posting frequent
2. Disclosure of attributes	Relatively difficult	Relatively easy
3. Property Rights	Limited [patents, trade secrets, copyright, etc.]	Broad
4. Item of Sale	License	Measurable units
5. Variety	Heterogeneous	Homogeneous
6. Unit of consumption	Often Unclear	$, Value, weight
Inherent tradability:	**Low**	**High**

Figure 1. Inherent tradeability of different assets.

industrial knowledge and away from scientific knowledge. While multiple use need not take away from knowledge—indeed it may well be augmented—the economic value may well decline with simultaneous use by multiple entities. This is saying little more than the obvious. Imitators can dramatically lower the market value of knowledge by augmenting its supply in the market.

Competition simply drives down the price of knowledge, even though its utility has not declined. Relatedly, while knowledge does not wear out as do most physical assets (like tractors, trucks, refrigerators, and disk drives), it is frequently exposed to rapid depreciation because of the creation of new knowledge. Thus leading edge products in the computer industry are often obsolete in a matter of months, not years. In fact, the depreciation may be so radical that a technological breakthrough drops the value of current practice technology to zero, or very nearly so.

An important difference between intangible and tangible assets is the availability and enforceability of property rights. Physical assets (land, cars, yachts, etc.) are generally well protected. Ownership is relatively easy to define, and the 'boundaries' of the property can be clearly delineated. Whether theft has occurred is relatively easy to ascertain, and in many jurisdictions there is a decent chance of getting police assistance in property recovery if the asset is of significant value. Not so with intangibles.

It may be natural to think that the different forms of intellectual property (patents, trade secrets, trademarks, copyrights, etc.) as providing similar ownership rights, with readily available protection against theft and misuse; but this is not so. There can be 'holes' and 'gaps' in intellectual property coverage,[8] and ascertaining whether trespass or theft has occurred can be difficult. Moreover, patents and copyrights eventually expire and cannot be extended. This is generally not so for physical assets.

Patents, trade secrets, trademarks provide protection for different mediums in different ways. The strongest form of intellectual property is the patent. The importance of patents for innovation was recognized by Mansfield (1986): 'The patent system', he noted, 'is at the heart of our nation's policies toward technological innovation. Consequently, it is of widespread interest to managers, management scientists, and economists, among others' (1986, p. 173).

A valid patent provides rights for exclusive use by the owner, although depending on the scope of the patent it may be possible to invent around it, albeit at some cost. Trade secrets do not provide rights of exclusion over any knowledge domain, but they do protect covered secrets in perpetuity. Trade secrets can well augment the value of a patent position. Different knowledge mediums quality for different types of intellectual property protection. The degree that intellectual property keeps imitators at bay may also depend on other external factors, such as regulations, which may block or limit the scope for invent-around alternatives.[9]

Replicability, imitability, and appropriability of knowledge

The economic value of knowledge depends not just on its ultimate utility, but on the case of transfer and replicability. If it can be replicated it can be 'scaled' and applied in new contexts. Replicability is closely related to transferability. If it can be transferred, from one geography to another, or from one product market context to a different one, then technology can potentially yield more value. But the catch is that if it can be readily transferred, it is often also prone to being lost to ones competitors through easy imitation (see Section 'Imitation').[10]

Replication

The replication of know how involves transferring or redeploying competences from one economic setting to another. Since productive knowledge is typically embodied, this cannot be accomplished by simply transmitting information. Only in those instances where all relevant knowledge is fully codified and understood can replication be collapsed into a simple problem of information transfer. Too often, the contextual dependence of original performance is poorly appreciated, so unless firms have replicated their systems of productive knowledge on many prior occasions, the act of replication is likely to be difficult (Teece,

1977a, 1993). Indeed, replication and transfer are often impossible without the transfer of people, though this can be minimized if investments are made to convert tacit knowledge to codified knowledge. However, this may not be possible.

In short, knowledge assets are normally rather difficult to replicate. Even understanding the relevant routines that support a particular competence may not be transparent. Indeed, Lippman and Rumelt (1982) have argued that some sources of competitive advantage are so complex that the firm itself, let alone its competitors, does not understand them.

Imitation can also be hindered by the fact that few routines work well in all contexts. Thus, imitating a part of what a competitor does may not enhance performance at all. Understanding the overall causal structure of processes, organization and superior performance is often critical to successful imitation and replication. This observation provides the foundation for the concept of uncertain immitability (Lippman and Rumelt, 1982). Because key performance factors in an organization are not understood (externally and possibly internally as well), replicating observable attributes is not guarantee of success.

At least two types of benefits flow to the firm from expertise in replication if it can be achieved. One is simply the ability to support geographic and product line expansion ('scalability'). To the extent that the organizational capabilities in question are relevant to the customer needs elsewhere, replication can confer value. Another is that the ability to replicate indicates that the firm has the foundations in place for learning and improvement.

Secondly, understanding processes, both in production and in management, is the key to process improvement; an organization cannot improve what it does not understand. Deep process understanding is often required to accomplish codification and replication. Indeed, if knowledge is highly tacit, it indicates that the phenomenon may not be well understood, except at an experiential level. When knowledge is tacit, the rate of learning may be limited because scientific and engineering principles cannot be systematically applied. Instead, learning is confined to proceeding through trial-and-error, and the amplification to learning that might otherwise come from the application of modern science is denied.

Imitation

Imitation is simply replication performed by a competitor. If self-replication is difficult, imitation is likely to be even harder. In competitive markets, it is the ease of imitation that determines the sustainability of competitive advantage. Easy imitation leads to the rapid dissipation of supernormal profits.

Factors that make replication difficult also make imitation difficult. Thus, the more tacit the firm's productive knowledge, the harder is replication by the firm itself, or by it's competitors. When the tacit component is high, imitation may well be impossible, absent the hiring away of key individuals and the transfer of key organizational processes.

In advanced industrial countries, intellectual property rights may impede imitation of certain capabilities.[11] Nevertheless, imitation lags tend to be short (Mansfield *et al.*, 1982, Chapter 2). These rights present a formidable imitation barrier in certain particular contexts. Several other factors, in addition to the patent system, cause there to be a difference between replication costs and imitation costs. The observability of the technology or the organization is one such important factor. As mentioned earlier, while insight into product technology can be obtained thorough strategies such as reverse engineering, this is not the case for process technology, as the firm need not expose its process technology to the outside in order to benefit from it. Firms with product technology, on the other hand, confront the unfortunate circumstances that they must expose what they have got in order to complete a sale. Secrets are thus more protectable if there is no need to expose them in contexts where competitors can learn about them.

Appropriability

Appropriability is a function of both the nature of knowledge, ease of replication, and the efficiency of intellectual property rights as a barrier to imitation. Appropriability is strong when a technology is both inherently difficult to replicate and intellectual property systems provides legal barriers to imitation. As shown in Figure 2, there are different layers of 'protection' which owners of valuable intangibles might enjoy. If technology is inherently easy to replicate and the intellectual property protection is either unavailable or ineffectual, then appropriability is weak.[12]

Much confusion has been caused by ignoring the significant distinction between an innovation and the intellectual property which embodies that innovation. The latter is merely a legal right (or, more precisely, a collection of various legal rights, some procedural, and some substantive).

An inventor develops say, a new technology for cracking petroleum. The technology exists when it has been developed and tested. But it only becomes covered by intellectual property once it is legally recognized as such—in the case of patents, when a particular country's patent office recognizes the inventor's application and grants a patent. An issued patent is presumed to be valid, but its ultimate validity is never established until it is challenged, and validity subsequently upheld in a court of law.

The distinction between the innovation and legal 'intellectual property' rights is most readily seen when the property right grant expires. Beethoven's copyright in his compositions has long since expired. But Beethoven's creations live on.

INTELLECTUAL PROPERTY RIGHTS		INHERENT REPLICABILITY	
		EASY	HARD
	LOOSE	WEAK	MODERATE
	TIGHT	MODERATE	STRONG

Figure 2. Appropriability regimes for knowledge assets.

An innovation may be just as valuable to society—in the sense that it represents an advance over the available alternative technologies—the day after the patent on that innovation expires as it was the day before the patent expires. But the legal rights of the innovator are radically different before and after the expiration date; after that date, the innovator has no right to exclude others from using the innovation. The private value falls, but the social value does not decline, and may in fact increase.

One other key distinction is that the innovation and the legal rights are often not coextensive. An innovator may only obtain legal rights over part of the totality of the innovation. Confusion can sometimes arise when individuals seek to assess the value of the 'technology' *per se*, rather than the value of the patent rights—namely, the right to exclude others from using the patented aspects of the technology. If the two are sold together it may not matter. When they are not, it does.

5. Capturing value from intellectual capital

As mentioned earlier, extracting value from intangible capital is a much more complicated and risky process than extracting value from tangible (physical) capital. Intellectual property, standing alone, generates little or no value to the final consumer. A patent, for instance is merely a piece of paper that conveys the right to exclude others. The vast majority of patents are never practiced. Rather, value typically arises only after inventions are embedded in devices which are then combined with other (complementary) assets to produce a product or service which is sold in a market.

To take a simple example: merely coming up with an idea for a new semiconductor device, or even obtaining a patent or copyright on a design for a better semi-conductor device, does not generate economic value. What generates value is when some entity combines an invention or a new design with the manufacturing, marketing, after sales support and other capabilities that are necessary to actually produce and sell semiconductors. Complementary assets typically assist in the extraction of value from intellectual property. Such assets generate a return which is analytically separate from the intellectual property itself.

In short, there are often significant hurdles that have to be cleared, and significant risks that must be undertaken, before an innovative idea can be successfully commercialized. Often, the individual(s) or firm(s) which supplies the necessary complementary assets and skills needed in order to commercialize the innovation, or which takes the necessary risks, are not the same as the inventor. When this is the case, the gains from innovation get split not only with the consumer, but also with the owners of the relevant complementary assets. Getting the commercialization strategy right is thus very important, as discussed in Teece (1986).

Appropriability regimes

One of the most fundamental reasons why innovators with good marketable ideas fail to open up markets successfully is that they are operating in an environment where appropriability is weak. This constrains their ability to capture the economic benefits arising from their ideas. As shown in Figure 2, the two most important environmental factors conditioning this are the efficacy of legal protection mechanisms and the nature of technology (including it's inherent replicability).

It is well known that patents do not generally block competitors. As Mansfield taught (1985, 1988), they can often (but not always) be worked around. Rarely, if ever, do patents confer perfect appropriability, although they do afford considerable advantage in some industries, such as with new chemical products, pharmaceuticals, and rather simple mechanical inventions (Levin *et al.*, 1987). They are especially ineffective at protecting process innovations. Often patents provide little protection because the legal and financial requirements for upholding their validity or for proving their infringement are high.

The degree of legal protection a firm enjoys is not necessarily a 'god given' attribute. The inventor's own intellectual property strategy itself enters the equation. The inventor of core technology need not only seek to patent the innovation itself, but can also seek complementary patents on new features and/or manufacturing processes, and possibly on designs.

Of course, the more fundamental the invention, the better the chances that a broad patent will be granted, and granted in multiple jurisdictions. It must be recognized that exclusionary rights are not fully secured by the mere issuance of a patent. While a patent is presumed to be valid in many jurisdictions, validity is never firmly established until a patent has been upheld in court. The strongest patents are those that are broad in scope, and have already been upheld in court.

In some industries, particularly where the innovation is embedded in processes, trade secrets are a viable alternative to patents. Trade secret protection is possible, however, only if a firm can put its product before the public and still keep the underlying technology secret. Usually only chemical formulas and industrial-commercial processes can be protected as trade secrets after they' are 'out'.

The degree to which knowledge about an innovation is tacit or easily codified also affects the ease of imitation. Tacit knowledge is, by definition, difficult to articulate and so is hard to pass on unless those who possess the know-how can demonstrate it to others. It is also hard to protect using intellectual property law. Codified knowledge is easier to transmit and receive and is more exposed to industrial espionage. On the other hand, it is often easier to protect using the instruments of intellectual property law.

As shown in Figure 2, appropriability regimes can be divided into 'weak' (innovations are difficult to protect because they can be easily codified and legal protection of intellectual property is ineffective) and 'strong' (innovations are easy to protect because knowledge about them is tacit and/or they are well protected legally). Despite recent efforts to strengthen the protection of intellectual property, strong appropriability is the exception rather than the rule. This has been so for centuries, and it will never be substantially different in democratic societies, where the migration of individuals and ideas face few governmental constraints.

Standards and timing issues

The success of the strategies, methods, and procedures by which innovators endeavor to develop new technology and capture value from it are frequently severely impacted by factors over which it may have little control. Standards and timing issues are amongst such factors.

Standard issues are particularly important when technologies must work closely together as a coupled or intertwined 'system'. Examples include telecommunications and computer equipment (interconnection is usually required) or even photocopiers—the 'aftermarket' products e.g. paper, toner must all conform to certain standards for the machine to work, or at least work well.

These factors lead to efforts by companies to promote proprietary standards (when they believe they have a good chance of success) or open standards when it's the success of a competitor's proprietary standard which is of greater concern. There are many factors which impact a firm's success, or lack thereof, in establishing standards. Achieving overall critical mass is frequently an issue, particularly when the phenomenon of two sided (or multisided) markets is at issue (Evans 2003; Rochet and Tirole, 2004). When standards are at issue, success may beget further success and dominant standards may emerge. When customers adopt a standard, they implicitly (and sometimes explicitly) abandon others. Inasmuch as innovations are often developed around existing or prospective standards, the rise and decline of certain standards is likely to have an impact on competitive outcomes, and possibly also on the value of technology.

6. Valuation issues: accounting and market metrics

It is undisputed that the creation of intangible assets and intellectual capital are sources of economic growth and productivity enhancement. It is also undisputed that private enterprise businesses in aggregate generate value from various investments, including investments aimed at creating valuable technological assets. Quite simply, firms would stop investing in R&D unless they continued to perceive that as a result they were generating an acceptable rate of return; and venture capitalists would not be able to raise money if they could not deliver the prospect of a positive return, commensurate with the risk.

Quantifying the value of intangibles and the returns they generate isn't easy.[13] However, as Mansfield recognized that it is a very important matter. One reason is that it's extremely hard to manage assets that you cannot describe or measure. Not only will one have difficulties in setting priorities, but one will also have difficulty determining success and/or failure in asset management activities. Also, if intangibles are not measured correctly, it might appear that an organization is doing poorly when in fact it is simply investing in intangibles. Accounting practices in the U.S. and elsewhere do not recognize many forms of intangibles, and this renders accounting data of limited value, and causes discrepancies to emerge between the market value and the book value of the business enterprise. Finally, to the extent that social returns exceed private, there is a case for government policies favoring innovation.

In recent decades, scholars have extended Mansfield's early work and embarked on inquiries as to the quantitative importance of intangibles, and their impact on the performance of the business enterprise. Four performance measures have received attention: (1) internal rates of return, as measured by the examination of the R&D portfolios of individual firms (2) market value, as established in (public) stock markets (3) gross margins (4) patents (5) direct measures of innovation such as innovation counts The latter is deeply imbued with judgmental assessment and will only be dealt with in a cursory fashion.

Internal (private and social) rates of return

Ed Mansfield was the pioneer in demonstrating empirically that private rates of return from investment in R&D were in the double digits for selected industrial enterprises, but that the social rates of return were many times the private rates of return. The latter findings were the first clear measurement of spillovers associated with R&D. These results have been cited extensively in the past. They undergrid the case for government support of R&D, and for policies that favor innovation. In an important paper, Mansfield and his co-authors (1977) made perhaps the first study of the social returns from a sample of industrial innovations, using a model that measured the social benefits from innovations. Social benefits included both the profits of the innovator plus the benefits to consumers due to reduction in prices due to the innovation. The median social rate of return for the 17 innovations reported in the study was 50%; about twice the median private return.

Two follow-up studies supported by the NSF supported and even strengthened Mansfield's findings. They showed that the median social rate of return to be 70% and the median private rate of return to be 36 (Mansfield himself reflected on these findings and subsequent studies in his 1991 paper, 'Social Returns from R&D: Findings, Methods and Limitations'). Moreover, several other studies have confirmed and extended the ideas and results, including two papers in the proceedings of the AER by Scherer (1983) and Piekarz (1983). Piekarz also discusses some of the policy issues associated with Mansfield's (and other's) findings. The findings support the case that government policy should favor innovation.

Stock market valuations

If the stock market is strongly efficient, the market value of a company is at all times equal to it's fundamental value, where fundamental value is defined as the expected present discounted value of future payments to shareholders. Assuming further the absence of market power, adjustment costs, and debt and taxes, then under the efficient market thesis a company's value as determined by investors pricing decisions will equate to enterprise value—that is, the replacement cost of it's assets. Put differently, the ratio of it's market value to the replacement cost of capital—known as Tobin's Q—should equal 1.

An inference is that if the market value of the firm is greater than the replacement cost of it's tangible assets, the difference must reflect the value of intangibles. Furthermore, since accounting standards require a very conservative treatment of intangibles, corporate balance sheets of publicly traded companies are believed to in the main capture tangible assets. Because intangibles are not properly reflected on balance sheets, researchers argue that the informativeness of financial information is compromised.

Nevertheless, the difference between market value and the replacement cost of tangible assets on the balance sheet has come to be used as a proxy for the value of intangibles. However, absent specification of what these intangibles are, it is very difficult to disaggregate and assign values to particular intangibles. Moreover, the inference that the difference between a firm's market value and the replacement cost of it's physical assets represents the value of it's intangibles require the assumption of 'strong form' market efficiency (where prices reflect all information, public as well as private)—but this may be difficult to accept if investors do not have good information about the firm's intangibles.

Nevertheless, researchers have begun to explore the empirical relevance of (stock) market values. For instance, studies have established that investors regard R&D expenditures as a significant value enhancing activity, presumable because they build (intangible) technological assets (Chan *et al.*, 1992). Also, econometric studies that explore relationships between market-to-book ratios and R&D-to-sales ratios show positive and statistically significant associations (see Hirchey and Weyganat, 1985). The evidence is clear that investors view R&D as on average as being value enhancing. Moreover, the magnitude of the contribution for the investing enterprise appears considerably higher than the cost of capital.

Gross margins

Another approach utilizes accounting data, and in particular gross margins (the differences between revenues and cost of goods sold), to assess how investment in intangibles affects performance. One basic approach offered by Hand (2003) is to use econometric analysis and to regress current year dollar gross margin on current and lagged R&D, advertising, and general and administrative expenses. Hand's analysis yielded several findings (p. 304): over the period 1980–2000, the mean yearly NPV of $1.00 spent on R&D, advertising, and personnel were $0.35, $0.24, and $0.14, respectively. Scale also mattered, at least for R&D and advertising activities. Based on his findings, Hand concludes:

'Overall, my findings support the view that R&D and advertising intangibles have emerged over the past 20 years to become a critical means by which firms today create value and that one mechanism of value creation is that of increasingly profitable returns-to-scale' (Hand, p. 304).

Patent and patent citation counts

The issuance of patents, and the size of a firm's patent portfolio, is also a measure, albeit a noisy one, of innovative output. Because of the skewness in patent values—many patents are quite worthless, but a few extremely valuable—it has turned out to be necessary to impose some at least crude measure of quality in order to make sense of the data. The most common measure of quality is the number of citations to a patent included in other subsequent patent applications. A number of studies have demonstrated that quality adjusted patents capture some element of the firm's R&D asset value. For instance, Hall *et al.* (2000) show that citation accepted patent counts help explain Tobin's Q values.

Innovation counts

Another way to measure innovative output is directly i.e. to map significant technological innovations, and then to assign them to particular firms responsible for their creation and commercialization. While this approach is at one level superior—it actually highlights innovation rather than say R&D expenditure (expenditure measures the cost of inputs into innovative activities)—it suffers from the lack of comparability i.e. there is no easy way to compare innovations, and to quantify their significance, except possibly through panels of experts who make qualitative judgments.[14]

Organizational capital

The primary focus in this very short survey of measurement issues has been on technological assets. However, it is well recognized that organizational innovation is as significant (if not more so) than technological innovation in creating value. Cole has asserted that 'if changes in business procedures and practices were patentable, the contribution of business change to the economic growth of the nation would be as widely recognized as the influence of mechanical inventions (1968, p. 61–62).

As an example, consider Henry Ford's invention of the moving assembly line. This was unquestionably one of the greatest innovations in the automobile industry, with ramification for other industries too. However, this invention was not technological, it was organizational. The Ford Motor Company's entire system of production had to be modified to accommodate it.

Another organizational innovation was the adaptation of the M-Form structure. The transition from corporations organized in a unitary structure to corporations organized in a decentralized profit center oriented multidivisional structure had a salutary effect on business performance. In a study of the adaptation of this new structure in the petroleum industry (Armour and Teece, 1978) the innovation was shown to produce a statistically significant improvement in return on equity of approximately two percentage points during the diffusion period 1955 – 1968. A subsequent study (Teece, 1981b) of the pair wise differential performance of the two leading firms in a number of industries yielded a similar finding. This study, which used a sample of the largest firms and most important U.S. industries, that the M-form innovation displayed a statistically significant improvement in firm performance amounting to 2.37 and 1.22% for return on equity and return on assets, respectively. These results held while the innovation was being diffused. Both studies support the insights from Chandler (1968) and Williamson (1975) on the importance of organizational innovation and organizational design on economic performance.

Also, the diffusion path of the M-Form innovation was not unlike diffusion paths that Mansfield identified for technological innovations. Teece (1980) argued that such similarities between the diffusion processes affecting technological and administration/organizational innovations indicates the broader potential of insights from the economics of technological change literature. Indeed, we may see recent work examining issues regarding the relationship between organizations and performance as contributors to this stream of ideas in the Mansfield/Teece tradition.

Recently, other (indirect) measures of the impact of organizational innovation have been attempted. Brynjolfsson and Yang (1999), have showed that a $1.00 investment in computers has about a $10.00 impact on market value. This has been interpreted to reflect positive results from new business processes which the installation of enterprise software frequently requires. The author's explanation is as follows:

'Our deduction is that the main portion of computer related intangible assets comes from the new business processes, new organizational structure and new market strategies—computer use is complementary to new workplace organization—Wal-Mart's main assets are not the computer software and hardware, but the intangible business processes they have built around those computer systems (1999, p. 30).

Furthermore, recent evidence (Morck and Yeung, 2003) supports earlier work (Teece, 1982) indicating that know-how transfer inside firms (across jurisdictions and product space) is value enhancing. In this earlier study, internal technology transfer processes were seen as more efficient and effective than arms length transfers across organizational boundaries. Morck and Yeung's recent work supports this analysis by showing a positive contribution of diversification to value when it is aimed at scaling intangibles.

7. The multinational firm, internalization, and R&D activity

My work with Ed on international technology transfer also helped lay the foundations for new thinking on the distinctive role of the multinational firm. While it is true that knowledge need not move freely inside the firm, as Szulanski and others have demonstrated, it generally does move easier inside firms than between unrelated entities. This is not only because it is easier to marshal the necessary transfer of people internally, but also because of common language and control the latter softening intellectual property ('leakage') concerns. Shared values and goals inside the firm also assist technology transfer, at least when they exist.

In the 1950s and 1960s, and 1970s, Hymer (1976) and others were trumpeting that the multinational firm was an instrument for exploiting monopoly power, in part through the manner in which it exploited technology developed at home. The Mansfield–Teece–Williamson tradition, focusing on the multinational firm as a relatively efficient mechanism for transferring technology,[15] was a significant counterpoint to the Hymer argument. Not only was Hymer's argument poor competition policy analysis (competition policy experts would not automatically conclude that a firm had (antitrust) market power simply because it had valuable intangibles and intellectual property, but the proper question to ask is whether it has market power in a *relevant* (antitrust) market, not simply whether it has hard to imitate assets.

In a series of articles (Teece 1981a, b, 1985, 1986a, b) I built upon insights by two of my teachers, Ed Mansfield and Oliver Williamson, to identify particular failures in the market for know-how. I used this in turn to explain the horizontal and vertical expansion of the multinational enterprise. While it is true that others had identified internalization efficiencies as the basis of multinational enterprise and foreign direct investment (e.g. Buckley and Casson, 1976) my work explicitly focused on technology transfer issues. These issues remain compelling explanations for the international scope of the firm. Indeed, subsequent work on appropriability (Teece, 1986) provided additional generality to these explanations. When combined with Mansfield's work on spillovers and high social rates of return to innovation, this broader body of work strongly supports the thesis that the multinational firm can be an instrument of economic development, not a tool for the extraction of monopoly rents and the amplification of poverty. While Mansfield chose to stay away from some of these broader policy issues, it is rather transparent that his work is relevant to many of the great policy debates we are currently experiencing.

With respect to the focus of R&D in the multinational firm, Mansfield *et al.* (1979) were amongst the first to examine the reasons why firms 'outsourced' R&D (i.e. conducted it abroad). This work is now an important historical benchmark, as it shows that in the 1970s most foreign R&D was aimed at adapting technology to local market conditions. However, even back then some firms performed R&D abroad to access particular R&D resources not otherwise available. While this early work did not ask the theoretical/design question around what R&D should be done offshore, Mansfield's early interest in offshore R&D did stimulate me and one of my former students (Chesbrough) to design a framework to help answer those questions (Teece and Chesbrough, 1996). Indeed, much of my subsequent work has involved trying to stitch together Mansfieldian issues and ideas with those of his early contemporaries (Oliver Williamson from Penn and Nelson and Winter from Yale).[16] Indeed, I for one have taken Ed's admonition to be interdisciplinary very seriously, and have found ways to weave his ideas and findings into the broader tapestry of innovation studies. I am most grateful to have had such helpful early guidance from a great master.

8. Closing

Much progress has been done in recent years in the area of the economics of technology, but it still builds on the foundational work that Schumpeter and Mansfield and others did. Ed was undoubtedly the pioneer in the study of industrial research, and one of a few leading scholars in the economics of technological change. In 1996 Medoff reports that Mansfield received the 26th highest number of citations from 1971 to 1992 among non Nobel prize winning economists younger than 70. But in the economics of technological innovation, Grandstrand (1994) reports Mansfield was the most cited author in each of the 4 years he examined (also see Grandstrand, 2004). But this only confirms what his friends and students already knew: Ed was second to none in his field, and he chose a field of compelling significance to understanding the business enterprise, economic growth, and the future of Western Civilization.

Acknowledgment

I wish to thank Mie Augier for many helpful comments and suggestions on early drafts. Patricia Lonergan supplied helpful technical support.

Notes

1. Mansfield acknowledges Schumpeter as founding the field of the economics of technology (1995, p. ix).
2. But see for instance Nelson and Winter's (1977) early critique of neoclassical theory of innovation: 'to the extent that technical advance is important [in neoclassical theory], the set of ideas built into the formulation that individual firms are maximizing profits over a common .. choice set and that the industry is in .. competitive equilibrium can be seen as serious structural misspecifications. It is exactly that some firms see alternatives that others do not, and that imitation is costly and takes time, that provides the incentive to try to innovate. It is a key structural characteristic ... of growth in a competitive market economy that there is a diversity of behavior (technologies used) by firms in the industry at any time. A chronic disequilibrium is what is driving the growth process. To assume .. equilibrium is to structurally misspecify the process'.
3. My doctoral work was referred to by Mansfield on several occasions; see for instance Mansfield (1975).
4. Ronald Coase admits to a similar benefit from his early field research in the U.S. in the 1930s (Coase, 1988).
5. The classical insights on the nature of tacit knowledge were provided by Hayek (1945) and Polyani (1966) and early applications to the study of technology include Mansfield (1975) and Teece (1981a).
6. Mansfield touched on some of the issues relating to imitation in an early paper 'Technical change and the Rate of Imitation' (1961).
7. Dosi's use of technological paradigms as a frame for understanding technological change can also accommodate the insights of dominating designs, technological regimes, etc. See Dosi (1982) for details.
8. In one of Ed's studies (Lee and Mansfield, 1996), it was established that the lack of intellectual property protection in certain host countries slowed technology transfer and direct foreign investment.
9. Contributions to the discussions of patent and patent protection include the early survey data from Mansfield *et al.*, 1981, Levin *et al.*, 1987. An extension and discussion of these studies can be found in Schankerman (1998). Moreover, Mansfield (1986) examined survey data of 100 manufacturing firms and found that—with the exceptions of the pharmaceutical and chemical industries—the firms found that most of their innovations would have been introduced even in the absence of patent protection.
10. Whether or not technology is exposed in this way is likely to depend in part on whether or not the technology enjoys intellectual property protection.
11. For Mansfield's most recent work on IP, see in particular Mansfield 1993, 1994.
12. A description of the results of an inquiry into appropriability conditions in manufacturing industries is found in Levin *et al.* (1987). Their data and discussion is consistent with the views discussed below.
13. As discussed below, Mansfield work on the private and social returns to innovation was pioneering (see in particular Mansfield 1977).
14. There have been very few studies of this kind. The most notable study was done by Mansfield (1968) where he

examined innovation in the petroleum industry. This study was extended and updated by Teece (1977b).

15. In Teece (1976, 1977a) I showed that internal transfer costs were generally less than the cost of transfer to unrelated entities.

16. In addition, much of Mansfield's early ideas on R&D originated while he was a consultant at the RAND Corporation—Winter and Nelson were also doing pioneering work on R&D there.

References

Armour, H. and D. Teece, 1978, 'Organizational Structure and Economic Performance: A Test of the Multidivisional Hypothesis,' *The Bell Journal of Economics* **9** (2), 106–122.

Brynjolfsson, E. and S. Yang, 1999, '*The Intangible Costs and Benefits of Computer Investments: Evidence From Financial Markets,*' Working Paper, Sloan School, Massachusetts Institute of Technology.

Buckley, P. and M. Casson, 1976, *The Future of the Multinational Enterprise,* London: MacMillan.

Coase, R.H., 1988, 'The Nature of the Firm: Origin, Meaning, Influence,' *Journal of Law, Economics, and Organization* **4** (1), 3–47.

Chan, S., J. Kesinger, and J. Martin, 1992, 'The Market Rewards Promising R&D,' *Journal of Applied Corporate Finance* **5**, 59–62.

Chandler, A., 1968, *Strategy and Structure*, Cambridge: Harvard University Press.

Cole, A.H., 1968, 'The Entrepreneur, Introductory Remarks,' *American Economic Review* **58** (2), 60–63.

Dosi, G., 1982, 'Technological Paradigms and Technological Trajectories. A Suggested Interpretation of the Determinants and Directions of technical Change,' *Research Policy* **11**, 147–162.

Evans, D., 2003, 'The Antitrust Economics of Multi Sided Platform Markets,' *Yale Journal of Regulation* (forthcoming).

Grandstrand, O., 1994, 'Economics of Technology: An Introduction and Overview,' in Idem (ed.), *Economics of Technology*, Amsterdam: North-Holland.

Grandstrand, O., 2004, *Economics, Law, and Intellectual Property*, Boson: Kluwer.

Grant, R.M., 1996, 'Prospering in Dynamically Competitive Environments: Organizational Capability as Knowledge Integration,' *Organization Science* **7** (4), 375–387.

Hall, B., A. Jaffee, and M. Trajtenberg, 2000, *Market Value and Patent Citations: A First Look,* Working Paper, National Bureau of Economic Research.

Hand, J. and B. Lev, 2003, *Intangible Asset Values, Measures, Risks*, Oxford: Oxford University Press.

Hayek, F.A., 1945, 'Economics and Knowledge,' in F.A. Hayek, (ed.), *Individualism and Economic Order,* University of Chicago Press.

Hymer, S., 1976, *International Operations of National Firms: A Study of Direct Investment*, Cambridge, MA: MIT Press.

Lee, J.Y. and E. Mansfield, 1996, 'Intellectual Property Protection and US Direct Investment,' *Review of Economics and Statistics* **78** (2), 181–186.

Kreps, D., 2004, 'Beliefs and Tastes: Confessions of an Economist,' in Augier and March (2004), *Models of a Man: Essays in Memory of Herbert Simon*, Cambridge: MIT Press.

Levin, R., C. Klevorick, R. Nelson, and S.G. Winter, 1987, 'Appropriating the Returns from Industrial Research and Development,' *Brookings Papers on Economic Activity* **3**, 783–820.

Lippman, S. and R. Rumelt, 1982, 'Uncertain Imitability: An Analysis of Interfirm Differences in Efficiency Under Competition,' *Bell Journal of Economics* **13**, 418–438.

Machlup, F., 1967, 'Theories of the Firm: Marginalist, Behavioral, Managerial,' *American Economic Review* **57**, 1–33.

Mansfield, E., 1961, 'Technical Change and the Rate of Imitation,' *Econometrica* **29** (40), 714–766.

Mansfield, E., 1968, *The Economics of Technological Change*, New York: W.W. Norton & Company Inc.

Mansfield, E., 1975, 'International Technology Transfer: Forms, Resource Requirements, and Policies,' *American Economic Review* **65** (2), 372–376.

Mansfield, E., 1986, 'Patents and Innovation: An Empirical Study,' *Management Science* **32**, 173–181.

Mansfield, E., 1991, 'Social Returns from R&D: Findings, Methods and Limitations,' *AAAS Science and Technology Policy Yearbook* **24**, 24–27.

Mansfield, E., 1993, 'Unauthorized use of Intellectual Property: Effects on Investment, Technology Transfer, and Innovation,' in M.B. Wallerstein *et al.* (eds.), *Global Dimensions of Intellectual Property Rights in Science and Technology*, Washington: National Academy Press.

Mansfield, E., 1994, *Intellectual Property Protection, Foreign Direct Investment, and Technology Transfer*, IFC discussion papers, No. 19, Washington, D.C.: The World Bank.

Mansfield, E., 1995, 'Introduction,' in Idem (ed.), *Innovation, Technology and the Economy: The Selected Essays of Edwin Mansfield, Vol. 1*, Brookfield: Edward Elgar.

Mansfield, E., A. Romeo, M. Schwartz, D. Teece, S. Wagner, and P. Brach, 1982, *Technology Transfer, Productivity, And Economic Policy*, New York: W.W. Norton & Company.

Mansfield E., D. Teece and A. Romeo, 1979, 'Overseas Research and Development by U.S.-Based Firms,' *Economica* **46** (May), 187–196.

Mansfield E., M. Schwartz and S. Wagner, 1981, 'Imitation Costs and Patents: An Empirical Study,' *The Economic Journal* **91**, 907–918.

Mansfield, E., R. John, R. Anthony, W. Samuel, and B. George, 1977, 'Social and Private Rates of Return from Industrial Innovations,' *Quarterly Journal of Economics* **91**, 221–240.

Medoff, M., 1996, 'A Citation-Based Analysis of Economists and Economics Programs,' *The American Economist* **40** (1), 46–49.

Morck, R. and B. Yeung, 2003, *Why Firms Diversify: Internalization v. Agency Behavior.*

Nelson, R. and S. Winter, 1977, 'In Search of a Useful Theory of Innovation,' *Research Policy* **6**, 36–76.

Piekarz, R., 1983, 'R&D and Productivity Growth: Policy Studies and Issues,' *American Economic Review* **73**, 210–214.

Polyani, M., 1966, *The Tacit Dimension*, New York: Doubleday.

Scherer, F., 1983, 'R&D and Declining Productivity Growth,' *American Economics Review* **73**, 215–218.

Schankerman, M., 1998, 'How Valuable Is Patent Protection? Estimates by Technology Field', *Rand Journal of Economics* **29** (1), 77–107.

Shannon, C. and W. Weaver, 1949, *A Mathematical Theory of Communication*, Urbana: University of Illinois Press.

Szulanski, G., 1996, 'Exploring Internal Stickiness: Impediments to the Transfer of Best Practice Within the Firm,' *Strategic Management Journal* **17**, 27–43.

Rochet, J. and J. Tirole, 2004, *Two Sided Markets: An Overview*, Mimeo, University of Toulouse.

Teece, D.J., 1976, *The Multinational Corporation and the Resource Cost of International Technology Transfer*, Cambridge, MA: Ballinger.

Teece, D.J., 1977a, 'Technology Transfer by Multinational Firms: The Resource Cost of Transferring Technological Know-how,' *The Economic Journal* **87**, 242–261.

Teece, D.J., 1977b, 'Time-Cost Tradeoffs: Elasticity Estimates and Determinants for International Technology Transfer Projects,' *Management Science* **23** (8), 830–837.

Teece, D.J., 1980, 'The Diffusion of an Administrative Innovation,' *Management Science* **26** (5) (May), 464–470.

Teece, D.J., 1981a, 'The Market for Know-how and the Efficient International Transfer of Technology,' *The Annals of the Academy of Political and Social Science* **458**, 81–196.

Teece, D.J., 1981b, The Multinational Enterprise: Market Failure and Market Power Considerations,' *Sloan Management Review* **22** (3), 3–17.

Teece, D.J., 1982, 'Towards an Economic Theory of the Multiproduct Firm,' *Journal of Economic Behavior and Organization* **3**, 39–33.

Teece, D.J., 1985, 'Multinational Enterprise, Internal Governance, and Industrial Organization,' *American Economic Review* **75** (2), 233–238.

Teece, D.J., 1986, 'Profiting from Technological Innovation,' *Research Policy* **15** (6), 285–305.

Teece, D.J. and H. Chesbrough, 1996, 'Organizing for Innovation: When is Virtual Virtuous?,' *Harvard Business Review*.

Williamson, O.E., 1975, *Markets and Hierarchies: Analysis and Antitrust Implications*, New York: The Free Press.

Part II

LICENSING AND CROSS-LICENSING

Profiting from technological innovation: Implications for integration, collaboration, licensing and public policy

David J. TEECE *

School of Business Administration, University of California, Berkeley, CA 94720, U.S.A.

Final version received June 1986

This paper attempts to explain why innovating firms often fail to obtain significant economic returns from an innovation, while customers, imitators and other industry participants benefit. Business strategy – particularly as it relates to the firm's decision to integrate and collaborate – is shown to be an important factor. The paper demonstrates that when imitation is easy, markets don't work well, and the profits from innovation may accrue to the owners of certain complementary assets, rather than to the developers of the intellectual property. This speaks to the need, in certain cases, for the innovating firm to establish a prior position in these complementary assets. The paper also indicates that innovators with new products and processes which provide value to consumers may sometimes be so ill positioned in the market that they necessarily will fail. The analysis provides a theoretical foundation for the proposition that manufacturing often matters, particularly to innovating nations. Innovating firms without the requisite manufacturing and related capacities may die, even though they are the best at innovation. Implications for trade policy and domestic economic policy are examined.

* I thank Raphael Amit, Harvey Brooks, Chris Chapin, Therese Flaherty, Richard Gilbert, Heather Haveman, Mel Horwitch, David Hulbert, Carl Jacobsen, Michael Porter, Gary Pisano, Richard Rumelt, Raymond Vernon and Sidney Winter for helpful discussions relating to the subject matter of this paper. Three anonymous referees also provided valuable criticisms. I gratefully acknowledge the financial support of the National Science Foundation under grant no. SRS-8410556 to the Center for Research in Management, University of California Berkeley. Earlier versions of this paper were presented at a National Academy of Engineering Symposium titled "World Technologies and National Sovereignty," February 1986, and at a conference on innovation at the University of Venice, March 1986.

Research Policy 15 (1986) 285–305
North-Holland

1. Introduction

It is quite common for innovators – those firms which are first to commercialize a new product or process in the market – to lament the fact that competitors/imitators have profited more from the innovation than the firm first to commercialize it! Since it is often held that being first to market is a source of strategic advantage, the clear existence and persistence of this phenomenon may appear perplexing if not troubling. The aim of this article is to explain why a fast second or even a slow third might outperform the innovator. The message is particularly pertinent to those science and engineering driven companies that harbor the mistaken illusion that developing new products which meet customer needs will ensure fabulous success. It may possibly do so for the product, but not for the innovator.

In this paper, a framework is offered which identifies the factors which determine who wins from innovation: the firm which is first to market, follower firms, or firms that have related capabilities that the innovator needs. The follower firms may or may not be imitators in the narrow sense of the term, although they sometimes are. The framework appears to have utility for explaining the share of the profits from innovation accruing to the innovator compared to its followers and suppliers (see fig. 1), as well as for explaining a variety of interfirm activities such as joint ventures, coproduction agreements, cross distribution arrangements, and technology licensing. Implications for strategic management, public policy, and international trade and investment are then discussed.

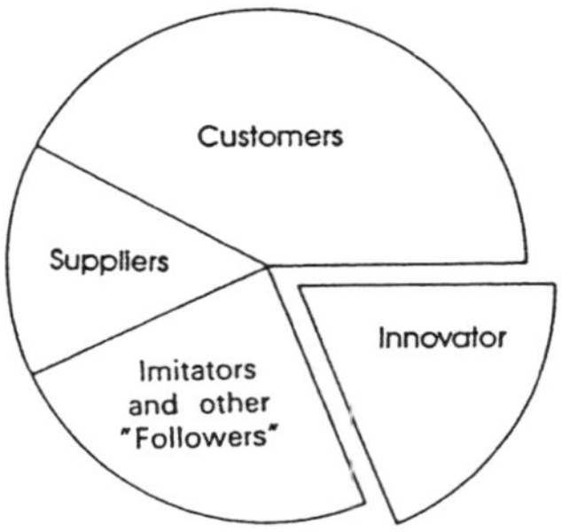

What determines the share of profits captured by the innovator?

Fig. 1. Explaining the distribution of the profits from innovation.

2. The phenomenon

Figure 2 presents a simplified taxonomy of the possible outcomes from innovation. Quadrant 1 represents positive outcomes for the innovator. A first-to-market advantage is translated into a sustained competitive advantage which either creates a new earnings stream or enhances an existing one. Quadrant 4 and its corollary quadrant 2 are the ones which are the focus of this paper.

The EMI CAT scanner is a classic case of the phenomenon to be investigated. [1] By the early 1970s, the UK firm Electrical Musical Industries (EMI) Ltd. was in a variety of product lines including phonographic records, movies, and advanced electronics. EMI had developed high resolution TVs in the 1930s, pioneered airborne radar during World War II, and developed the UK's first all solid-state computers in 1952.

In the late 1960s Godfrey Houndsfield, an EMI senior research engineer engaged in pattern recognition research which resulted in his displaying a scan of a pig's brain. Subsequent clinical work established that computerized axial tomography (CAT) was viable for generating cross-sectional "views" of the human body, the greatest advance in radiology since the discovery of X rays in 1895.

While EMI was initially successful with its CAT scanner, within 6 years of its introduction into the US in 1973 the company had lost market leadership, and by the eighth year had dropped out of the CT scanner business. Other companies successfully dominated the market, though they were late entrants, and are still profiting in the business today.

Other examples include RC Cola, a small beverage company that was the first to introduce cola in a can, and the first to introduce diet cola. Both Coca Cola and Pepsi followed almost immediately and deprived RC of any significant advantage from its innovation. Bowmar, which introduced the pocket calculator, was not able to withstand competition from Texas Instruments, Hewlett Packard and others, and went out of business. Xerox failed to succeed with its entry into the office computer business, even though Apple succeeded with the MacIntosh which contained many of Xerox's key product ideas, such as the mouse and icons. The de Havilland Comet saga has some of the same features. The Comet I jet was introduced into the commercial airline business 2 years or so before Boeing introduced the 707, but de Havilland failed to capitalize on its substantial early advantage. MITS introduced the first personal computer, the Altair, experienced a burst of sales, then slid quietly into oblivion.

If there are innovators who lose there must be followers/imitators who win. A classic example is IBM with its PC, a great success since the time it was introduced in 1981. Neither the architecture nor components embedded in the IBM PC were considered advanced when introduced; nor was the way the technology was packaged a significant departure from then-current practice. Yet the IBM PC was fabulously successful and established MS-DOS as the leading operating system for 16-bit PCs. By the end of 1984, IBM has shipped over 500 000 PCs, and many considered that it had irreversibly eclipsed Apple in the PC industry.

3. Profiting from innovation: Basic building blocks

In order to develop a coherent framework within which to explain the distribution of outcomes illustrated in fig. 2, three fundamental building blocks must first be put in place: the appropriability regime, complementary assets, and the dominant design paradigm.

[1] The EMI story is summarized in Michael Martin, *Managing Technological Innovation and Entrepreneurship*, (Reston Publishing Company, Reston, VA, 1984).

	Innovator	Follower-Imitator
Win	1 • Pilkington (Float Glass) • G.D. Searle (NutraSweet) • Dupont (Teflon)	2 • IBM (Personal Computer) • Matsushita (VHS video recorders) • Seiko (quartz watch)
Lose	4 • RC Cola (diet cola) • EMI (scanner) • Bowmar (pocket calculator) • Xerox (office computer) • DeHavilland (Comet)	3 • Kodak (instant photography) • Northrup (F20) • DEC (personal computer)

Fig. 2. Taxonomy of outcomes from the innovation process.

3.1. Regimes of appropriability

A regime of appropriability refers to the environmental factors, excluding firm and market structure, that govern an innovator's ability to capture the profits generated by an innovation. The most important dimensions of such a regime are the nature of the technology, and the efficacy of legal mechanisms of protection (fig. 3).

It has long been known that patents do not work in practice as they do in theory. Rarely, if ever, do patents confer perfect appropriability, although they do afford considerable protection on new chemical products and rather simple mechanical inventions. Many patents can be "invented around" at modest costs. They are especially ineffective at protecting process innovations. Often patents provide little protection because the legal requirements for upholding their validity or for proving their infringement are high.

In some industries, particularly where the innovation is embedded in processes, trade secrets are a viable alternative to patents. Trade secret protection is possible, however, only if a firm can put its product before the public and still keep the underlying technology secret. Usually only chemical formulas and industrial-commercial processes (e.g., cosmetics and recipes) can be protected as trade secrets after they're "out".

The degree to which knowledge is tacit or codified also affects ease of imitation. Codified knowledge is easier to transmit and receive, and is more exposed to industrial espionage and the like. Tacit knowledge by definition is difficult to articulate, and so transfer is hard unless those who possess the know how in question can demonstrate it to others (Teece [9]). Survey research indicates that methods of appropriability vary markedly across industries, and probably within industries as well (Levin et al. [5]).

The property rights environment within which a firm operates can thus be classified according to the nature of the technology and the efficacy of the legal system to assign and protect intellectual property. While a gross simplification, a dichotomy can be drawn between environments in which the appropriability regime is "tight" (technology is relatively easy to protect) and "weak" (technology is almost impossible to protect). Examples of the former include the formula for Coca Cola syrup; an example of the latter would be the Simplex algorithm in linear programming.

3.2. The dominant design paradigm

It is commonly recognized that there are two stages in the evolutionary development of a given branch of a science: the preparadigmatic stage when there is no single generally accepted conceptual treatment of the phenomenon in a field of study, and the paradigmatic stage which begins when a body of theory appears to have passed the canons of scientific acceptability. The emergence of a dominant paradigm signals scientific maturity and the acceptance of agreed upon "standards" by which what has been referred to as "normal" scientific research can proceed. These "standards" remain in force unless or until the paradigm is overturned. Revolutionary science is what overturns normal science, as when the Copernicus's theories of astronomy overturned Ptolemy's in the seventeenth century.

Abernathy and Utterback [1] and Dosi [3] have provided a treatment of the technological evolution of an industry which appears to parallel

• Legal instruments	• Nature of technology
- Patents	- Product
- Copyrights	- Process
- Trade secrets	- Tacit
	- Codified

Fig. 3. Appropriability regime: Key dimensions.

Kuhnian notions of scientific evolution. [2] In the early stages of industry development, product designs are fluid, manufacturing processes are loosely and adaptively organized, and generalized capital is used in production. Competition amongst firms manifests itself in competition amongst designs, which are markedly different from each other. This might be called the preparadigmatic stage of an industry.

At some point in time, and after considerable trial and error in the marketplace, one design or a narrow class of designs begins to emerge as the more promising. Such a design must be able to meet a whole set of user needs in a relatively complete fashion. The Model T Ford, the IBM 360, and the Douglas DC-3 are examples of dominant designs in the automobile, computer, and aircraft industry respectively.

Once a dominant design emerges, competition shifts to price and away from design. Competitive success then shifts to a whole new set of variables. Scale and learning become much more important, and specialized capital gets deployed as incumbent's seek to lower unit costs through exploiting economies of scale and learning. Reduced uncertainty over product design provides an opportunity to amortize specialized long-lived investments.

Innovation is not necessarily halted once the dominant design emerges; as Clarke [2] points out, it can occur lower down in the design hierarchy. For instance, a " v" cylinder configuration emerged in automobile engine blocks during the 1930s with the emergence of the Ford V-8 engine. Niches were quickly found for it. Moreover, once the product design stabilizes, there is likely to be a surge of process innovation as producers attempt to lower production costs for the new product (see fig. 4).

The Abernathy–Utterback framework does not characterize all industries. It seems more suited to mass markets where consumer tastes are relatively homogeneous. It would appear to be less characteristic of small niche markets where the absence of scale and learning economies attaches much less of a penalty to multiple designs. In these instances, generalized equipment will be employed in production.

[2] See Kuhn [4].

The existence of a dominant design watershed is of great significance to the distribution of profits between innovator and follower. The innovator may have been responsible for the fundamental scientific breakthroughs as well as the basic design of the new product. However, if imitation is relatively easy, imitators may enter the fray, modifying the product in important ways, yet relying on the fundamental designs pioneered by the innovator. When the game of musical chairs stops, and a dominant design emerges, the innovator might well end up positioned disadvantageously relative to a follower. Hence, when imitation is possible and occurs coupled with design modification before the emergence of a dominant design, followers have a good chance of having their modified product annointed as the industry standard, often to the great disadvantage of the innovator.

3.3. Complementary assets

Let the unit of analysis be an innovation. An innovation consists of certain technical knowledge about how to do things better than the existing state of the art. Assume that the know-how in question is partly codified and partly tacit. In order for such know-how to generate profits, it must be sold or utilized in some fashion in the market.

In almost all cases, the successful commercialization of an innovation requires that the know-how in question be utilized in conjunction with other capabilities or assets. Services such as marketing, competitive manufacturing, and after-sales support are almost always needed. These services are often obtained from complementary assets which are specialized. For example, the commercialization of a new drug is likely to require the dissemination of information over a specialized information channel. In some cases, as when the innovation is systemic, the complementary assets may be other parts of a system. For instance; computer hardware typically requires specialized software, both for the operating system, as well as for applications. Even when an innovation is autonomous, as with plug compatible components, certain complementary capabilities or assets will be needed for successful commercialization. Figure 5 summarizes this schematically.

Whether the assets required for least cost production and distribution are specialized to the

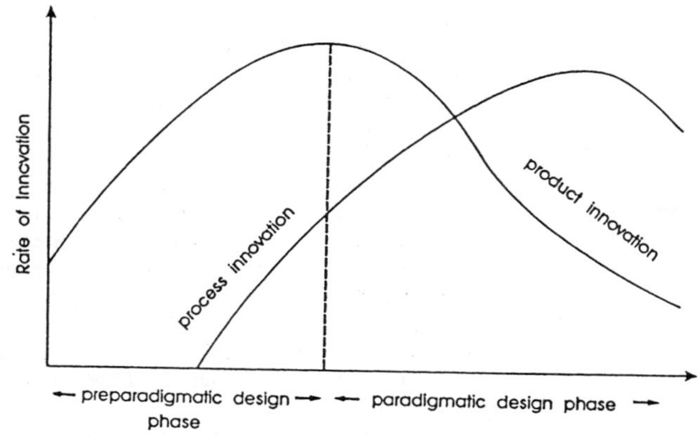

Fig. 4. Innovation over the product/industry life cycle.

innovation turns out to be important in the development presented below. Accordingly, the nature of complementary assets are explained in some detail. Figure 6 differentiates between complementary assets which are generic, specialized, and cospecialized.

Generic assets are general purpose assets which do not need to be tailored to the innovation in question. Specialized assets are those where there is unilateral dependence between the innovation and the complementary asset. Cospecialized assets are those for which there is a bilateral dependence. For instance, specialized repair facilities were needed to support the introduction of the rotary engine by Mazda. These assets are cospecialized because of the mutual dependence of the innovation on the repair facility. Containerization similarly required the deployment of some cospecialized assets in ocean shipping and terminals. However, the dependence of trucking on containerized shipping was less than that of containerized shipping on trucking, as trucks can convert from containers to flat beds at low cost. An example of a generic asset would be the manufacturing facilities needed to make running shoes. Generalized

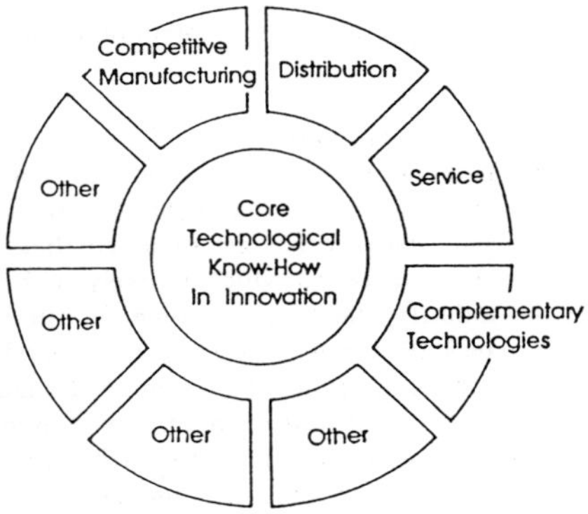

Fig. 5. Complementary assets needed to commercialize an innovation.

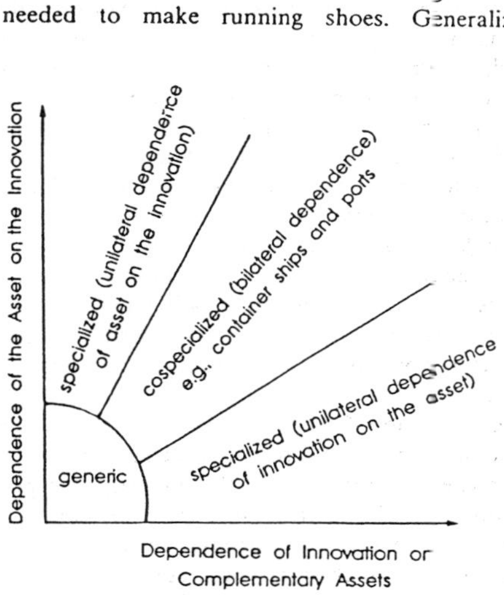

Fig. 6. Complementary assets: Generic, specialized, and cospecialized.

equipment can be employed in the main, exceptions being the molds for the soles.

4. Implications for profitability

These three concepts can now be related in a way which will shed light on the imitation process, and the distribution of profits between innovator and follower. We begin by examining tight appropriability regimes.

4.1. Tight appropriability regimes

In those few instances where the innovator has an iron clad patent or copyright protection, or where the nature of the product is such that trade secrets effectively deny imitators access to the relevant knowledge, the innovator is almost assured of translating its innovation into market value for some period of time. Even if the innovator does not possess the desirable endowment of complementary costs, iron clad protection of intellectual property will afford the innovator the time to access these assets. If these assets are generic, contractual relation may well suffice, and the innovator may simply license its technology. Specialized R&D firms are viable in such an environment. Universal Oil Products, an R&D firm developing refining processes for the petroleum industry was one such case in point. If, however, the complementary assets are specialized or cospecialized, contractual relationships are exposed to hazards, because one or both parties will have to commit capital to certain irreversible investments which will be valueless if the relationship between innovator and licensee breaks down. Accordingly, the innovator may find it prudent to expand its boundaries by integrating into specialized and cospecialized assets. Fortunately, the factors which make for difficult imitation will enable the innovator to build or acquire those complementary assets without competing with innovators for their control.

Competition from imitators is muted in this type of regime, which sometimes characterizes the petrochemical industry. In this industry, the protection offered by patents is fairly easily enforced. One factor assisting the licensee in this regard is that most petrochemical processes are designed around a specific variety of catalysts which can be kept proprietory. An agreement not to analyze the catalyst can be extracted from licensees, affording extra protection. However, even if such requirements are violated by licensees, the innovator is still well positioned, as the most important properties of a catalyst are related to its physical structure, and the process for generating this structure cannot be deduced from structural analysis alone. Every reaction technology a company acquires is thus accompanied by an ongoing dependence on the innovating company for the catalyst appropriate to the plant design. Failure to comply with various elements of the licensing contract can thus result in a cutoff in the supply of the catalyst, and possibly facility closure.

Similarly, if the innovator comes to market in the preparadigmatic phase with a sound product concept but the wrong design, a tight appropriability regime will afford the innovator the time needed to perform the trials needed to get the design right. As discussed earlier, the best initial design concepts often turn out to be hopelessly wrong, but if the innovator possesses an impenetrable thicket of patents, or has technology which is simply difficult to copy, then the market may well afford the innovator the necessary time to ascertain the right design before being eclipsed by imitators.

4.2. Weak appropriability

Tight appropriability is the exception rather than the rule. Accordingly, innovators must turn to business strategy if they are to keep imitators/followers at bay. The nature of the competitive process will vary according to whether the industry is in the paradigmatic or preparadigmatic phase.

4.2.1. Preparadigmatic phase

In the preparadigmatic phase, the innovator must be careful to let the basic design "float" until sufficient evidence has accumulated that a design has been delivered which is likely to become the industry standard. In some industries there may be little opportunity for product modification. In microelectronics, for example, designs become locked in when the circuitry is chosen. Product modification is limited to "debugging" and software modification. An innovator must begin the design process anew if the product doesn't fit the market well. In some respects, however, selecting designs is dictated by the need to meet certain compatibility standards so that new hardware can interface with existing applications software. In one sense, therefore, the design issue for the microprocessor industry today is relatively straightforward: deliver greater power and speed while meeting the the computer industry standards of the existing software base. However, from time to time windows of opportunity emerge for the introduction of entirely new families of microprocessors which will define a new industry and software standard. In these instances, basic design parameters are less well defined, and can be permitted to "float" until market acceptance is apparent.

The early history of the automobile industry exemplifies exceedingly well the importance for subsequent success of selecting the right design in the preparadigmatic stages. None of the early producers of steam cars survived the early shakeout when the closed body internal combustion engine automobile emerged as the dominant design. The steam car, nevertheless, had numerous early virtues, such as reliability, which the internal combustion engine autos could not deliver.

The British fiasco with the Comet I is also instructive. De Havilland had picked an early design with both technical and commercial flaws. By moving into production, significant irreversibilities and loss of reputation hobbled de Havilland to such a degree that it was unable to convert to the Boeing design which subsequently emerged as dominant. It wasn't even able to occupy second place, which went instead to Douglas.

As a general principle, it appears that innovators in weak appropriability regimes need to be intimately coupled to the market so that user needs can fully impact designs. When multiple parallel and sequential prototyping is feasible, it has clear advantages. Generally such an approach is simply prohibitively costly. When development costs for a large commercial aircraft exceed one billion dollars, variations on a theme are all that is possible.

Hence, the probability that an innovator – defined here as a firm that is first to commercialize a new product design concept – will enter the paradigmatic phase possessing the dominant design is problematic. The probabilities will be higher the lower the relative cost of prototyping, and the more tightly coupled the firm is to the market. The later is a function of organizational design, and can be influenced by managerial choices. The former is embedded in the technology, and cannot be influenced, except in minor ways, by managerial decisions. Hence, in industries with large developmental and prototyping costs – and hence significant irreversibilities – and where innovation of the product concept is easy, then one would expect that the probability that the innovator would emerge as the winner or amongst the winners at the end of the preparadigmatic stage is low.

4.2.2. Paradigmatic stage

In the preparadigmatic phase, complementary assets do not loom large. Rivalry is focused on trying to identify the design which will be dominant. Production volumes are low, and there is little to be gained in deploying specialized assets, as scale economies are unavailable, and price is not a principal competitive factor. However, as the leading design or designs begin to be revealed by the market, volumes increase and opportunities for economies of scale will induce firms to begin gearing up for mass production by acquiring specialized tooling and equipment, and possibly specialized distribution as well. Since these investments involve significant irreversibilities, producers are likely to proceed with caution. Islands of specialized capital will begin to appear in an industry, which otherwise features a sea of general purpose manufacturing equipment.

However, as the terms of competition begin to change, and prices become increasingly unimportant, access to complementary assets becomes absolutely critical. Since the core technology is easy to imitate, by assumption, commercial success swings upon the terms and conditions upon which the required complementary assets can be accessed.

It is at this point that specialized and cospecialized assets become critically important. Generalized equipment and skills, almost by definition, are always available in an industry, and even if they are not, they do not involve significant irreversibilities. Accordingly, firms have easy access to this type of capital, and even if there is insufficient capacity available in the relevant assets, it can easily be put in place as it involves few risks. Specialized assets, on the other hand, involve significant irreversibilities and cannot be easily accessed by contract, as the risks are significant for the party making the dedicated investment. The firms which control the cospecialized assets, such as distribution channels, specialized manufacturing capacity, etc. are clearly advantageously positioned relative to an innovator. Indeed, in rare instances where incumbent firms possess an airtight monopoly over specialized assets, and the innovator is in a regime of weak appropriability, all of the profits to the innovation could conceivably innure to the firms possessing the specialized assets who should be able to get the upper hand.

Even when the innovator is not confronted by situations where competitors or potential competitors control key assets, the innovator may still be disadvantaged. For instance, the technology embedded in cardiac pacemakers was easy to imitate, and so competitive outcomes quickly came to be determined by who had easiest access to the complementary assets, in this case specialized marketing. A similar situation has recently arisen in the United States with respect to personal computers. As an industry participant recently observed: "There are a huge numbers of computer manufacturers, companies that make peripherals (e.g. printers, hard disk drives, floppy disk drives), and software companies. They are all trying to get marketing distributors because they cannot afford to call on all of the US companies directly. They need to go through retail distribution channels, such as Businessland, in order to reach the marketplace. The problem today, however, is that many of these companies are not able to get shelf space and thus are having a very difficult time marketing their products. The point of distribution is where the profit and the power are in the marketplace today". (Norman [8, p.438])

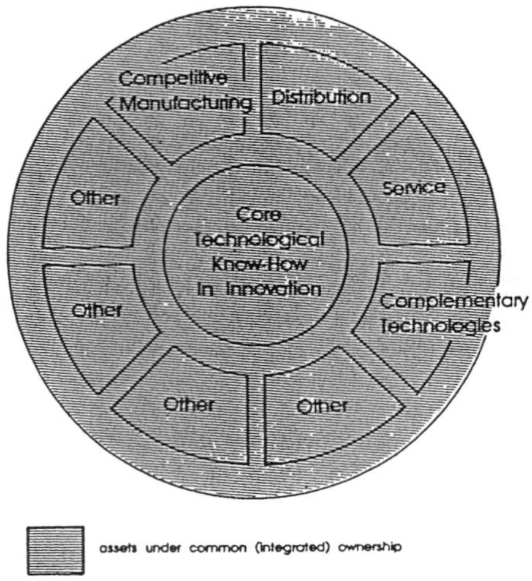

Fig. 7. Complementary assets internalized for innovation: Hypothetical case #1 (innovator integrated into all complementary assets).

There are a myriad of possible channels which could be employed. At one extreme the innovator could integrate into all of the necessary comple-

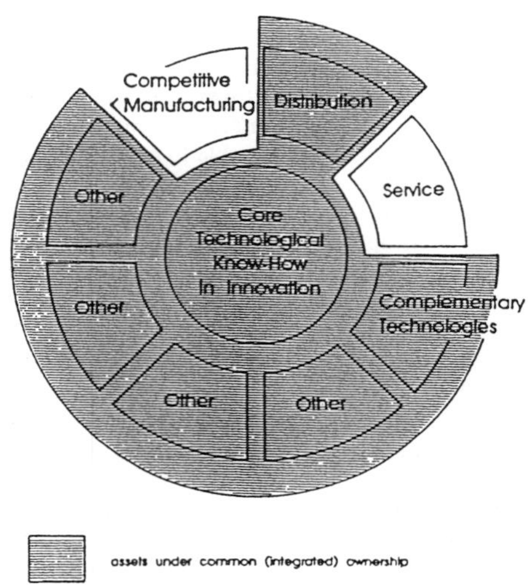

Fig. 8. Complementary assets internalized for innovation: Hypothetical case #2 (innovator subcontracts for manufacturing and service).

5. Channel strategy issues

The above analysis indicates how access to complementary assets, such as manufacturing and distribution, on competitive teams is critical if the innovator is to avoid handling over the lion's share of the profits to imitators, and/or to the owners of the complementary assets that are specialized or cospecialized to the innovation. It is now necessary to delve deeper into the appropriate control structure that the innovator ideally ought to establish over these critical assets.

mentary assets, as illustrated in fig. 7, or just a few of them, as illustrated in fig. 8. Complete integration (fig. 7) is likely to be unnecessary as well as prohibitively expensive. It is well to recognize that the variety of assets and competences which need to be accessed is likely to be quite large, even for only modestly complex technologies. To produce a personal computer, for instance, a company needs access to expertise in semiconductor technology, display technology, disk drive technology, networking technology, keyboard technology, and several others. No company can keep pace in all of these areas by itself.

At the other extreme, the innovator could attempt to access these assets through straightforward contractual relationships (e.g. component supply contracts, fabrication contracts, service contracts, etc.). In many instances such contracts may suffice, although it sometimes exposes the innovator to various hazards and dependencies that it may well wish to avoid. In between the fully integrated and full contractual extremes, there are a myriad of intermediate forms and channels available. An analysis of the properties of the two extreme forms is presented below. A brief synopsis of mixed modes then follows.

5.1. Contractual modes

The advantages of a contractual solution – whereby the innovator signs a contract, such as a license, with independent suppliers, manufacturers or distributors – are obvious. The innovator will not have to make the upfront capital expenditures needed to build or buy the assets in question. This reduces risks as well as cash requirements.

Contracting rather than integrating is likely to be the optimal strategy when the innovators appropriability regime is tight and the complementary assets are available in competitive supply (i.e. there is adequate capacity and a choice of sources).

Both conditions apply in petrochemicals for instance, so an innovator doesn't need to be integrated to be a successful. Consider, first, the appropriability regime. As discussed earlier, the protection offered by patents is fairly easily enforced, particularly for process technology, in the petrochemical industry. Given the advantageous feedstock prices available in hydrocarbon rich petrochemical exporters, and the appropriability regime characteristic of this industry, there is no incentive or advantage in owning the complementary assets (production facilities) as they are not typically highly specialized to the innovation. Union Carbide appears to realize this, and has recently adjusted its strategy accordingly. Essentially, Carbide is placing its existing technology into a new subsidiary, Engineering and Hydrocarbons Service. The company is engaging in licensing and offers engineering, construction, and management services to customers who want to take their feedstocks and integrate them forward into petrochemicals. But Carbide itself appears to be backing away from an integration strategy.

Chemical and petrochemical product innovations are not quite so easy to protect, which should raise new challenges to innovating firms in the developed nations as they attempt to shift out of commodity petrochemicals. There are already numerous examples of new products that made it to the marketplace, filled a customer need, but never generated competitive returns to the innovator because of imitation. For example, in the 1960s Dow decided to start manufacturing rigid polyurethene foam. However, it was imitated very quickly by numerous small firms which had lower costs. [3] The absence of low cost manufacturing capability left Dow vulnerable.

Contractual relationships can bring added credibility to the innovator, especially if the innovator is relatively unknown when the contractual partner is established and viable Indeed, arms-length contracting which embodies more than a simple buy-sell agreement is becoming so common, and is so multifaceted, that the term strategic partnering has been devised to describe it. Even large companies such as IBM are now engaging in it. For IBM, partnering buys access to new technologies enabling the company to "learn things we couldn't have learned without many years of trial and error." [4] IBM's arrangement with Microsoft to use the latter's MS-DOS operating system software on the IBM PC facilitated the timely introduction of IBM's personal computer into the market.

[3] Executive V.P. Union Carbide, Robert D. Kennedy, quoted in *Chemical Week*, Nov. 16, 1983, p. 48.

[4] Comment attributed to Peter Olson III, IBM's director of business development, as reported in The Strategy Behind IBM's Strategic Alliances, *Electronic Business*, October 1 (1985) 126.

Smaller less integrated companies are often eager to sign on with established companies because of the name recognition and reputation spillovers. For instance Cipher Data Products, Inc. contracted with IBM to develop a low-priced version of IBM's 3480 0.5 inch streaming cartridge drive, which is likely to become the industry standard. As Cipher management points out, "one of the biggest advantages to dealing with IBM is that, once you've created a product that meets the high quality standards necessary to sell into the IBM world, you can sell into any arena." [5] Similarly, IBM's contract with Microsoft "meant instant credibility" to Microsoft (McKenna, 1985, p. 94).

It is most important to recognize, however, that strategic (contractual) partnering, which is currently very fashionable, is exposed to certain hazards, particularly for the innovator, when the innovator is trying to use contracts to access specialized capabilities. First, it may be difficult to induce suppliers to make costly irreversible commitments which depend for their success on the success of the innovation. To expect suppliers, manufacturers, and distributors to do so is to invite them to take risks along with the innovator. The problem which this poses for the innovator is similar to the problems associated with attracting venture capital. The innovator must persuade its prospective partner that the risk is a good one. The situation is one open to opportunistic abuses on both sides. The innovator has incentives to overstate the value of the innovation, while the supplier has incentives to "run with the technology" should the innovation be a success.

Instances of both parties making irreversible capital commitments nevertheless exist. Apple's Laserwriter – a high resolution laser printer which allows PC users to produce near typeset quality text and art department graphics – is a case in point. Apple persuaded Canon to participate in the development of the Laserwriter by providing subsystems from its copiers – but only after Apple contracted to pay for a certain number of copier engines and cases. In short, Apple accepted a good deal of the financial risk in order to induce Canon to assist in the development and production of the Laserwriter. The arrangement appears to have been prudent, yet there were clearly hazards for both sides. It is difficult to write, execute, and enforce complex development contracts, particularly when the design of the new product is still "floating." Apple was exposed to the risk that its co-innovator Canon would fail to deliver, and Canon was exposed to the risk that the Apple design and marketing effort would not succeed. Still, Apple's alternatives may have been rather limited, inasmuch as it didn't command the requisite technology to "go it alone."

In short, the current euphoria over "strategic partnering" may be partially misplaced. The advantages are being stressed (for example, McKenna [6]) without a balanced presentation of costs and risks. Briefly, there is the risk that the partner won't perform according to the innovator's perception of what the contract requires; there is the added danger that the partner may imitate the innovator's technology and attempt to compete with the innovator. This latter possibility is particularly acute if the provider of the complementary asset is uniquely situated with respect to the complementary asset in question and has the capacity to imitate the technology, which the innovator is unable to protect. The innovator will then find that it has created a competitor who is better positioned than the innovator to take advantage of the market opportunity at hand. *Business Week* has expressed concerns along these lines in its discussion of the "Hollow Corporation." [6]

It is important to bear in mind, however, that contractual or partnering strategies in certain cases are ideal. If the innovator's technology is well protected, and if what the partner has to provide is a "generic" capacity available from many potential partners, then the innovator will be able to maintain the upper hand while avoiding the costs of duplicating downstream capacity. Even if the partner fails to perform, adequate alternatives exist (by assumption, the partners' capacities are commonly available) so the innovator's efforts to successfully commercialize its technology ought to proceed profitably.

[5] Comment attributed to Norman Farquhar, Cipher's vice president for strategic development, as reported in *Electronic Business*, October 1 (1985) 128.

[6] See *Business Week*, March 3 (1986) 57–59. *Business Week* uses the term to describe a corporation which lacks in-house manufacturing capability.

5.2. Integration modes

Integration, which by definition involves ownership, is distinguished from pure contractual modes in that it typically facilitates incentive alignment and control. If an innovator owns rather than rents the complementary assets needed to commercialize, then it is in a position to capture spillover benefits stemming from increased demand for the complementary assets caused by the innovation.

Indeed, an innovator might be in the position, at least before its innovation is announced, to buy up capacity in the complementary assets, possibly to its great subsequent advantage. If futures markets exist, simply taking forward positions in the complementary assets may suffice to capture much of the spillovers.

Even after the innovation is announced, the innovator might still be able to build or buy complementary capacities at competitive prices if the innovation has iron clad legal protection (i.e. if the innovation is in a tight appropriability regime). However, if the innovation is not tightly protected and once "out" is easy to imitate, then securing control of complementary capacities is likely to be the key success factor, particularly if those capacities are in fixed supply – so called "bottlenecks." Distribution and specialized manufacturing competences often become bottlenecks.

As a practical matter, however, an innovator may not have the time to acquire or build the complementary assets that ideally it would like to control. This is particularly true when imitation is easy, so that timing becomes critical. Additionally, the innovator may simply not have the financial resources to proceed. The implications of timing and cash constraints are summarized in fig. 9.

Accordingly, in weak appropriability regimes innovators need to rank complementary assets as to their importance. If the complementary assets are critical, ownership is warranted, although if the firm is cash constrained a minority position may well represent a sensible tradeoff.

Needless to say, when imitation is easy, strategic moves to build or buy complementary assets which are specialized must occur with due reference to the moves of competitors. There is no point moving to build a specialized asset, for instance, if one's imitators can do it faster and cheaper.

It is hopefully self evident that if the innovator is already a large enterprise with many of the relevant complementary assets under its control, integration is not likely to be the issue that it might otherwise be, as the innovating firm will already control many of the relevant specialized and cospecialized assets. However, in industries experiencing rapid technological change, technologies advance so rapidly that it is unlikely that a single company has the full range of expertise needed to bring advanced products to market in a timely and cost effective fashion. Hence, the integration issue is not just a small firm issue.

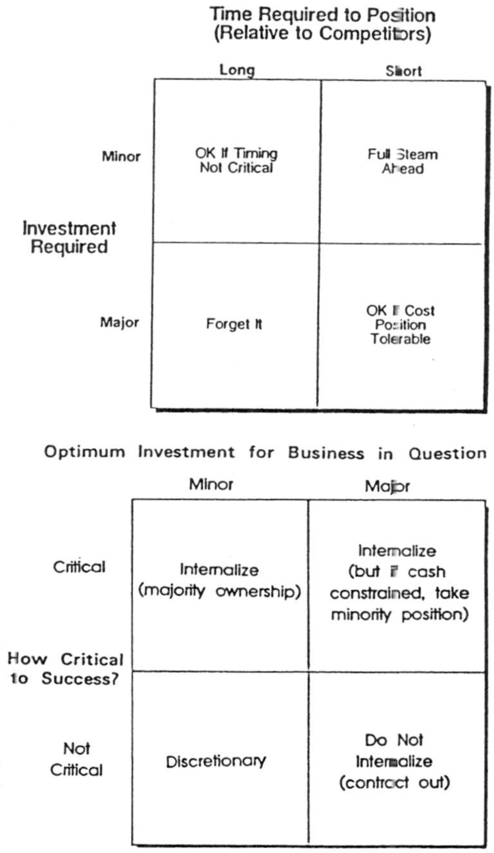

Fig. 9. Specialized complementary assets and weak appropriability: Integration calculus.

5.3. Integration versus contract strategies: An analytic summary

Figure 10 summarizes some of the relevant considerations in the form of a decision flow chart. It indicates that a profit seeking innovator, confronted by weak intellectual property protection and the need to access specialized complementary assets and/or capabilities, is forced to expand its activities through integration if it is to prevail over imitators. Put differently, innovators who develop new products that possess poor intellectual property protection but which requires specialized complementary capacities are more likely to parlay their technology into a commercial advantage, rather than see it prevail in the hands of imitators.

Figure 10 makes it apparent that the difficult strategic decisions arise in situations where the appropriability regime is weak and where specialized assets are critical to profitable commercialization. These situations, which in reality are very common, require that a fine-grained competitor analysis be part of the innovator's strategic assessment of its opportunities and threats. This is carried a step further in fig. 11, which looks only at situations where commercialization requires certain specialized capabilities. It indicates the appropriate strategies for the innovators and predicts the outcomes to be expected for the various players.

Three classes of players are of interst: innovators, imitators, and the owners of cospecialized assets (e.g. distributors). All three can potentially benefit or lose from the innovation process. The latter can potentially benefit from the additional business which the innovation may direct in the asset owners direction. Should the asset turn out

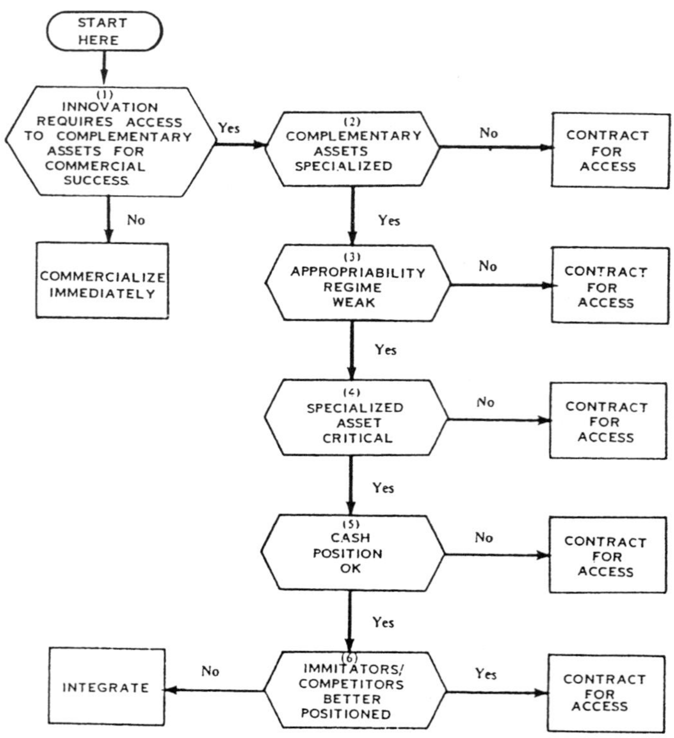

Fig. 10. Flow chart for integration versus contract decision.

to be a bottleneck with respect to commercializing the innovation, the owner of the bottleneck facilities is obviously in a position to extract profits from the innovator and/or imitators.

The vertical axis in fig. 11 measures how those who possess the technology (the innovator or possibly its imitators) are positioned vis à vis those firms that possess required specialized assets. The horizontal axis measures the "tightness" of the appropriability regime, tight regimes being evidence by iron clad legal protection coupled with technology that is simply difficult to copy; weak regimes offer little in the way of legal protection and the essence of the technology, once released, is transparent to the imitator. Weak regimes are further subdivided according to how the innovator and imitators are positioned vis à vis each other. This is likely to be a function of factors such as lead time and prior positioning in the requisite complementary assets.

Figure 11 makes it apparent that even when firms pursue the optimal strategy, other industry participants may take the jackpot. This possibility is unlikely when the intellectual property in question is tightly protected. The only serious threat to the innovator is where a specialized complementary asset is completely "locked up," a possibility recognized in cell 4. This can rarely be done without the cooperation of government. But it frequently occurs, as when a foreign government closes off access to a foreign market, forcing the innovators to license to foreign firms, but with the government effectively cartelizing the potential licensees. With weak intellectual property protection, however, it is quite clear that the innovator will often loose out to imitators and/or asset holders, even when the innovator is pursuing the appropriate strategy (cell 6). Clearly, incorrect strategies can compound problems. For instance, if innovators integrate when they should contract,

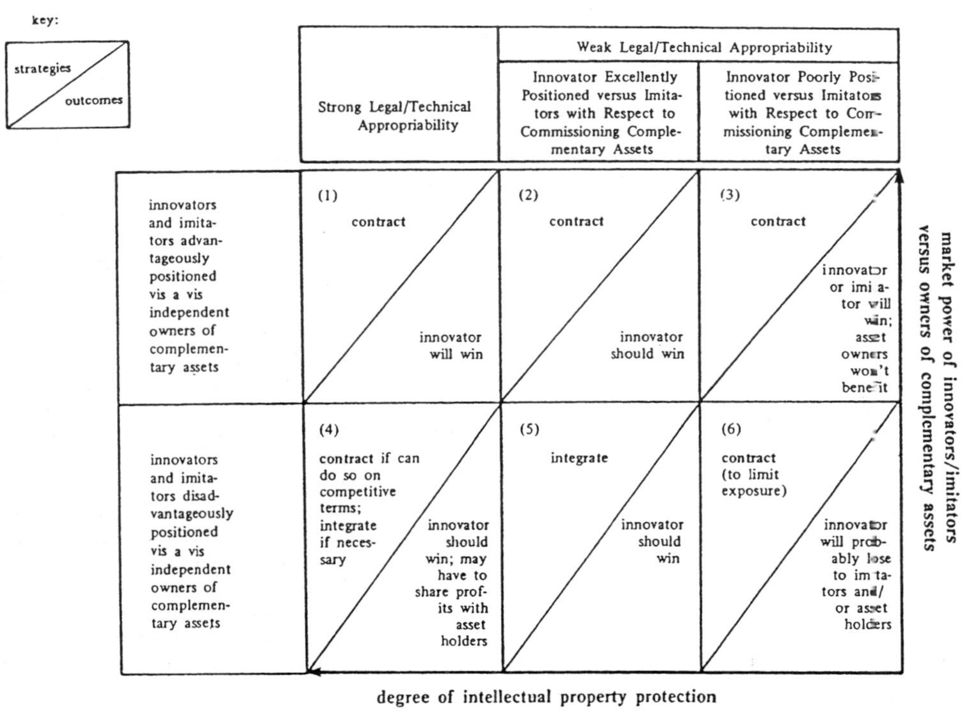

Fig. 11. Contract and integration strategies and outcomes for innovators: Specialized asset case.

a heavy commitment of resources will be incurred for little if any strategic benefit, thereby exposing the innovator to even greater losses than would otherwise be the case. On the other hand, if an innovator tries to contract for the supply of a critical capability when it should build the capability itself, it may well find it has nutured an imitator better able to serve the market than the innovator itself.

5.4. Mixed modes

The real world rarely provides extreme or pure cases. Decisions to integrate or• license involve tradeoffs, compromises, and mixed approaches. It is not surprising therefore that the real world is characterized by mixed modes of organization, involving judicious blends of integration and contracting. Sometimes mixed modes represent transitional phases. For instance, because of the convergence of computer and telecommunication technology, firms in each industry are discovering that they often lack the requisite technical capabilities in the other. Since the technological interdependence of the two requires collaboration amongst those who design different parts of the system, intense cross-boundary coordination and information flows are required. When separate enterprises are involved, agreement must be reached on complex protocol issues amongst parties who see their interests differently. Contractual difficulties can be anticipated since the selection of common technical protocols amongst the parties will often be followed by transaction-specific investments in hardware and software. There is little doubt that this was the motivation behind IBM's purchase of 15 percent of PBX manufacturer Rolm in 1983, a position that was expanded to 100 percent in 1984. IBM's stake in Intel, which began with a 12 percent purchase in 1982, is most probably not a transitional phase leading to 100 percent purchase, because both companies realized that the two corporate cultures are not very compatible, and IBM may not be as impressed with Intel's technology as it once was.

5.5. The CAT scanner, the IBM PC, and NutraSweet: Insights from the framework

EMI's failure to reap significant returns from the CAT scanner can be explained in large measure by reference to the concepts developed above. The scanner which EMI developed was of a technical sophistication much higher than would normally be found in a hospital, requiring a high level of training, support, and servicing. EMI had none of these capabilities, could not easily contract for them, and was slow to realize their importance. It most probably could have formed a partnership with a company like Siemens to access the requisite capabilities. Its failure to do so was a strategic error compounded by the very limited intellectual property protection which the law afforded the scanner. Although subsequent court decisions have upheld some of EMI's patent claims, once the product was in the market it could be reverse engineered and its essential features copied. Two competitors, GE and Technicare, already possessed the complementary capabilities that the scanner required, and they were also technologically capable. In addition, both were experienced marketers of medical equipment, and had reputations for quality, reliability and service. GE and Technicare were thus able to commit their R&D resources to developing a competitive scanner, borrowing ideas from EMI's scanner, which they undoubtedly had access to through cooperative hospitals, and improving on it where they could while they rushed to market. GE began taking orders in 1976 and soon after made inroads on EMI. In 1977 concern for rising health care costs caused the Carter Administration to introduce "certificate of need' regulation, which required HEW's approval on expenditures on big ticket items like CAT scanners. This severely cut the size of the available market.

By 1978 EMI had lost market share leadership to Technicare, which was in turn quickly overtaken by GE. In October 1979, Godfrey Houndsfield of EMI shared the Nobel prize for invention of the CT scanner. Despite this honor, and the public recognition of its role in bringing this medical breathrough to the world, the collapse of its scanner business forced EMI in the same year into the arms of a rescuer, Thorn Electrical Industries, Ltd. GE subsequently acquired what was EMI's scanner business from Thorn for what amounted to a pittance. [7] Though royalties continued to flow to EMI, the company had failed to capture the

[7] See GE Gobbles a Rival in CT Scanners, *Business Week*, May 19, 1980, issue no. 2637.

lion's share of the profits generated by the innovation it had pioneered and successfully commercialized.

If EMI illustrates how a company with outstanding technology and an excellent product can fail to profit from innovation while the imitators succeeded, the story of the IBM PC indicates how a new product representing a very modest technological advance can yield remarkable returns to the developer.

The IBM PC, introduced in 1981, was a success despite the fact that the architecture was ordinary and the components standard. Philip Estridge's design team in Boca Raton, Florida, decided to use existing technology to produce a solid, reliable micro rather than state of the art. With a one-year mandate to develop a PC, Estridge's team could do little else.

However, the IBM PC did use what at the time was a new 16-bit microprocessor (the Intel 8088) and a new disk operating system (DOS) adapted for IBM by Microsoft. Other than the microprocessor and the operating system, the IBM PC incorporated existing micro "standards" and used off-the-shelf parts from outside vendors. IBM did write its own BIOS (Basic Input/Output System) which is embedded in ROM, but this was a relatively straightforward programming exercise.

The key to the PC's success was not the technology. It was the set of complementary assets which IBM either had or quickly assembled around the PC. In order to expand the market for PCs, there was a clear need for an expandable, flexible microcomputer system with extensive applications software. IBM could have based its PC system on its own patented hardware and copyrighted software. Such an approach would cause complementary products to be cospecialized, forcing IBM to develop peripherals and a comprehensive library of software in a very short time. Instead, IBM adopted what might be called an "induced contractual" approach. By adopting an open system architecture, as Apple had done, and by making the operating system information publicly available, a spectacular output of third part software was induced. IBM estimated that by mid-1983, at least 3000 hardware and software products were available for the PC. [8] Put differently, IBM pulled together the complementary assets, particularly software, which success required, without even using contracts, let alone integration. This was despite the fact that the software developers were creating assets that were in part cospecialized with the IBM PC, at least in the first instance.

A number of special factors made this seem a reasonable risk to the software writers. A critical one was IBM's name and commitment to the project. The reputation behind the letters I.B.M. is perhaps the greatest cospecialized asset the company possesses. The name implied that the product would be marketed and serviced in the IBM tradition. It guaranteed that PC-DOS would become an industry standard, so that the software business would not be solely dependent on IBM, because emulators were sure to enter. It guaranteed access to retail distribution outlets on competitive terms. The consequences was that IBM was able to take a product which represented at best a modest technological accomplishment, and turn into a fabulous commercial success. The case demonstrates the role that complementary assets play in determining outcomes.

The spectacular success and profitability of G.D. Searle's NutraSweet is an uncommon story which is also consistent with the above framework. In 1982, Searle reported combined sales of $74 million for NutraSweet and its table top version, Equal. In 1983, this surged to $336 million. In 1985, NutraSweet sales exceeded $700 million [9] and Equal had captured 50 percent of the U.S. sugar substitute market and was number one in five other countries.

NutraSweet, which is Searle's tradename for aspartame, has achieved rapid acceptance in each of its FDA approved categories because of its good taste and ability to substitute directly for sugar in many applications. However, Searle's earnings from NutraSweet and the absence of a strategic challenge can be traced in part to Searle's clever strategy.

It appears that Searle has managed to establish an exceptionally tight appropriability regime around NutraSweet – one that may well continue for some time after the patent has expired. No competitor appears to have successfully "invented around" the Searle patent and commercialized an alternative, no doubt in part because the FDA

[8] F. Gens and C. Christiansen, Could 1,000,000 IBM PC Users Be Wrong, *Byte*, November 1983, 88.

[9] See *Monsanto Annual Report, 1985*.

approval process would have to begin anew for an imitator who was not violating Searle's patents. A competitor who tried to replicate the aspartame molecule with minor modification to circumvent the patent would probably be forced to replicate the hundreds of tests and experiments which proved aspartame's safety. Without patent protection, FDA approval would provide no shield against imitators coming to market with an identical chemical and who could establish to the FDA that it is the same compound that had already been approved. Without FDA approval on the other hand, the patent protection would be worthless for the product would not be sold for human consumption.

Searle has aggressively pushed to strengthen its patent protection. The company was granted U.S. patent protection in 1970. It has also obtained patent protection in Japan, Canada, Australia, U.K., France, Germany, and a number of other countries. However, most of these patents carry a 17-year life. Since the product was only approved for human consumption in 1982, the 17-year patent life was effectively reduced to five. Recognizing the obvious importance of its patent, Searle pressed for and obtained special legislation in November 1984 extending the patent protection on aspartame for another 5 years. The U.K. provided a similar extension. In almost every other nation, however, 1987 will mark the expiration of the patent.

When the patent expires, however, Searle will still have several valuable assets to help keep imitators at bay. Searle has gone to great lengths to create and promulgate the use of its NutraSweet name and a distinctive "Swirl" logo on all goods licensed to use the ingredient. The company has also developed the "Equal" tradename for a table top version of the sweetener. Trademark law in the U.S. provides protection against "unfair" competition in branded products for as long as the owner of the mark continues to use it. Both the NutraSweet and Equal trademarks will become essential assets when the patents on aspartame expire. Searle may well have convinced consumers that the only real form of sweetener is NutraSweet/Equal. Consumers know most other artificial sweeteners by their generic names – saccharin and cyclamates.

Clearly, Searle is trying to build a position in complementary assets to prepare for the competition which will surely arise. Searle's joint venture with Ajinomoto ensures them access to that company's many years of experience in the production of biochemical agents. Much of this knowledge is associated with techniques for distillation and synthesis of the delicate hydrocarbon compounds that are the ingredients of NutraSweet, and is therefore more tacit than codified. Searle has begun to put these techniques to use in its own $160 million Georgia production facility. It can be expected that Searle will use trade secrets to the maximum to keep this know-how proprietary.

By the time its patent expires, Searle's extensive research into production techniques for L-phenylalanine, and its 8 years of experience in the Georgia plant, should give it a significant cost advantage over potential aspartame competitors. Trade secret protection, unlike patents, has no fixed lifetime and may well sustain Searle's position for years to come.

Moreover, Searle has wisely avoided renewing contracts with suppliers when they have expired.[10] Had Searle subcontracted manufacturing for NutraSweet, it would have created a manufacturer who would then be in a position to enter the aspartame market itself, or to team up with a marketer of artificial sweeteners. But keeping manufacturing inhouse, and by developing a valuable tradename, Searle has a good chance of protecting its market position from dramatic inroads once patents expire. Clearly, Searle seems to be astutely aware of the importance of maintaining a "tight appropriability regime" and using cospecialized assets strategically.

6. Implications for R&D strategy, industry structure, and trade policy

6.1. Allocating R&D resources

The analysis so far assumes that the firm has developed an innovation for which a market exists. It indicates the strategies which the firm must

[10] Purification Engineering, which had spent $5 million to build a phenylalanine production facility, was told in January 1985 that their contract would not be renewed. In May, Genex, which claimed to have invested $25 million, was given the same message, A Bad Aftertaste, *Business Week*, July 15, 1985, issue 2903.

follow to maximize its share of industry profits relative to imitators and other competitors. There is no guarantee of success even if optimal strategies are followed.

The innovator can improve its total return to R&D, however, by adjusting its R&D investment portfolio to maximize the probability that technological discoveries will emerge that are either easy to protect with existing intellectual property law, or which require for commercialization cospecialized assets already within the firm's repertoire of capabilities. Put differently, if an innovating firm does not target its R&D resources towards new products and processes which it can commercialize advantageously relative to potential imitators and/or followers, then it is unlikely to profit from its investment in R&D. In this sense, a firm's history – and the assets it already has in place – ought to condition its R&D investment decisions. Clearly, an innovating firm with considerable assets already in place is free to strike out in new directions, so long as in doing so it is cognizant of the kinds of capabilities required to successfully commercialize the innovation. It is therefore rather clear that the R&D investment decision cannot be divorced from the strategic analysis of markets and industries, and the firm's position within them.

6.2. Small firm versus large firm comparisons

Business commentators often remark that many small entrepreneurial firms which generate new, commercially valuable technology fail while large multinational firms, often with a less meritorious record with respect to innovation, survive and prosper. One set of reasons for this phenomenon is now clear. Large firms are more likely to possess the relevant specialized and cospecialized assets within their boundaries at the time of new product introduction. They can therefore do a better job of milking their technology, however meager, to maximum advantage. Small domestic firms are less likely to have the relevant specialized and cospecialized assets within their boundaries and so will either have to incur the expense of trying to build them, or of trying to develop coalitions with competitors/owners of the specialized assets.

6.3. Regimes of appropriability and industry structure

In industries where legal methods of protection are effective, or where new products are just hard to copy, the strategic necessity for innovating firms to integrate into cospecialized assets would appear to be less compelling than in industries where legal protection is weak. In cases where legal protection is weak or nonexistent, the control of cospecialized assets will be needed for long-run survival.

In this regard, it is instructive to examine the U.S. drug industry (Temin [10]). Beginning in the 1940s, the U.S. Patent Office began, for the first time, to grant patents on certain natural substances that involved difficult extraction procedures. Thus, in 1948 Merck received a patent on streptomycin, which was a natural substance. However, it was not the extraction process but the drug itself which received the patent. Hence, patents were important to the drug industry in terms of what could be patented (drugs), but they did not prevent imitation [10, p.436]. Sometimes just changing one molecule will enable a company to come up with a different substance which does not violate the patent. Had patents been more all-inclusive – and I am not suggesting they should – licensing would have been an effective mechanism for Merck to extract profits from its innovation. As it turns out, the emergence of close substitutes, coupled with FDA regulation which had the de facto effect of reducing the elasticity of demand for drugs, placed high rewards on a product differentiation strategy. This required extensive marketing, including a sales force that could directly contact doctors, who were the purchasers of drugs through their ability to create prescriptions.[11] The result was exclusive production (i.e., the earlier industry practice of licensing was dropped) and forward integration into marketing (the relevant cospecialized asset).

Generally, if legal protection of the innovator's profits is secure, innovating firms can select their

[11]. In the period before FDA regulation, all drugs other than narcotics were available over-the-counter. Since the end user could purchase drugs directly, sales were price sensitive. Once prescriptions were required, this price sensitivity collapsed; the doctors not only did not have to pay for the drugs, but in most cases they were unaware of the prices of the drugs they were prescribing.

boundaries based simply on their ability to identify user needs and respond to those through research and development. The weaker the legal methods of protection, the greater the incentive to integrate into the relevant cospecialized assets. Hence, as industries in which legal protection is weak begin to mature, integration into innovation-specific cospecialized assets will occur. Often this will take the form of backward, forward and lateral integration. (Conglomerate integration is not part of this phenomenon.) For example, IBM's purchase of Rolm can be seen as a response to the impact of technological change on the identity of the cospecialized assets relevant to IBM's future growth.

6.4. Industry maturity, new entry, and history

As technologically progressive industries mature, and a greater proportion of the relevant cospecialized assets are brought in under the corporate umbrellas of incumbents, new entry becomes more difficult. Moreover, when it does occur it is more likely to involve coalition formation very early on. Incumbents will for sure own the cospecialized assets, and new entrants will find it necessary to forge links with them. Here lies the explanation for the sudden surge in "strategic partnering" now occurring internationally, and particularly in the computer and telecommunications industry. Note that it should not be interpreted in anti-competitive terms. Given existing industry structure, coalitions ought to be seen not as attempts to stifle competition, but as mechanisms for lowering entry requirements for innovators.

In industries in which technological change of a particular kind has occurred, which required deployment of specialized and/or cospecialized assets at the time, a configuration of firm boundaries may well have arisen which no longer has compelling efficiencies. Considerations which once dictated integration may no longer hold, yet there may not be strong forces leading to divestiture. Hence existing firm boundaries may in some industries – especially those where the technological trajectory and attendent specialized asset requirements has changed – be rather fragile. In short, history matters in terms of understanding the structure of the modern business enterprise. Existing firm boundaries cannot always be assumed to have obvious rationales in terms of today's requirements.

6.5. The importance of manufacturing to international competitiveness

Practically all forms of technological know-how must be embedded in goods and services to yield value to the consumer. An important policy for the innovating nation is whether the identity of the firms and nations performing this function matter.

In a world of tight appropriability and zero transactions cost – the world of neoclassical trade theory – it is a matter of indifference whether an innovating firm has an in-house manufacturing capability, domestic or foreign. It can simply engage in arms-length contracting (patent licensing, know-how licensing, co-production, etc.) for the sale of the output of the activity in which it has a comparative advantage (in this case R&D) and will maximize returns by specializing in what it does best.

However, in a regime of weak appropriability, and especially where the requisite manufacturing assets are specialized to the innovation, which is often the case, participation in manufacturing may be necessary if an innovator is to appropriate the rents from its innovation. Hence, if an innovator's manufacturing costs are higher than those of its imitators, the innovator may well end up ceding the lion's share of profits to the imitator.

In a weak appropriability regime, low cost imitator-manufacturers may end up capturing all of the profits from innovation. In a weak appropriability regime where specialized manufacturing capabilities are required to produce new products, an innovator with a manufacturing disadvantage may find that its advantage at early stage research and development will have no commercial value. This will eventually cripple the innovator, unless it is assisted by governmental processes. For example, it appears that one of the reasons why U.S. color TV manufacturers did not capture the lion's share of the profits from the innovation, for which RCA was primarily responsible, was that RCA and its American licenses were not competitive at manufacturing. In this context, concerns that the decline of manufacturing threatens the entire economy appear to be well founded.

A related implication is that as the technology gap closes, the basis of competition in an industry will shift to the cospecialized assets. This appears to be what is happening in microprocessors. Intel is no longer out ahead technologically. As Gordon Moore, CEO of Intel points out, "Take the top 10 [semiconductor] companies in the world... and it is hard to tell at any time who is ahead of whom.... It is clear that we have to be pretty damn close to the Japanese from a manufacturing standpoint to compete." [12] It is not just that strength in one area is necessary to compensate for weakness in another. As technology becomes more public and less proprietary through easier imitation, then strength in manufacturing and other capabilities is necessary to derive advantage from whatever technological advantages an innovator may possess.

Put differently, the notion that the United States can adopt a "designer role" in international commerce, while letting independent firms in other countries such as Japan, Korea, Taiwan, or Mexico do the manufacturing, is unlikely to be viable as a long-run strategy. This is because profits will accrue primarily to the low cost manufacturers (by providing a larger sales base over which they can exploit their special skills). Where imitation is easy, and even where it is not, there are obvious problems in transacting in the market for know-how, problems which are described in more detail elsewhere [9]. In particular, there are difficulties in pricing an intangible asset whose true performance features are difficult to ascertain ex ante.

The trend in international business towards what Miles and Snow [7] call "dynamic networks" – characterized by vertical disintegration and contracting – ought thus be viewed with concern. (*Business Week*, March 3, 1986, has referred to the same phenomenon as the Hollow Corporation.) "Dynamic networks" may not so much reflect innovative organizational forms, but the disassembly of the modern corporation because of deterioration in national capacities, manufacturing in particular, which are complementary to technological innovation. Dynamic networks may therefore signal not so much the rejuvenation of American enterprise, but its piecemeal demise.

6.6. How trade and investment barriers can impact innovators' profits

In regimes of weak appropriability, governments can move to shift the distribution of the gains from innovation away from foreign innovators and towards domestic firms by denying innovators ownership of specialized assets. The foreign firm, which by assumption is an innovator, will be left with the option of selling its intangible assets in the market for know how if both trade and investment are foreclosed by government policy. This option may appear better than the alternative (no renumeration at all from the market in question). Licensing may then appear profitable, but only because access to the complementary assets is blocked by government.

Thus when an innovating firm generating profits needs to access complementary assets abroad, host governments, by limiting access, can sometimes milk the innovators for a share of the profits, particularly that portion which originates from sales in the host country. However, the ability of host governments to do so depends importantly on the criticality of the host country's assets to the innovator. If the cost and infrastructure characteristics of the host country are such that it is the world's lowest cost manufacturing site, and if domestic industry is competitive, then by acting as a de facto monopsonist the host country government ought to be able to adjust the terms of access to the complementary assets so as to appropriate a greater share of the profits generated by the innovation. [13]

If, on the other hand, the host country offers no unique complementary assets, except access to its own market, restrictive practices by the government will only redistribute profits with respect to domestic rather than worldwide sales.

6.7. Implications for the international distribution of the benefits from innovation

The above analysis makes transparent that innovators who do not have access to the relevant specialized and cospecialized assets may end up ceding profits to imitators and other competitors, or simply to the owners of the specialized or cospecialized assets.

Even when the specialized assets are possessed by the innovating firm, they may be located abroad. Foreign factors of production are thus

[12] Institutionalizing the Revolution, *Forbes*, June 16, 1986, 35.

[13] If the host country market structure is monopolistic in the first instance, private actors might be able to achieve the same benefit. What government can do is to force collusion of domestic enterprises to their mutual benefit.

likely to benefit from research and development activities occurring across borders. There is little doubt, for instance, that the inability of many American multinationals to sustain competitive manufacturing in the U.S. is resulting in declining returns to U.S. labor. Stockholders and top management probably do as well if not better when a multinational accesses cospecialized assets in the firm's foreign subsidiaries; however, if there is unemployment in the factors of production supporting the specialized and cospecialized assets in question, then the foreign factors of production will benefit from innovation originating beyond national borders. This speaks to the importance to innovating nations of maintaining competence and competitiveness in the assets which complement technological innovation, manufacturing being a case in point. It also speaks to the importance to innovating nations of enhancing the protection afforded worldwide to intellectual property.

However, it must be recognized that there are inherent limits to the legal protection of intellectual property, and that business and national strategy are therefore likely to the critical factors in determining how the gains from innovation are shared worldwide. By making the correct strategic decision, innovating firms can move to protect the interests of stockholders; however, to ensure that domestic rather than foreign cospecialized assets capture the lion's share of the externalities spilling over to complementary assets, the supporting infrastructure for those complementary assets must not be allowed to decay. In short, if a nation has prowess at innovation, then in the absence of iron clad protection for intellectual property, it must maintain well-developed complementary assets if it is to capture the spillover benefits from innovation

7. Conclusion

The above analysis has attempted to synthesize from recent research in industrial organization and strategic management a framework within which to analyze the distribution of the profits from innovation. The framework indicates that the boundaries of the firm are an important strategic variable for innovating firms. The ownership of complementary assets, particularly when they are specialized and/or cospecialized, help establish who wins and who loses from innovation. Imitators can often outperform innovators if they are better positioned with respect to critical complementary assets. Hence, public policy aimed at promoting innovation must focus not only on R&D, but also on complementary assets, as well as the underlying infrastructure. If government decides to stimulate innovation, it would seem important to clear away barriers which impede the development of complementary assets which tend to be specialized or cospecialized to innovation. To fail to do so will cause an unnecessary large portion of the profits from innovation to flow to imitators and other competitors. If these firms lie beyond one's national borders, there are obvious implications for the internal distribution of income.

When applied to world markets, results similar to those obtained from the "new trade theory" are suggested by the framework. In particular, tariffs and other restrictions on trade can in some cases injure innovating firms while simultaneously benefiting protected firms when they are imitators. However, the propositions suggested by the framework are particularized to appropriability regimes, suggesting that economy-wide conclusions will be illusive. The policy conclusions derivable for commodity petrochemicals, for instance, are likely to be different than those that would be arrived at for semiconductors.

The approach also suggests that the product life cycle model of international trade will play itself out very differently in different industries and markets, in part according to appropriability regimes and the nature of the assets which need to be employed to convert a technological success into a commercial one. Whatever its limitations, the approach establishes that it is not so much the structure of markets but the structure of firms, particularly the scope of their boundaries, coupled with national policies with respect to the development of complementary assets, which determines the distribution of the profits amongst innovators and imitator/followers.

References

[1] W.J. Abernathy and J.M. Utterback, Patterns of Industrial Innovation, *Technology Review* 80(7) (January/July 1978) 40–47.

[2] Kim B. Clarke, The Interaction of Design Hierarchies and Market Concepts in Technological Evolution, *Research Policy* 14 (1985) 235–251.

[3] G. Dosi, Technological Paradigms and Technological Trajectories, *Research Policy* 11 (1982) 147–162.

[4] Thomas Kuhn, *The Structure of Scientifc Revolutions*, 2nd ed (University of Chicago Press, Chicago, 1970).

[5] R. Levin, A. Klevorick, N. Nelson, and S. Winter, Survey Research on R&D Appropriability and Technological Opportunity, unpublished manuscript, Yale University, 1984.

[6] Regis McKenna, Market Positioning in High Technology, *California Management Review*, XXVII (3) (spring 1985).

[7] R.E. Miles and C.C. Snow, Network Organizations: New Concepts for New Forms, *California Management Review* (spring 1986) 62–73.

[8] David A. Norman, Impact of Entrepreneurship and Innovations on the Distribution of Personal Computers, in: R. Landau and N. Rosenberg (eds.), *The Positive Sum Strategy* (National Academy Press, Washington, DC, 1986).

[9] D.J. Teece, The Market for Know how and the Efficient International Transfer of Technology, *Annals of the American Academy of Political and Social Science*, November 1981.

[10] P. Temin, Technology, Regulation, and Market Structure in the Modern Pharmaceutical Industry, *The Bell Journal of Economics* (autumn 1979) 429–446.

Managing Intellectual Capital:

Licensing and Cross-Licensing in Semiconductors and Electronics

Peter C. Grindley
David J. Teece

One of the most significant emerging business developments in the last decade has been the proactive management of intellectual capital by innovating firms. While firms have for decades actively managed their physical and financial assets, until quite recently intellectual property (IP) management was a backwater. Top management paid little attention and legal counsel did not participate in major managerial decisions. This is changing. High-technology firms now often have "IP" managers as well as "IT" managers.[1] In some firms considerations of intellectual capital management have expanded from the mere licensing of residual technology to become a central element in technology strategy. This development is spurred by the increasing protection afforded IP worldwide and by the greater importance of technological know-how to competitive advantage. These developments herald a new era for management.

Patents and trade secrets have become a key element of competition in high-technology industries. In electronics and semiconductors, firms continually make large investments in R&D in their attempts to stay at the frontier and to utilize technological developments external to the firm. Fierce competition has put a premium on innovation and on defending IP from unlicensed imitators. As IP owners have taken a more active stance regarding their patent portfolios, industry participants increasingly find it necessary to engage in licensing and cross-licensing.[2] Moreover, and relatedly, royalty rates have risen. The effect has been positive for firms with strong portfolios, who are now able to capture considerable benefit from their patent estates. Firms that are high net users of others' patents have a choice. They must increasingly pay royalties, or they must develop their own portfolios so as to bring something to the table in cross-licensing negotiations.

The new environment affords new challenges. If a firm is to compete with advanced products and processes, it is likely to utilize not only its own technology, but also the patents of others. In many advanced products, the range of technology is too great for a single firm to develop its entire needs internally. In cumulative technology fields such as electronics and semiconductors, one innovation builds on another. There are inevitably overlapping developments and mutually blocking patents. It is likely that firms will need to cross-license patents from others to ensure that they have freedom to manufacture without infringement. Thus in many industries today, firms can generate value from their innovation not only by embedding it in new products and processes, but also through engaging in licensing and cross-licensing.

In electronics and semiconductors, cross-licensing is generally more complex than the exchange of individual patent rights. The size of the patent portfolios of some firms is often too great for it to be feasible to identify individual infringements. Companies may own thousands of patents, used in literally tens of thousands of products, and may add hundreds more each year. With this degree of overlap of technology, companies protect themselves against mutual infringement by cross-licensing portfolios of all current and future patents in a field-of-use, without making specific reference to individual patents. It is simply too cumbersome and costly to license only the specific patents you need for specific products. The portfolio approach reduces transactions costs and allows licensees freedom to design and manufacture without infringement.[3]

An important dimension of field-of-use cross-licensing is the calculation of balancing royalty payments, according to the relative value of the patent portfolios of each party. This calculation is made prospectively, based on a sample of each firm's leading patents. Weight is given to the quality and market coverage of the patents. Desirable portfolios have excellent patents covering technology widely used in the industry. A quality portfolio is a powerful lever in negotiating access to required technology and may lead to significant royalty earnings or, at a minimum, to reduced payments to others. Obviously, a firm which is a large net user of other firms' patents, without contributing comparable IP in exchange, is likely to have to pay significant royalties.

Many managers now understand the use of licensing and cross-licensing as part of business strategy as well as the importance of a valuable patent portfolio. The key to successful cross-licensing is a portfolio of quality patents that covers large areas of the partner's product markets. Significantly, for the balancing process, the firm should not necessarily emulate the portfolio of its cross-licensing partner. Rather it should concentrate R&D in those areas in which it does best and has the most comparative advantage to develop patents that its partners need. This will give maximum leverage in negotiating access and balancing royalties. This might be in product design, software, or manufacturing processes, wherever the firm's R&D is most effective and its IP most widely used. In this sense, cross-licensing has a double positive effect on innovation. It allows firms greater means of earning a return on innovation (to help fund further

R&D), while allowing firms to concentrate their innovation and patenting activities according to their comparative advantage. In this way, firms can develop complementary rather than duplicative technology, thereby benefiting the public interest.

The unprecedented rates of technological development in the electronics industries have been made possible by a combination of the ability to capture value from innovation and the freedom to design and manufacture. Cross-licensing has been crucial. A key lesson for managers is to be aware of the value of developing a strong, high-quality IP portfolio and the effect this can have on licensing and cross-licensing strategies. This protects the firm's innovations and may significantly reduce royalty payments and fund further R&D.

The Licensing Legacy

Background—The Formation of RCA

Cross-licensing is not a new phenomenon in electronics; it goes back almost to the beginning of the industry. Cross-licensing is typical of industries involved in "cumulative systems technologies," where one innovation builds on another and products may draw on several related technologies. Multiple firms develop patented innovations in the same technological fields, and the "state of the art" of the technology tends to be covered by a large number of different patents held by different firms. Because of the potential for mutually blocking patents, firms typically cross-license all patents in a field-of-use to ensure adequate access to technology. The strongest examples of cumulative systems technologies are in electronics, including computers and semiconductors, where extensive cross-licensing ensures "design freedom" or "freedom-to-manufacture."[4] Note that this is a different situation than in some other industries not characterized by cumulative systems technologies, such as chemicals and pharmaceuticals, where cross-licensing, or, rather, reciprocal licensing, is typically aimed at exchanging technology rather than avoiding patent interference.[5]

An important instance of field-of-use cross-licensing is the development of radio in the first quarter of this century.[6] It epitomizes the complexities surrounding intellectual property arrangements that may be encountered with cumulative systems technologies. Also, many of the cross-licensing ideas used later by the electronics industry were pioneered during the early days of radio.

The commercialization of radio required a number of basic inventions. The scientific basis for wireless was developed by university scientists such as Maxwell, Hertz, and Lodge in the 19th century. Their discoveries were first applied to practical communication with the development of wireless telegraphy by Marconi in Britain in 1896. The first speech transmissions were made in the U.S. by Fessenden in 1900, using a high-frequency alternator. Further basic innovations were made over the next two decades.[7]

Many of these inventions were initially developed by individuals working independently of each other. Indeed, many carry the name of the inventor, such as the Poulsen arc, the Fleming valve, and the de Forest triode.[8] As the potential for radio became apparent, and the need for large-scale R&D and investment grew, large corporations entered the field. The pace of development accelerated and the number of patents multiplied. The companies involved included Marconi, General Electric (GE), Westinghouse, AT&T, Telefunken, and others. In addition to their considerable R&D effort, these corporations also acquired key patents where appropriate.[9] There was considerable competition, and with research teams in different companies working in parallel, patent interferences were common.[10] By 1918, it was apparent that several technologies were needed to manufacture radio systems, and each of these technologies itself involved multiple patents from different firms. In the words of Armstrong, one of the pioneers of radio, "It was absolutely impossible to manufacture any kind of workable apparatus without using practically all of the inventions which were then known."[11]

The result was deadlock. A number of firms had important patent positions and could block each other's access to key components. They refused to cross-license each other. It was a "Mexican standoff," with each firm holding up the development of the industry.[12] The situation arose in large part as a result of the way radio had developed. Key patent portfolios had been developed by different individuals and corporations, who were often adamant about refusing to cross-license competitors. Also, in a new industry in which large scale interference was a novel problem, there was no well developed means of coordinating cross-licensing agreements between these groups.

The situation was resolved in the U.S. only when, under prompting by the U.S. Navy, the various pioneers formed the Radio Corporation of America (RCA) in 1919.[13] This broke a key source of the deadlock. RCA acquired the U.S. rights to the Marconi patents, and cross-licensed the U.S. rights for other major patent portfolios.[14] The major U.S. patent holders became shareholders in RCA. In this way, RCA acquired the U.S. rights to all the constituent radio patents under one roof—amounting to over 2,000 patents.[15] It established RCA as the technical leader in radio, but also enabled the other cross-licensees to continue their own development of the technology for use in other fields or as suppliers to RCA. The RCA cross-licensing agreements became a model for the future.[16]

The case shows that because of the reluctance of the parties to cross-license, technological progress and the further commercialization of radio was halted. In this case, the debacle was resolved only by the formation of RCA, a rather radical organizational solution. However, it became clear from the experience that the same ends—namely design freedom—may be achieved more simply, without such fundamental reorganization, by cross-licensing alone. This helped set the stage for further development of cross-licensing in electronics.

AT&T's Cross-Licensing Practices

The need to achieve design freedom was soon experienced in other fields of electronics and resulted in patent cross-licensing agreements. One of the most influential firms in shaping the industry practices was AT&T, whose licensing and cross-licensing policy, especially from the 1940s until its breakup in 1984, has been crucial to the development of similar practices in U.S. electronics and semiconductor industries.

Over its long history, AT&T's licensing policy has had three phases, reflecting changes in its overall business strategy. First, from AT&T's establishment in 1885 until its first antitrust-related commitment in 1913, it used its IP rights in a forthright fashion to establish itself in the service market.[17] In the second phase, from 1914 until 1984, AT&T became a regulated monopoly. Its policy (as a matter of law under the 1956 antitrust consent decree) was to openly license its IP to everyone for minimal fees. Reasons of technology access similar to those in radio led to patent cross-license agreements between the major producers of telephone equipment, starting in the 1920s. This developed into a more widespread policy. It was during this period that the transistor was invented at Bell Labs. This and other breakthroughs laid the foundation for the semiconductor industry and shaped the development of the telecommunications, computer, and electronics industries. In the current phase, dating from divestiture in 1984, AT&T is no longer bound by the consent decree, and its IP licensing can be aligned with its proprietary needs.[18]

The 1956 antitrust consent decree required AT&T to openly license all patents controlled by the Bell System to any applicant at "reasonable royalties," provided that the licensee also grant licenses at reasonable royalties in return. AT&T was also required to provide technical information with the licenses on payment of reasonable fees; licensees had the right to sublicense the technology to their associates.[19] The impact of AT&T's liberal licensing on the industry was considerable, especially when considered in parallel with that at IBM.[20]

To a large extent, the licensing terms in AT&T's 1956 decree simply codified what was already AT&T policy. As an enterprise under rate-of-return regulation, it had little reason to maximize royalty income from its IP. Instead, it used its technology and IP to promote new services and reduce costs. It procured a tremendous amount of equipment and materials on the open market and apparently figured that its service customers would be better off if its technologies were widely diffused amongst its actual and potential suppliers, as this would lower prices and increase the performance of procured components.[21] It was the first company we are aware of to have "design freedom" as a core component of its patent strategy. It did not see licensing income as a source of funds for R&D, as Bell Labs research was largely funded by the "license contract fee," assessed on the annual revenues of the Bell operating companies. This very stable source of research funding supported a constant stream of basic innovations.[22] Using its own portfolio as leverage, AT&T was able to obtain the (reciprocal) rights it needed to continue to innovate, unimpeded by the IP of others. It successfully accomplished this limited objective.

An interesting aspect of AT&T's IP strategy was that technologies (though not R&D programs) were often selected for patent protection based on their potential interest to other firms generating technology of interest to AT&T. Since the legal requirement for open licensing specifically did not extinguish all of AT&T's intellectual property rights, the company was able to gain access to the external technology that it needed, while contributing enormously to innovation in telecommunications, computers, and electronics worldwide.[23]

The terms of AT&T's licenses set a pattern that is still commonplace in the electronics industries. The "capture model" was defined in the consent decree.[24] Under this arrangement, the licensee is granted the right to use existing patents and any obtained for inventions made during a fixed capture period of no more than five years, followed by a survivorship period until the expiration of these patents and with subsequent agreement renewals. The open licensing regimes this led to were persistent, since with the long survivorship period on many of the basic patents, there was limited scope to introduce more stringent conditions for new patents.

AT&T's licensing policy had the effect of making its tremendously large IP portfolio available to the industry worldwide for next to nothing. This portfolio included fundamental patents such as the transistor, basic semiconductor technology, and the laser, and included many other basic patents in telecommunications, computing, optoelectronics, and superconductivity. Shaped under antitrust policy reflecting the needs and beliefs of an era in which U.S. firms did not have to worry much about foreign competition, such a liberal policy appears quite anachronistic today. However, there is no doubt that it provided a tremendous contribution to world welfare. It remains as one of the most unheralded contributions to economic development—possibly far exceeding the Marshall Plan in terms of the wealth generation capability it established abroad and in the United States.

The traditional cross-licensing policy of AT&T was greatly extended following the invention of the transistor. Widespread "field-of-use" licenses in the semiconductor industry is a legacy, as the industry was founded on the basic semiconductor technology developed by AT&T. In the early days of semiconductor technology, AT&T controlled most of the key patents in the field. It soon realized that, given the importance of semiconductor technology, other electronics companies were developing their own technologies and could eventually invent around the AT&T patents. Cross-licensing ensured that AT&T would have reciprocal access to this technology and be able to develop its own technology without risking patent interference.[25]

AT&T's liberal licensing allowed the semiconductor industry to grow rapidly, and members of the industry did not care much about individual patents. The culture of the industry still reflects this, with a tradition of spin-outs and new ventures, open communications and frequent job changes.[26] The continued speed of technological progress in the industry and the difficulty of monitoring technological use are reasons why there is still a need for the transactional simplicity associated with "lump-sum" or bundled licensing.[27] With individual product life cycles short compared with the long patent lives, any new innovation is likely to infringe several existing patents. Licensing thus typically involves clusters of patents.

Not surprisingly, AT&T now uses its IP more strategically. No longer bound by the consent decree, and with R&D funding no longer guaranteed by the telephone subscribers, its IP policy is necessarily linked more closely to individual business opportunities. This is especially true of trade secret licensing, which is often a key component of international joint ventures, involving omnibus IP agreements combining patents, trademarks, and know-how.

Cross-Licensing in the Computer Industry—IBM

A second major influence on licensing practice across the electronics industry has been IBM. It has long been heavily involved in licensing and cross-licensing its technology, both as a means of accessing external technology and to earn revenues. In many ways, it has been in a similar position to AT&T in that it has been a wellspring of new technology but was also subject to a consent decree in 1956 that had certain compulsory licensing terms. Under the IBM consent decree, IBM was required to grant non-exclusive, non-transferable, worldwide licenses for any or all of its patents at reasonable royalties (royalty free for existing tabulating card/machinery patents) to any applicant—provided the applicant also offered to cross-license its patents to IBM on similar terms. The provision covered all existing patents at the time of the decree (i.e., as of 1956) plus any that were filed during the next five years. The rights lasted for the full term of the patents.[28]

IBM's cross-licensing activity continues today. IBM states that it is "exploiting our technology in the industry through agreements with companies like Hitachi, Toshiba, Canon, and Cyrix." Patent and technology licensing agreements earned $640 million in cash for IBM in 1994.[29] IBM is one of the world's leading innovators, with more U.S. patents granted in each of the three years from 1993 to 1995 than any other company (see Table 1).

The central importance IBM attaches to its patent portfolio in providing an arsenal of patents for use in cross-licensing and negotiating access to outside technology has been borne out in public statements by the company.[30] For IBM, the main object of its licensing policy has been "design freedom," to ensure "the right to manufacture and market products." To be able to manufacture products, IBM needs rights to technology owned by others:

> Market driven quality demands that we shorten our cycle times. This means we have to speed up the process of innovation. And that means there is less time to invent everything we need. We can't do everything ourselves. IBM needs to have access to the inventions of others.[31]

TABLE 1. Top Ten U.S. Patent Recipients (1990-1995)

Company	US Patents Received					
	1990	1991	1992	1993	1994	1995
IBM	608	684	851	1,088	1,305	1,383
Canon	868	831	1,115	1,039	1,100	1,088
Motorola	396	614	662	731	839	1,012
NEC	448	441	462	602	901	1,005
Mitsubishi	862	964	977	944	998	971
Toshiba	891	1,031	1,036	1,064	985	970
Hitachi	902	962	973	949	1,002	909
Matsushita	351	467	616	722	782	852
Eastman Kodak	720	863	778	1008	890	772
General Electric	785	818	943	942	973	757

Source: IFI/Plenum Data Corp., USPTO

It acquires these rights "primarily by trading access to its own patents, a process called 'cross-licensing'."[32] IBM has often had the reputation of being a "fast follower" in some areas of technology, and it has used the power of its patent portfolio to negotiate the access needed. The company notes that:

> You get value from patents in two ways: through fees, and through licensing negotiations that give IBM access to other patents. Access is far more valuable to IBM than the fees it receives from its 9,000 active [U.S.] patents. There is no direct calculation of this value, but it is many times larger than the fee income, perhaps an order of magnitude larger.[33]

The effect of the consent decree for IBM, as for AT&T, was in large part to formalize policies that were already partly in effect. While IBM already used cross-licensing for design freedom where appropriate, the consent decree expanded the scope and in a sense prodded IBM into treating licensing and cross-licensing as a central aspect of its business.

Impact of Consent Decrees on Industry Development

The combined cross-licensing of basic technology by the technologically leading firms—AT&T, IBM, and others—had a profound influence on the development of the post-war electronics industry. The effect of the 1956 AT&T and IBM consent decrees was to make a huge range of basic semiconductor and telecommunications technology widely available for next to nothing to domestic and foreign firms. Even so, for AT&T and its existing cross-licensing partners, the AT&T 1956 consent decree merely formalized what was already established corporate policy. This was exchanged for rights to related technology where this was available; otherwise it was offered at low royalty payments. The availability of the basic technology formed the basis for the rapid growth of the semiconductor industry. Given the common technological base, firms relied on the rapid development and introduction of new products to succeed.

Yet the very prevalence of AT&T, IBM, and others in licensing at low royalties also created a mind set in the industry that became accustomed to artificially low royalties. This contributed to some initial agitation, if not outrage, in some quarters when in the 1980s some intellectual property owners such as Texas Instruments began to seek market returns on their IP.[34]

Licensing Practice at a Semiconductor Company—Texas Instruments[35]

Licensing Objectives

In the semiconductor industry, IP licensing is an integral and essential element of competition, and a corollary of innovation. As noted above, the industry was launched with the invention of the transistor by Bell Laboratories in 1947. First commercial transistor production took place in 1952. By 1995, worldwide sales of the industry were over $150 billion. Like other parts of the electronics industry, the semiconductor industry is characterized by wide use of cross-licensing. The main purpose of cross-licensing is to ensure "freedom-to-operate" or "design freedom" in an industry where there are likely to be large numbers of overlapping patents. Given rapid technological development and many industry participants, the probability is high that any new product or process will overlap technology developed by other firms pursuing parallel paths. Also, the technology often overlaps that developed in related industries, such as computers and telecommunications.

The licensing procedures and royalty rate determination process at Texas Instruments (TI) illustrates the ways in which cross-licensing agreements are used in practice. TI has two main licensing objectives. The first and primary objective is to ensure freedom to operate in broad areas of technology supporting given product markets, without running the risk of patent infringement litigation by other firms with similar technology. Agreements cover groups of patents within designated "fields-of-use," including existing and new patents developed within the fixed term of the agreement. The second objective is to obtain value from the firm's IP, in the form of its patent portfolio, by generating royalty income. The purpose and result of royalty payments received under cross-licensing agreements is "competitive re-balancing," which equalizes the net cost and profit advantage for imitators who otherwise might free-ride on technology TI developed.

Buying "freedom-to-operate" is vital in the semiconductor industry, with its rapid innovation, short product life cycles, and ubiquity of patents. In a typical technological field, there may be as many as a half dozen other firms with patents that an innovator could potentially infringe while implementing its independent research strategy. In semiconductor devices and manufacture, there are huge numbers of patents to consider, with many more generated each year, as seen in Table 2. Bear in mind that a particular product can utilize technology from several other technology fields, such as computers, software, materials, communications, and general systems, each with large patent establishments.

At the start of an R&D program, possible infringements cannot be easily predicted, as firms are quite ignorant of the R&D and product development plans of competitors. Yet a firm investing in R&D and product development needs to be confident that patents developed through independent R&D efforts by others will not hinder commercialization of its technology. Consider that a wafer production facility now costs $1 billion.[36] The facility may have a five-year life or longer, and it is not known in advance what products will be developed for manufacture during that time. R&D is similarly becoming more expensive. Companies need to be able to develop new products to fill the wafer fabrication facilities without being concerned that startup may be blocked by patents owned by competitors and other companies inside and outside the industry.

TABLE 2. U.S. Patents Granted in Semiconductor Devices and Manufacture (1969-1994)

Company*	Patents granted (1969-94)	Patents granted (1994)
IBM	3,435	220
Toshiba	2,492	245
Texas Instruments	2,366	231
AT&T	2,342	110
Hitachi	2,218	170
Motorola	1,882	210
Mitsubishi	1,691	275
RCA	1,601	0
Siemens	1,518	46
U.S. Philips	1,482	61
General Electric	1,446	48
NEC	1,360	261
Fujitsu	1,335	125

* Companies with over 1,000 semiconductor patents granted (1969-94).
Source: USPTO, 1995

One approach for a developer to deal with the IP rights of others would be simply to identify all infringements as they arise, and negotiate separate licenses for each. However, the transactions costs of such an approach would be inordinate.[37] Moreover, it would expose the potential licensee to large risks.

A typical cross-license includes all patents that licensees may own in a given field-of-use, giving each firm the freedom to infringe the other's existing and future patents for a given period, typically five years. Such licenses are typically non-exclusive and rarely include any trade-secret or know-how transfer or sublicensing rights.[38]

In a cross-license, technology is not usually transferred, as the parties each are capable of using the technology in question without assistance. Firms will usually gain access to the relevant technology either by developing it themselves, or by other means such as reverse engineering, hiring consultants, other technical agreements, or technical publications.[39] In either case, the cross-license primarily confers the right to use the patented technology without being sued for infringement. This avoids monitoring costs and adjusts royalty payments to reflect overall contributions to the stock of IP currently in use.[40]

In the semiconductor industry, licensing agreements sometimes go further, and may include transfer of trade secrets and know-how. However, trade secret licenses are quite different, typically involve technology transfer, and often accompany a joint venture or strategic alliance. Technology transfer involves significant costs and managerial effort, and often "creates competitors", as it frequently transfers to the licensee important technological capabilities otherwise inaccessible.[41]

Types of Cross-Licenses

There are two main models for cross-licensing agreements in the semiconductor industry: "capture" and "fixed period." In the "capture" model the licensee has rights to use, in a given field-of-use, all patents within a technological field which exist or are applied for during the license period, usually five years, and, importantly, retains "survivorship" rights to use the patents until they expire, up to 20 years later. The agreement does not generally list individual patents, but some patents of particular strategic importance to the licensor may be excluded. In the "fixed period" model the licensee has similar rights to use patents existing or applied for during the license period, but with no survivorship rights once the license period has expired. This requires full renegotiation of the cross-license for succeeding periods.

TI has been a leader in the use of fixed period licensing, which is becoming more widely used. The capture model became widespread through the industry following its use by AT&T and IBM. It gives broad rights to patents for a long period. The fixed period model allows more flexible commercialization of patent portfolios, since licensing terms can be periodically adjusted to account for changes in competitive conditions and the value of the technology. This increases strategic flexibility and allows the parties more freedom to negotiate royalty terms so that they more closely mirror the value of the patents. It is a logical evolution of licensing practices reflecting the difficulties and changes in the market for know-how.

"Proud List" Royalty Valuation Process

Balancing payments are negotiated as part of the agreement, to account for the relative value of the IP contributed by two firms. Each firm's contribution is evaluated by estimating the value of a firm's patent portfolio to its licensing partner, with the net royalty payment to the one with the greater contribution. Where both firms contribute similar portfolio values, the net payment will be small or zero. Where one firm has developed little technology and the other a great deal, the payments may be significant. Occasionally, cross-licenses are royalty-free because contributions are either very close or difficult to assess. However, even in royalty-free agreements it should not be assumed that a detailed patent balancing process has not taken place. Also, the cross-license may be included as part of a larger joint venture.

Royalty balancing is performed according to a "proud list" procedure. In this procedure, each firm identifies a sample list of its most valuable patents and this is used as a representative proxy group for estimating the value of the entire portfolio. There is a great deal of preparation before the negotiations. Having identified a potential cross-licensing candidate, TI first performs extensive reverse engineering of the other's products to assess the extent of any infringement—called "reading" the patents on the infringer's products—and identifies product market sizes involved. This may take a year of effort.[42] As part of this effort, it generates the proud list of about 50 of its major patents which it believes are being infringed, and which apply over a large product base of the other firm. The other firm also prepares a proud list of its own strongest patents.

In the negotiations, each of the sample patents is evaluated by both sides according to its quality and coverage. Quality measures include: the legal validity and enforceability of the patent; the technological significance of this feature to the product compared with other (non-infringing) ways of achieving the same end; and the similarity between the infringing features and the patent. These determine quality weighting factors for each patent so that a legally strong patent, which is hard to invent-around and is close to the infringing feature, has a high relative weight. The coverage is the size of the infringer's product market using the patent. Each patent is assigned a nominal royalty rate, which is then multiplied by its quality weighting factor and the annual sales of the affected product base to arrive at a dollar amount. Certain patents of particular strategic significance to the technology are assigned a flat rate as a group and do not go through the weighting process.

The dollar amounts are summed for all the listed patents and expressed as a royalty rate percentage of the licensee's total sales. Typically, the values of each side's estimated royalty payments are netted out to give a single royalty rate paid by the firm with the less valuable portfolio.[43] This royalty rate applies to the licensee's sales for the term of the license. When the license expires the same procedure will be used to reevaluate the relative portfolio values for the next five years.[44]

Strategic Considerations

TI's procedures provide a formal mechanism for determining royalty rates based on best estimates of the economic and technological contribution of the patent portfolios of the two firms. These procedures have been applied to a wide variety of relative IP contributions, both where these are roughly in balance and where not. Even so, there are often other considerations to include in final negotiations of a licensing agreement. Much depends on the individual needs of the parties, their negotiating strength, and the broader strategic considerations of each firm. Individual rates and the overall rates also tend to recognize overall competitive effects of the royalty payments, as well as "what the market will bear."[45]

There is obviously an upper limit on royalties, since royalties that are too high will cripple the competitive capacities of the licensee, causing royalty payments to decline. If a potential problem in this respect exists, it is usually not with an individual agreement, which is likely to be set at reasonable royalty rates. Rather, problems may arise when a licensee is subject to claims from several licensors and the cumulative royalty payments become onerous. This can create serious problems in negotiating agreements with would-be licensees. There does not seem to be an easy solution to this problem, given that agreements are negotiated individually.[46]

Royalty rates may also be affected by longer-term strategic considerations. For one thing, both parties are likely to need to renew the agreement in future, and an aggressive royalty rate now may make negotiations more difficult later, when the balance of IP may have shifted in a different direction. The firms may have, or expect to have, overlapping interests in other market areas, which will also condition negotiations. Licenses often may also be part of a cooperative venture of some kind. Patents can often be traded for know-how, or used as an entry ticket to a joint development arrangement. For example, rather than seek royalties, TI has had technology development agreements with Hitachi. It also has several manufacturing joint ventures around the world.

Strategic considerations may also affect the usual licensing process where the technology is intended to become part of an industry standard. Industry standards bodies sometimes require that patent holders agree to license their patents with low or zero royalty fees, often on a non-discriminatory basis. Similarly, when trying to establish a *de facto* market standard, a firm may charge low royalty rates.[47] The aim is to ensure the wide adoption of the technology as an industry-wide standard. Value from the technology may then be earned through product sales in an expanded market. The "reasonable rate" royalty involved is likely to be low, though need not be zero.[48]

Impact of TI's Licensing Strategy

TI has led industry moves to take a more active stance on licensing and cross-licensing. The impact of its licensing strategy on its capability to compete and innovate is of particular interest. TI instituted its current licensing strategy in 1985. Cumulative royalty earnings of over $1.8 billion had been achieved during the period from 1986 to 1993. Among other effects, this enabled TI to maintain a high level of R&D spending during 1989-91, when the semiconductor market was in a downturn, as shown in Figure 1. However, moving to a more active licensing strategy and the aggressive assertion of its IP rights was a major step for the company—and the industry—and involved considerable risk.[49] TI's strategy was enhanced by the stronger U.S. treatment of IP after 1982.

FIGURE 1. Texas Instruments: Royalty Earnings, Net Income and R&D

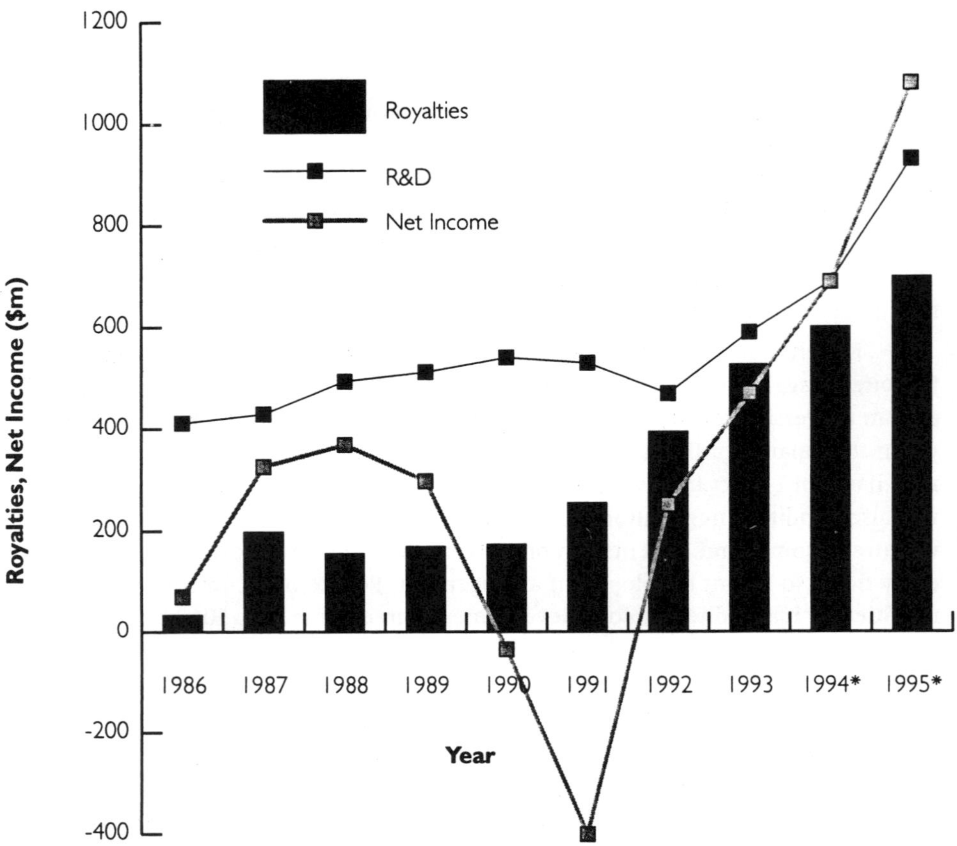

*Royalties for 1994 and 1995 estimated
Source: Annual Reports

TI's IP portfolio has been valuable in negotiating R&D cooperation. For example, TI has had a series of ventures with Hitachi for the joint technological development of DRAM memory chips. TI's ability to supply technology, supported by its IP rights, was a crucial component in making these agreements.[50] TI's changed IP strategy has allowed it to implement new product market strategies to expand its manufacturing capacity by means of joint ventures, based partly on the negotiating value of its IP portfolio, and expanding its development of high value added components. It has been a partner in a number of international manufacturing joint ventures to set up production facilities for memory

TABLE 3. Top 10 Merchant Semiconductor Firms: 1980-1995

	1980			1990			1995		
	Company	**$m**	**%**	**Company**	**$m**	**%**	**Company**	**$m**	**%**
1	Texas Inst.	1,580	12.2	NEC	4,952	8.6	Intel	13,830	8.9
2	Motorola	1,110	8.5	Toshiba	4,905	8.5	NEC	11,360	7.3
3	Philips	935	7.2	Hitachi	3,927	6.8	Toshiba	10,190	6.6
4	NEC	787	6.1	Intel	3,135	5.5	Hitachi	9,420	6.1
5	National	747	5.7	Fujitsu	3,019	5.3	Motorola	9,170	5.9
6	Toshiba	629	4.8	Motorola	3,692	6.4	Samsung	8,340	5.4
7	Hitachi	622	4.8	Texas Inst.	2,574	4.5	Texas Inst.	8,000	5.2
8	Intel	575	4.4	Mitsubishi	2,476	4.3	Fujitsu	5,510	3.6
9	Fairchild	566	4.4	Matsushita	1,945	3.4	Mitsubishi	5,150	3.3
10	Siemens	413	3.2	Philips	1,932	3.4	Philips	4,040	2.6
	Others	5036	38.7	Others	24,943	43.4	Others	69,990	45.2
	Total	**13,000**	**100.0**	**Total**	**57,500**	**100.0**	**Total**	**155,000**	**100.0**

Source: Dataquest

chip production.[51] TI and Hitachi also entered a joint venture in 1996 to manufacture DRAMs in Texas.

These changes have had a major impact on TI's performance, helping the company to grow and to increase its world market share since the mid 1980s. This helped reverse a relative decline in its position beginning in the mid-1970s due to inroads made in world markets by foreign producers, as seen in Table 3.

IP Management and Cross-Licensing in an Electronics Company—Hewlett-Packard[52]

Innovation Strategy

Many aspects of licensing elsewhere in electronics are similar to those described for semiconductors. The electronics industry shares many of the basic features of the semiconductor industry: rapid technological innovation, short product life cycles, and significant patenting. The computer, telecommunications, electronics, and semiconductor industries also use many of the same technologies and have been influenced by the practices of AT&T and other major corporations. Field-of-use cross-licensing is used widely.

However, a difference between many electronics firms outside of semiconductors is the breadth of technologies that are practiced. In addition to semiconductor technology, product development may involve integrating many aspects of computing, telecommunications, software, systems design, mechanical engineering, ergonomics and so forth. There are also likely to be complex manufacturing and marketing requirements. Thus, IP strategies in such firms are likely to involve broader considerations.

Hewlett-Packard (HP) produces many different types of products, from laser printers and computers to hand-held calculators and electronic instruments. HP is currently organized into Computer Products, Systems, Measurement Systems, and Test and Measurement organizations.

To maintain its high rate of innovation, a high priority for HP in its IP strategy is maintaining "design freedom." It has two principal objectives: ensuring that its own technology is not blocked by competitors' patents; and ensuring that it has access to outside technology. HP's products include complex systems that typically involve several different technologies, some of which may be developed by other firms and other industries. HP alone can not develop the complete range of technologies used in its products. To obtain access to needed technologies, Hewlett-Packard needs patents to trade in cross-licensing agreements. The company has a huge portfolio of patents and know-how in leading-edge technologies, developed as part of its extensive R&D programs. This IP portfolio is the basis for protecting HP's own products; it is also invaluable as leverage to ensure access to outside technology.

Licensing Objectives

One type of HP cross-licensing takes place as "program licensing," which is aimed at acquiring access to specific technologies. The company identifies firms with technologies of interest. There may be several different technologies at a given firm so the strategic overlaps must be considered in assessing each licensing opportunity.

HP's licensing activities are not focused primarily on cash income. With a wide range of products, the company's interests in one area are likely to overlap with those in other areas. It may encounter licensing partners in several different markets in a variety of circumstances—a competitor in one field may be a supplier or customer in another. HP does not want negotiations in one product group to interfere with those in another. This leads to a long-term bias towards meaningful cross-licensing agreements and a soft approach to royalties. HP recognizes that it is likely to deal with the same partners repeatedly and therefore normally does not require high royalty rates that could be used as a precedent against it in the future.

There are some exceptions in that some strategic patents are only licensed at high royalty rates, or more likely are not licensed at all. In products where HP has a strong leadership position (e.g. printers), it is unlikely to license out its core IP rights. HP's IP policy in this area is aimed, as it must be, at the aggressive protection of a key source of competitive advantage. The company would normally consider licensing such IP rights only as part of a specific strategic alliance and would normally exclude such technology from cross-licensing agreements.

The form of the cross-license agreements is quite standard, with a limited capture period, usually with survivorship rights. The objective is to estimate the relative value of the infringements that are likely to take place over a five-year period. Other inputs to the licensing decision include the expected R&D spending in the field by each firm, the number of patents held by each party in the particular field, and determination of the value to the infringer of a limited number of pertinent patents. Each side to the agreement may select a limited number of patents which it has determined are being infringed by the other party's products. This may be as few as six to twelve patents each. The imputed royalty fee for these patents over the next five years becomes one of the inputs to the negotiation. In general, this balancing process is not unlike that which exists in the semiconductor industry.

Royalties are often paid as a lump sum. Agreements almost never include sublicensing rights, since the company could lose control of its own technology if sublicensing were permitted. Exclusive licensing is also rare, partly because of potential antitrust concerns, but also because the historical practice of non-exclusive cross-licensing leaves fewer innovations that could be treated as exclusive.

Even after a patent cross-license agreement is concluded, HP policy is not to over-use the technology of the other party to the agreement. This is again related to a long-term view of licensing. The agreement will probably need to be renewed in the future and the more of the other party's technology HP uses, the greater the leverage the other party would have the next time around. Also, patents are lagging indicators of research, so that to be at the forefront of technology each party will need to have developed its own application of the technology well before the patents are issued. One purpose of the agreement is to be able to use the technology in the development of new products without worrying about "accidental infringement."

Licensing is only secondarily seen as a source of royalty earnings. Royalty earnings are significant but not material, given the overall size of HP's operations. However, there are some cases where licensing for revenue is pursued. One is where the company has world-class technology and is approached by others seeking a license. If the technology is not of strategic importance to HP, the company may license it out for profit. Another is the "rifle shot" license, where a single patent may be licensed, if it has specific value to a licensee. Licensing terms in either case are usually very simple, amounting to an agreement to allow use of the innovation for a royalty payment or lump sum without being subject to an infringement claim.

IP Management

Given the importance of IP to Hewlett-Packard, a formal IP strategy has been developed for managing its large and diverse IP portfolio. Since products combine many technologies, IP may need to be even more closely integrated with business strategy than at a single product corporation. HP has a series of

FIGURE 2. Intellectual Property and Patenting Decision Process at HP

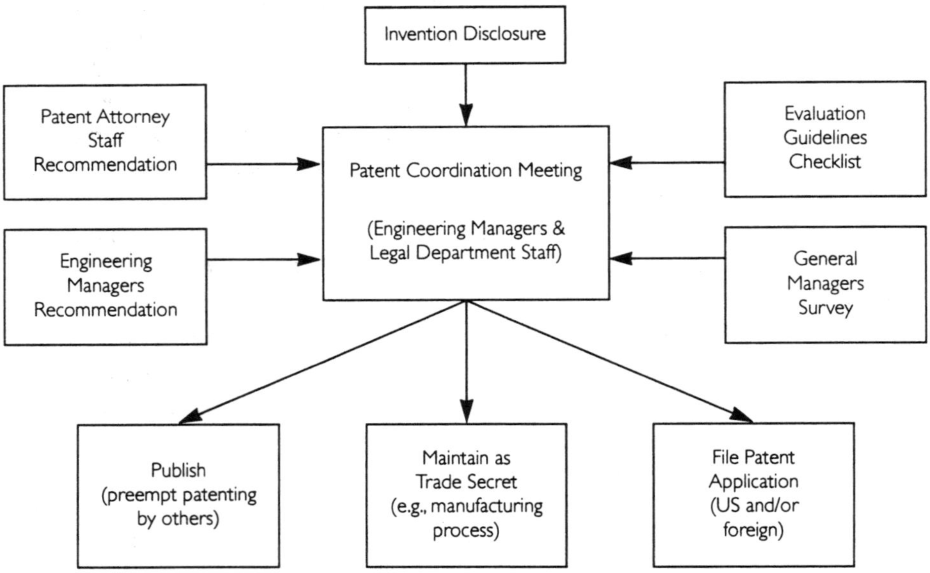

procedures for identifying technological areas to stress for patent protection and for making individual decisions about the best method of protecting innovations. Obtaining and maintaining patent protection is costly, and hence only selected innovations are patented. This process starts with "templates" to guide what IP should be protected. The templates are updated each year to protect technologies that will be strategically important to the company in the future. These templates are developed by a process that rates and prioritizes products and technologies and reviews patent needs throughout the world. This does not go as far as targeting R&D programs at innovations that will be useful in negotiating cross-licenses; rather it aims to make maximum use of innovations by creating patent portfolios that will be strategically valuable. This supports rather than directs corporate strategy.

The IP protection decision process for individual innovations is shown in Figure 2. When a product or process innovation is developed, a determination is made whether to patent it, to keep it as a trade secret, or, if it not believed worthwhile to patent, to publish it. The inputs to this decision take place in an internal committee process, with inputs from engineering management and the legal (IP) department. Innovations that are likely to be of strategic value are either patented immediately or, if they are not yet completed or proven, are reviewed again at a later time. If the innovation is valuable but its use by an imitator would be undetectable (such as for some process innovations), then the innovation may be kept as a trade secret. Marginal ideas are published immediately to preempt patenting by a competitor who might later block their use by HP. "Vanity publishers" for publicly disclosing the results of research exist for this purpose.[53]

Managing Intellectual Capital in the Electronics Industry

Contrasting IP Management Objectives

The case studies indicate several similarities in the way firms in the electronics industry use licensing and cross-licensing to ensure design freedom as well as some level of licensing earnings. They also illustrate how differences in management objectives are reflected in cross-licensing strategies.

RCA represents a rather complex organizational response to the problem of design freedom, in which a single company acquired exclusive cross-licensed rights to all the patents needed for radio manufacture. It then licensed out these rights to other manufacturers. Partly as a result, RCA was able to dominate the radio market for many years.

AT&T, as a regulated monopoly before 1984, was primarily interested in the dissemination of technology to as many producers as possible, to develop technologies that would be useful in its telecommunications services—as purchased components or in its own systems development. It was barred from competing in product markets, so it cross-licensed on liberal terms with the aim of stimulating development and obtaining access to new technology.

A primary concern of IBM in cross-licensing has been design freedom. As one of the world's leading innovators it has been very active in using its IP for competitive advantage, both in products and to obtain the widest possible access to other technology. IBM's interests have spanned a wide range of computer related markets and it has needed broad access to many different technologies. It also obtains significant income from its licenses.

TI's interests have generally been more specific to the semiconductor industry, although it also has interests in other areas of electronics. Its concerns have been to obtain freedom-to-operate given the dense patent concentration in semiconductors, and to obtain cash from cross-licensing its IP, to help fund R&D and to equalize any advantage it would otherwise be allowing competitors using its IP.

Finally, HP is in a somewhat similar position to IBM in having a broad range of interests in different markets and being especially interested in design freedom for products spanning many technologies. HP's breadth of interests—in which a competitor in one field may be a customer, supplier, or venture partner in another—moderates its approach to seeking high royalties. IP is central to its business, needed to support its rapid product innovation and to trade for technology access. It has well developed procedures for developing and protecting IP across its diverse fields.

Changing IP Modes in the Semiconductor Industry

The strengthening of IP rights and increased licensing and cross-licensing have extended the ability of the innovator to earn a reward from R&D. In addition to providing better IP protection for new products, there are greater opportunities for earning value via access to technology, joint ventures, technology exchanges, and R&D collaboration. Royalty earnings have become more significant. Much of this is a recent development and there are many questions as to how much strategic emphasis firms should place on licensing and cross-licensing compared with manufacturing, and on the importance of licensing revenue earning compared with freedom-to-operate.

It may help put these questions in context by reviewing the changing modes of competition in semiconductors, where firms have gradually needed to place increased stress on innovation, IP protection, and licensing and cross-licensing as a basis for product competition. There have been major changes in the way firms have obtained value from innovation as the industry has developed. The weak IP regime in effect during the first two or three decades of the industry was not a barrier to R&D and investment, and the liberal licensing practices used by AT&T and others accelerated the initial diffusion of the technology. This nurtured the early growth of a new industry. However, firms could not operate successfully in today's technological and competitive environment with the strategies and policies in place in the 1950s and 1960s. Competition to stay at the forefront of innovation is sharper and R&D and investment take place on a much bigger scale. AT&T no longer has a franchise monopoly, the market power of other industry participants is at best a phantom, and the industry is global.

Initial Growth Phase

From 1950 until the late 1970s, semiconductor and electronics firms used technology to open up new markets. Semiconductor technology was new and developing rapidly, and was too big and too important to be developed and commercialized adequately within one organization. There were benefits from having multiple sources of innovation. This was epitomized by AT&T's policy. As a major consumer of semiconductors, it wished to spread the use of the technology as widely as possible. Elements of this reasoning applied to other firms, who benefited from the rapid expansion of technology and markets. And, given the weak protection of IP afforded by the courts at this time, patents were not seen as a major factor in building competitive advantage.[54]

At that time, firms relied primarily on time-to-market advantages to keep ahead. The basic semiconductor patents were already widely licensed, so any individual patent had limited power.[55] Product life cycles were short and often firms would simply not bother to patent inventions, believing that there was no point in patenting products and processes that would soon be obsolete. The fragmented structure of the new "merchant" semiconductor industry (which had grown up around spin-offs from Bell Labs and others), the rapidity of innovation, and the high level of competition reflected the fact that not much attention was paid to protecting IP.[56] The predominant strategy for capturing value from technology involved "riding the experience curve"—reducing prices rapidly as unit costs fell with the hope of earning enough to fund the next round of development.[57]

Second sourcing, licensed or not, was often required by many of the large customers to ensure continuous and competitive supplies. There was significant cross-licensing (often associated with second sourcing), but it rarely involved significant royalty payments.[58] Customers like the Department of Defense (DOD) had sufficient clout to force small suppliers like Intel to second source. During this period, licenses were mainly used to get some residual value from an innovation when it could not be recovered via the product market because of investment restrictions or trade restrictions. An example is the difficulty U.S. firms had selling products in Japan. Faced with effective trade protection, most U.S. firms' only recourse was to license technology to Japanese firms.[59]

At this time, TI was one of the first firms to make strategic use of its IP. It established a production plant in Japan in 1968, one of the very few foreign firms to do so. It achieved these rights from MITI by using the power of its patent portfolio.[60] This heralded a new role for IP in global commerce and firm competitiveness.

Increased Global Competition

The competitive environment began to change during the 1970s. The complexity of the technology and the scale of investment in R&D and capacity were rising, increasing the business risk of each new development. Moreover, as requirements for specialized investment increased, the business risk associated with a patent holder's ability to obtain an injunction (in the case of inadvertent or intentional infringement) increased.

Managers were at first distracted by the increasing size of the total market when new mass markets opened up in the 1970s for consumer electronics (including calculators, watches, and later personal computers) and computer memories. By the early 1980s, new competitors from Japan (and later Korea) had entered the world markets and were challenging the U.S. firms, using technology largely developed in the United States. Changes were most dramatic in the manufacture of "commodity" DRAM memory chips, in which U.S. manufacturers' share of the world market fell from 75% in 1980 to 17% in 1986, while over the same period Japanese memory share rose from 25% to 79%.[61] U.S. firms could no longer rely on success in the product market alone to obtain returns from innovation.

The new entrants to the industry depended on access to existing technology and often sought to cross-license it. Yet nominal or royalty-free cross-licenses, which had been common in the industry prior to the 1980s, came to be seen as unfair when the entrants from outside the industry offered to pay the nominal cross-license fees, but with no balancing portfolios of patents to offer. Royalty fees also reflect payment for access to technology accumulated in prior years, often at great expense. TI and others realized that more detailed evaluation of relative contributions to cross-licenses were required.

Innovation Leadership

The situation today is that, with numerous qualified competitors, competitive advantage requires more emphasis on strong IP rights. Stronger IP protection calls for dual strategies for capturing value from technology—the simultaneous use of product manufacturing using the IP in question together with IP licensing. Market developments have put more emphasis on chip design, developed close to the customer, and on being able to protect this and leading-edge process technology from imitation by fully able competitors. The increase in cooperative R&D and manufacturing joint ventures, often underpinned by IP rights, represent a market response to increased costs and the risks of development.

A regime shift occurred when many of the once small semiconductor firms such as Intel could no longer be forced into second sourcing their products. The demise of contractually required second sourcing suddenly made the value of IP more significant. The successful blunting of buyers' demand for second sourcing made IP more important—so much so that many companies, such as Intel, now have designated IP managers.

Many in the semiconductor industry have been opposed to stronger assertion of IP rights, having grown accustomed to a relatively open exchange of ideas and personnel. Not surprisingly, advocates of this view include start-ups, who claim that if they pay the full price of technology, it would limit their ability to compete. This may be true, but it is also trite. We observe that supporters of open ideas often become more protective once they have invested heavily in R&D. Most significantly, there has been a change in the global competitive reality. What may have been a useful model in the early days of the industry (in which it may be argued that all firms in a local market benefit from mutual exchange of ideas), becomes a different equation when firms are global.[62]

Lessons for Innovation Management

To an extent, management today has little choice but to adopt a more active IP and licensing stance. IP rights have been strengthened and, not surprisingly, firms have become more strategic about commercializing IP. Cross-licensing enables firms to protect their IP while at the same time obtaining freedom to manufacture. The new IP and licensing circumstances have increased incentives to build IP portfolios and to innovate. In these new circumstances, there are some key lessons for innovation management.

Using IP to Support Core Business

Despite, or because of, the growing importance of licensing and cross-licensing, IP strategy should still be designed primarily to support technological developments and strategies surrounding the firm's core business. The global marketplace still rewards firms primarily for developing and commercializing products and processes as such, not for developing IP. Accordingly, few firms target technologies primarily for their value in earning royalties or for trading IP rights in future cross-licensing agreements.

Furthermore, for long-term success, firms typically need to be closely involved with the markets in which they operate and to develop core capabilities (in manufacture and design) closely linked to the products and processes. Maintaining a stream of valuable innovations requires extensive, up-to-date information about market demand and technological possibilities, especially in industries where technology is changing rapidly. Although this depends on the nature of the product, it usually also calls for close functional links between design, production, and marketing. These needs are typically best served by active participation in the product market.[63]

The alternative—becoming a pure "licensing company" not directly involved in the product market and increasingly remote from the manufacture and design of the product itself—can be a risky strategy. Such a strategy, on its own, not only risks the erosion of the dynamic capabilities of the firm to continue innovating, it also is likely to be less financially rewarding than developing and commercializing products.[64]

Importance of Developing a Valuable Patent Portfolio

Developing a valuable patent portfolio is an increasingly important part of strategy. In the electronics industry, patents are valuable because they provide protection from imitation for new proprietary products and services; they provide bargaining chips in negotiating access to other firms' technology (to avoid patent blocking and ensure freedom-to-operate); and patents may be an additional source of earnings or of reduced royalty fees the firm might otherwise have to pay.

The value of a portfolio is greatest when it has a high proportion of high-quality patents that cover significant product markets. These patents affect each of the reasons for holding a portfolio, but are seen most directly in the effect on cross-licensing. Patents have greatest cross-licensing value when they give the firm maximum leverage to obtain a favorable cross-license. This means that the patents should be legally and technically strong and should cover key aspects of the licensee's product base.

Concentrate R&D Where the Firm Is Strongest

In developing its patent portfolio, the firm can concentrate its R&D in those areas where it has the greatest competitive advantage in developing valuable innovations, provided these are also areas needed by other firms. It need not focus on those technological areas where its cross-licensee is strongest in an attempt to duplicate or avoid the licensee's patents—a hopeless task with complex cumulative technology, such as electronics, where infringement is almost inevitable.[65] This might be in the same fields that it wishes to cross-license from its partners, or it might be in a more specialized area. For cross-licensing with a multidivisional corporation with interests in several markets, it might be in a different business area or field-of-use than the one from which it wishes to access technology. As argued above, a firm is most likely to create valuable IP where it is actively involved in the market, i.e., its core business. Provided this is also a commercially important field to cross-licensing partners, the firm can concentrate on developing and protecting IP in this field, rather than seeking another.

Licensing and cross-licensing enable firms to capture value from technology so long as they contribute to the common pool of industry knowledge. Innovators who are contributors have every incentive to avoid duplicative R&D investments, since a contribution to an industry's useful stock of proprietary knowledge is recognized no matter what the precise domain of applicability. Firms are advised to focus on innovating where they can best make a contribution to the development of quality patents they and other firms are likely to need. Cross-licensing thus enables firms to play to their technological strengths.

Although the number of patents a firm holds is important, of even greater importance is their quality. A single key patent is often worth more than a portfolio of questionable ones when it comes to assessing the ability of a patent owner to stop an infringer. The most effective way to acquire a portfolio of valuable patents is likely to be through in-house R&D. Occasionally, firms can purchase a portfolio of patents with which to establish cross-licensing relationships; but quality patents often are not available in this fashion.

In summary, the reality of the global marketplace today indicates that firms should proactively develop IP portfolios with an eye towards value in the market for know-how. A corollary is that to create a valuable patent portfolio for cross-licensing, it matters little where R&D is aimed, so long as it creates quality patents in a field that one's competitors need to license.

Policy Issues

Intellectual property is more critical than ever to competitive advantage and, as a result, is being given increasing attention by strategists and policy makers. IP protection has been strengthened and firms are more actively defending and exploiting their IP. Coincident with the increased importance of patents is the increased importance of licensing and cross-licensing. Cross-licensing has become a significant dimension of competition. Absent the ability to offer an equivalent IP portfolio, licensees must incur considerable costs. This in and of itself is a spur to innovation.

Cross-licensing outcomes do not, however, tilt towards the large firm at the expense of the small. Rather, they favor firms with significant IP regardless of size. In a particular market niche where patents from two firms overlap, a small firm may have as many patents as a large firm, and as much bargaining power as the large firm. It may have sufficient IP leverage to block a larger competitor by pursuing a claim in court (or credibly threatening to do so). Indeed, in the evaluation process, a small innovator with a strong patent may be the net gainer, if the patent applies to a high-volume product of a large corporation.[66] Some competitors may possess "equal patents but unequal products." Nor need the licensing process disadvantage a new entrant firm. If a new entrant has significant relevant technology, it can in principle be a beneficiary of the cross-licensing regime.

Those investing in R&D need to ensure that they earn an adequate return, and royalties from licensing are an increasingly significant part. A company that develops technology will be at a competitive disadvantage in the market if its competitors are free to use its technology without incurring any expenses. Licensing fees on patented technology help ensure that the innovator earns an adequate return, which helps support future R&D. Cross-licensing helps balance the costs for developers and imitators. Thus, products manufactured by imitators who have not performed R&D do not have a competitive advantage merely by virtue of engaging in "copycat" imitation. If both parties to a licensing agreement have contributed similarly to a product field-of-use—in terms of the number, quality, product base coverage, and commercial significance of the patents included in the agreement—then the net royalty payments will be small, or possibly zero. In short, royalty payments help level the playing field, thereby ensuring competition on the merits.

The result is that IP now often has great value, both as a lever to obtain design freedom and as a vehicle to assist innovators in capturing value from innovation. This is of considerable consequence to firms without much IP—they must expect to pay—and also for firms with significant IP portfolios. IP and other knowledge assets are the core assets of many high-technology companies.

However, and perhaps because IP rights have become more valuable, infringers do not always step forward and offer to pay royalties. Accordingly, patent owners must often be proactive in obtaining royalty payments. Litigation or the threat of it may sometimes be necessary to enforce one's rights. Unfortunately, at least in the U.S., litigation is often slow and costly, and antitrust and patent misuse defenses are often raised, sometimes frivolously. The archaic state of the law on patent misuse may further handicap the chances of efficient and socially desirable outcomes.[67] Moreover, antitrust attorneys are often ready to argue that a package license is a tying arrangement with anticompetitive effects, and/or that cross-licensing is a front for collusion. However, the truth of the matter is that such arguments are out of step with the new competitive order.

Such arrangements are pro-innovation and pro-competitive. There would appear to be a significant knowledge gap in some circles with respect to the nature, purposes, and effects of cross-licensing. For instance, the field-of-use cross-licensing of patents in widespread use today is quite different from the traditional practice of licensing and cross-licensing involving individual patents. In the electronics industries, it is simply too cumbersome and transactionally costly to license specific patents for specific products, and so licensing commonly proceeds on a portfolio basis. Yet patent misuse and patent antitrust arguments often assume a world where infringement is easy to detect and costless to enforce. This is rarely the case in the electronics industry today.

At the most elementary level, licensing and cross-licensing involve merely the sale or exchange of property rights. Indeed, it often involves precisely that and no more. However, such arrangements ensure that firms have freedom-to-operate in developing and using innovations, without risking infringement claims from holders of patents in the same field of technology. In industries experiencing rapid technological innovation, patents, even when developed independently, will inevitably overlap technological domains worked by other firms. Cross-licensing agreements provide firms active in R&D with protection against inadvertent infringement and the rights to use the licensee's patents. Cross-licensing arrangements provide a mechanism for recognizing contributions through the establishment of balancing royalty payments. Royalty flows thus recognize the relative contributions to the product technology of the parties, thereby providing a mechanism for net takers to compensate net contributors. The arrangements thereby provide some limited protection against "free riders" who wish to use an industry's stock of proprietary knowledge without contributing. Balancing royalty payments are part of most cross-licenses, even when the main purpose is freedom-to-operate. "Pure" royalty free cross-licenses are rare for some companies and nowadays tend only to apply where the patent portfolios of both firms are large and the overall technological balance is both hard to assess and roughly equal.

Conclusion

Licensing is no longer a marginal activity in semiconductors and electronics. Whereas the management of patents and other forms of IP have always been of great importance in some industries like chemicals and pharmaceuticals, the ascendancy of IP in electronics is relatively recent. This is not just because the industry is new, but because regulatory and judicial distortions which impaired the value of IP have now been substantially rectified. The U.S. Department of Justice (DOJ) and the Courts forced AT&T, and to a lesser extent IBM, to license their technologies way below market value.[68] Not surprisingly, the electronics industry worldwide grew up with a distorted view of the value of intellectual property. This was reinforced by second sourcing requirements imposed by the DOD and other large buyers of integrated circuits that could, and did, insist on licensing for second sourcing purposes at low or zero royalties. Moreover, AT&T itself, being a significant purchaser of telecommunications and electronic equipment, and with protected service markets, had private incentives to diffuse technology rather than use it to build competitive advantage.

This confluence of very special factors has ended. The AT&T consent decree is gone, and AT&T must now be far more proprietary with its technology. The IBM patent provisions ended in 1961. Intel, TI, and other integrated circuit producers are no longer forced to second source. Moreover, the courts are more inclined to enforce IP rights than ever before. In these respects, hopefully the DOJ/FTC 1995 Antitrust Guidelines for the Licensing of IP, which include statements regarding the potential efficiency benefits of licensing and cross-licensing, are an important step in the right direction and reflect more modern thinking about IP.[69] However, these guidelines are non-binding in litigation, though one would of course hope that the courts would take them into account.

The old regime—whereby the antitrust authorities pressed major IP owners to give up whatever rights they held, where the courts were reluctant to enforce IP rights and were eager to see IP as a barrier to competition rather than as an instrument of it—has faded away. Meanwhile, the ability of the buyers of electronic componentry to bargain for and achieve second source arrangements (which indirectly lowered the value of IP by causing owners to create their own competition) has declined. As a result of these developments, a new order has emerged in which IP rights are valuable. Firms must either invest in R&D and develop patentable technology, or pay to license the patent portfolios of others. The free ride appears to be coming to an end, and IP management is now critical to the success of new entrants and incumbents alike.

Notes

1. By "IT," we refer of course to information technology.
2. In cross-licensing, two or more firms license their IP to each other.
3. Cross-licensing is not the same as a patent pool, in which member firms contribute patents to a common pool and each member accesses them on the same conditions. In cross-licensing, firms agree one-on-one to license their IP to each other and retain control over their proprietary technology, which is used for competitive advantage via product manufacturing and further licensing.
4. Other examples of "cumulative systems" include aircraft and automobiles. In aircraft, problems of blocking patents, stemming from different approaches by pioneers such as the Wright Brothers and Curtiss, were only resolved during World War II when automatic cross-licensing was introduced. In automobiles, the Association of Licensed Automobile Manufacturers (although formed to exploit the Selden patent) developed means for automatic cross-licensing of patents early this century. In both cases, the lack of cross-licensing probably held up industry development. R. Merges and R. Nelson, "On the Complex Economics of Patent Scope," *Columbia Law Review*, 90 (1990): 839-916.
5. In chemicals and pharmaceuticals, although patenting is extensive, individual technology development paths are less likely to overlap, and cross-licensing may be used to ensure broad product lines. For licensing strategy in the chemicals industry, see P. Grindley and J. Nickerson, "Licensing and Business Strategy in the

Chemicals Industry," in R. Parr and P. Sullivan, eds., *Technology Licensing Strategies* (New York, NY, NY: Wiley, 1996), pp. 97-120.

6. The early history of radio is described in: G. Archer, *History of Radio to 1926* (New York, NY: American Historical Society, 1938); W. Maclaurin, *Invention and Innovation in the Radio Industry* (New York, NY: Macmillan, 1949); J. Jewkes, D. Sawers, and R. Stillerman, *The Sources of Innovation* (New York, NY: Norton, 1969), pp. 286-288; G. Douglas, *The Early Days of Radio Broadcasting* (Jefferson, NC: McFarland, 1987); Merges and Nelson, op. cit., pp. 891-896.
7. These included the high-frequency alternator, high-frequency transmission arc, magnetic amplifier, selective tuning, crystal detector, heterodyne signal detection, diode valve, triode valve, high vacuum tube, and directional aerials.
8. Not all early inventors were independent. Alexanderson—who improved the Fessenden alternator, invented a magnetic amplifier, electronic amplifier, and multiple tuned antenna, and co-invented the "Alexanderson-Beverage static eliminator"—was a General Electric employee.
9. AT&T acquired the de Forest triode and feedback patents in 1913-1914 for $90,000, and his remaining feedback patents in 1917 for $250,000; Westinghouse cross-licensed the Fessenden heterodyne interests in 1920, and acquired the Armstrong super heterodyne patents in 1920 for $335,000. Archer, op. cit., p. 135; Maclaurin, op. cit., p. 106.
10. The fact that GE and AT&T alone were each devoting major research attention to the vacuum tube led to no less than twenty important patent interferences in this area. Maclaurin, op. cit., p. 97.
11. Federal Trade Commission, *The Radio Industry* (Washington DC: FTC, 1923); Maclaurin, op. cit., p. 99.
12. To cite one important example, Marconi and de Forest both had critical valve patents. Marconi's diode patent was held to dominate de Forest's triode patent. Both technologies were vital to radio, yet the interests refused to cross-license. [Archer, op. cit., pp. 113-114; Douglas, op. cit., p. 12.] The application of the triode (audion) to feedback amplification was also the subject of a long-running patent priority dispute between de Forest and Armstrong (finally resolved in de Forest's favor by the Supreme Court in 1934). Its use in transmission oscillation was the subject of four-way patent interference between Langmuir, Meissner, Armstrong, and de Forest. [Maclaurin, op. cit., p. 77.] These problems held up the use of the triode—a crucial component of signal transmission, detection, and amplification, which has been called "the heart and soul of radio" [Douglas, op. cit., p. 8], and "so outstanding in its consequences it almost ranks with the greatest inventions of all time" [Nobel Prize physicist Rabi, quoted in Maclaurin, op. cit., p. 70].
13. A main concern of the U.S. Navy was that international wireless communications were dominated by the British firm Marconi, and the patent impasse helped perpetuate this. It favored the establishment of an "All American" company in international communications. RCA was formed by GE in 1919, and simultaneously acquired the American Marconi Corp. Major shareholders included GE, AT&T (1920) and Westinghouse (1921). Archer, op. cit., pp. 176-189; Maclaurin, op. cit., p. 105.
14. As part of its role in the formation of RCA, the U.S. Navy also initiated cross-licensing to resolve the patent situation in radio manufacture. It wished to have clear rights to use the radio equipment it purchased, without risking litigation due to the complex patent ownership—noting in 1919 that "there was not a single company among those making radio sets for the Navy which possessed basic patents sufficient to enable them to supply, without infringement, . . . a complete transmitter or receiver." A formal letter suggesting "some agreement between the

several holders of permanent patents whereby the market can be freely supplied with [vacuum] tubes," sent from the Navy to GE and AT&T in January 1920, may be seen as an initiating point for cross-licensing in the industry. Archer, op. cit., pp. 180-186; Maclaurin, op. cit., pp. 99-110.

15. RCA concluded cross-license agreements with firms including GE, Westinghouse, AT&T, United Fruit Company, Wireless Specialty Apparatus Company, Marconi (Britain), CCTF (France), and Telefunken (Germany). Archer, op. cit., p. 195; Maclaurin, op. cit., p. 107.
16. A distinction was that the RCA cross-licenses typically granted (reciprocal) exclusive rights to use the patents in given territories or markets, compared with the non-exclusive cross-licenses that became the norm later. The cross-license with GE (and later Westinghouse) included provisions for the supply of components to RCA. The RCA cross-licenses were for very long terms—many for 25 years, from 1919 to 1945. They covered current and future patents. Other radio manufacturers took licenses with RCA, starting in the late 1920s. Some of RCA's cross-licensing policies were later questioned on antitrust grounds, and modified following a consent decree in 1932. Archer, op. cit., pp. 381-387; Maclaurin, op. cit., pp. 107-109, 132-152.
17. Historical perspective on competition in the telecommunications industry is given in: M. Irwin, "The Telephone Industry," in W. Adams, ed., *The Structure of American Industry*, 5th ed. (New York, NY: Macmillan, 1977), pp. 312-333; G. Brock, *The Telecommunications Industry: The Dynamics of Market Structure* (Cambridge, MA: Harvard University Press, 1981); Office of Technology Assessment, *Information Technology Research and Development: Critical Trends and Issues* (New York, NY: Pergamon Press, 1985); R. Noll and B. Owen, "The Anticompetitive Uses of Regulation: United States *v.* AT&T," in J. Kwoka and L. White, eds., *The Antitrust Revolution* (New York, NY: Macmillan, 1989); G. Rosston and D. Teece, "Competition and "Local" Communications: Innovation, Entry, and Integration," *Industrial and Corporate Change*, 4/4 (1995).
18. OTA, op. cit.; M. Noll, "Bell System R&D Activities: The Impact of Divestiture," *Telecommunications Policy*, 11 (1987): 161-178; R. Harris, "Divestiture and Regulatory Policies," *Telecommunications Policy*, 14 (1990): 105-124.
19. The two substantive provisions of the 1956 consent decree were that (a) it confined AT&T to providing regulated telecommunications services, and its manufacturing subsidiary Western Electric to making equipment for those services (effectively prohibiting it from selling semiconductors in the commercial market), and (b) all patents controlled by the Bell System should be licensed to others on request. Licenses for the 8,600 patents included in existing cross-licensing agreements were royalty free to new applicants, and licenses to all other existing or future patents were to be issued at a non-discriminatory "reasonable royalty" (determined by the court if necessary). AT&T was also to provide technical information along with the patent licenses for reasonable fees. Licenses were unrestricted, other than being non-transferable. [*USA v. Western Electric Co. Inc. and AT&T*, Civil Action, 17-49, Final Judgment, January 24, 1956; Brock, op. cit., pp. 166, 191-194; R. Levin, "The Semiconductor Industry," in R. Nelson, ed., *Government and Technical Progress* (New York, NY: Pergamon, 1982), pp. 9-101.] In fact, AT&T went beyond the Consent Decree in its efforts to diffuse transistor technology, including symposia and direct efforts to spread know-how. [Levin, op. cit., pp. 76-77.]
20. See section later in this article on "Lessons for Innovation Management."
21. "We realized that if [the transistor] was as big as we thought, we couldn't keep it to ourselves and we couldn't make all the technical contributions. It was to our interest to spread it around." AT&T executive, quoted in Levin, op. cit., p. 77, after

J. Tilton, *International Diffusion of Technology: The Case of Semiconductors* (Washington, D.C.: The Brookings Institution, 1971).

22. By 1983, Bell Labs had received 20,000 patents. This may be compared to about 10,000 currently at IBM and 6,000 at Texas Instruments.
23. W. Kefauver, "Intellectual Property Rights and Competitive Strategy: An International Telecommunications Firm," in M. Wallerstein, M. E. Mogee, and R. Schoen, eds., *Global Dimensions of Intellectual Property Rights in Science and Technology* (Washington, D.C.: National Academy Press, 1993), pp. 236-240.
24. For the capture model, see section below on "Policy Issues." The survivorship period could be as much as 17 years from the grant date (possibly several years after filing), under U.S. patent rules prior to 1995, or 20 years from the filing date, after 1995.
25. In the U.S., during 1953-1968, 5,128 semiconductor patents were awarded. Bell Laboratories was granted 16% of these; the next five firms were RCA, General Electric, Westinghouse, IBM, and Texas Instruments. Tilton, op. cit.
26. E. von Hippel, "Cooperation Between Rivals: Informal Know-How Trading," Research Policy, 16 (1987): 416-424; A. Saxenian, "Regional Networks and the Resurgence of Silicon Valley," California Management Review, 33/1 (Fall 1990): 89-112.
27. There are also transactions costs reasons for using bundled licensing, as noted previously.
28. If the parties could not agree on a reasonable royalty rate, the court could impose one. Patent rights could be very long lived, since, at that time, patent life was 17 years from the grant date, which might be some years after the filing date. The patent licensing provisions ended in 1961. The decree also included other provisions related to the sale of IBM products and services. USA v. International Business Machines Corporation, CCH 1956 Trade Cases para. 68, 245, SDNY 1956.
29. This increased from $345 million in 1993 [IBM Annual Report, 1994]. IBM initiated a more active approach to licensing in 1988, when it increased the royalty rates sought on its patents from 1% of sales revenue on products using IBM patents to a range of 1% to 5%. Computerworld, April 11, 1988, p. 105.
30. R. Smith, "Management of a Corporate Intellectual Property Law Department," *AIPLA Bulletin* (April/June 1989), pp. 817-823; C. Boyer, "The Power of the Patent Portfolio," *Think*, 5 (1990): 10-11.
31. Gary Markovits, IBM patent process manager, in Boyer, op. cit., p. 10.
32. Jim McGrody, IBM VP and director of research, in Boyer, op. cit.
33. Roger Smith, IBM assistant general counsel, in Boyer op. cit. In all, IBM has about 11,000 active inventions, with about 35,000 active patents around the world. Smith, op. cit.
34. Many firms in the U.S. semiconductor industry were reported to be "dismayed" and "outraged" over the higher royalties and more active IP strategies of TI and others. [S. Weber, "The Chip Industry is Up in Arms Over TI's Pursuit of Intellectual Property Rights at the ITC," *Electronics* (February 1991), p. 51.] For example, T. J. Rodgers, CEO of Cypress Semiconductor described the practice of increased litigation over patent rights as a "venture capital investment." [*Upside* (December 1990).] Others have questioned whether the strengthening of patent rights might be hindering innovation, by enabling IP holders to demand "crippling royalties from young companies." Several small Silicon Valley semiconductor firms, including Cypress Semiconductor, LSI Logic, and VLSI Technology, formed a consortium to defend themselves against patent suits. [B. Glass, "Patently Unfair: The System Created to Protect the Individual Inventor May be Hindering Innovation," *InfoWorld*, October 29, 1990, p. 56.] Although some Japanese manufacturers reportedly described royalty demands as "possibly exorbitant," the Japanese

response has generally been to increase their own patenting effort. [*Computergram*, September 14, 1990; Weber, op. cit.] Similar objections to increased patent strength and licensing activity have also been evident in resistance to the growing use of patents for computer software, which it has been claimed may restrict innovation by small enterprises. [B. Kahin, "The Software Patent Crisis," *Technology Review* (April 1990), pp. 53-58.] However, here too, many software firms who at first resisted the trend have now accepted the need to build their own patent portfolios. [M. Walsh, "Bowing to Reality, Software Maker Begins Building a Patent Portfolio," *The Recorder*, August 17, 1995, p. 1.]

35. This section is based in part on discussions with Texas Instruments executives. However, the views expressed here are those of the authors and should not be seen as necessarily reflecting those of Texas Instruments.
36. The costs of manufacturing facilities have risen dramatically. A new wafer fabrication plant cost $10-20 million in 1975 (4-kilobit DRAM), $300-400 million in 1990 (16-megabit DRAM) and over $1 billion in 1991 (256-megabit DRAM). SEMATECH, *Annual Report, 1991*; "Foreign Alliances Which Make Sense," *Electronic Business*, September 3, 1990, p. 68.
37. Without field-of-use cross-licenses, a typical semiconductor firm might need to reverse engineer an average of two or three competitors' products a day, as each is introduced over the course of a five-year license, to ascertain whether these are infringing its patents. It must do the same for its own products. This would be prohibitively expensive. Tracking sales by each of hundreds of affected products, on a patent by patent basis, to ascertain royalties, would be virtually impossible.
38. In some cases, where there are only a few very specific overlaps between two firms' technology needs, firms may choose to license single patents. Although an option, it is rarely convenient compared with field-of-use cross-licensing when there are substantial technology overlaps.
39. R. Levin, A. Klevorick, R. Nelson, and S. Winter, "Appropriating the Returns to Industrial R&D," *Brookings Papers on Economic Activity*, 3 (1987): 783-820 Of course, even reading the patent is a helpful guide to someone knowledgeable in the field.
40. The most powerful threat to enforce a patent is an injunction to close down the infringer's production line. This could be ruinous for a manufacturing corporation, especially in fast developing markets such as electronics and semiconductors. The threat of damages may also be important, but as these are often based on projected royalties (and hence may be little worse than freely negotiated licensing terms) they are less potent, unless multiplied by the court.
41. For the economics of technology transfer see D. Teece, "The Market for Know-How and the Efficient International Transfer of Technology," *Annals of the American Academy of Political and Social Science*, 458 (1981): 81-96.
42. Reverse engineering a semiconductor product is not a simple matter, involving as it does decapping and microscopic examination at the submicron level. Although the process is by now largely automated, it can take 400-500 man-hours per device.
43. For cross-licenses with firms outside the semiconductor industry, such as the personal computer industry, the process used is simpler. In this case, there may be few patents to balance against the proffered patents. Licensing follows precedents long established in the computer industry, primarily under the leadership of IBM, as the holder of many of the patents used in the industry. The negotiations are similar, but the weighting process is not involved. Royalty rates are influenced by industry norms.
44. In some cases licensees may only wish to license a few selected patents, rather than all patents in a field-of-use. For this reason licenses are generally also offered

for individual or specific patents, as well as for all patents in a given field. However, there are significant transactions savings to both sides from a field-of-use license, and the cost per patent is likely to be higher when only a few patents are licensed.

45. For general considerations affecting royalty rates, see M. Lee, "Determining Reasonable Royalty," *Les Nouvelles*, 27 (1992): 124-128; R. Parr, *Intellectual Property Infringement Damages: A Litigation Support Handbook* (New York, NY: Wiley, 1993).
46. To an extent this may be a transitional problem. As licensing becomes more widespread, individual licenses are more likely to be negotiated in the knowledge that other licenses, potential or actual, must be taken into account.
47. For strategies to establish standards see R. Hartman and D. Teece, "Product Emulation Strategies in the Presence of Reputation Effects and Network Externalities," *Economics of Innovation and New Technology*, 1 (1990): 157-182; L. Gabel, *Competitive Strategies and Product Standards* (London: McGraw-Hill, 1991); P. Grindley, *Standards, Strategy, and Policy: Cases and Stories* (Oxford: Oxford University Press, 1995).
48. However liberal the licensing terms, the patent holder should not inadvertently assign away IP rights beyond those specifically needed to operate the standard, and may need to condition rights over its IP to uses related to the standard. The innovator might otherwise be deterred from participating in standards setting. There is a balance to be drawn between committing to an open standard and limiting that commitment to what is needed for the standard and to keep access open in future.
49. Risks include the likelihood that the validity of the patents would be challenged in court, that firms—and nations—would retaliate, and that the corporate image with customers would suffer. Patent assertion against customers and partners is an especially sensitive area.
50. R&D agreements with Hitachi have ranged from a 4-megabit DRAM know-how exchange in 1988 to a 256-megabit DRAM co-development agreement in 1994. According to Yasutsugu Takeda of Hitachi, "You can't create [a successful cooperative venture] just because you sign up a lot of companies that are barely committed and don't have anything to bring." The Hitachi-TI collaboration on 256-megabit memory chips has been successful because it is a "meeting of equals" [*Business Week*, June 27, 1994, p. 79]. Complementary capabilities are generally considered important factors in selecting international collaborative venture partners. D. Mowery, "International Collaborative Ventures and the Commercialization of New Technologies," in N. Rosenberg, R. Landau, and D. Mowery, *Technology and the Wealth of Nations* (Stanford, CA: Stanford University Press, 1992), pp. 345-380.
51. TI entered joint ventures during 1989-1990 to build manufacturing plants with total investments over $1 billion: with the Italian government; Acer (Taiwan); Kobe Steel (Japan); and the Singapore government, HP, and Canon (Singapore).
52. This section is based in part on discussions with Hewlett-Packard executives. However, the views expressed here are those of the authors, and should not be seen as necessarily reflecting those of Hewlett-Packard.
53. Examples include *Research Disclosure* and other publications. Such journals charge fees to authors, yet often have large circulations for reference libraries and research laboratories.
54. Surveys of executives in a range of industries taken in the early 1980s typically rated methods such as lead time and superior sales and service effort as the most effective means of protecting innovations, rather than patent protection, which was considered relatively ineffective. Levin et al., op. cit.
55. The original transistor process patents were held by AT&T, so that all transistor manufacturers needed to cross-license their own patents at least with AT&T.

Similarly, the key patents for the integrated circuit (IC) technology were held by two firms, TI and Fairchild, ensuring that these too were widely licensed. With the critical patents widely available, the cumulative nature of innovation guaranteed broad cross-licensing. Levin, op. cit., pp. 79-82.

56. The first commercial producers of transistors in the 1950s, using AT&T licenses, included Shockley Labs, Fairchild, Motorola and TI. These gave rise to a wave of spin-off companies in the 1960s, such as National Semiconductor, Intel, AMD, Signetics and AMI, which in turn gave rise to subsequent waves of new companies, such as, Cypress Semiconductor, Cyrix, LSI Logic, Chips and Technologies, Brooktree Semiconductor, and others.
57. At TI this approach was formalized in the Objectives, Strategies, and Tactics (OST) product development management process, including "design to cost" methods formalizing experience curve pricing procedures. *Business Week*, September 18, 1978; B. Uttal, "TI Regroups," *Fortune*, August 9, 1982, p. 40; M. Martin, *Managing Technological Innovation and Entrepreneurship* (Reston, VA: Reston, 1984).; R Burgelman and M. Maidique, *Strategic Management of Technology and Innovation* (Homewood, IL: Irwin, 1988).
58. Tilton, op. cit.; M. Borrus, J. Millstein, and J. Zysman, *International Competition in Advanced Industrial Sectors: Trade and Development in the Semiconductor Industry* (Washington, D.C.: U.S. Department of Commerce, 1982).
59. Borrus et al., op. cit.
60. The same is broadly true of IBM's entry into Japan.
61. Dataquest figures, quoted in United Nations Organization (UNO), *The Competitive Status of the U.S. Electronics Sector* (New York, NY: United Nations Organization, 1990). For comments on the U.S. recovery since the late 1980s, see W. Spencer and P. Grindley, "SEMATECH After Five Years: High-Technology Consortia and U.S. Competitiveness," *California Management Review*, 35/4 (Summer 1993): 9-32; P. Grindley, D. Mowery, and B. Silverman, "SEMATECH and Collaborative Research: Lessons in the Design of High-Technology Consortia," *Journal of Policy Analysis and Management*, 13 (1994): 723-758.
62. For contrasting views on the responses of Silicon Valley to international competition, see R. Florida and M. Kenney, "Why Silicon Valley and Route 128 Can't Save Us," *California Management Review*, 33/1 (Fall 1990): 66-88; Saxenian, op. cit.
63. Hazards for innovation when a firm is remote from business transactions, and hence from the technological frontier, are outlined in J. de Figueiredo and D. Teece, "Strategic Hazards and Safeguards in Competitor Supply," *Industrial and Corporate Change*, vol. 5.2 (1996). The similar vulnerability of the "virtual corporation," which contracts out development and manufacturing, is discussed in H. Chesbrough and D. Teece, "When Is Virtual Virtuous: Organizing for Innovation," *Harvard Business Review* (January/February 1996), pp. 65-73.
64. For the nature of dynamic capabilities of firms and their relationship to innovation, see D. Teece and G. Pisano, "The Dynamic Capabilities of Firms: An Introduction," *Industrial and Corporate Change*, 3.3 (1994): 537-556; D. Teece, G. Pisano, and A. Shuen, "Dynamic Capabilities and Strategic Management," *Strategic Management Journal* (forthcoming in 1997). For the role of complementary assets in commercializing innovation, see D. Teece, "Profiting from Technological Innovation," *Research Policy*, 15 (1986): 285-305.
65. Indeed, in some cases the firm might conceivably do better if it has strengths in an area where the licensee is relatively weak, since it will have greatest difficulty avoiding their patents in those areas, whereas where it is strongest it may have more ability to invent around the patents.

66. An example is Brooktree Corporation, a small semiconductor design company in San Diego, which concluded a favorable cross-licensing agreement with TI in 1993.
67. See E. Sherry and D. Teece, "The Patent Misuse Doctrine: An Economic Reassessment," in *Antitrust Fundamentals*, ABA Section of Antitrust Law, Chicago (forthcoming).
68. IP rights to the transistor were given away to U.S. and foreign firms for very small amounts. Levin, op. cit.
69. DOJ/FTC, *Antitrust Guidelines for the Licensing of IP*, April 6, 1995 (Washington, DC: U.S. Department of Justice and the Federal Trade Commission, 1995).

Patents, Licensing, and Entrepreneurship: Effectuating Innovation in Multi-invention Contexts

DEEPAK SOMAYA AND DAVID J. TEECE

1. Introduction

In recent years, patents have become more significant as mechanisms to capture value from innovation. Simultaneously, the innovation context has become more complicated, not only because many patents are implicated in "multi-invention" or "systemic" innovation, but also because there is great dispersal in the ownership of patents. Accordingly, many new products require the use of patents owned by several different entities. This creates both challenges and opportunities for entrepreneurs. Sometimes these challenges are easily worked through; on other occasions it requires "new combinations" not just of patents, but of tangible assets, components, and technologies in order to enable innovations to proceed to market. This paper maps opportunities and challenges, identifies possible solutions, and discusses how (entrepreneurial) firms have responded to these opportunities and challenges.

2. Intellectual Property and Entrepreneurship

The patent system grants the inventor/patent owner time-bound exclusive rights to practice the technology that is covered by the patent. In exchange for this right, the inventor recognizes that the invention will be disclosed when the patent is published. Absent the patent system, inventions could be more readily imitated. It is frequently argued that patents are needed to help provide incentives not only for invention, but also for the commercial application of inventions (Kitch 1977). Small inventors and entrepreneurs are among the most enthusiastic supporters of the patent system. Their support stems in part from the perception that the system provides safeguards for the inventor, who might otherwise be subject to use of the invention without compensation. This is corroborated by a survey in the U.S. semiconductor

Reprinted with permission from *Entrepreneurship, Innovation, and the Growth Mechanism of the Free-Enterprise Economies* (Princeton University Press, 2007), edited by Eytan Sheshinski, Robert J. Strom, and William J. Baumol, pp. 185-212.

industry (Hall and Ziedonis 2001), where small start-up firms were found to be champions of patents, who recognize that patents can assist them in acquiring investment funds from venture capitalists. And, it is in sharp contrast to the perception of some commentators that patents may be harmful to small firms and the entrepreneurship process.

In industries where innovation requires the combination of a very large number of inventions to create new products and services, what we term here the multi-invention or systemic innovation[1] context, additional opportunities and constraints arise from patenting. In these situations, patents may provide leverage to the entrepreneur or individual inventor, while at the same time requiring new entrants and incumbents alike to navigate patent thickets where the relevant patents required for a particular innovation have distributed ownership; that is, the patents needed to design and manufacture a product may be held by several unrelated entities. These environments may require the exercise of entrepreneurial skills as well as the establishment of potentially quite different organizational arrangements (modes) to enable an innovation to proceed to market.

According to Schumpeter (1934), the entrepreneur drives economic growth. The function of the entrepreneur is to innovate, to "carry out new combinations." In Schumpeter's treatment, the entrepreneur is not the inventor. Rather, the entrepreneur exploits the invention to effectuate innovation in the marketplace. Nor is the entrepreneur a risk bearer, as that function is performed by the capitalist (the venture capitalists in today's vernacular). The Schumpeterian entrepreneur has an organizing and decision-making role. Typically that role is exercised under considerable uncertainty, and usually with quite limited information.

An important function for the entrepreneur is to assemble assets and exploit complementarities among them. By exploring and exploiting co-specialization, rent streams can be generated (Teece 1986, 2003a). Competitive advantage is obtained when tangible and intangible assets are assembled that yield joint returns that are more than the sum of the parts (Lippman and Rumelt 2003).

Although possibly amplified in the context of untested patents,[2] the condition whereby factor/asset owners and entrepreneurs have different beliefs about the value of different factor/asset combinations is not uncommon in the economy. It is the role of the entrepreneur to perceive value that no one else can; or even if others perceive it, the entrepreneur is able to organize resources to achieve ends that other (nonentrepreneurial) individuals (or managers) are unable or unwilling to achieve. Entrepreneurial opportunities inherently depend upon asymmetries of information, belief, and individual and organizational capability.[3] As Baumol (1993) has explained, entrepreneurship is certainly not an optimization process by which people make mechanical calculations in response to a given set of alternatives imposed upon them. Clearly, entrepreneurship is a scarce resource. History is replete with examples where inventors and incumbents alike did not initially see commercial opportunities resulting from the invention of new technologies (Rosenberg 1994).

The particular opportunities and challenges afforded by multi-invention (systemic) innovation require particular entrepreneurial responses. These can be of two kinds. First, entrepreneurs must understand and address the challenges and opportunities of combining inventions from a variety of disparate sources in a multi-invention or systemic context. Organizational barriers and transaction costs can be a significant hurdle in unlocking the value hidden in a new invention or combination of inventions. Moreover, patents owned by others may appear to block the road ahead and amplify organizational challenges. Second, entrepreneurs must evaluate how best to appropriate value from the unique combinations that they create. There is no benefit to engaging in entrepreneurial efforts if another enterprise can simply appropriate all the returns. Entrepreneurs must therefore pay attention to how they will appropriate returns while at the same time creating the factor/asset combinations necessary to effectuate innovation.

In this essay we explain that workable solutions usually exist to both these challenges. A variety of organizational arrangements can help manage the challenges of combining inventions in multi-invention settings. Some solutions require action in technology and component markets, and some require actions in the market for corporate control (i.e., mergers). In other instances, the lowering of transaction costs in patent licenses may require certain bargaining and negotiating skills. Further, in each instance, there are implications for the role of patents in helping to appropriate returns for the entrepreneur. In other words, the organizational responses used to effectuate innovation in a multi-invention or systemic context need to be supported by an appropriate patent strategy. To our knowledge, the literature has not explored these issues in any systematic way, and certainly not with attention to entrepreneurial solutions that are in our view central to the issues at hand.

3. The Multi-invention (Systemic Innovation) Context

In many high-tech sectors of the economy, multi-invention contexts are the norm; that is, very large numbers of inventions are combined to develop end products and services. In semiconductors, increasing miniaturization has made it feasible to manufacture large and complex electronics systems on a single chip (so-called systems-on-a-chip, or SOCs). This creates the need for large numbers of patented inventions to be combined in any single product (Teece 1998; Linden and Somaya 2003). In biotechnology, increasingly large portfolios of inventions in genomics, research tools, and other areas need to be assembled to bring new medical solutions to fruition. Similarly, hundreds or thousands of patentable software inventions may be combined in contemporary software programs, and this trend is intensifying with the increasing size and complexity of software products.

Innovation in these industries comes from multiple sources—from within large firms, from start-ups and specialized players, from firms outside the industry, and even from universities and other research establishments—creating phenomenal entrepreneurial opportunities to combine knowledge in innovative ways and create valuable new products. But these multiple sources of invention also imply a tangled web of patent rights, which must be navigated for commercial success. A central challenge for entrepreneurs in such multi-invention contexts is to determine how the production of end products from large numbers of potential inventions can be most effectively organized, and what role the entrepreneurial firm will play in this organizational structure.

Broadly, one can think of two types of organizational arrangements or modes (or business "models") by which inventions may be combined—integrated modes and nonintegrated modes. Integrated modes arise when firms innovate by using their own internal technologies and resources, without relying on external access. Nonintegrated modes can broadly be separated into licensing and component modes, where access to external technologies is obtained in abstract and product-embodied forms, respectively. Of course, these distinctions are somewhat stylized. In any given multi-invention context, entrepreneurs may choose to develop some technologies internally in an integrated fashion, and use market (nonintegrated) arrangements to access others. Furthermore, actual organizational arrangements (or business model choices) may exhibit hybrid integrated and nonintegrated characteristics—for example, interfirm alliances—or hybrid licensing and component characteristics—for example, transfer of highly flexible components (like a programmable chip) or highly codified product designs (which would enable component manufacture, but not transfer any know-how).

Which organizational mode should be chosen by the (entrepreneurial) firm for effectuating innovation depends on the associated organizational costs and benefits in each mode. Generally, integrated modes are considered to be advantageous for overcoming transaction costs of various kinds, whereas nonintegrated modes are considered advantageous in terms of incentives and access to best-of-class inventions or components.

When transaction costs in know-how, licensing, or component markets are low, it makes sense for entrepreneurs to innovate by transacting for complementary assets and inventions through these markets. Otherwise, the costs of developing the required technologies and capabilities in-house and the added cost of internal bureaucracy produce a significant drag on commercialization. However, when these transaction costs are high, it makes sense to seek more integrated solutions, either through internal development or through the market for corporate control (mergers and acquisitions).

The study of transaction costs has become a significant research enterprise (Williamson 1985, 1996; Shelanski and Klein 1995), to which we surely cannot do justice in the limited space available. However, we draw attention to some types of transaction costs that are particularly relevant in technology-related transactions. First, there are barriers that arise due to the *technological interconnectedness* (Linden and Somaya 2003) or *the systemic nature* (Teece 1996) of innovation in some multiproduct contexts. In essence, the difficulty of partitioning the problem domain in these contexts makes it very costly to transact because of the various technologies that must work together as a whole. Another source of transaction costs is the potential *leakage of know-how* through transactions in technology markets (Arrow 1971; Teece 1982). When (entrepreneurial) firms either buy or sell technologies and components, they may end up disclosing elements of inventions to their partners, which (despite the existence of nondisclosure agreements) subsequently undermines their own ability to appropriate returns. In addition, there are often disagreements between firms over the contribution to value that is created by each of their technologies when used together in a particular product. These *value allocation* problems can lead to significant delays in negotiating contractual arrangements, and are particularly exacerbated by the idiosyncratic nature of each invention (Merges and Nelson 1994; Somaya 2005), and by the fact that there may be uncertainty over patent validity and infringement.

The transaction cost issues surrounding valuation and know-how leakage tend to be somewhat greater with "know-how" and intellectual property markets compared to "product" or "component" markets. Components are tangible products with measurable performance characteristics. Components are often easier to value against competing alternatives; also, it is likely to be more difficult for the technologies embedded in a component to inadvertently leak out. In addition, component markets also have lower *monitoring and metering* costs than licensing markets because each use of the technology is limited to a single well-defined physical artifact. With know-how and intellectual property, it is often difficult to ascertain how, where, and how often a technology is being used by the licensee and whether patents are valid. Both questions can lead to uncertainties and disputes about value and royalty payments.

Ultimately, entrepreneurs must evaluate the potential costs and benefits arising from each organizational mode, and choose that which has the best performance characteristics. In other words, careful attention must be paid to the business model and the organizational challenges of innovation if entrepreneurship is to succeed in multi-invention contexts. In the case of Kentron, discussed below, the firm encountered high transaction costs in licensing and had to quickly modify its strategy to a component-focused one. Each organizational strategy in turn must be complemented by a suitable patent strategy so as to ensure that the firm is also able to appropriate returns from its innovation.

4. Understanding Patent Strategy

One can think of patent strategy as occurring in three related domains of activity—patenting, licensing, and enforcement. "Patenting" refers to the gamut of actions whereby patent rights are obtained, renewed, maintained, and protected, including through the purchase of others' patents in the secondary market. "Licensing" involves the provision of exclusive and nonexclusive rights to use the patent. Distinctions can be made between the instances in which patent rights are licensed along with know-how transfer, and those in which only patent rights are licensed. "Enforcement" entails the use or threatened use of litigation to persuade infringers to desist or pay royalties. Since no one would take a naked patent license absent fear of a court sanctions at some level, patent licensing (as distinct from pure know-how licensing) always takes place in the shadow of court-enforced sanctions against infringement.

Patent strategy ought to be formulated in the broader context of the business strategies required for establishing and maintaining competitive advantage at the enterprise level. While there are no doubt specific issues that arise in each domain of patent strategy—patenting, licensing, or enforcement—some important commonalities cut across all of them. Three generic patent strategies are presented below—namely, proprietary use, defensive use, and royalty generation.

4.1. *Proprietary Use (No Licensing)*

Patents and other forms of intellectual property have long been recognized as tools that can in some cases protect technologies from imitation by rivals. Put differently, patents are "isolating mechanisms" (Rumelt 1984) that can help protect "rent" streams. Indeed, in the popular literature, the role of patents in enabling firms to "stake out and defend a proprietary market advantage" has been characterized as "their most powerful benefit" (Rivette and Kline 2000, 4). The central insight here is that the ownership of IP conveys the right to *exclude* others from the use of patented invention. In most circumstances, of course, this does not convey the ability to exclude competitors from a market. Such power is only conveyed with very fundamental patents, which cannot be worked around for one reason or another. Even then, the period of exclusion is of course limited by the length of time the patent has to run. In the real world, situations where patents confer market power are quite rare. Furthermore, the use of patents to protect fundamental new areas of technology has been acknowledged as one of the critical functions of the patent system, without which firms might be reluctant to make additional investments to commercialize their inventions (Kitch 1977; Mazzoleni and Nelson 1998).

If a business enterprise has a fundamental interest in a particular opportunity, and seeks to control the technology, there are implications for how the enterprise will need to conduct its patent-related activities (Somaya 2003). One implication is that the firm would most likely need to invent follow on technologies and also patent these. Another implication, naturally, is that such patents will generally not be licensed. It is of course the patent owner's choice not to license. As discussed in Teece (1986), the strategy of eschewing licensing is likely to be preferred only if the enterprise's patent portfolio is strong, the enterprise does not need access to anyone else's patent, and the enterprise is well positioned in the complementary assets required to successfully commercialize the innovation. Furthermore, to sustain a proprietary strategy with respect to select patents, infringement of these patents by others would need to be prosecuted aggressively.

4.2. "Design Freedom" (Defensive) Patent Strategies

Defensive strategies relate to the actions of business enterprises to protect themselves against the use of patents by their rivals in the marketplace. In fast-paced high-technology industries, enterprises often desire the freedom to design, innovate, and manufacture without being too constrained by the patent rights, present and future, of other firms. In part, this desire for design and operating freedom may be motivated by irreversible investments that they have made or expect to make, including investments in highly capital intensive manufacturing facilities, as in the case of the semiconductor industry. These investments, and the firm's commercial interests in general, can potentially be put at risk by others' patents, including those that had not issued at the time investment or commercialization decisions are made. As a result, these patents pose a significant threat to the firm—entire lines of business may be put at risk and significant royalties may have to be paid to license necessary patents. These may reflect the firms' much higher ex post willingness to pay (Sherry and Teece 2003). Needless to say, these issues are more common in multi-invention contexts, where the likelihood of infringing one or more patents among hundreds or thousands is quite high.

In some cases, the enterprise's own patents can be used as bargaining chips. Attempts by rivals to assert their own patents can be countered with threats to enforce the firm's own patents against them. This situation of "mutual holdup" can facilitate the negotiation of reasonable terms between the parties, and the effective removal of patent barriers (Grindley and Teece 1997; Somaya 2003). Research in the semiconductor industry has demonstrated that firms often engage in reciprocal cross-licensing as part of their patent strategy. Such firms accumulate large portfolios of patents in part because they are desirous of achieving design and operating freedom (Hall and Ziedonis 2001). It is important to bear in mind, however, that defensive patenting may not be effective under all circumstances, especially because it assumes that the threat of reciprocal patent enforcement is effective. When this is not the case (for example, with individual inventors or universities who have few commercial interests that can be held up), this defensive strategy may not work and a license may need to be taken. This may be comparatively costly in many multi-invention contexts, particularly if the invention is important and good alternatives do not exist.

4.3. Royalty Generation Strategies

Licensing on an exclusive or nonexclusive basis is the other obvious strategy for capturing value from a patent. The licensing of technologies is a much-studied phenomenon in the management and economics literature (Teece 1986, 2000, 2003b; Arora 1995). Much of the literature has tended to assume that patents and know-how are always bundled together; however, patent-only licensing has grown considerably in recent years. Firms like IBM and Texas Instruments have earned very substantial licensing revenues from licensing patent rights (an average of $580 million a year over 1999–2001 for IBM alone).[4] While some know-how transfer may accompany patent-licensing deals, TI's and IBM's primary focus is the granting of rights to use patents to companies that are already using (i.e., infringing) the technology in question. When patents are strong and provide utility to an entire industry, they are natural candidates for licensing in this manner. If the incumbent firms are already infringing the patent(s) and have invested substantially in using patented technologies, they may have few practical alternatives to licensing the patent(s). Royalty generation in this manner is supported by the business enterprise's efforts to identify potential licensees, its negotiating and bargaining skills, and the implicit threat of sanctions obtained from a court (or the International Trade Commission if imported goods implicate the patent).

With respect to enforcement of property rights, there are significant differences between intangible and tangible goods. When the input is a tangible good, it is impossible for the manufacturer of the final product to produce it unless the physical input is delivered. Stolen goods are not acceptable. However, when the use of patented technology already known to the user is the input,[5] production can commence and sale of the final product can be completed without "delivery" of the IP rights. This is because the manufacturer can simply go ahead and infringe the patent. The only barrier is the prospect of a court-ordered injunction and the court's determination that damages should be paid. Therefore, legal enforcement of property rights (patents) plays a critical role with respect to intangible property, when compared with tangible property, for the collection of monies for the use of the input.

5. Entrepreneurship and Patent Strategy in Multi-invention Contexts

What, then, are the implications for entrepreneurship emerging from our understanding of multi-invention (or systemic) contexts and patent strategy? Interest in the role of patents in multi-invention settings goes at least as far back as Kitch (1977), who contended that patent rights should facilitate coordination between owners of related inventions. Subsequent research has focused considerable attention on a particular type of multi-invention context, namely sequential innovation (Merges and Nelson 1990, 1994; Scotchmer 1991, 1996; Chang 1995; Green and Scotchmer 1995). These studies have primarily addressed the desirable scope of patents, implicitly taking entrepreneurship as given (at what we think is a low level). More recent work has focused on the transactional challenges posed by patents in multi-invention contexts, leading—according to the authors—to the potential underutilization of innovative resources, a so-called "tragedy of the anticommons" (Heller and Eisenberg 1998). These transaction costs may arise from *diffuse ownership* and associated *royalty stacking* problems in patent licensing (Teece 2000, 208–9; Somaya 2005), or from *valuation disputes* that are due at least in part to the *fuzzy boundaries* of patents (Teece 2000, 149–50; Somaya 2005).

However, in our view, Eisenberg and Heller may have exaggerated the problems associated with the so-called "anticommons." They provide no compelling evidence of the scope of the "problem." It is well known that patents can also facilitate transactions in technology, for example by facilitating transactions in know-how without the fear of misappropriation (Teece 1982; Arora 1995; Oxley 1999). Ultimately, transaction costs, both in general and those induced (or remedied) by patent rights, speak to the need for entrepreneurship in multi-invention settings. Entrepreneurs play an important role in figuring out the right organizational arrangements (or business model) for innovation. In addition, entrepreneurs and managers must choose appropriate patent strategies to support their innovative efforts. These strategies are likely to depend in large part on the precise organizational arrangements chosen for commercialization. We assess the implications for both integrated and nonintegrated modes below, using mini case studies drawn from past multi-invention contexts.

5.1. Integrated Modes

Entrepreneurs and managers should choose an integrated mode to innovate when the transaction costs in licensing and component markets for complementary technologies are relatively high. Given this choice, the main challenge for the entrepreneur becomes how to assemble all the required assets and technologies within a single firm. Given the nature of innovation in multi-invention contexts, it would be highly unlikely that a single firm has invented and patented all the technologies necessary to commercialize the end product, and will continue to do so in the future. Integrated approaches therefore employ different ways of obtaining both the technologies and the patents rights needed for commercialization.

One alternative is to develop all the technologies needed in-house, but rely on patent licensing to obtain access to the patent rights *owned* by other enterprises. Access to patents could be obtained in a number of ways, including patent pools, cross-licenses, and other patent-sharing arrangements. Since this form of integrated innovation implicitly acknowledges the existence of patents that may be infringed by the enterprise, defensive patent strategies are especially important for ensuring freedom to design and innovate. In industries like semiconductors, electronics, and computers, these defensive strategies are often pursued through the building up of large patent portfolios, and the proactive development of cross-licensing relationships.

Beyond the firm's defensive needs, however, patents may also be used in this context to generate licensing revenues. Later entrants and noninnovators in such a market may have weaker patent portfolios, reflecting their limited contribution to technological advances in the industry. Firms with a more robust history of innovation and patenting need not license their own patents to these firms on a purely reciprocal basis. In other words, offsetting royalty payments can be negotiated to reflect the asymmetry among the patent portfolios. The case of AT&T, IBM, and Texas Instruments in electronics and semiconductors provides a graphic illustration of the entrepreneurial creation of patent exchange mechanisms to deal with defensive concerns, and the use of strong patent portfolios to generate royalty income.

CROSS-LICENSING IN ELECTRONICS AND SEMICONDUCTORS

In the electronics and semiconductor industries, the multi-invention context is frequently the norm. In many advanced products, the range of technology is simply too great for a single firm to develop its entire needs internally. The "state of the art" of the technology tends to be covered by a large number of different patents held by different firms. Companies may produce hundreds of products, which use literally thousands of patents, and many hundreds more may be added each year. One innovation builds on another. Overlapping developments and mutually blocking patents are inevitable. To solve these problems, the business practice of cross-licensing has emerged. Enterprises cross-license patents from others to ensure that they themselves have the freedom to innovate and manufacture without inadvertent infringement. Cross-licenses typically cover portfolios of all current and future patents in a field of use, without making specific reference to individual patents. It is simply too cumbersome and costly to license only specific patents needed for specific products. The portfolio approach reduces transactions costs and allows licensees freedom to design and manufacture without triggering infringement, inadvertent or otherwise.

Cross-licensing has developed in a quite sophisticated fashion (Grindley and Teece 1997).[6] An important feature is the calculation of balancing royalty payments, according to the relative value of the patent portfolios of each party. This calculation is made prospectively, based on a sample of each firm's leading patents. Weight is given to the quality and market coverage of the patents. The key to successful cross-licensing is a portfolio of quality patents that covers large portions of the licensing partner's product markets. A quality portfolio is a powerful lever in negotiating access to required technology and may lead to significant royalty generation or, at a minimum, to reduced payments to others. Obviously, a firm that is a large net user of other firms' patents, without contributing comparable IP in exchange, is likely to have to pay significant royalties. Significantly, for the balancing process, the firm should concentrate its patenting in those areas where it does best and has a comparative advantage to develop patents that its cross-licensing partners need. In this way, firms can develop complementary rather than duplicative technology, thereby also benefiting the public interest.

In patent cross-licenses, technology is not usually transferred, as the parties are often capable of using the technology in question without assistance. Rather, these licenses confer the right to use the intellectual property without being sued for infringement. The licensing agreements sometimes include transfer of trade secrets and know-how. However, these licenses are quite different, as they involve technology transfer, and may accompany a joint venture or strategic alliance.

AT&T'S CROSS-LICENSING PRACTICES

Cross-licensing is not a new phenomenon in electronics; it goes back almost to the beginning of the industry.[7] One of the most influential firms was AT&T, whose licensing and cross-licensing practices, especially from the 1940s until its breakup in 1984, were the initial templates for the development of similar programs by other firms. Over its long history, AT&T's licensing policy has had three phases, reflecting changes in its overall business strategy. First, from AT&T's establishment in 1885 until its first antitrust-related commitment in 1913, it used IP rights in a forthright exclusive fashion to establish itself in the service market.[8] In the second phase, from 1914 until 1984, AT&T was a regulated monopoly. The need for access to patents led to cross-license agreements between the major producers of telephone equipment, starting in the 1920s, which soon developed into a more widespread policy. In the last phase, dating from divestitures in 1984, AT&T was no longer bound by the consent decree, and its IP licensing has been increasingly aligned with its commercial needs (OTA 1985; Noll 1987, 161–78; Harris 1990, 105–24).

AT&T's policy was to openly license its IP to everyone for minimal fees. The 1956 consent decree required AT&T to license all patents at "reasonable royalties," provided that the licensee also grants licenses at reasonable royalties in return. AT&T was also required to provide technical information in exchange for the payment of reasonable fees, and licensees had the right to sublicense the technology to their associates.[9] The impact of AT&T's liberal licensing on the industry was considerable, especially when considered in parallel with that at IBM.

To a large extent, the licensing terms in AT&T's 1956 decree simply codified what was already AT&T policy. As an enterprise under rate-of-return regulation, it had little reason to maximize royalty income from its IP. It perhaps figured that its service customers would be better off if its technologies were widely diffused among suppliers, as this would lower the prices and increase the performance of procured components

(Levin 1982, 77). It appears to be the first company to have had "design freedom" as a core component of its patent strategy. However, it did not see licensing income as a source of funds for R&D, as Bell Laboratories' research was largely funded by the "license contract fee," assessed on the annual revenues of the Bell operating companies. By 1983, Bell Laboratories had received 20,000 patents, as compared to about 10,000 held by IBM in 1995 and 6,000 by Texas Instruments. AT&T's portfolio was fundamental, and included patents such as the transistor, basic semiconductor technology, and the laser, and indeed many other basic patents in telecommunications, computing, optoelectronics, and superconductivity.

Using its own portfolio as leverage, AT&T was able to obtain the (reciprocal) rights it needed to continue to innovate, unimpeded by the IP of others. An interesting aspect of AT&T's IP strategy was that technologies (though not R&D programs) were often selected for patent protection based on their potential value to other firms generating technology of interest to AT&T. Since the legal pressures by the regulators for open licensing did not extinguish all of AT&T's intellectual property rights, the company was able to gain access to the external technology that it needed (Kefauver 1993).

The terms of AT&T's licenses set a pattern that is still commonplace in the electronics industry through the "capture model," which was defined in the 1956 consent decree. Under this arrangement, the licensee is granted the right to use existing patents plus any obtained for inventions made during a fixed future capture period of no more than five years, followed by a survivorship period until the expiration of the patents. The licensing regimes this led to were persistent, since the long survivorship period on many of the basic patents provided only limited scope to introduce more stringent conditions for new patents.

The traditional cross-licensing policy of AT&T was greatly extended following the invention of the transistor in 1947. Widespread "field-of-use" licensing in the semiconductor industry is one of AT&T's legacies, as the industry was founded on the basic semiconductor technologies developed by the company. AT&T soon realized that other electronics companies were developing their own semiconductor technologies and obtaining patents, which led to its policy of cross-licensing by field of use.[10] These cross-licenses ensured that the company had reciprocal access to patents and was able to develop its own technology without risking patent infringement.

Not surprisingly, AT&T/Lucent Technologies has subsequently used its IP more strategically. No longer bound by the consent decree, with R&D facilities mainly in Lucent Technologies (which has legacy connections back to AT&T's manufacturing arm known as Western Electric), its IP policy has necessarily been linked more closely to particular business opportunities.

IBM'S CROSS-LICENSING PRACTICES

A second major influence on licensing practice across the electronics industry has been IBM. The company has long been involved in licensing and cross-licensing its technology, both as a means of accessing external technology and to gain profit (generate royalties). In many ways, it has been in a similar position to AT&T in that it has been a wellspring of new technology, and was subject to a 1956 consent decree that contained certain compulsory licensing terms. Under the consent decree, IBM was required to grant nonexclusive, nontransferable, worldwide licenses for all of its patents at reasonable royalties (royalty free for existing tabulating card/machinery patents)—provided the applicant also offered to cross-license its patents to IBM on similar terms.[11]

The importance IBM attaches to its patents for use in cross-licensing and negotiating access to outside technology is reflected in its public statements (Smith 1989, 817–23; Boyer 1990). The main object of its licensing policy has been "design freedom," and to ensure "the right to manufacture and market products" by obtaining rights to use technologies and patents owned by others. IBM acquires these rights primarily by trading access to its own patents, that is, through cross-licensing.[12] IBM has often had the reputation of being a "fast follower" in some areas of technology, and it has used the power of its patent portfolio to negotiate access. The company has noted that "You get value from patents in two ways; through fees, and through licensing negotiations that give IBM access to other patents. Access is far more valuable to IBM than the fees it receives from its 9,000 active (U.S.) patents. There is no direct calculation of this value, but it is many times larger than the fee income, perhaps an order of magnitude larger."[13]

IBM's cross-licensing activity continues today. But, the company has complemented this essentially defensive policy with a strategy to generate royalty income from its licenses. IBM initiated this more active approach to licensing in 1988, when it increased the royalty rates sought on its patents from 1 percent of sales revenue (on products using IBM patents) to a range of 1–5 percent.[14] The company has also adopted a proactive strategy for identifying potential patent infringement and negotiating royalty-yielding licenses with them. Cash revenues earned from IBM's patent and technology licensing agreements increased from $345 million in 1993 to $640 million in 1994, and were well over $1 billion per year by the end of the decade.[15] It is important to bear in mind that these revenues carry low incremental costs, and accrue in large measure to the company's bottom line. In terms of their profit impact, these licensing operations are equivalent to a multi-billion-dollar business for IBM. IBM is one of the world's leading innovators, with more U.S. patents granted to it than any other company in every year since 1993. The company's licensing strategy has enabled IBM to appropriate some of the returns to its inventions by essentially charging users for access to these technologies.

LICENSING PRACTICES AT TEXAS INSTRUMENTS

Like other parts of the electronics industry, the semiconductor industry is characterized by widespread use of cross-licensing.[16] The licensing procedures at Texas Instruments (TI) illustrate the ways in which cross-licensing is used in the modern electronics industry. TI has two main licensing objectives. The first and primary objective is to ensure freedom to operate in broad areas of technology, without running the risk of patent infringement litigation by other firms in given product markets. Thus the first strategic goal is fundamentally a defensive one. The second objective is to obtain value from the firm's IP, in the form of its patent portfolio, by generating royalty income. The purpose and result of royalty generation through cross-licensing agreements is "competitive re-balancing," which offsets the advantage for imitators who might otherwise free-ride on technology TI developed.

Establishing "freedom to operate" is vital in the semiconductor industry, with its rapid innovation, short product life cycles, and ubiquity of patents. At the start of an R&D program, possible patent infringements cannot be easily predicted, as firms are quite ignorant of the R&D and patenting plans of competitors. Yet when it invests in R&D and product development, TI needs to be confident that patents developed by others through independent R&D efforts will not hinder commercialization of its technology. This need is heightened by the significant investments TI makes in capital-intensive semiconductor manufacturing facilities. TI has responded to this challenge by building a robust portfolio of semiconductor patents, which it essentially uses to defend against other's patents and to negotiate preemptive cross-licenses. It has also divested from DRAM (**dynamic random access memory**) fabrication in part to minimize its potential infringement of others' patents, and to enhance the royalty generation capacities of its IP strategy.

There are two main models for cross-licensing agreements in the semiconductor industry: "capture" and "fixed period." In the "capture" model discussed earlier, the licensee retains "survivorship" rights to use the patents until they expire, sometimes up to 20 years later. In the "fixed period" model the licensee has similar rights to use patents existing or applied for during the license period, but with no survivorship rights once the license period has expired. Full renegotiation of the cross-license is required for succeeding periods. Texas Instruments (TI) has been a leader in the use of fixed period licensing, which has gained in popularity in the industry. The fixed period model allows more flexible commercialization of patent portfolios, since licensing terms can be periodically adjusted to account for changes over time. For example, it mitigates the possibility of being locked into a cross-license even though the licensing partner has stopped contributing new inventions to the focal field of use. Thus, fixed period licensing allows TI to carefully calibrate the "openness" of its cross-licensing policy to avoid potential abuse by noninventors. Like IBM, TI has also been successful in generating royalties for access to its pioneering patents through its licensing efforts and enforcement actions. In the latter half of the 1980s, when the company was facing stiff competition in the DRAM market, licensing royalties sometimes exceeded the net profits of the company. Absent these revenues, TI would not have been profitable.

Licensing arrangements, including the cross-licensing of patents, may not always be easy to achieve in multi-invention contexts. Reliance on such licensing assumes a willingness to license on the part of others, an assumption that will be strongly challenged if some patent owners have somewhat different strategic goals or a different appreciation of the value of their own technology. In these cases, entrepreneurs can sometimes overcome the barriers in the licensing market by using the markets for corporate control to acquire technologies and patents. Firms like Cisco Systems have championed this approach, typically acquiring smaller innovative firms with technologies (and patents) that the company needs. Similarly, in the agricultural biotechnology industry, a number of firms have consolidated to bring together germplasm, genomic, and plant variety patents relating to specific crops.

From the perspective of a small entrepreneurial firm, setting oneself up as an acquisition target would be an appropriate strategy when organizational costs dictate an integrated mode, but there are also significant barriers to developing complementary technologies in-house and to obtaining access to the necessary patents. A strong set of blocking patents, which has the potential for generating patent exclusivity for the merged firm, while simultaneously dissuading potential merger partners from a go-it-alone strategy, would increase the attractiveness of the entrepreneurial firm in the market for corporate control. Even among larger firms, entrepreneurial opportunities exist to agglomerate businesses across firms so as to pool technologies and patents, and overcome transaction costs in patent licensing. For example, in 1998, when a long-running patent dispute between Digital and Intel (relating to Digital's Alpha processor patents) had reached a stalemate, Intel was able to break the impasse by simply buying out Digital's semiconductor business. Perhaps one of the earliest examples of entrepreneurship to consolidate inventions into a single firm, in this case spurred by a major customer (the U.S. Navy), was the creation of RCA for the development of radio in the first quarter of the twentieth century.[17]

THE FORMATION OF RCA

Early developments in wireless radio epitomize the complexities surrounding intellectual property arrangements that may be encountered with systems innovation (or multi-invention) technologies. The commercialization of radio required a number of basic inventions. The scientific basis for wireless was developed by university scientists such as Maxwell, Hertz, and Lodge in the nineteenth century. Their discoveries were first applied to practical communication with the development of wireless telegraphy by Marconi in Britain in 1896. The first speech transmissions were made in the United States by Fessenden in 1900, using a high-frequency alternator. Further basic innovations were made over the next two decades.[18]

Many of these inventions were initially developed by individuals working independently of each other. Indeed, many carry the name of the inventor, such as Poulsen arc, the Fleming valve, and the de Forest triode.[19] As the potential for radio became apparent, and the need for large-scale R&D and investment grew, large corporations entered the field. The pace of development accelerated, and the number of patents multiplied. The companies involved included Marconi, General Electric (GE), Westinghouse, AT&T, Telefunken, and others. In addition to their considerable R&D effort, these corporations also acquired key patents (Archer 1938, 135; Maclaurin 1949, 106).[20] There was considerable competition, and with research teams in different companies working in parallel, patent interferences were common (Maclaurin 1949, 97).[21] By 1918, it was apparent that several technologies were needed to manufacture radio systems, and each of these technologies itself involved multiple patents from different firms. In the words of Armstrong, one of the pioneers of radio, "It was absolutely impossible to manufacture any kind of workable apparatus without using practically all of the inventions that were then known" (Federal Trade Commission 1923; Maclaurin 1949, 99).

The result was deadlock. A number of firms had important patent positions and could block each other's access to key components. They refused to cross-license. This held up the development of the industry (Archer 1938, 113–14; Douglas 1987, 8, 12; Maclaurin 1949, 77).[22] The situation arose in large part as a result of the way radio had developed.

Key patent portfolios had been developed by different individuals and corporations, who were often adamant about refusing to cross-license competitors. In addition, in a new industry in which large-scale patent overlaps were a novel problem, there was no well-developed means of coordinating licensing agreements between these groups.

The situation was resolved in the United States only when, under prompting by the U.S. Navy, the various pioneers formed the Radio Corporation of America (RCA) in 1919 (Archer 1938, 176–89; Maclaurin 1949, 105).[23] This broke a key source of the deadlock. RCA acquired the U.S. rights to the Marconi patents, and the other major U.S. patent holders became shareholders in RCA.[24] In this way, RCA acquired the U.S. rights to all the constituent radio patents under one roof—amounting to over 2,000 patents (Archer 1938, 195; Maclaurin 1949, 107).[25] It established RCA as the technical leader in radio, but also granted cross-licenses to the other firms to continue their own development of the technology for use in other fields or as suppliers to RCA.[26]

The RCA example highlights the perils to the economy when patent owners pursue exclusivity too vigorously in multi-invention contexts. Without the willingness to allow others access to one's own patents, there is virtually no prospect for reciprocal access. Because of the high transactions costs reflected in this reluctance to cross-license, technological progress and the further commercialization of radio was halted. In this case, the debacle was resolved only by the formation of RCA. However, it is now clear that the same ends—namely design freedom—may often be achieved more simply, without such fundamental reorganization, by cross-licensing alone. The wireless patent-licensing deadlock and the formation of RCA helped set the stage for further development of cross-licensing in electronics.

5.2. Nonintegrated Modes

Nonintegrated modes use licensing and component markets to supply intellectual property bundled-in with their technologies or component products. In this way, specialized firms can avoid the need to develop all the complementary technologies in-house. The use of such nonintegrated modes or "business models" usually occurs when the transaction costs in the associated (licensing or component) market are not especially high. One fundamental challenge for entrepreneurs is to recognize opportunities for commercializing inventions in this fashion, instead of resorting to integrated commercialization by default.

It is important to understand that when technologies are licensed or components are sold, there is definite risk that knowledge may leak out and the transaction partner may learn too much about the firm's technology. Attempts to barricade the company's know-how may fail because successful commercialization may require joint problem solving and the exchange of technical information. This is a very serious problem because uniqueness of the company's technology is often the primary business proposition of enterprises employing nonintegrated modes of innovation. Patents can play a vital role by limiting misappropriation of the firm's technology, and facilitating transactions between the entrepreneurial firm and its business partners. Naturally, enterprises relying on patent protection to sustain a licensing or component-product business model would take a more proprietary view in their IP strategy. We illustrate the type of entrepreneurship entailed in nonintegrated modes by reviewing two case studies, which highlight the differences in strategy not only between integrated and nonintegrated modes, but also among nonintegrated modes.

SYSTEM ON A CHIP (SOC) AND ARM, LTD.

In semiconductors, relentless miniaturization has made it possible to put entire electronic systems on a single semiconductor chip (Teece 1998; Linden and Somaya 2003). Market demand for the advantages in size, power consumption, and production cost that such systems-on-a-chip (SOCs) promise has also been growing rapidly. Previously, it was common for semiconductor technologies to be transferred between firms through the sale of various component integrated circuits (ICs), where the technology itself was transferred in "embodied" form. However, when entire electronic systems needed to be put on a single chip or IC, it became practically impossible to conduct transactions in technology by conducting transactions in components.

One solution to this problem in the component market was pioneered by new "chip-less" firms, who went about creating a licensing market for design modules that other firms can license and integrate into their own system-on-a-chip designs. ARM, Ltd., based in Cambridge in the United Kingdom, is one of the leading firms that adopted this approach. ARM's RISC (reduced instruction set computer) processor designs are used in literally tens of millions of cell phones and handheld devices sold around the world. ARM is a **spin-off** from the Power-PC consortium assembled by Apple Computer, IBM, and Motorola to design and manufacture microprocessors. The company resisted the temptation to be acquired by a large semiconductor firm, or to expand the scope of its own technological domain. Instead, it specialized in developing processor designs, which it then sought to license to other firms. As the SOC revolution took root, ARM's licensing-based strategy turned out be extremely successful.

ARM appears to have taken patent protection of its technologies very seriously and accumulated a portfolio of over 80 patents by 2000. When picoTurbo, a rival firm based in Milpitas, California, came up with a product that could essentially run any software written for ARM's processor, the company immediately filed suit.[27] PicoTurbo's technology threatened the exclusive position that ARM had built up in many ways. First and foremost, it threatened to invade the installed base of complementary ARM-related software and software programmers, and thus in effect invent around ARM's technology. Given the importance of an exclusive position for ARM's commercial success, it is no surprise that the company enforced its patents so aggressively. Eventually, the suit was settled with ARM simply acquiring picoTurbo with all its product designs and IP assets.

Dolby, which licenses its noise reduction designs for high-fidelity sound systems, and Rambus from the semiconductor industry are other prominent examples of firms that have pursued a licensing mode to commercialize their inventions. In both these cases, well-developed patent strategies are important to the viability of the licensing option, and indeed their business model more generally.

COMPONENTIZATION AT KENTRON TECHNOLOGIES

Sometimes licensing and component sales are transparent alternatives, in the sense that after pursing one strategy, the entrepreneur changes gears and adopts another. Kentron Technologies, a semiconductor firm, developed a technology to (effectively) double the bandwidth of DRAM modules by interleaving signals from two slower "single data rate" DRAMs, rather than using the patented "double data rate" technology developed by Rambus.[28] Kentron originally offered to license its technology for a 5 percent royalty, but did not get any takers. Kentron subsequently changed its patent strategy to offer "royalty free" licenses if DRAM users would buy special switches from Kentron that enable the interleaving to occur.[29]

Put differently, Kentron took the payment for the use of its technology and patents in the form of a premium price for the switches. Robert Goodman, Kentron's CEO, indicated that (1) the price of the special switch was set at a level that yielded Kentron the same revenue as a 5 percent royalty would have yielded, (2) users resisted taking a license from Kentron.[30] However, they were willing to pay Kentron for the use of its technology. Quite simply, the users may have perceived the transaction costs in the licensing market to be too high for Kentron's technology. The value of the technology may have seemed uncertain absent a physical component that could be evaluated, and potential licensees may have discounted its importance. There may have also been concerns about how well this technology would work with their own DRAM designs, in other words, about *technological interconnectedness*. Firms may also have wanted to avoid the monitoring and compliance costs associated with a license (tracking infringing sales, and calculating and paying royalties), and preferred to have the price of Kentron's technology built into the price of the physical switches instead. Because Kentron had a strong patent position built up around its pioneering technology, DRAM manufacturers could not simply appropriate its technology for internal use, nor could they effectively invent around Kentron's patent position.

The ARM and Kentron examples show how licensing a technology or selling components in which the technology is embedded are alternative strategies for commercialization. One of the functions of the entrepreneur is to recognize which organizational approach is most appropriate for generating value from an innovative idea, and to implement this organizational strategy. A second important function of the entrepreneur is to ensure that the firm captures value from its innovation, and does not simply dissipate it to other firms. In part, this implies designing the appropriate patent strategies for a given organizational mode.

6. Conclusion

In this chapter, we have described some of the challenges presented to entrepreneurs by multi-invention contexts, and analyzed implications for entrepreneurship and patent strategy in these contexts. Our analysis provides two main insights. First, to maximize chances of success, entrepreneurs must assess the relative organizational costs and benefits of different organizational modes or business "models," and commercialize inventions by using the most effective mode in a given multi-invention context. Second, patent strategies must be chosen to complement the choice of business model.

Entrepreneurship is critical for effectuating new combinations of assets, resources, and technologies in multi-invention contexts. Often incumbent firms face significant inertia to change their existing modes of behavior and organization. In many cases, they are unable to perceive the potential value that can be generated by developing a new invention or combining existing inventions in a new way. However, entrepreneurship is also likely to fail if the business models adopted are too cumbersome or flawed. In addition, it is not enough to unlock the potential value hidden in multi-invention or systemic contexts. It is equally important to devise strategies that appropriate some of this value to compensate for the entrepreneur's efforts. Appropriate patent strategy can play a useful role in capturing this value.

In response to transactional problems in multi-invention contexts, the instinctive public policy remedy sought by some is often a weakening of patent rights. While this solution may address certain transaction cost problems, it also means a weakening of incentives for innovation and the reinforcement of integrated modes of innovation and production. Licensing and component modes, on the other hand, benefit from patents because of stronger incentives, and in some cases, because of lower transaction costs. Evaluation of these alternative organizational forms and business models is essential in any attempt to address concerns about patents in multi-invention contexts. Often, astute management and entrepreneurial efforts are sufficient to allow technology and component markets to work effectively. Thus entrepreneurship and patent rights may have a symbiotic relationship that can be undermined by policy responses that do not account for all possible organizational arrangements through which innovation can take place.

Notes

1. For a discussion of systemic innovation, see Teece 2000.
2. By untested, we mean that the validity of the patent has not yet been tested in court.
3. The treatment of entrepreneurship here is sympathetic to the work of Hayek (1945), Casson (1982), Kirzner (1973), as well as Schumpeter himself.
4. IBM 2001 annual report.
5. This could be due to the user reading the patent, or it could be due to the user's independent invention of the patent, or hearing about it from some source.
6. Cross-licensing is not the same as "patent pooling," in which member firms contribute patents to a common pool and each member accesses them on the same terms and conditions. In cross-licensing, firms agree one-on-one to license their IP to each other and retain control over their proprietary technology, which is used for competitive advantage via product manufacturing and further licensing.
7. Note that the situation is different in other industries not characterized by cumulative systems technologies, such as chemicals and pharmaceuticals, where cross-licensing or, rather, reciprocal licensing, is typically aimed at exchanging technology rather than avoiding patent conflicts. In chemicals and pharmaceuticals, although patenting is extensive, individual technology development paths are less likely to overlap, and cross-licensing may be used to ensure broad product lines. For licensing strategy in the chemicals industry, see Grindley and Nickerson 1996, 97–120.
8. Historical perspective on competition in the telecommunications industry is given in Irwin 1977, 312–33; Brock 1981; OTA 1985; Noll and Owen 1989; and Rosston and Teece 1997.

9. The two substantive provisions of the 1956 consent decree were that (*a*) it confined AT&T to providing regulated telecommunications services, and its manufacturing subsidiary Western Electric to making equipment for those services (effectively prohibiting it from selling semiconductors in the commercial market), and (*b*) all patents controlled by the Bell System should be licensed to others on request. Licenses for the 8,600 patents included in existing cross-licensing agreements were royalty-free to new applicants, and licenses to all other existing or future patents were to be issued at a nondiscriminatory "reasonable royalty" (determined by the court if necessary). AT&T was also to provide technical information along with the patent licenses for reasonable fees. Licenses were unrestricted, other than being nontransferable (*USA v. Western Electric Co. Inc., and AT&T*, Civil Action, 17–49, Final Judgment, January 24, 1956; Brock 1981, 166, 191–94; Levin 1982, 9–101).

10. "We realized that if [the transistor] was as big as we thought, we couldn't keep it to ourselves and we couldn't make all the technical contributions. It was to our interest to spread it around" (AT&T executive, quoted in Levin 1982, 77, after Tilton 1971). The strategy appears to have been prescient. In the United States, during 1953–68, 5,128 semiconductor patents were awarded. Bell Laboratories was granted only 16 percent of these; the next five firms were RCA, General Electric, Westinghouse, IBM, and Texas Instruments, all AT&T cross-licensees (Tilton 1971).

11. The provision covered all existing patents at the time of the decree (i.e., as of 1956) plus any that were filed during the next five years. The rights lasted for the full term of the patents. If the parties could not agree on a reasonable royalty rate, the court could impose one. Patent rights could be very long lived, since, at that time, patent life was 17 years from the grant date, which might be some years after the filing date. The patent-licensing provisions ended in 1961. The decree also included other provisions related to the sale of IBM products and services (*USA v. International Business Machines Corporation*, CCH 1956 Trade Cases par. 68, 245, SDNY 1956).

12. Jim McGrody, IBM VP and director of research, in Boyer 1990, 10–11.

13. Roger Smith, IBM assistant general counsel, in Boyer 1990.

14. *Computerworld*, April 11, 1988, 105.

15. IBM Annual Report, 1993, 1994, 2000, 2001.

16. This section is based in part on discussion with Texas Instruments executives. However, the views expressed here are those of the authors, and should not be seen as reflecting those of Texas Instruments.

17. The early history of radio is described in Archer 1938; Jewkes, Sawers, and Stillerman 1969; Douglas 1987; and Merges and Nelson 1990, 891–96.

18. These included the high-frequency alternator, high-frequency transmission arc, magnetic amplifier, selective tuning, crystal detector, heterocyne signal detection, diode valve, triode valve, high vacuum tube, and directional aerials.

19. Not all early inventors were independent. E.F.W. Alexanderson—who improved the Fessenden alternator, invented a magnetic amplifier, electronic amplifier, and multiple tuned antenna, and co-invented the "Alexanderson-Beverage static eliminator"—was a General Electric employee.

20. AT&T acquired the de Forest triode and feedback patents in 1913–14 for $90,000, and his remaining feedback patents in 1917 for $250,000; Westinghouse cross-licensed the Fessenden heterodyne interests in 1920, and acquired the Armstrong super heterodyne patents in 1920 for $335,000 (Archer 1938, 135; Maclaurin 1949, 106).

21. The fact that GE and AT&T alone were each devoting major research attention to the vacuum tube led to no less than 20 important patent interferences in this area (Maclaurin 1949, 97).

22. To cite one important example, Marconi and de Forest both had critical valuable patents. Marconi's diode patent was held to dominate de Forest's triode patent. Both technologies were vital to radio, yet the interests refused to cross-license (Archer 1938, 113–14; Douglas 1987, 12). The application of the triode (audion) to feedback amplification was also the subject of a long-running patent priority dispute between de Forest and Armstrong (finally resolved in de Forest's favor by the Supreme Court in 1934). Its use in transmission oscillation was the subject of four-way patent interference between Langmuir, Meissner, Armstrong, and de Forest (Maclaurin 1949, 77). These problems held up the use of the triode—a crucial component of signal transmission. Detection and amplification, which has been called "the heart and soul of radio" (Douglas 1987, 8) and "so outstanding in its consequences it almost ranks with the greatest inventions of all time" (Nobel Prize physicist Rabi, quoted in Maclaurin 1949, 70).

23. RCA was formed by GE in 1919, and simultaneously acquired the American Marconi Corporation. Major shareholders included GE, AT&T (1920), and Westinghouse (1920) (Archer 1938, 176–89; Maclaurin 1949, 105). A major concern of the U.S. Navy was that international wireless communications were dominated by the British firm Marconi, and the patent impasse helped perpetuate this. The Navy favored the establishment of an "All American" company in international communications. A similar concern on the eve of U.S. entry into World War I prompted the U.S. government (the secretaries of war and the navy) to also intervene in the case of aircraft patents, and create a patent pooling arrangement in January 1917 (Bittlingmayer 1988).

24. As part of its role in the formation of RCA, the U.S. Navy also initiated cross-licensing to resolve the patent situation in radio manufacture. It wished to have clear rights to use the radio equipment it purchased, without risking litigation due to complex patent ownership—noting in 1919 that "there was not a single company among those making radio sets for the Navy which possessed basic patents sufficient to enable them to supply, without infringement, . . . a complete transmitter or receiver." A formal letter suggesting "some agreement between the several holders of permanent patents whereby the market can be freely supplied with [vacuum] tubes," sent from the navy to GE and AT&T in January 1920, may be seen as an initiating point for cross-licensing in the industry (Archer 1938, 180–86; Maclaurin 1949, 99–110).

25. RCA concluded cross-license agreements with firms including GE, Westinghouse, AT&T, United Fruit Company, Wireless Specialty Apparatus Company, Marconi (Britain), CCTF (France), and Telefunken (Germany) (Archer 1938, 195; Maclaurin 1949, 107).

26. A distinction was that the RCA cross-licenses typically granted (reciprocal) exclusive rights to use the patents in given territories or markets; compared with the nonexclusive cross-licenses that became the norm later. The cross-license with GE (and later Westinghouse) included provisions for the supply of components to RCA. The RCA cross-licenses were for very long terms—many for 25 years, from 1919 to 1945. They covered current and future patents. Other radio manufacturers took licenses with RCA, starting in the late 1920s. Some of RCA's cross-licensing policies were later questioned on antitrust grounds, and modified following a consent decree in 1932 (Archer 1938, 381–87; and Maclaurin 1949, 107–9, 132–52).

27. See online: http://www.reed-electronics.com/electronicnews/article/CA186719.html (accessed October 27, 2005).

28. One of the authors (Teece) is familiar with the Kentron example from his work as an expert in *In the Matter of Rambus, Inc.*, FTC Docket No. 9302. In this paper, we are relying only on the public testimony given by Mr. Robert Goodman, Kentron's CEO, and not on any information that is subject to the protective order in that case. See *In the Matter of Rambus, Inc.*, June 19, 2003, 6020–29, 6041, 6078–87.

29. As a news story indicates: "Kentron makes the special QBM module switches, which is [*sic*] used in its modules and those of licensees. Bob Goodman, Kentron's CEO, said the firm licenses its QBM technology on a royalty-free basis, and derives revenues from the sale of its QBM switches." See http://siliconstrategies.com/article/printableArticle.jhtmlarticle4ID=10806590 (accessed September 16, 2003).

30. At the public hearing *In the Matter of Rambus, Inc.*, FTC Docket No. 9302.

References

Archer, G. 1938. *History of Radio to 1926*. New York: American Historical Society.

Arora, A. 1995. "Licensing Tacit Knowledge: Intellectual Property Rights and the Market for Know-How." *Economics of Innovation and New Technology* 4: 41–59.

Arrow, K. J. 1971. *Essays in the Theory of Risk-Bearing*. Chicago: Markham.

Baumol, W. 1993. "Formal Entrepreneurship Theory in Economics: Existence and Bounds." *Journal of Business Venturing* 8: 197–210.

Bittlingmayer, G. 1988. "Property Rights, Progress, and the Aircraft Patent Agreement." *Journal of Law and Economics* 31 (1): 227–48.

Boyer, C. 1990. "The Power of the Patent Portfolio." *Think* 5: 10–11.

Brock, G. 1981. *The Telecommunications Industry: The Dynamics of Market Structure*. Cambridge: Harvard University Press.

Casson, M. 1982. *The Entrepreneur*. Totawa, N.J.: Barnes and Noble.

Chang, H. F. 1995. "Patent Scope, Antitrust Policy, and Cumulative Innovation." *Rand Journal of Economics* 26 (1): 34–57.

Douglas, G. 1987. *The Early Days of Radio Broadcasting*. Jefferson, N.C.: McFarland.

Federal Trade Commission. 1923. *The Radio Industry*. Washington, D.C.: Government Printing Office.

Green, J. R., and S. Scotchmer. 1995. "On the Division of Profit in Sequential Innovation." *Rand Journal of Economics* 26 (1): 20–33.

Grindley, P. C., and J. Nickerson. 1996. "Strategic Objectives Supported by Licensing." In *Technology Licensing: Corporate Strategies for Maximizing Value*, ed. R. L. Parr and P. H. Sullivan. New York: John Wiley and Sons.

Grindley, P. C., and D. J. Teece. 1997. "Managing Intellectual Capital: Licensing and Cross-Licensing in Semiconductors and Electronics." *California Management Review* 39 (2): 8–41.

Hall, B. H., and R. M. Ziedonis. 2001. "The Patent Paradox Revisited: An Empirical Study of Patenting in the Semiconductor Industry, 1979–1995." *Rand Journal of Economics* 32 (1): 101–28.

Harris, R. 1990. "Divestiture and Regulatory Policies." *Telecommunications Policy* 14: 105–24.

Hayek, F. A. 1945. "The Use of Knowledge in Society." *American Economic Review* 35 (4): 519–30.

Heller, M. A., and R. S. Eisenberg. 1998. "Can Patents Deter Innovation? The Anticommons in Biomedical Research." *Science* 280 (5364): 698–701.

Irwin, M. 1977. "The Telephone Industry." In *The Structure of American Industry*, ed. W. Adams. 5th ed. New York: Macmillan.

Jewkes, J., D. Sawers, and R. Stillerman. 1969. *The Sources of Innovation*. New York: Norton.

Kefauver, W. 1993. "Intellectual Property Rights and Competitive Strategy: An International Telecommunications Firm." In *Global Dimensions of Intellectual Property Rights in Science and Technology*, ed. M. Wallerstein, M. E. Mogee, and R. Schoen. Washington, D.C.: National Academy Press.

Kitch, E. W. 1977. "The Nature and Function of the Patent System." *Journal of Law and Economics* 20: 265–90.

Kirzner, I. 1973. *Competition and Entrepreneurship*. Chicago: University of Chicago Press.

Levin, R. 1982. "The Semiconductor Industry." In *Government and Technical Progress*, ed. R. Nelson. New York: Pergamon.

Linden, G., and D. Somaya. 2003. "System-on-a-Chip Integration in the Semiconductor Industry: Industry Structure and Firm Strategies." *Industrial and Corporate Change* 12 (3): 545–76.

Lippman, S. A., and R. P. Rumelt. 2003. "A Bargaining Perspective on Resource Advantage." *Strategic Management Journal* 24: 1069–86.

Maclaurin, W. 1949. *Invention and Innovation in the Radio Industry*. New York: Macmillan.

Mazzoleni, R., and R. R. Nelson. 1998. "The Benefits and Costs of Strong Patent Protection: A Contribution to the Current Debate." *Research Policy* 27 (3): 273–84.

Merges, R. P., and R. R. Nelson. 1990. "On the Complex Economics of Patent Scope." *Columbia Law Journal* 90 (4): 839–916.

———. 1994. "On Limiting or Encouraging Rivalry in Technical Progress: The Effect of Patent Scope Decisions." *Journal of Economic Behavior and Organization* 25 (1): 1–24.

Office of Technology Assessment (OTA). 1985. *Information Technology Research and Development: Critical Trends and Issues*. New York: Pergamon.
Oxley, J. E. 1999. "Institutional Environment and the Mechanisms of Governance: The Impact of Intellectual Property Protection on the Structure of Interfirm Alliances." *Journal of Economic Behavior and Organization* 38 (3): 283–309.
Noll, M. 1987. "Bell System R&D Activities: The Impact of Divestiture." *Telecommunications Policy* 11: 161–78.
Noll, R., and B. Owen. 1989. "The Anticompetitive Uses of Regulation: United States v. AT&T." In *The Antitrust Revolution*, ed. J. Kwoka and L. White. New York: Macmillan.
Rivette, K. G., and D. Kline. 2000. "Discovering New Value in Intellectual Property." *Harvard Business Review*, January–February, 2–12.
Rosenberg, N. 1994. *Exploring the Black Box*. Cambridge: Cambridge University Press.
Rosston, G. L., and D. J. Teece. 1997. "Competition and 'Local' Communications: Innovation, Entry, and Integration." In *Globalism and Localism* in Telecommunications, ed. E. M. Noam and A. J. Wolfson. North Holland: Elsevier.
Rumelt, R. P. 1984. "Towards a Strategic Theory of the Firm." In *Competitive Strategic Management*, ed. R. B. Lamb. Englewood Cliffs, N.J.: Prentice-Hall.
Schumpeter, J. 1934. *Capitalism, Socialism, and Democracy*. New York: Harper and Row.
Scotchmer, S. 1991. "Standing on the Shoulders of Giants: Cumulative Research and the Patent Law." *Journal of Economic Perspectives* 3: 29–41.
———. 1996. "Protecting Early Innovators: Should Second-Generation Products be Patentable?" *Rand Journal of Economics* 27 (2): 322–31.
Shelanski, H., and P. G. Klein. 1995. "Empirical Research in Transaction Cost Economics: A Review and Assessment." *Journal of Law, Economics, and Organization* 11 (2): 335–61.
Sherry, E., and D. J. Teece. 2003. "Standards Setting and Antitrust." *Minnesota Law Review* 87 (6): 1913–94.
Smith, R. 1989. "Management of a Corporate Intellectual Property Law Department." *AIPLA Bulletin*, April–June, 817–23.
Somaya, D. 2003. "Strategic Determinants of Decisions Not to Settle Patent Litigation." *Strategic Management Journal* 24: 17–38.
———. 2005. "Combining Inventions in Multi-invention Products: Patents, Organizational Alternatives, and Public Policy." Working paper, University of Maryland.
Teece, D. J. 1982. "Towards an Economic Theory of the Multiproduct Firm." *Journal of Economic Behavior and Organization* 3 (March): 39–64.
———. 1986. "Profiting from Technological Innovation: Implications for Integration, Collaboration, Licensing, and Public Policy." *Research Policy* 15 (6): 285–305.
———. 1996. "Firm Organization, Industrial Structure, and Technological Innovation." *Journal of Economic Behavior and Organization* 31 (2): 193–224.
———. 1998. "Capturing Value from Knowledge Assets: The New Economy, Markets for Know-how, and Intangible Assets." *California Management Review* 40 (3): 55–79.

———. 2000. *Managing Intellectual Capital*. New York: Oxford University Press.
———. 2003a. "Explicating Dynamic Capabilities: Asset Selection, Cospecialization, and Entrepreneurship in Strategic Management Theory." Working paper.
———. 2003b. *Essays in Technology Management and Policy*. Hackensack, N.J.: World Scientific Publishing.
Tilton, J. 1971. *International Diffusion of Technology: The Case of Semiconductors*. Washington, D.C.: Brookings Institution.
Williamson, O. E. 1985. *The Economic Institutions of Capitalism*. New York: Free Press.
———. 1996. *Mechanisms of Governance*. New York: Free Press.

Available online at www.sciencedirect.com
SCIENCE DIRECT
ELSEVIER
Research Policy 33 (2004) 179–191
research policy
www.elsevier.com/locate/econbase

Royalties, evolving patent rights, and the value of innovation

Edward F. Sherry[1], David J. Teece*,[2]

Institute of Management, Innovation and Organization, Haas School of Business, University of California, 1930 554 Barrows Hall, Berkeley, CA 94720-1930, USA

Received 2 September 2002; received in revised form 17 April 2003; accepted 12 May 2003

Abstract

The value of an innovation to the innovator can change over time, especially in response to changes in the legal protection (such as patent rights) afforded the innovator. A proven-valid-and-infringed patent is a more valuable economic commodity than is an untested patent. The increase in value can be estimated using the success rate of patent lawsuits. Using a database of the outcomes of U.S. patent litigation, we find that plaintiffs win patent litigation some 45% of the time at the trial court level. This has implications both for patent damages awards and for the incentives to innovate.

Keywords: Intellectual property; Patent rights; Value of innovation

1. Introduction: value and the stages of intellectual property (IP)

It is widely acknowledged that the value of an innovation can change (often dramatically) over time, for a number of reasons. Some are technological. New technology may come along to supplant the old, rendering the old partially or possibly totally obsolete and hence less valuable. Conversely, complementary technology may be developed which enables an old innovation which previously could not be implemented to now be commercialized, rendering the old innovation more valuable than it previously had been. Other reasons are commercial. The market for the services of the product in which the innovation is embedded may simply disappear because of a change in consumer taste. Or the complementary assets used to commercialize the innovation could become more (or less) competitive in supply, raising (or lowering) the value of the innovation in the process. Moreover, innovators differ in their ability to commercialize their innovations, and the value that the innovator can obtain from commercialization depends not only on the appropriability regime but also on the commercialization strategy the innovator chooses (Teece, 1986).

But discussions of innovation and its value often fail to differentiate between two very different (albeit related) concepts. The first is the innovation itself (say, a technological breakthrough). The second are the intellectual property rights (patents, copyrights, trademarks and/or trade secrets)[3] associated with that innovation. The two are not synonymous. Yet in

* Corresponding author. Tel.: +1-510-642-4041.
E-mail address: murphy@hass.berkeley.edu (D.J. Teece).
[1] Edward F. Sherry is Senior Managing Economist, LECG, LLC and a member of the California Bar.
[2] David J. Teece is Mitsubishi Bank Professor, Haas School of Business, University of California (Berkeley) and Chairman of LECG, LLC.

[3] For simplicity and concreteness, the remaining discussion will focus on patentable innovations.

0048-7333/$ – see front matter © 2003 Elsevier B.V. All rights reserved.
doi:10.1016/S0048-7333(03)00088-X
With permission from Elsevier.

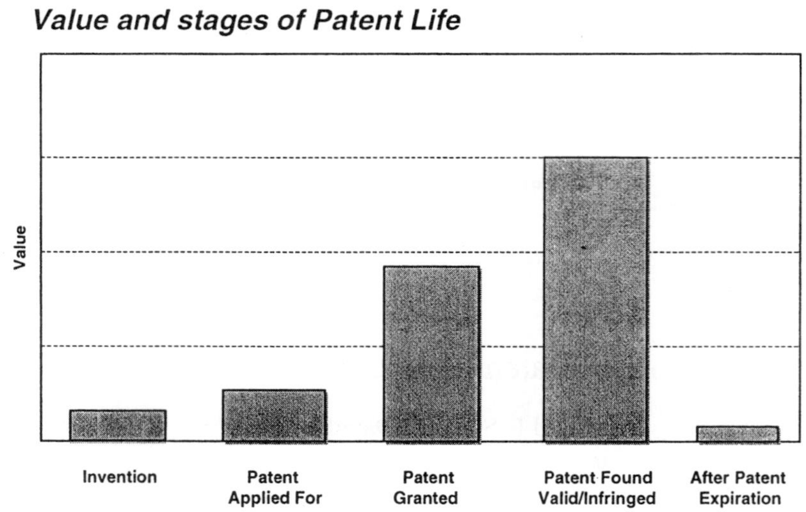

Fig. 1. Value and stages of patent life.

our experience, many analysts and some managers fail to properly differentiate between the two, and the resulting confused thinking often leads to poor managerial decisions and/or inappropriate policy recommendations.[4] In particular, it is all too common to fall into the trap of talking about "the value of the innovation", when what is meant is the value of the intellectual property rights associated with the innovation, or possibly the value of the complementary assets.

Consider the various stages in the evolution of an invention, as shown in Fig. 1. Suppose an inventor comes up with an idea for an invention. That idea may have some economic value. But it may not be patentable. Research scientists, patent lawyers, and venture capitalists are all well aware of the difference between coming up with an innovation and obtaining a patent on that innovation. Few innovations are patentable; even fewer are patented.[5]

The inventor searches the prior art and determines whether or not the invention may be patentable. Suppose the conclusion is that it is worth filing a patent application. Because the idea passes the initial prior art screening, and because filing an application involves incurring additional cost, the patent application incorporating the invention is likely to be more valuable than the invention alone.[6]

If and when the Patent Office grants the patent, the value of "the invention" is now significantly greater.

[4] Those responsible for applying for patents, or for in- or out-licensing intellectual property, are likely to be intimately familiar with the distinction between innovations and intellectual property rights associated with those innovations. But this awareness need not extend to other individuals less involved in the process. And in our experience, an appreciable number of academic studies of innovation and the innovation process fail to appreciate the significance of the distinction.

[5] It is true that, in recent years, the Patent and Trademark Office in the US has granted the large majority of patent applications. But not all innovations result in a patent application. A recent estimate is that manufacturing firms seek patent protection on only 49% of their product innovations and 31% of their process innovations. The propensity to patent varies widely across industries (see Table A.1 of Cohen et al., 2000).

[6] Filing a patent application also "tolls" the date for determining what can count as "prior art" against which the patentability of the invention is measured. As such, an early filing date makes it less likely that subsequent material will count as "prior art", and thus increases the likelihood that the patent will ultimately issue.

The Patent Office has, in essence, put its "seal of approval" on the idea that the invention is worthy of patent protection. It has granted the inventor the right to exclude others from using the now-patented invention for a period of years.

This legal transformation, from a patent application to an issued patent, has nothing to do with "the invention" in a technological sense. But it can have extremely high economic value to the (new) patent holder.[7] For example, in late October 1989, the Japanese Patent Office granted to Texas Instruments (TI) a patent, known as the Kilby patent (after the inventor), some 30 years after TI had applied for the patent. The Kilby patent is widely acknowledged[8] to be a fundamental patent covering virtually all integrated circuits. When the news was released, TI's stock price jumped 15%, reflecting a US$ 387 million increase in TI's market capitalization,[9] an extremely large amount given that the newly issued patent applied only in Japan (its US counterpart had already expired).[10]

But this does not end the inquiry. Under the law, an issued patent is presumed in many jurisdictions (including the US) to be valid. But both validity and infringement can be disputed. If a patent is found invalid, then the patent holder no longer can exclude others from using the invention, and its private value falls to zero. Less far-reaching rulings, such as a judicial interpretation of the patent claims that narrows the effective scope of the patent, or a finding that a particular firm does not infringe the patent, can likewise have significant effects on the value of the patent.

This again is demonstrated by real-world events. When a Japanese trial court ruled in August 1994 against TI's claim that Fujitsu had infringed the Kilby patent, TI's stock price fell by 5.6%, a loss in

[7] As such, we are to some extent addressing a different question from the question whether there is a relationship between a firm's overall patent portfolio, and/or the issuance of a new patent, and the *firm's* value. There are a number of studies of the former relationship, including Pakes and Schankerman (1984), Lanjouw (1998) and others.

Obviously, an increase in the value of a given patent will, ceteris paribus, increase the value of the firm that owns the patent. But for publicly traded firms, one would expect it to be rare for the value of a single patent to comprise a material portion of the overall value of the firm, and as a result it would be rare that events related to a single patent (such as a finding of validity or invalidity) would have a measurable effect on the firm's stock price. Because of this factor, we are not aware of any general empirical studies of the relationship between stages in IP protection and firm value; all of the analyses we have seen are anecdotal.

More significantly, to a significant extent our analysis can be seen as making the *normative* claim that, in awarding patent infringement damages, the courts should treat proven-valid-and-infringed patents as economically more valuable (and thus deserving of higher damages awards) than patents that have not yet been tested. This is not a testable empirical proposition. We discuss this issue later.

[8] For his part in the invention of the integrated circuit, Jack Kilby was awarded the National Medal of Science in 1970 and the Nobel Prize in physics in 2000.

[9] The price of TI stock increased from US$ 31.75 to 36.50 per share. TI's 1989 Annual Report indicated (p. 43) that TI had an average of just under 81.5 million shares of common stock outstanding in the fourth quarter of 1989. The US$ 4.75 per share increase translates to a US$ 387 market capitalization increase.

[10] Indeed, the change in market value of the company in response to such news will represent only a portion of the total value of the issued patent. Economic theory suggests that the price of the stock prior to the issuance of the patent (or rejection of the application) will already reflect the market's assessment of the probability that the patent will issue. The market's reaction to news of issuance (or denial) will reflect the elimination of this ex ante uncertainty, not the full value of the patent.

To see this, consider a company whose sole asset is a pending patent application. Suppose that, if it issues, the company will be worth US$ 100 million; if it does not, the company will be worth nothing. Suppose further that ex ante there is a 60% chance that a patent will issue, and thus a 40% chance that it will not. That is, before the uncertainty about whether the patent will issue is resolved, the company will be worth (in an expected value sense) US$ 60 million. Investors will rationally bid up the stock of the company to that level.

Suppose that the patent then issues. Then the value of the company increases from US$ 60 million to 100 million. The US$ 40 million increase represents, not the *full* value of the patent (which, by assumption, is US$ 100 million), but only the resolution of the remaining uncertainty (the ex post elimination of the 40% ex ante prospect that the patent would not issue). Conversely, if it is announced that the patent will not issue, the market value of the company falls from US$ 60 million to zero, which again understates the total value of the patent. The same economic logic applies when the firm has other assets.

Mathematically, with B the baseline value of the company (excluding the patent), H the value of the patent if issued and p the ex ante probability that the patent will issue, the ex ante value of the company will be $B+pH$. If the patent issues, the ex post value of the company will rise to $B+H$; conversely, if it is announced that the patent will not issue, the ex post value of the company falls to B. For values of p strictly between 0 and 1, both the rise on issuance, $(1-p)H$, and the fall on denial, pH, understate the total value H of an issued patent.

market capitalization of some US$ 426.5 million.[11] The stock of Rambus, a designer of high-speed memory chips, fell some 54% (a loss in market capitalization of over US$ 1.9 billion) over a 2-day period in March 2001 in response to news that a judge overseeing a patent infringement case brought by Rambus intended to interpret the claims in some of Rambus' patents in a narrow fashion.[12]

Conversely, if the patent is found valid and infringed, then from an economic standpoint its value is significantly *greater* than it was before, while the issues of validity and infringement are still disputed. One can think of it as though a proven-valid-and-infringed patent has received a *second* "seal of approval", this time from a court that has found that the patent is valid and infringed.[13]

Finally, consider what happens when the patent expires. The innovation itself may still have significant advantages over the alternative technologies. But the patent holder no longer has the right to exclude others from using the innovation.[14] The value of the patent rights effectively drops to zero.[15] Note that, throughout this evolution, "the invention" itself has not changed. What has changed are the legal rights of the inventor.

The fundamental reality is that the different stages in the legal evolution of patent rights transform the nature of the underlying asset. An issued patent is more valuable than a pending patent application. Likewise, a patent for which validity and infringement has been established is a different, and more valuable, economic commodity than a patent for which validity and infringement have not been established. Talk of "the value of the invention" that fails to recognize these key distinctions largely misses the point.

2. Estimating value at different stages

The value transformation noted above becomes significant when one tries to ascertain "the value" of the rights to use a patented invention at different stages in the legal/economic evolution of the patent.[16] This issue frequently arises in patent infringement lawsuits, where the relevant question is how much in the way of damages should be awarded the patent holder for

[11] See Edmund Andrews, "Texas instrument loses in Japanese ruling", New York Times, September 1, 1994 (US$ 4.625 per share price drop); CRSP (92.2 million shares outstanding as of August 31, 1994).

[12] See Michael Kanellos, "Rambus shares plunge on talk of trial ruling", CNET News, March 15, 2001; http://news.cnet.com/news/0-1003-200-5148431.html (30% 1-day fall).

[13] Conversely, if a patent is found invalid, then its ability to exclude others drops to zero even before the patent expires.

Only a small fraction of patents are ever litigated, and only a fraction of litigated patents result in a finding of validity (or, indeed, of invalidity) (see Lanjouw and Schankerman, 2001). As such, one should not interpret Fig. 1 as a *chronological* figure so much as a *conceptual* depiction of "stages" of value that are reached in only a fraction of all cases.

[14] This is easiest to see in the context of copyrights. Shakespeare's plays and Beethoven's music still inspire and delight audiences, though any copyright protection has long since expired.

[15] This does not mean, however, that the (former) patent holder may not continue to receive benefits in the future, in two main ways. First, even after the patent expires, the patent owner may continue to have damages claims against other firms for their infringement of the patent during the pre-expiration period. Second, it may take time for others to introduce their own versions of the (formerly) patented product and to obtain distribution and market share. In the interim, the (former) patent holder may continue to earn supra-competitive returns.

One example involved Searle's patent on the sweetener aspartame, marketed under the trademark Nutrasweet®. While the patent was still in force, Searle consciously marketed the product under its trade name, and contractually required its customers to feature the trademark on their own products, so that consumers would come to associate the product with the trademark, thereby providing Searle with a continuing competitive advantage once the patent (but not the trademark) expired.

Another example involves brand-name prescription drugs. After patent expiration, others can introduce non-branded (generic) versions of the drug, but the (former) patent holder often retains significant market share for some period, despite the fact that the generic drug is therapeutically equivalent to the branded drug and typically sells for a significant discount relative to the branded version.

In these cases, however, we believe that a reasonable interpretation of the continuing benefits is that the patent enabled the (former) patent holder to obtain other, valuable complementary assets (such as a strong brand name or market presence), which continue to generate value even after the patent has expired and the patent per se no longer provides any protection against competition.

[16] Economists have estimated the value of patents using patent renewal data. In many countries, a patent holder must pay a renewal fee at intervals in order to keep the patent in force. A patent holder may elect not to pay the fee, in which case the patent lapses. A rational patent holder will elect to pay the renewal fee if the private value of maintaining the patent in force exceeds the renewal fee. The size of the fee thus puts a lower bound on the private value of those patents that are renewed. See, e.g., Lanjouw et al. (1998) and Schankerman (1998).

the infringer's unauthorized past use of the patented invention.[17] The law in the US says that, in such cases, the patent holder is entitled to "damages adequate to compensate for the infringement, but in no event less than a reasonable royalty for the use of the invention by the infringer . . . ."[18]

But the patent holder is entitled to an award of damages only if the patent is found to be valid and infringed. If the patent is found invalid or not infringed, the patent holder receives nothing. Consequently, the "reasonable royalty" rate is the rate appropriate for a proven-valid-and-infringed patent.[19]

How does one go about determining what such a "reasonable royalty" rate is? In many such cases, the patent holder will have entered into licenses granting others the right to use their patented invention in exchange for payment of royalties. The royalty rates in those licenses can be thought of as the economic price for the right to use the patented invention.

But such licenses are typically negotiated *before* the issues of validity and infringement have been resolved. The question then becomes: how informative are such negotiated royalty rates in determining the appropriate economic price for a proven-valid-and-infringed patent?

The key insight flows from the earlier conclusion that a proven-valid-and-infringed patent is a different, and more valuable, economic commodity than "the same" patent for which the issues of validity and infringement have not yet been resolved The easiest way to see this is to conduct a "thought experiment", in which one starts with an assumption about what one ultimately wants to determine (the appropriate economic price for a proven-valid-and-infringed patent) and work "backwards" to take account of uncertainty.

---

[17] One aspect of the scope of intellectual property rights is the remedy that the patent holder can receive should others infringe the patent. The patent holder's effective rights are stronger in a legal regime that awards high damages than in a regime that awards low damages. As such, the proper damages award can be thought of as going to the question of the optimal strength of patents. (We thank an anonymous referee for this suggestion.)

There is a reasonably large economic literature on that topic (see Besen, 1998 and the references cited therein). However, based on our review of that literature, it does not appear that the topic discussed in the present article has received much (if any) attention. The tacit premise that appears to underlie most of that literature is that, once a patent has been granted, the patent holder's intellectual property rights are costlessly enforceable, and that there will be no unauthorized use (so that the issue of the appropriate measure of damages for infringement does not arise). In those few examples in the literature where the issue of damages is addressed (see Lanjouw and Schankerman, 2001), the authors assume that the courts somehow costlessly know the amount of damages. By contrast, our analysis focuses on the pragmatic difficulties associated with trying to *determine* the appropriate level of damages for a proven-valid-and-infringed patent.

[18] 35 U.S.C. § 284. In appropriate contexts (notably when the patent holder makes and sells the patented product in competition with the infringer), the patent holder may also obtain damages in the form of: (a) lost profits on lost sales as a result of the infringement; and/or (b) "price erosion" due to the lower price resulting from the infringement. In many contexts, these sorts of damages will exceed the damages that would be appropriate under a "reasonable royalty" approach. As such, plaintiffs will prefer to sue for lost profit damages, and will use a claim for "reasonable royalty" damages only as a fallback position, as the patent statute recognizes ("in no case less than a reasonable royalty"). Similarly, because "reasonable royalty" damages will be less than "lost profit" damages, we expect (ceteris paribus) that the likelihood of infringement will be greater when only reasonable royalty damages are available than when lost profits damages are available.

For ease of presentation, our discussion is couched only in terms of a "reasonable royalty", but similar conceptual considerations apply to other damages claims.

An anonymous referee suggested that, "there may be good policy reasons why the legal system keeps patentee damages on the low side when only a royalty is available. For example, this form of damages is generally all the patentee can rely on of the patentee does not also manufacture a product under its own patent. Perhaps the low royalty rate is an incentive for a patentee to also be a manufacturer."

We believe that accepting this argument would lead to an inappropriate element of "double counting". As the referee acknowledges, a non-manufacturing patentee can only receive "reasonable royalty" damages, not the higher amount of "lost profit" damages that it could receive if it elected to manufacture and sell the patented product. This implies that a non-manufacturing patent holder is already disadvantaged (in the form of receiving lower patent infringement damages) relative to a manufacturing patent holder.

To suggest that, in addition, "reasonable royalty" damages should intentionally be kept low so as to provide a *further* incentive for firms to engage in manufacturing would appear to be inefficient. There is, and should be, no economic presumption that the most societally efficient way for firms to commercialize their innovations is for the innovating firm to also engage in manufacturing (see Teece, 1986 for an extended discussion of this issue). To the contrary, the legal system should encourage innovators to use other commercialization strategies if those strategies are superior.

[19] Strictly speaking, this is a normative claim, not an empirical one. We believe it to be uncontroversial as a matter of economics. But it is not entirely clear that the legal system fully accepts this proposition. This issue is discussed in more detail in footnote 24.

To take a numerical example, assume that everyone agrees that, if the patent in question is valid and if that patent is infringed by the product in question, a reasonable royalty would be 10% of sales.[20] But further assume that both parties also believe that the patent only has a 50% chance of being upheld as valid.[21] (Assume for now that there is no dispute about infringement if the patent is found to be valid.) Clearly, from an economic and managerial perspective, the licensee would not be willing to pay the full 10% royalty, given the uncertainty about validity. Rather, in real-world negotiations, people who face a (known) 50% uncertainty about the question whether a patent is valid might agree to a 5% negotiated royalty (50% of 10%), even though they all agree that the patent, if known to be valid, would be worth a 10% royalty.

Similarly, one can have an adjustment for uncertainty about whether or not the patent is in fact being infringed (i.e. whether the patent "reads on" the product being made or the process being used). For example, suppose that, even assuming there is no question that the patent is valid, there is only a 60% chance that the patent is being infringed by the product in question. (This might be due, for example, to some ambiguity in the language of the patent claims.) Standing alone, this uncertainty too could affect negotiated royalty. In particular, if everyone agreed that in this case a patent known to be both valid and infringed patent is worth 10%, they might negotiate a 6% royalty rate (60% of 10%) to reflect the 40% chance that patent is not in fact being infringed.

Obviously, one can also combine both types of uncertainty. For example, a 50% uncertainty about the patent's validity, coupled with a 60% chance of infringement if the patent is indeed valid, could lead to a 3% actual negotiated royalty rate (50% of 60% of 10%), even if everyone agrees that a valid infringed patent is worth 10%.

The point here is that an economically appropriate price for a valid-and-infringed patent—which we believe is the "reasonable royalty" concept in patent infringement damages cases[22]—may be significantly different from (and, in particular, higher than) the actual negotiated royalty for the same patent, because the actual negotiated royalty reflects what might be termed an "uncertainty discount" because of uncertainty about validity and infringement issues.[23]

It is important to note the nature of the adjustment here. In the "thought experiment" examples above, we

[20] The problem is somewhat more complicated in the more realistic situation in which the parties disagree about what an appropriate royalty rate would be if the patent in question were known to be valid and infringed. But the general principle discussed in the text is still applicable.

[21] The problem is significantly more complicated if the parties disagree about the likely probabilities of a finding of validity and/or infringement, but the same general principles apply.

[22] This proposition is not uncontroversial. Patent damages awards apply only to past infringement. (If the patent holder prevails on its liability claim, the patent holder typically will be able to obtain an injunction against continuing infringement.) It might be argued that the appropriate damages rate for infringement prior to a finding of validity and infringement should not be the rate for a proven-valid-and-infringed patent (which, by assumption, had not yet been established at the time of the actual infringement), but rather the rate for an untested patent (which was the status of the patent during the period of actual infringement).

We believe that such an approach would be economically inappropriate. Allowing an infringer to pay damages calculated at the rate for an untested patent on the grounds that, until the court ruled, the patent had not yet been shown to be valid and infringed, amounts to a court-sanctioned "heads I win, tails I break-even" scenario, as discussed below. It makes no more economic sense than an argument that, because the patent had not yet been proven to be infringed during the pretrial period of infringement, that therefore *no* damages should be awarded.

We also believe that such an approach is inconsistent with the "hypothetical negotiation" approach to assessing "reasonable royalty" damages, where the assumption underlying the hypothetical negotiation is that (unlike in the real world) the parties are assumed to agree that the patent is known to be valid and infringed.

[23] Similar logic is laid out in Kalos and Putnam (1997), where the authors explained the difference between a negotiated royalty and what they term an "infringer's royalty". Their analysis focuses on the loss to the patent holder if, after a finding of validity and infringement, the infringer must pay only what others paid. As discussed below, our analysis is different: we focus on the economic value-added from a finding of validity and infringement and on deterrence issues, and we provide an empirical estimate of the potentially appropriate adjustment. (We developed our analysis independently of Kalos and Putnam.)

We should stress, however, that this argument does not imply that "infringer's royalties" are always necessarily larger than actual negotiated royalty rates. Depending on the factual circumstances of the case, it may well be that actual negotiated royalty rates are a good estimate of the rates that would have been agreed to in a "hypothetical negotiation". For example, in some contexts validity is not controverted, and infringement is easy to establish. Under these circumstances, actual royalty rates are not likely to reflect much if any "discount" for uncertainty about validity or infringement, and thus provide a good measure of the rates that would have been agreed to in the "hypothetical negotiation".

started with a given commonly accepted (and assumed known) "reasonable royalty" rate, and then noted that, in actual negotiations, the actual royalty rate that the parties would agree to would be adjusted downward, to reflect the (assumed) uncertainty about validity and infringement. In practice, of course, the analysis goes the other direction. One has to work "forwards" from what one can observe (license rates negotiated in the real world, in which validity and/or infringement may be disputed), and the task is to try to infer the appropriate rate for a proven-valid-and-infringed patent, by estimating how significant the "uncertainty discount" might be and adjusting upward to offset that discount. In other words, our "thought experiment" by its nature does not represent what actually happens; nevertheless, its artificiality is precisely what enables us to make our theoretical point.

To return to the example above, if one observes an actual negotiated rate of 3%, and one knows (or has good reason to believe) that the parties believed that there was only a 50% chance that the patent would be found valid, and one further knows (or has good reason to believe) that the parties believed that, even if the patent was found to be valid, there was only a 60% chance that the patent would be found to be infringed, then one can infer that the parties must have likewise believed that the patent, if known to be valid and infringed, would command a 10% royalty.

The conclusion to be drawn from this is that, in order to infer the value for a valid-and-infringed patent, or to award the economically appropriate amount of patent infringement damages,[24] it is typically necessary to make what might be termed a "certainty adjustment" to negotiated rates. Otherwise, the infringer gets to play a "heads I win, tails I break-even" game. If the patent holder is unable to prove validity or infringement, the infringer does not have to pay anything (the "heads I win" side of the coin). If, following a finding that the patent is valid and infringed, the infringer is merely required to pay what everyone else (who in fact negotiated a royalty) pays, then the infringer faces no downside risk; it pays only what it would have had to pay anyway (the "tails I break-even" side of the coin).[25]

In conclusion, an "infringer's royalty" (or a "reasonable royalty") can (depending on factual circumstances) be significantly different from, and higher than, the rate actually negotiated by those who took a license. This is the main reason why we earlier said that market-negotiated rates must be interpreted with caution.[26]

[24] After all, the intellectual property owner will be able to recover damages only if it shows that: (a) it owned a valid piece of intellectual property; and (b) the defendant infringed it.

As used here, the term "economically appropriate" refers to the appropriate valuation for a proven-valid-and-infringed patent. There is an unrelated theme in the literature, which focuses on the relationship between patent damages awards (and other factors such as the rules regarding which party should bear the costs of litigation) and the incentives both: (a) to litigate cases (rather than settle); and (b) to bring lawsuits in the first place.

It is likely that the approach proposed in this paper, if adopted by the courts, would lead to higher damages awards than would otherwise be the case, thus (ceteris paribus) raising the return to innovation, the likelihood of bringing a suit, the stakes in litigation and the likely settlement amount. We make no suggestion here that the "economically appropriate" damages we propose will lead either to the optimal level of litigation or to the optimal incentives to litigate versus settle. That is a complex issue best left for another time.

[25] This has been noted by several courts, notably in *Panduit Corp. versus Stahlin Bros. Fibre Works*, 575 F. 2d 1152 (6th Cir. 1978). Some courts have refused to award what they refer to as a "*Panduit* kicker", on the grounds that an award of what we have termed "infringer's royalties" amounts to an improper "penalty" on an accused infringer that asserted its legal right to compel the patent holder to *prove* that its patent is valid and infringed. Other courts have noted that there are other means (notably an award of additional damages for "willful infringement") that can be used to address the issue raised in this paper.

We believe that the discussion in the text establishes that this concern is to some extent conceptually misplaced. The issue is not whether the defendant should have to pay a "penalty". Rather, the issue is whether the court will require the defendant to pay damages at a level that *is* appropriate for a proven-valid-and-infringed patent, rather than damages calculated using as a reference point licenses entered into for patents that have *not* been proven to be valid-and-infringed. This issue is conceptually different from the question whether infringement was "willful" in the legal sense. In particular, in our experience courts are reluctant to impose higher damages for "willful infringement" if the accused infringer had reasonable (though ultimately unsuccessful) arguments of invalidity or non-infringement. But it is precisely in such cases that the "infringer's royalty" discussed in the this article is *most* necessary, as a deterrent against the "heads I win, tails I break-even" argument.

[26] An anonymous referee has suggested that our proposal has the potential for abuse. A company could create internal memos (falsely) suggesting that the royalty rates that it had actually negotiated in its outstanding licenses reflected a significant discount to reflect uncertainty about validity and infringement, and then use those memos in later litigation to suggest that the court should award high damages to offset the "discount".

3. Quantifying the adjustment

Of course, to perform this sort of adjustment in practice, we must estimate the probability of a finding of validity and infringement. Given the uncertainty of litigation, these probabilities are clearly not 100%. To illustrate, one can make use of data from US patent lawsuits. Data regarding the frequency with which plaintiffs have prevailed in patent infringement suits was collected. While this is not precisely the same as the question of whether challenged patents are found invalid and/or not infringed (as not all patent infringement suits involve defenses of invalidity or non-infringement, and parties sometimes stipulate to validity or infringement), it provides a useful benchmark.

We examined a large and comprehensive database collected by the Administrative Office of the US Courts and administered by the US Federal Judicial Center[27] on lawsuits filed in US District Courts[28] over the 1979–1995 period,[29] and their disposition. Cases are coded on the basis of the field of litigation involved; we examined the data for patent-related suits. We examined the data for information on the number of lawsuits resolved at each stage in the litigation process (via settlement or trial), focusing on: (1) cases resolved by pretrial motions (largely summary judgment motions); and (2) cases resolved at trial (whether by a jury verdict, a directed verdict, or a court verdict).[30] Of these cases,[31] we determined the percentage which resulted in a judgment for the plaintiff[32] and the percentage which resulted in a

---

We acknowledge the potential for such abuse. But this issue has nothing to do with our fundamental points: that a proven-valid-and-infringed patent is a more valuable economic commodity than an untested patent, and that in appropriate cases it may be necessary to upwardly adjust observed royalty rates to reflect the difference. Instead, it goes to the sorts of proof that the courts should require before awarding damages. The litigation process is inherently susceptible to manipulation by false testimony or self-serving documents prepared in anticipation of litigation. But that proposition is true for *all* litigated issues. We know of no reason why the potential for abuse is particularly problematic here. Moreover, if courts were concerned about the issue, one possible solution would be to rely less on such documents and more on empirical evidence on success rates generally, as discussed in the next Section. Such broad-based data is not subject to "abuse" by individual litigants.

[27] The data are available from the Inter-University Consortium for Political and Social Research (ICPSR), located at the University of Michigan.

There have been numerous empirical studies of litigation generally, and patent litigation in particular, that use the FJC data. Of special note is Lanjouw and Schankerman (2001), who combine the FJC data with information about the nature of the disputed patent (obtained from another data source) to examine various propositions about the effect of patent characteristics on the likelihood and outcomes of litigation.

[28] Because patents are a creation of federal law, patent cases are tried in the Federal Courts, not the State Courts.

[29] While the entire database covered the 1970–1995 period, the information we needed was coded only for the period 1979–1995. We examined whether there was any trend in the data indicating that a different period would be appropriate, but identified no trends which were relevant to our analysis. Consequently, we used the entire 1979–1995 sample in our analysis.

[30] Because our concern is with the uncertainty associated with litigation, we disregarded the large majority of cases which settle. The database does not (except in isolated cases) code for whether a settlement favored the plaintiff or the defendant.

[31] There is a large literature establishing that cases selected for trial are not likely to be a representative sample of all litigated cases. Many scholars have sought to identify factors that differentiate between cases litigated to a verdict and cases settled before a verdict (see Waldfogel, 1998 and the references cited therein).

The FJC database does contain information on "judgment(s) on consent", and indicates whether the judgment was for the plaintiff, the defendant, or "both". In our experience, and based on discussions with Federal Judicial Center personnel about what this category includes, we believe that such "judgment(s) on consent" are much more akin to settlements than they are to judgments on the merits (or on summary judgment motions). We consequently chose not to include such "judgment(s) on consent" in our totals. We were left with 2586 cases.

Though cases litigated to an outcome are not representative of all filed cases, we believe that they are the relevant universe of comparison for our purposes. The task is to determine what damages should be awarded for a proven-valid-and-infringed patent. But damages will be awarded only in those cases which go to a verdict. Thus, the universe of cases that are "comparable" to cases in which the issue of damages arises are precisely those cases which go to a verdict. As such, the fact that such cases are not representative of all filed cases is, in our view, immaterial.

[32] The Federal Judicial Center database identifies whether the case involves patent issues, and whether the result favors the plaintiff or the defendant. The "plaintiff" is identified as the party which brought the original suit. In an ordinary patent infringement case, the patent holder is the original plaintiff, so that a verdict for the "plaintiff" is a verdict for the patent holder.

Obviously, there are cases: (a) where a firm sues a patent holder on some grounds (such as breach of contract) and the patent holder defendant countersues for patent infringement; (b) where a patent holder sues another firm for patent infringement, and the second firm countersues alleging that the first firm infringes a patent held by the second firm; and (c) where a firm sues a patent holder seeking a declaratory judgment that the patent is invalid.

judgment for the defendant.[33] The overall size of the database, the number of patent-related suits, and the size of each of these categories are all very large, all of which would make our results highly statistically reliable.[34]

The results are shown in Table 1. Of cases resolved by pretrial motions, 34.2% were resolved in favor of the plaintiff. Of cases resolved by a trial verdict, 56.6% were resolved in favor of the plaintiff.[35] If we combine these two categories, of all cases resolved either by pretrial motions or by a trial verdict, 45.0% were resolved in favor of the plaintiff.[36]

That is, if one were to consider a "typical" patent infringement case brought in a US Federal District Court during the 1979–1995 period and taken through to disposition either by summary judgment or by a trial court verdict, the plaintiff would have an 45.0% probability of prevailing at the District Court level.

Our results are generally consistent with results of earlier published studies, many of which also used samples drawn from the Federal Judicial Center database to estimate plaintiffs' win rates in patent litigation. Waldfogel (1995) estimates plaintiffs' win rate at trial in patent cases at 55.8%.[37] Kessler et al. (1996) estimate plaintiffs' win rates at trial in patent cases at 53.3%.[38] Because our study uses the entire Federal Judicial Center database over a longer period, it is more comprehensive than their studies. Allison and Lemley (1998) use a different data source and conclude that patents are held valid 51.7% of the time.[39]Combining the results of the various studies, it is reasonable to infer that patent holders[40] win

---

We discussed this matter with the staff of the Federal Judicial Center, and concluded that, because of the coding system used by the FJC, cases of type (a) would not be coded as "patent" cases and would thus fall outside the set of data we have examined. Consequently, whether a type (a) plaintiff does or does not prevail in its original suit is irrelevant to our analysis.

Under cases of type (b), it is possible that both patent holders would win, or neither would win. The database lacks the necessary information to enable us to fully resolve these concerns.

Under cases of type (c), a win for the plaintiff is a finding of patent invalidity, and thus a finding against the patent holder. Conversely, a win for the defendant is a finding that the patent is valid, and thus a finding in favor of the patent holder. The data in the FJC dataset provide no way of separating out these cases, as it does not code for which party owns the patent. To resolve this issue for the FJC dataset would require a significant research effort, essentially involving physically going to the federal courthouses and manually recoding the data.

Consequently, our empirical result—that *plaintiffs* win at trial in roughly 45% of patent cases—does not directly measure the probability that *patent holders* win at trial. We discuss this issue later.

33 The database indicates that some cases involved a verdict for "both" plaintiff and defendant. We investigated this issue with the Federal Judicial Center, and determined that these cases largely involve counter-claims (or cross-claims) brought by a defendant accused of infringing a patent against the patent holder. Unfortunately, the data available do not allow us to identify these outcomes further.

In order to be conservative, we have interpreted "both" verdicts as findings for the patent holder on the questions of validity and infringement. This leads to the numbers given in the text. If "both" verdicts are disregarded, then 30.5% of cases resolved by pretrial motions, 53.8% of cases resolved by a trial verdict, and 41.7% of all resolved cases involve findings for the plaintiff. This would lead to a higher adjustment factor.

34 Technically, because we have used the entire database rather than a sample, the issues of statistical significance and confidence intervals do not arise.

35 The likelihood that the plaintiff would obtain a favorable verdict at trial depends on whether the verdict was a court (bench) verdict, a directed verdict, or a jury verdict. The figures reported in the text aggregate across all three categories.

36 Obviously, trial court verdicts can be appealed. The Federal Judicial Center database has some information on whether trial court outcomes were appealed, and the outcome on appeal. Unfortunately, given the sheer size of the database and the difficulty in matching trial court outcomes with appellate outcomes (especially since the Courts of Appeals will often remand a case for subsequent District Court proceedings), it has proven impracticable to trace cases through to final disposition after appeal.

37 This estimate was based on 170 cases filed in the southern district of New York during 1984–1987.

38 This estimate was based on 30 cases decided on the merits by the Seventh Circuit Court of Appeals between 1982 and 1987. The probability that the plaintiff wins after appeals falls to 36.7%; the probability that an appellant prevails is 30%.

39 Allison and Lemley found that the success rate varied depending on how the case was resolved. Patents were found valid in 67.1% of the time in jury trials, 57.3% of the time bench trials, but only 28.1% of the time in cases resolved by pretrial motion. The percentage reported in the text aggregates across all three categories.

40 As noted earlier, the empirical results show the probability that a *plaintiff* will prevail at the trial court level, which is not the same thing as whether the *patent holder* will prevail at the trial court level. The issue then becomes: (a) whether it is reasonable to use the former as a *proxy* for the latter, or (b) whether it is possible to use our empirical estimate, and the other estimates in the literature, to put a rough bound on the latter.

Analytically, denote by N the percentage of "patent" cases that are ordinary infringement actions, so that the plaintiff is the patent holder. Of those cases, the plaintiff (patent holder) will win at

Table 1
Frequency of plaintiff winning a patent case in Federal District Courts 1979–1995

	Court verdict	Directed verdict	Jury verdict	All verdicts	Disposed of in pretrial motion	All
Number of cases						
Plaintiff wins	350	27	254	631	386	1017
Defendant wins	379	26	137	542	880	1422
Both win	57	0	19	76	71	147
Percent win by plaintiff						
Percent plaintiff wins (excluding "both")	48.0	50.9	65.0	53.8	30.5	41.7
Percent plaintiff wins ("both" = win for plaintiff)	51.8	50.9	66.6	56.6	34.2	45.0

Sources: (1) Federal Court Cases: Integrated Database, 1970–1994 (Parts 82–89: Appellate and Civil Data, 1992–1994) (Computer file), ICPSR version. Federal Judicial Center, Washington, DC (producer), 1995. Inter-University Consortium for Political and Social Research, Ann Arbor, MI (distributor), 1996. (2) Federal Court Cases: Integrated Database, 1970–1995 (Parts 76–81: Criminal Data, 1992–1995; Parts 96 and 97: Appellate Data, 1995; Parts 98 and 99: Civil Data, 1995) (Computer file), ICPSR version. Federal Judicial Center, Washington, DC (producer), 1995. Inter-University Consortium for Political and Social Research, Ann Arbor, MI (distributor), 1996.

roughly 45–55% of the time in cases resolved at the trial court level.[41]

Thus, for example, the analysis above suggests that, if the parties agreed that a particular valid-and-infringed patent was worth a 10% roy-

---

trial some fraction of the time; denote that fraction by $X\%$. The remaining $(1-N)\%$ of "patent" cases are declaratory judgment cases (or other cases in which the patent holder is the defendant). Of those cases, the plaintiff (the accused infringer) will again win at trial some fraction of the time; denote that by $Y\%$. Then the patent holder wins the remaining $(1-Y)\%$ of the time.

Basic probability theory shows that the total probability that the *plaintiff* wins, which we will denote by P_{π} (using the standard lawyer's mnemonic of "π" for plaintiff) is then:

$$P_{\pi} = NX + (1-N)Y.$$

(This is the probability that we have estimated at 45%, using the Federal Judicial Center data.)

Similarly, the total probability that the *patent holder* wins, which we denote by P_{H} (where "H" is the mnemonic for "holder"), is:

$$P_{\mathrm{H}} = NX + (1-N)(1-Y).$$

The question becomes: how significant is the discrepancy between P_{π} and P_{H}? One way to measure that discrepancy is to look at the *difference* between these two numbers, which can readily be shown to equal:

$$P_{\pi} - P_{\mathrm{H}} = (1-N)(2Y-1).$$

Note that, when $Y = 50\%$, the difference $(P_{\pi} - P_{\mathrm{H}})$ is zero. The closer that Y is to 50%, the smaller the difference. When Y is greater than 50%, $(2Y-1)$ is positive, so P_{π} exceeds P_{H}.

Similarly, holding Y constant, the greater that $(1-N)$ is (or, equivalently, the smaller that N is), the more that P_{π} differs from P_{H}.

Using a somewhat different data set, a recent empirical study by Lanjouw and Schankerman (2001) found that, "infringements account for the bulk of litigation, about 84% if unclassified cases are excluded and 90% if they are treated as infringement suits." (They conclude that their "unclassified category is primarily infringement suits brought by an exclusive licensee or the parent company of the patent holder.") That is, Lanjouw and Schankerman estimate N at 84–90%. Suppose for concreteness that N is 85%, and that $Y = 45\%$. Then,

$$P_{\pi} - P_{\mathrm{H}} = 15\% \times (90\% - 100\%) = 15\% \times (-10\%) = -1.5\%.$$

This implies that P_{H} (the probability that the patent holder wins) $= P_{\pi} + 1.5\%$. With P_{π} estimated at 45%, this implies that P_{H} is 46.5%. This suggests that, in practice, P_{π} is likely to be a reasonable proxy for P_{H}.

Another plausible way to approach this issue is to see if it is possible to put reasonable "bounds" on the likely outcome. Suppose that 100% of cases coded as "patent" cases were brought by the patent holder. Then our empirical results (that plaintiffs win 45% of the time) would mean that patent holders won 45% of the time. At the other extreme, suppose that 100% of cases coded as "patent" cases were declaratory judgment cases, so that the patent holder was the defendant. Then our empirical results (that plaintiffs win 45% of the time, or, conversely, that defendants win 55% of the time) would mean that patent holders win 55% of the time.

The reality falls somewhere between these two extremes., We believe that it is plausible (though not, we must stress, strictly mathematically necessary) that the probability that the patent holder would prevail at the trial court level will thus fall within the 45–55% range.

[41] One interpretational issue is whether these results reflect the Priest–Klein hypothesis (Priest and Klein, 1984), which suggests that self-selection of cases for litigation will lead to plaintiffs' success rates at trial of 50%. For a general discussion of the Priest–Klein hypothesis and various alternatives, see the survey by Waldfogel (1998).

alty, but they discounted for the fact that plaintiff would prevail (if the case went to trial) with only a 45–55% probability, then the parties would be likely to agree to a negotiated royalty in the range of 4.5–5.5%.

Conversely, if one were to start with an (observed) negotiated royalty of (say) 3% for a given patent, and believed that it was discounted to reflect uncertainty about validity or infringement, one might conclude that a reasonable royalty for that patent, assuming that it was shown to be both valid and infringed, would be 3% divided by 45–55%, or 5.45–6.67%. That is, given the estimates above of the uncertainty about validity and infringement, the appropriate royalty rate for a known-valid-and-infringed patent will on average[42] likely be roughly twice the rates in negotiated licenses for the same or equivalent patents.

This does *not* mean, however, that in patent infringement litigation it is always appropriate to calculate the "infringer's royalty" simply by multiplying the existing royalty rates by the inverse of the large-sample probability of a finding of validity and infringement. There are numerous complications—ranging from the fact that the probability of validity and infringement varies across patents and can vary over time even for a single patent, to the fact that the value of being able to practice a patent can vary over time, to the dynamics of the licensing process in the context of multiple potential licensees, to the fact that many licenses (and many patent infringement lawsuits) involve multiple patents, to the presence of terms in licenses (such as a "most favored nations" clause) that can affect the royalties paid over the term of the license—that make it necessary in practice to proceed with caution.[43]

We have searched for real-world examples comparing royalty rates negotiated before the issues of validity and infringement have been resolved with royalty rates negotiated after a finding of validity and infringement has been made. This has proven difficult, because royalty rates are typically treated as confidential business information and are rarely publicly disclosed.

However, one well-known example illustrates the point. Prior to suing Toshiba for patent infringement on a SIMM patent, Wang Laboratories had only been able to license the patent to one firm, for a 1% royalty. After a trial court decision upholding the validity of the patent, Wang was able to enter into over forty licenses for a 3% royalty.[44]

4. Implications and conclusions

Our analysis has a number of managerial and policy implications. First, as stressed at the outset, it can often be important to differentiate between the innovation and the intellectual property rights to that innovation. Second, any analysis of the value of intellectual property rights needs to take account the various stages in the evolution of those rights, and the impact on economic value of changes in the legal status. While the discussion and analysis here relates only to patents, it is also applicable at least in part to other form of intellectual property too.

Third, an uncritical attitude to "the value of a patent license" that fails to differentiate between an untested patent and a proven-valid-and-infringed patent can dramatically understate the difference in the economic value between these two economic commodities.

Fourth, litigation is costly. Sound public policy seeks to encourage parties to resort to market transactions rather than litigation. If (when awarding patent infringement damages) courts fail to acknowledge the economic difference in value between an untested patent and a proven-valid-and-infringed patent, the

[42] The issue of whether a given patent will be found valid and infringed by a particular product turns on the particulars of the patent, on the prior art, on the product, and (most significantly) on how the features of the product relate to the claims of the patent. This suggests, on the one hand, that estimates (such as those we have calculated) drawn from a range of cases are likely to be less relevant than the particulars of any given case. On the other hand, in litigation the parties are likely to dispute the case-specific features. And one does not want to use the ex post outcome to assess the ex ante uncertainty; an argument along the lines of "because the court ultimately found that the patent was valid and infringed, therefore it must be the case that the ex ante probability of a finding of validity and infringement was 100%" is clearly invalid. Balancing case-specific factors against empirical results drawn from a range of cases can be a difficult task.

[43] For a more extensive discussion, see Ingberman, Sherry, and Teece, "The law and economics of 'infringer's royalties'", unpublished working paper (2003), available from the authors.

[44] Most of this information is taken from Wang versus Mitsubishi, 860 F. Supp. 1448, 1450-51 n. 2 (C.D. Cal. 1993); the rest is from unpublished sources.

(claimed) infringer, facing a "heads I win, tails I break-even" situation, has little incentive to negotiate a license, but instead has a strong incentive to litigate and take its chances.[45] This in turn leads to excessive and inefficient litigation.[46]

Fifth, licenses are negotiated "in the shadow of the law". Once the patent is granted, it is published; the information contained in the patent is available for inspection by all. Obviously, a prospective licensee would not willingly pay anything for this information, in the absence of a "threat" (generally implicit) of injunction or damages for a refusal to accept a license, unless the license also carried with it something else (like a transfer of related technology from the patent holder to the licensee).

Of course, the "threat" in question is rarely made overtly, far less baldly; in modern capitalist societies, firms realize that they must pay for what they use, and will voluntarily negotiate such payments, even if there is not the equivalent of a policeman with a loaded gun standing guard to prevent them from taking it without payment. But in the background is always the (implied) "threat" of litigation; if the infringer knew for certain that the patent holder could not sue, or that the court would award little or no damages even if the patent holder successfully sued, it is unlikely that the patent holder would be able to get much if anything for its patented technology.

Similarly, the parties' expectations of the damages that will be awarded should the case go to trial will affect negotiated rates. If both parties know, for example, that a court will award (say) a 10% royalty if the patent is found valid and infringed, then that will in effect put a cap on the rates that the parties will agree to, whether in settling an already-filed lawsuit or in a negotiating a license in the shadow of the law.

[45] Against this must be set the cost of litigation, which the (prospective) licensee would avoid if it took a license.

[46] This does not necessarily mean that more cases will be taken to trial. The patent holder, knowing that the courts will only award low damages, may settle for low royalty rates. In other words, low expected damages will reduce both the plaintiff's settlement demand and willingness to take the case to trial (or, indeed, to bring the case in the first place) and the defendant's settlement offer. For an analysis of the economics of suit and settlement, see the survey article by Hay and Spier (1998) and the references cited therein.

But this in turn implies that the standards used to set intellectual property damages awards will have an impact far out of proportion to the (relatively tiny) number of cases that actually go to trial. Expectations of damages awards will affect both settlements and licensing rates, which in turn will affect the return that successful innovators can achieve.

As such, while at first glance it might appear that our analysis is primarily of interest to lawyers and legal scholars, we believe that it has broader implications of interest to economists and others studying technical change and innovation. Incentives matter; greater incentives encourage more innovation. And the manner in which the legal system treats patent infringement damages can have significant incentive effects on the propensity to engage in R&D in the first place.

The value of innovation is clearly a complex and multifaceted subject. Private value is quite distinct from social value.[47] Moreover, the private value of innovation can be quite different from the private value of the intellectual property associated with that innovation, and vice versa. The value of intellectual property is itself dynamic, varying according to many factors, not least of which is the legal status/evolutionary state of the intellectual property, a distinction we have illustrated here in the case of patents. We believe that this is but a first step in the better appreciation of one of the most important classes of intangible assets in the global economy today.

Acknowledgements

We wish to thank Alan Marco for research assistance, Dan Ingberman and Mark Schankerman for useful discussions, and Gary Pisano and three anonymous referees for helpful comments on earlier drafts. All errors remain our own.

References

Allison, J., Lemley, M., 1998. Empirical evidence on the validity of litigated patents. AIPLA Quarterly Journal 26, 185–275.

[47] Empirical studies show that innovators generally receive only a fraction of the total social benefits resulting from their innovations. See Levin et al. (1987).

Besen, S., 1998. Intellectual property. In: Newman, P. (Ed.), The New Palgrave Dictionary of Economics and the Law, vol. 2, Macmillan Reference Ltd., pp. 348–351.

Cohen, W., Nelson, R., Walsh, J., 2000. Protecting their intellectual assets: appropriability conditions and why US manufacturing firms patent (or not). NBER Working Paper no. 7552.

Hay, B., Spier, K., 1998. Settlement of litigation. In: Newman, P. (Ed.), The New Palgrave Dictionary of Economics and the Law, vol. 3, Macmillan Reference Ltd., pp. 442–450.

Kalos, S., Putnam, J., 1997. On the incomparability of 'comparables': an economic interpretation of 'infringer's royalties'. Journal of Proprietary Rights 9 (4), 2–5.

Kessler, D., Meites, T., Miller, G., 1996. Explaining deviations from the 50-percent rule—a multimodal approach to the selection of cases for litigation. Journal of Legal Studies 25, 233–259.

Lanjouw, J., 1998. Patent value in the shadow of infringement: simulation estimations of patent value. Review of Economic Studies 65, 671–710.

Lanjouw, J., Schankerman, M., 2001. Characteristics of patent litigation: a window on competition. RAND Journal of Economics 32, 129–151.

Lanjouw, J., Pakes, A., Putnam, J., 1998. How to count patents and value intellectual property: the uses of patent renewal and application data. Journal of Industrial Economics 46, 405–432.

Levin, R., Klevorick, A., Nelson, R., Winter, S., 1987. Appropriating the returns from industrial research and development. Brookings Papers on Economic Activity 3, 783–820.

Pakes, A.S., Schankerman, M., 1984. The rate of obsolescense of patents, research gestation lags, and the private rate of return to research resources. In: Griliches, Z. (Ed.), R&D, Patents and Productivity, University of Chicago Press, Chicago.

Priest, G., Klein, B., 1984. The selection of disputes for litigation. Journal of Legal Studies 22, 1–55.

Schankerman, M., 1998. How valuable is patent protection: estimates by technology field. RAND Journal of Economics 29, 77–107.

Teece, D., 1986. Profiting from technological innovation: implications for integration, collaboration, licensing, and public policy. Research Policy 15, 285–305.

Waldfogel, J., 1995. The selection hypothesis and the relationship between trial and plaintiff victory. Journal of Political Economics 103, 229–260.

Waldfogel, J., 1998. Selection of cases for trial. In: Newman, P. (Ed.), The New Palgrave Dictionary of Economics and the Law, vol. 3, Macmillan Reference Ltd., pp. 419–424

Minnesota Law Review, 87 (June 2003), 1913–1994

Standards Setting and Antitrust

David J. Teece[†] and Edward F. Sherry[††]

In recent years, several high-profile cases have raised questions about the appropriate relationship between antitrust policy and enforcement and the standards-setting process. This Article uses the tools of economic analysis to address several aspects of this debate.

In Part I, we identify certain different types of standards, differentiate between standards and regulations, and offer a recent example of intellectual property issues arising in the standards-setting process. In Part II, we identify several ways in which intellectual property and antitrust regulations interact during the standards-setting process. Part III analyzes the interplay between antitrust and intellectual property, focusing on the various roles that firms play in standards-setting organizations (SSOs). In Part IV, we discuss SSO policies regarding intellectual property, including search, disclosure, and licensing rules. Finally, in Part V we argue that, to capture the social and economic benefits made available through standards, antitrust authorities must regulate sparingly, with no presumption that a "one size fits all" antitrust policy is appropriate. In our view, a better approach is for SSOs to establish clearly articulated rules or policies governing both members' obligations and the SSO's own actions.

I. TYPES OF STANDARDS AND REGULATIONS DIFFERENTIATED

A. Different Types Of Standards

Standards have become an increasingly significant aspect of many industries. But the term "standard" has a number of different meanings or uses, with fundamentally different implications for antitrust policy. In their recent treatise on *IP and Antitrust*, Professors Hovenkamp, Janis, and Lemley "define a standard rather broadly as any set of technical specifications which either does, or is intended to, provide a common design for a product or process."[1] They note that "standards" are not restricted to today's high-tech fields, but include such mundane matters as electrical plugs and outlets.[2] Other common examples of standardized products include nuts and bolts (Society of Automotive Engineering versus metric) and electrical voltage (110 volts versus 220 volts).

The Hovenkamp, Janis, and Lemley definition can be misleading in some contexts. Many standards (including many interoperability standards) do not specify a "design for a product" so much as they identify certain *features* of the product that are standardized, leaving many if not most other product features unspecified and unstandardized. For example, the design of a toaster includes the design of the plug at the end of the power cord. While the *plug* design needs to be standardized so that the cord fits into the electrical socket, and the toaster needs to be designed to run on common household current, the remainder of the design of the toaster per se does not need to be, and typically is not, standardized.

Similarly, in the semiconductor industry, one common standardization issue for new chips involves the physical chip size and shape (the "form factor") and pin assignment configuration[3] of the new chip.[4] Once these (and other)

---

1. Herbert Hovenkamp et al., IP and Antitrust: An Analysis of Antitrust Principles Applied to Intellectual Property Law § 35.1a (2002 & Supp. 2003).

2. *Id.*

3. Chips (such as DRAMs or microprocessors) have numerous "pins" (small metal wires) that plug into receptacles in chip sockets. Different functions are assigned to different pins. Common functions include power, electrical ground, and various inputs and outputs. For chips from different manufacturers to be compatible with one another (and with the equipment in which they are installed), they must share the same pin assignment.

standards are set, different firms can compete among themselves to design different versions of the chip, all of which share the standardized features but which can differ significantly in cost and performance. Another common example involves automobile tires. Tire sizes are standardized so that tires from different manufacturers can be used interchangeably on (suitably sized) wheel rims.

In such cases, the standard serves to promote what is often termed "compatibility" or "interoperability."[5] Products that comport with the standard share common features so that they are compatible with other complementary products (e.g., tires must be compatible with wheels). In particular, interoperability standards govern how products interact with other products, not how they perform their functions. Many different designs can have the same interoperability features.

In many such cases, manufacturers compete vigorously *within* the standard along a number of dimensions such as non-standardized features, quality, and price.[6] It is all too easy to slip into the habit of assuming that, because certain product *features* have been standardized, the *product* itself has been standardized.

Adoption of a standard limits "standards wars" (in which different firms compete by offering mutually incompatible products, such as VHS and Beta VCRs), but frequently promotes competition *within* a given standard.[7] To take a well-known example, the IBM personal computer (PC) versus Macintosh "standards war" involved competition between Apple's Macintosh standard (which Apple kept proprietary, so that "Mac clones" were generally not available) and the "open" PC standard, which fostered competition between IBM and numerous brand-name (e.g., Compaq) and no-name "PC clone" manufacturers.[8]

Other standards do not involve "design" issues. One well-

---

4. *See, e.g.*, JEDEC SOLID STATE TECHNOLOGY ASSOCIATION, DOUBLE DATA RATE (DDR) SDRAM SPECIFICATION 3-7, http://www.jedec.org/download/search/JESD79R2.pdf (May 2002).

5. *See, e.g.*, PETER GRINDLEY, STANDARDS, STRATEGY, AND POLICY: CASES AND STORIES 22-23 (1995).

6. For example, there are certain technical standards for VHS VCRs to ensure their compatibility. *See* GRINDLEY, *supra* note 5, at 75-98. But manufacturers compete with one another by offering VCRs with a wide range of features at a wide range of prices. *See id.* at 86.

7. *See id.* at 75-98.

8. *Id.* at 131-55.

known example is product gradation standards for non-manufactured goods. Familiar examples include United States Department of Agriculture (USDA) size categories (Large, Extra Large, Jumbo) and quality categories (Grades AA, A, B) for eggs, USDA grade categories for beef (Select, Prime, Choice), and grading standards for gemstones such as diamonds (color and clarity). Of course, other product gradation standards (such as gradation standards for plywood, tire tread wear, or gasoline octane ratings) do involve manufactured products and, thus, affect certain aspects of product design.[9] The principal purpose of such product gradation standards is to provide customers with a way of comparing product features (such as size and quality) across different sellers.[10] As long as such gradation standards are provided for information purposes only, such gradation standards would not appear to raise competition policy concerns.[11]

In his 1982 book, now-Justice Stephen Breyer drew another key distinction between *performance standards* (which specify that goods must achieve certain performance characteristics, but which allow firms to achieve those performance levels in whatever fashion they can) and *design standards* (which specify particular features which goods must have if they are to conform with the standard).[12] Performance standards do not have the Hovenkamp, Janis, and Lemley "common design" feature.

Nevertheless, their definition captures the most common situation involving standardization disputes. And such "compatibility" (or "interoperability") standards represent perhaps the most important economic examples of standardization. Significant economic benefits accrue if the products of different manufacturers are compatible, even if they are not identical. Such compatibility or interoperability promotes consumers' ability to substitute among the products of different suppliers, avoiding customer "lock-in" to a

9. STEPHEN BREYER, REGULATION AND ITS REFORM 97 (1982).

10. *See, e.g.*, *id.* (explaining that the purpose behind tire standards is to help consumers make rational decisions).

11. One possible exception is when participants in the SSOs manipulate the outcome so that the SSO refuses to draw a distinction that is important to consumers. *See* HOVENKAMP ET AL., *supra* note 1, § 35.5.

12. BREYER, *supra* note 9, at 105-06.

particular supplier.[13] For example, standardizing the physical configuration and input-output workings of computer chips allows computer makers to design their motherboards to accommodate chips from different suppliers. Standardization also helps create markets by enabling firms to achieve scale economies in production. For example, by standardizing on a relatively small number of different tire sizes, auto manufacturers facilitate the development of competition among tire manufacturers, competition that would be lessened (or absent entirely) if each car was custom designed to work only with a particular tire (and vice versa).

1. Formal Standards Versus De Facto Standards

Technical specifications for standards are typically formally adopted by an SSO.[14] Some SSOs are long-lived formal bodies and may have delegated authority. For example, the American National Standards Institute (ANSI) is a quasi-governmental, non-profit, "umbrella" organization comprising hundreds of different special-purpose SSOs. ANSI is the official United States representative to both the International Electrotechnical Commission (IEC) and the International Organization for Standardization (ISO).[15] Such organizations often have numerous committees (or subcommittees) that adopt standards in particular fields.[16] Other SSOs are less structured, often ad hoc consortia of interested parties formed for a particular purpose. There are hundreds of such ad hoc consortia SSOs, especially in high-tech industries.[17]

---

13. For an extended discussion of "lock-in" and its competitive and economic significance, see CARL SHAPIRO & HAL R. VARIAN, INFORMATION RULES: A STRATEGIC GUIDE TO THE NETWORK ECONOMY 103-34 (1999).

14. *See* GRINDLEY, *supra* note 5, at 61-70.

15. Am. Nat'l Standards Inst., *Annual Report*, 6, 17, http://public.ansi.org/ansionline/Documents/News%20and%20Publications/Brochures/ar_2001.pdf (last visited Mar. 13, 2003).

16. *Id.* at 9-10. For example, JEDEC (formerly the Joint Electron Device Engineering Council) is "the semiconductor engineering standardization body of the Electronic Industries Alliance (EIA), a trade association that represents all areas of the electronics industry," JEDEC, *About JEDEC*, http://www.jedec.org/Home/about_jedec.cfm (last visited Mar. 10, 2003). It has eleven committees and thirty-eight subcommittees (including "letter subcommittees"). JEDEC, *Roster of Committees*, http://www.jedec.org/service_members/Rosters/committee_roster.cfm (June 2001).

17. A recent survey by Professor Mark Lemley identified forty-three SSOs and consortia in the "telecommunications and computer-networking industries" alone. Mark A. Lemley, *Intellectual Property Rights and*

In addition, many "standards" are not set by SSOs at all. Rather, they reflect the market success of a particular product in competition with other competing products. Such "de facto" or "market" standards are common in what economists term "network industries" in which consumers benefit by adopting products or processes adopted by others.[18] Well-known examples include VHS VCRs (which "won" a "market standards" war with Sony's Betamax VCRs) and Microsoft's DOS and Windows operating systems.[19]

2. Standards Versus Regulations

Another important distinction is between private standards and government regulations. While many regulations are often described as "standards" (such as "clean air standards" or "health and safety standards"), in practice regulations often serve to limit the range of goods or services that can be sold, in ways that privately adopted standards do not. The principal antitrust concern with regulatory standards is that interested parties may be able to co-opt the regulatory process to protect their market position against potential competitors. For example, many local building codes routinely adopt (as regulations) standards promulgated by otherwise-private SSOs such as the National Fire Protection Association (NFPA).[20] The rationale for such delegation is that the SSO has specialized expertise in the area (especially with respect to safety issues involving changing technology) that local government bodies are likely to lack.

But by manipulating the actions of those SSOs, private parties can achieve an effect that would otherwise require lobbying thousands of different city councils. The best known example is *Allied Tube & Conduit Corp. v. Indian Head, Inc.*, where manufacturers of metal electrical conduit "stuffed the ballot box" (by recruiting new members to the NFPA) in order

---

Standards Setting Organizations, 90 CAL. L. REV. 1889, 1903, 1973-80 (2002). He acknowledged that his list is "by no means comprehensive, even within the telecommunications and computer-networking industries" and that he "consolidated multiple working SSOs operating under the aegis of the same SSO and subject to the same policy." *Id.* at 1903 & n.45.

18. For a discussion of network industries, see SHAPIRO & VARIAN, *supra* note 13, at 173-225.

19. *See* GRINDLEY, *supra* note 5, at 75, 97, 151-52.

20. *See* National Fire Protection Association, *About NFPA*, *at* http://www.nfpa.org/catalog/home/AboutNFPA/index.asp (last visited Mar. 10, 2003).

to prevent the SSO from authorizing the use of competing polyvinyl chloride (PVC) conduit for electrical wiring.[21]

Similarly, product gradation standards have sometimes been used, in conjunction with government regulations, to limit competition, often by prohibiting the sale of (purportedly) "unsafe" products (e.g., PVC electrical conduit) or of products below a certain size or "quality" level. The latter is especially common in agricultural commodities. For example, certain agricultural marketing programs, relics of the New Deal, allow a marketing board to prevent the sale of "low quality" products.[22] Similarly, professional licensing requirements are often supported on the grounds that they prevent "unqualified" individuals from practicing the profession, albeit at the (often unstated) cost of restricting entry and raising prices.[23]

Just as "market" or de facto standards raise issues that do not arise in the context of formally adopted standards, regulations (or regulatory standards) also raise a host of new issues. For example, government agencies (unlike private entities) enjoy governmental immunity from suit.[24] Conversely, government agencies (unlike many private entities) are constrained by the requirements of due process generally and the strictures of administrative law, and often cannot adopt the sorts of streamlined procedures that private SSOs are able to adopt.

Perhaps the most significant difference between regulations and standards involves the coercive power of regulations. Most privately-set standards are voluntary; individuals and firms can, if they choose, elect not to adhere to the standard.[25] Unlike private standards, many regulations

---

21. Allied Tube & Conduit Corp. v. Indian Head, Inc., 486 U.S. 492, 495-97 (1988).

22. *See* Agricultural Marketing Service, U.S. Dep't of Agric., *Fruit and Vegetable Programs*, http://www.fred.ifas.ufl.edu/courses/AEB6383/Lectures 2001/mktorders.pdf (last visited Mar. 10, 2003).

23. For a discussion, see J.A.H. Maks and N.J. Philipsen, *An Economic Analysis of the Regulation of Professions*, EUROREGIONAL J. OF SCI. 17-19, *at* http://www.fdewb.unimaas.nl/Eurecom/professionalregul.pdf (last visited Mar. 10, 2003).

24. For a discussion of the state-action and petitioning immunity issues associated with regulatory agencies, see 1 PHILLIP E. AREEDA & HERBERT HOVENKAMP, ANTITURST LAW ¶¶ 200-231 (2d. ed. 2000).

25. This clearly does not mean that doing so is without cost. To the contrary, the benefits associated with being compatible with the standard, and the costs associated with being "the odd man out," are often so great that firms frequently have little practical choice but to comply with the standard. But

are not "voluntary." Since compliance with government regulations is frequently not voluntary (unlike many privately-set standards), individuals and firms often have little choice but to try to influence whether a regulation will be adopted, the content of that regulation, or both. As such, individuals and firms have immunity under the *Noerr-Pennington* doctrine from antitrust liability for "petitioning" conduct designed to influence what regulations are adopted.[26]

B. A CASE STUDY: REFORMULATED GASOLINE

A recent well-known and controversial example of how a regulatory "standard" raises intellectual property issues involves reformulated gasoline.[27] It is widely recognized that air pollution is a significant social and environmental problem, and a major source of air pollution is automotive exhaust. In the early 1990s, the California Air Resources Board (CARB), a government entity, was under a legislative mandate to adopt regulations to reduce the level of pollution.[28]

While initial efforts to address the problem focused on changing the *car* (e.g., mandating the use of catalytic converters and requiring cars to pass periodic "smog checks"), it became apparent that progress could also be made by changing

---

the source of the constraint is economic, not regulatory.

26. E. R.R. President's Conf. v. Noerr Motor Freight, 365 U.S. 127, 135-40 (1961); United Mine Workers v. Pennington, 381 U.S. 657, 670 (1965).

27. In March 2003, the Federal Trade Commission (FTC) filed an administrative complaint against Unocal Corporation, alleging that Unocal had violated section 5 of the FTC Act in connection with Unocal's patents on reformulated gasoline. Compl., *In re* Union Oil Co. of Cal. (Fed. Trade Comm'n Mar. 4, 2003) (No. 9305) [hereinafter FTC Unocal Compl.], http://www.ftc.gov /os/2003/03/unocalcmp.htm (last visited Mar. 10, 2003).

The following factual discussion is drawn from a number of sources, including the FTC Complaint. For background information and an extensive discussion of various issues, including links to various briefs filed in the cases, see Unocal's patent website, http://www.unocal.com/rfgpatent/index.htm (last updated Mar. 4, 2003). See also the various reported opinions, including: Union Oil Co. of Cal. v. Atl. Richfield Co., No. CV-95-2379-KMW, 1998 U.S. Dist. LEXIS 22847 (C.D. Cal. Mar. 10, 1998); Union Oil Co. of Cal. v. Atl. Richfield Co., 34 F. Supp. 2d 1208, 1222 (C.D. Cal. 1998); Union Oil Co. of Cal. v. Atl. Richfield Co., 208 F.3d 989 (2000), *cert. denied*, 531 U.S. 1183 (2001), and sources cited therein. Further factual discussion (and additional citations) is found in Janice M. Mueller, *Patenting Industry Standards*, 34 J. MARSHALL L. REV. 897 (2001), and Scott H. Segal, *Fuel For Thought: Clean Gasoline and Dirty Patents*, 51 AM. U. L. REV. 49 (2001). (We should note that, based on our first-hand familiarity with the matter, we disagree with many of the factual contentions and conclusions reached by those authors.)

28. CAL. HEALTH & SAFETY CODE § 43018(a) (West 1996).

the nature of the *gasoline* that cars burned.[29] CARB was considering adopting strict regulations requiring that gasoline sold in California meet certain technical criteria governing gasoline composition and/or characteristics.[30]

A research group (Auto/Oil) comprised of representatives of the gasoline industry (refiners) and the automobile industry, conducted research on how to change gasoline formulations to reduce pollution.[31] One participant in Auto/Oil was Unocal Corporation, at the time a major gasoline refiner and marketer in California.[32] Unocal scientists believed that the Auto/Oil research was too limited in scope (and, in the Unocal scientists' opinions, aimed in the wrong direction) and proposed a more extensive research project.[33]

In particular, Unocal believed that CARB's proposed "strict limits" approach to permissible gasoline reformulations was unduly restrictive and that it would be possible to adopt a more flexible approach involving what became known as a "predictive model" (whereby one would measure certain gasoline characteristics and predict the level of pollution that would result from burning gasoline with those characteristics).[34] In Justice Breyer's terminology (discussed above), CARB was proposing a *design* standard, requiring that gasoline meet certain fixed limits on various characteristics, while Unocal was advocating a (limited) *performance* standard, whereby refiners would be allowed to make and sell other gasoline formulations that did *not* meet the fixed limits so long as the predictive model suggested that the other formulations would generate no more pollution than the fixed-limit gasoline formulations.

Auto/Oil rejected the Unocal proposed research agenda.[35]

---

29. Both the federal Environmental Protection Agency and CARB ultimately adopted regulations for reformulated gasoline. *See* Segal, *supra* note 27, at 51-54.

30. *Id.* at 55-56.

31. *See id.* at 55; Unocal, *Auto/Oil Study Provisions*, *at* http://www.unocal.com/rfgpatent/rfgao.htm (last visited Mar. 10, 2003).

32. Unocal subsequently sold its refining and marketing operations to Tosco. *See* Press Release, Unocal, Tosco Sign Agreement for Sale of Unocal's West Coast Downstream Assets, http://www.unocal.com/uclnews/96htm/121696a.htm (Dec. 16, 1996).

33. Unocal, *RFG Emissions Research*, *at* http://www.unocal.com/rfgpatent/rfgresch.htm (last visited Apr. 17, 2003).

34. *See* FTC Unocal Compl., *supra* note 27, ¶¶ 37, 44, 47.

35. Unocal, *RFG Emissions Research*, *at* http://www.unocal.com/rfgpatent/

The terms of the Auto/Oil research collaboration agreement made it clear that, while group-sponsored research would be made freely available to all participants, participants were free to conduct their own proprietary research and to patent the results.[36] Unocal funded a significant research effort, which demonstrated both (1) that air pollution could be reduced significantly by reformulating gasoline in previously unexplored ways and (2) that it was possible to generate a predictive model.[37]

Unocal scientists filed for a patent on their invention, and then disclosed the results of their research (but *not* the fact that they had applied for a patent) both to Auto/Oil and CARB.[38] Some four months later, CARB adopted regulations (to go into effect several years later) requiring California gasoline refiners to make and sell reformulated gasoline.[39] In response to the regulations, California refiners had to spend billions of dollars to upgrade their refineries to make it possible to produce adequate quantities of reformulated gasoline.[40]

Notably, the Unocal research was apparently instrumental in persuading CARB to use a flexible—"predictive model"—approach.[41] The key economic benefit of such a "predictive model" approach was that it would give refiners more flexibility to produce acceptable gasoline, thus significantly reducing the cost of complying with the new regulations.

In February 1994, between the date that the CARB regulations were adopted (in November 1991) and the date that the regulations went into effect (in March 1996), Unocal's first patent issued.[42] The claims of the patent were such that many (though *not* all) of the gasoline formulations that satisfied the

---

rfgresch.htm (last updated June 15, 2001).

36. Unocal, *Auto/Oil Study Provisions*, *at* http://www.unocal.com/rfgpatent/rfgao.htm (last visited Mar. 10, 2003).

37. *See* Unocal, *RFG Emissions Research*, *at* http://www.unocal.com/rfgpatent/rfgresch.htm (last updated June 15, 2001); FTC Unocal Compl., *supra* note 27, ¶ 37.

38. FTC Unocal Compl., *supra* note 27, ¶¶ 32, 61-62, 66.

39. *Id.* ¶ 44; Unocal, *Timeline of RFG Patent Issues*, *at* http://www.unocal.com/rfgpatent/rfgtime.htm (last visited Mar. 4, 2003).

40. Western States Petroleum Association, *California Petroleum Industry Overview*, *at* http://www.wspa.org/factcpio.htm (last visited Mar. 10, 2003); FTC Unocal Compl., *supra* note 27, ¶ 93.

41. *See* FTC Unocal Compl., *supra* note 27, ¶¶ 37, 39-40, 43, 45.

42. Five such patents have now been issued. Unocal, *Unocal's RFG Patents*, *at* http://www.unocal.com/rfgpatent/patents.htm (last visited Mar. 10, 2003).

CARB regulations were also covered by the patent.[43] Unocal announced that it would license other refiners to use its patent.[44] Before the licensing terms were announced, a group of California refiners sued Unocal, seeking a declaratory judgment that the Unocal patent was invalid.[45] After a jury trial, Unocal's patent was upheld, and Unocal was awarded significant damages for patent infringement.[46]

One commentator cites Unocal's conduct as an example of "patenting industry standards."[47] This claim ignores the fact that what was at issue in the Unocal example was not an "industry" or SSO setting a "standard," but a *government* body adopting a *regulation* that imposed billions of dollars of costs on the regulated California refiners (including Unocal). Unocal's disclosure of its research apparently played a significant role in CARB's adoption of a "predictive model" that reduced refiners' costs of complying with the regulations by millions of dollars per year.[48] Moreover, the regulations at issue did not implicate the sorts of "compatibility" (or "interoperability") standardization issues that are at the core of many standards.

As such, the CARB/Unocal situation raises a wide range of issues not present in the context of private SSOs. In particular, the CARB/Unocal situation raised constitutional takings

43. The jury concluded that only 29% of the reformulated gasoline made and sold by California refiners infringed the Unocal patent. Unocal, *Data on Patent Infringement*, *at* http://www.unocal.com/rfgpatent/rfgdata.htm (last visited Mar. 10, 2003).

44. Unocal, *Timeline of RFG Patent Issues*, *at* http://www.unocal.com/rfgpatent/rfgtime.htm (last visited Mar. 10, 2003).

45. Union Oil Co. of Cal. v. Atl. Richfield Co., 208 F.3d 989, 994 (Fed. Cir. 2000), *cert. denied*, 531 U.S. 1183 (2001). The refiners also claimed that the invention arose out of the Auto/Oil research, thus challenging Unocal's claim that its scientists were responsible for the patented invention. *Id.* at 995-96. The refiners also claimed that Unocal had failed to adequately disclose the patented invention, thus not complying with the patent law's "written disclosure" and "enablement" requirements. *Id.* at 996-1001. The trial court and court of appeals both rejected the refiners' arguments. *Id.* at 991.

46. Unocal, *Testing the Patent in Court*, *at* http://www.unocal.com/rfgpatent/rfgsuit.htm (last visited Mar. 10, 2003). David Teece testified as Unocal's damages expert at that trial. The discussion in this paper is based solely on publicly available information (including the public transcript of that trial) and does not reflect any confidential or proprietary information learned during the course of that engagement. The views expressed in this paper are those of the authors, and do not necessarily reflect Unocal's position.

47. *See* Mueller, *supra* note 27, at 897-901.

48. *See* FTC Unocal Compl., *supra* note 27, ¶¶ 37, 39-40, 43, 45.

concerns,[49] *Noerr-Pennington* immunity issues,[50] and issues of administrative law regarding the obligations (if any) of participants in a regulatory process to disclose confidential proprietary information (i.e., the fact that Unocal had pending patent applications[51]) to the regulatory agency. Moreover, the CARB regulations were *not* coextensive with the scope of the Unocal patent. Indeed, the jury concluded that only some 29% of the reformulated gasoline made and sold by California refiners infringed the Unocal patent.[52] As such, firms could and did comply with the regulations without thereby infringing the patent.[53]

Finally, as noted above, the first Unocal patent issued (and Unocal's plans to license the patent were announced) before the CARB regulations were slated to go into effect. CARB continues to maintain its regulations even after a jury ordered the other California refiners to pay Unocal substantial damages for patent infringement.[54] If CARB believes that Unocal improperly "captured" the CARB regulations, it has

---

49. Professor Mueller proposes,

> When government mandates a technological standard, particularly a standard pertaining to public health and safety, any entity holding patent rights in the subject matter of the standard should be required to license all users at reasonable commercial terms. If the patent owner fails to meet this requirement, the government *should consider the exercise of eminent domain* over the patent.

Mueller, *supra* note 27, at 945 (emphasis added).

50. E. R.R. President's Conference v. Noerr Motor Freight, 365 U.S. 127, 138 (1961); United Mine Workers v. Pennington, 381 U.S. 657, 670 (1965).

51. At the time, pending patent applications were confidential. Congress has since amended the patent statute to publish most (though not all) patent applications eighteen months after filing. *See* 35 U.S.C. § 122(b)(1) (2001).

52. Unocal, *Patent Infringement Data*, *at* http://www.unocal.com/rfgpatent/rfgdata.htm (last visited Mar. 24, 2003). The percentage varied significantly by refiner. Because the patent claims were different for higher-octane gasoline than for lower-octane gasoline, the infringement percentage was also significantly higher for premium gasoline than for regular gasoline. *See* Unocal, *Infringement Data from the Trial*, *at* http://www.unocal.com/rfgpatent/rfgsuit.htm (last visited Mar. 10, 2003).

53. A subsequently adopted industry-wide reduction in the octane level of premium gasoline in California is likely to reduce the infringement percentage significantly from the 29% figure on a going-forward basis, making the claim that the patent "captured" the standard even less credible. *See Lowered Octane Fuels Debate*, *at* http://www.geocities.com/njvorc/loweroctan.html (last visited Mar. 10, 2003). Similarly, a number of refiners have contended that they are able to "blend around" the Unocal patents and thus do not need to take a license.

54. *See* Unocal, *Testing the Patent in Court*, *at* http://www.unocal.com/rfgpatent/rfgsuit.htm (last visited Mar. 24, 2003).

clearly had the opportunity to address the issue by rescinding its regulations. The fact that it has not done so casts doubt on any suggestion that CARB would have adopted different regulations had it known ex ante that Unocal had applied for a patent.[55]

These issues, though significant, are not implicated in most standards-setting contexts and are thus tangential to our main concerns. Therefore, this Article will focus on formal standards set by private (non-governmental) SSOs, with particular focus on compatibility/interoperability standards.

One point is worth noting, however. There is no dispute that Unocal disclosed its research results to CARB and Auto/Oil.[56] But research results do not grow on trees. The revelation of the results ipso facto implied that research was done to discover those results. Such research is costly. For-profit firms such as Unocal presumably conduct research, despite the cost, because they believe that they will benefit by doing so. One common way to benefit from research is to seek (and hopefully receive) a patent on the results of that research. By granting the Unocal patents, the Patent and Trademark

---

55. One possible caveat here is that, in response to the adoption of the CARB regulations, California refiners spent over four billion dollars upgrading their refineries to make reformulated gasoline. *See* Western States Petroleum Ass'n, *California Petroleum Industry Overview*, *at* http://www.wspa.org/factcpio.htm (last visited Mar. 10, 2003). Since those costs have already been incurred and cannot be recovered (i.e., they are what economists term "sunk" costs), CARB *may* have reasoned that there was no point in rescinding the regulations once the investment had already been made.

This argument ignores the fact that CARB anticipated that the adoption of the CARB regulations would add twelve to seventeen cents per gallon to the cost of gasoline, while the actual price increase following the implementation of the regulations was only five to eight cents. CARB, *Final Statement of Reasons to Rulemaking, Including Summary of Comments and Agency Response*, 85 (Nov. 1991) (document on file with authors). In other words, CARB was willing to adopt the regulations in the belief that the benefits to the public as a whole (in terms of reduced air pollution) exceeded the costs to automobile users (in the form of higher prices), even when it anticipated that the costs would be significantly higher than they in fact turned out to be.

As such, it is unlikely that the relatively small amount of royalties collected (of 5.75 cents per gallon on the infringing 29% of gasoline production, which amounts to 1.67 cents per gallon on all gallons) would have been the "straw that broke the camel's back" in CARB's decision to adopt the regulations. *See* Unocal, *Testing the Patent in Court*, *at* http://www.unocal.com/rfgpatent/rfgsuit.htm (last visited Mar. 24, 2003).

56. *See* Segal, *supra* note 27, at 58-59; Unocal, *Unocal Patent Issues Timeline*, *at* http://www.unocal.com/rfgpatent/rfgtime.htm (last visited Mar. 24, 2003).

Office (PTO) agreed that the Unocal's scientists' invention was worthy of being patented. And in upholding Unocal's patent against a variety of challenges to validity and inequitable conduct, the trial court judge ruled (in effect) that Unocal had acted properly in obtaining its patent.[57]

The Auto/Oil rules explicitly allowed members (including Unocal) to conduct their own independent research and to patent the results.[58] The other Auto/Oil members presumably knew or should have known that the Unocal research results were the result of a Unocal-funded research effort—indeed, the research program that Auto/Oil had been offered but had previously rejected. No one should have been surprised that Unocal, having invested in the research and having discovered a better way to make reformulated gasoline, applied for a patent on that invention.

Absent some explicit CARB rule or regulation requiring disclosure of pending patent applications—and there was *no* such rule; indeed, CARB did *not* ask Unocal (nor, apparently, any other firm) whether it had relevant patents or patent applications—Unocal's "failure" to disclose its patent application (especially given that pending applications were confidential under then-current PTO rules) should *not* have deluded anyone into believing that Unocal would *not* seek to patent its invention.[59] Others may have believed or assumed that no patent would ever *issue*, but that mistaken belief was not due to any conduct by Unocal.

---

57. *See* Union Oil Co. v. Atl. Richfield Co., 34 F. Supp. 2d 1208, 1210 (C.D. Cal. 1998); Union Oil Co. v. Atl. Richfield Co., No. CV-93-2379-KMW, 1998 U.S. Dist. LEXIS 22847, at *5 (C.D. Cal. Mar. 6, 1998); Segal, *supra* note 27, at 66-67.

58. *See* Unocal, *Independent Research Provision of the Auto/Oil Study*, *available at* http://www.unocal.com/rfgpatent/rfgsuit.htm (last visited Mar. 10, 2003).

59. In its Complaint, the FTC (correctly) notes that Unocal agreed with CARB to make its research *data* and *equations* public and non-proprietary. *See* FTC Unocal Compl., *supra* note 27, ¶¶ 39-42. The FTC alleges that this meant that Unocal "created the materially false and misleading impression that Unocal had relinquished or would not enforce any proprietary interests in its *emissions research results*." *Id.* ¶ 78.b (emphasis added).

The Unocal patents cover various low-emissions formulations of gasoline. Gasoline formulations are not *data* (or equations). The FTC's Complaint effectively transforms Unocal's statement that its *data* was not proprietary into a statement that *patent claims* derived from the research program that yielded that data were non-proprietary.

II. ANTITRUST ISSUES IN STANDARDS SETTING

There are two main classes of antitrust issues in the context of standards setting: *procedural* issues, which address the *method* by which the standard was set, and *substantive* issues, which go to the *content* of the standard adopted.

Procedural issues include both the prospect that the SSO will exclude interested parties who are thus denied a stake in setting a standard that affects them, and concerns about manipulation of the standards-setting process, including "stuffing the ballot box."[60] One common class of issues involves SSO procedural rules that give some participants undue influence over the outcome of the standards-setting process, or that bias the outcomes of the process in particular ways.[61] Substantive issues tend to arise less frequently, if only because the antitrust authorities rarely have the expertise to evaluate whether the "appropriate" standard was chosen.

A. STANDARDS, INTELLECTUAL PROPERTY, AND MARKETS

One area that has received much attention of late has been the interplay between standards and intellectual property rights that "read on" the standard.[62] Before addressing this issue in detail, it is worth drawing some important background distinctions.

Perhaps the most common context in which standards are set involves the development of new products (or aspects of products, such as interfaces between different products) for which standardization is seen to be desirable. Often, the SSO contains representatives from both those who will make the product and those who will use the product. For example, when setting a pin-assignment standard for a new generation of computer chips, the SSO will often involve representatives of both chipmakers and computer (or motherboard) makers.[63] In other words, both the *supply side* and the *demand side* of the potential market for the standardized product will typically be represented. From an economic perspective, this is not

60. *See* HOVENKAMP, ET AL., *supra* note 1, §§ 35.3-35.4, at 35-18 to 35-33.

61. *See id.*, § 35.5a, at 35-34 to 35-37.

62. *See id.*, *supra* note 1, § 35.1b, at 35-6 to 35-7. See also the articles cited in Lemley, *supra* note 17, at n.11.

63. The JEDEC website indicates, "Presently there are about 300 member companies in JEDEC including both manufacturers and users of semiconductor components and others allied to the field." *See* JEDEC, *at* http://www.jedec.org/Home/about_jedec.cfm (last visited Mar. 10, 2003).

surprising, because the standard the SSO adopts will often affect both suppliers and customers. Both sides of the market strive to ensure that the chosen standard reflects their concerns.

But when considering the relationships between standards and intellectual property, there are two fundamentally different "markets" at issue. The first is the *product* market: the market for the products (e.g., the computer chips whose features are being standardized) themselves. The second is the *technology* market: the market for the technology (whether patented or unpatented) used to design and manufacture the product. In the technology market, patent holders license their technology to manufacturing firms, which then use that licensed technology (and other inputs) to make and sell the standardized products.[64] Economists call demand for technology a "derived demand," derived from the demand for the products.[65] As noted above, many SSOs include both suppliers (manufacturers) and customers in the *product* market.[66] And the fact that the manufacturers may need to use intellectual property belonging to others in order to make and sell the standardized product means that the *demand* side of the *technology* market is represented in the standard-setting process. But there is often little or no assurance that the *supply* side of the technology market—the *owners* of the relevant intellectual property—will be adequately represented in the standard-setting process.

In some contexts—especially in some high-tech industries such as semiconductors—many, but not all, patent holders are also manufacturers. Such firms "wear three hats," often participating in the relevant markets in three fundamentally different roles. First, as patent holders, they may *out*-license their patented technology to others, and, as such, they are sellers in the technology market. Second, as manufacturers they manufacture and sell the standardized products (i.e., they are sellers in the product market). Third, they may also *in-*

64. For a discussion of the distinction between product markets and technology markets, see, for example, U.S. DEP'T OF JUSTICE & FED. TRADE COMM'N, ANTITRUST GUIDELINES FOR THE LICENSING OF INTELLECTUAL PROPERTY 8 (1995).

65. For a discussion of the concept of derived demand, see the survey article, J.K. Whitaker, *Derived Demand*, *in* 1 THE NEW PALGRAVE DICTIONARY OF ECONOMICS 813-14 (John Eatwell et al. eds., 1998) [hereinafter PALGRAVE].

66. *See supra* note 63 and accompanying text.

license other firms' patented technology (i.e., they are buyers in the technology market).[67]

Moreover, there are some firms in many industries that do not fit this three-role model. For example, some manufacturing firms have no significant patented technology of their own. As such, they are not sellers in the technology market. If (as is common in the semiconductor industry) they need to use patented technology belonging to others in order to make and sell products in the product market, they will be buyers in the technology market, needing to obtain the necessary in-licenses. Firms without patented technology of their own to "swap" (barter) with other patent-holding manufacturers (in the form of out-licenses), often pay royalties that are significantly *greater* than the *net* "balancing payment" royalties paid by other firms that *do* have valuable patents.[68] Finally, some firms are what might be termed "pure play" technology companies and do not participate in the product market. They out-license their patented technology for cash to manufacturing firms. Because they are not manufacturers, they do not need in-licenses for other firms' patents.

B. STANDARDS SETTING AND DIVERGENT ROLES

Many of the tensions involved in standard-setting, and many of the most complicated public policy issues, can best be understood by recognizing two key implications of the multiplicity of roles. First, because different firms play different roles in the technology and product markets, the interests of different participants in the standard-setting

---

67. It is rare that such firms also play a fourth role, as a buyer in the product market, for the simple reason that firms are rarely both buyers and sellers of the same commodity. Some buyers in the product market (e.g., computer manufacturers who buy chips), however, may also own relevant patents and, thus, may be sellers in the technology market.

In some standards-setting contexts, the SSO is trying to develop standards for the next-generation of products. As such, the products at issue do not yet exist and thus are not yet part of the product market. But some of the technology to make and sell the next-generation products presumably already exists. What is being standardized are certain aspects of that technology (e.g., certain design features of the next-generation of computer chips). What is at issue in connection with the standards-intellectual property issue is manufacturers' access to the technology that they will need to make the next generation of products.

68. *See* Peter C. Grindley & David J. Teece, *Managing Intellectual Capital: Licensing and Cross-Licensing in Semiconductors and Electronics*, 39 CAL. MANAGEMENT REV. 8, 9, 18-19, 33 (1997).

process are frequently *not* aligned. Second, because some firms play multiple roles (e.g., a firm can be a seller in the technology market and a buyer in the product market), there are often internal conflicts within the firm when the firm's different roles conflict with one another.

Further, because of the technical nature of many standards, SSO participants tend to be technical people, often engineers, who rarely are familiar with their firms' patent portfolios.[69] Moreover, in our experience, engineers tend to have a negative attitude toward intellectual property (IP) rights generally, seeing IP claims as "getting in the way" of choosing the "best technological solution." To some extent, this may be a consequence of three key facts. First, the patent prosecution process takes time. Professors Allison and Lemley estimate that the mean time lag between the application and the issuance of patents issued between 1996-98 was 2.77 years, and the median was 2.22 years.[70] Second, inventors often disclose their inventions to the public (whether by publishing papers or by marketing products incorporating the innovation) after filing their patent applications but while the applications are still pending. Consequently, others know of the invention before the patent issues; they know of the technology but not of the patent claims associated with that technology. Third, many patented innovations are independently discovered by others. The patent system grants the patent to the first to invent,[71] but another inventor may have already independently discovered the technology before the patent issues.

The confluence of these three factors, coupled with the engineering mentality of many SSO participants, suggests that SSOs often downplay the technological contributions that presumably led to the issuance of the patent in the first place. By the time the patent issues, the patented technology is often seen as "old hat" or is already widely known in the industry, and the patent's grant of exclusivity is thus seen as merely taking away others' ability to do what they already know how

---

69. The authors have obtained this information from conversations with senior management at several major semiconductor manufacturers. *See also* Lemley, *supra* note 17, at 1907 ("A company's representative to such an SSO is normally an engineer with little or no understanding of patent law.").

70. John R. Allison & Mark A. Lemley, *The Growing Complexity of the United States Patent System*, 82 B.U. L. REV. 77, 98 (2002).

71. *See* ROBERT MERGES, PATENT LAW AND POLICY: CASES AND MATERIALS 37-38 (2d ed. 1997).

to do.[72] This attitude is exacerbated by two commonly held beliefs: (1) patents are frequently issued for trivial inventions and (2) the scope of the issued patent is often much broader than the contribution of the inventor.

One major public policy issue thus involves *balancing* the interests of intellectual property owners and the users of that intellectual property. Almost by definition, the latter are likely to outnumber the former; a patent has only one owner, but multiple manufacturers may need to use the patented technology. Hence, SSOs tend to be dominated by the demand side of the technology market, and they are likely to adopt procedural and substantive rules that favor IP users over IP owners. Moreover, some SSOs, especially in the Internet field, have a policy of flatly refusing to adopt a standard that implicates any proprietary technology, insisting on "open" standards.[73]

C. PATENTED STANDARDS AND SOCIAL EFFICIENCY

Indeed, economics suggests that SSOs have a strong tendency to act in a socially *inefficient* fashion when determining whether to adopt a standard on which a firm has a patent. Royalty payments for the use of a patented technology are a *transfer payment* from the users of the patent to the owner of the patent. Thus, the royalty payments per se[74]

---

72. This suggests that the commonly held view that patent infringement inherently involves improperly and knowingly "copying" others' innovations is incorrect. In our experience, a significant amount of patent infringement is "inadvertent" in the sense that the infringer did not learn of the patented technology directly from the patent holder, but developed the technology independently.

73. The best-known example is the World Wide Web Consortium (W3C). Though W3C recently contemplated changing its policies to allow the use of patented technology in its standards, it decided to retain its earlier policy of refusing to adopt any standard that implicated patented technology. *See* Margaret Kane, *W3C Bows to Royalty Free Pressure*, CNET News.com, *available at* http://news.com.com/2100-1001-965863.html (Nov. 22, 2002).

74. The fact that the patent holder is charging for the use of its patent, rather than letting others use the patents for free, leads to some degree of what economists term "dead weight loss," in the sense that there are some consumers who would buy the patented product (and obtain some degree of consumer surplus) if no royalty was charged, but who will elect not to purchase (and thus will forego the consumer surplus) when a royalty is charged. For a discussion of the concept of dead-weight loss and its economic significance, see HAL R. VARIAN, MICROECONOMIC ANALYSIS 229-30 (3d ed. 1992).

This is not unique to the standards-setting situation, however. Some represent no *net* cost to society: the users have less money, but the patent owner has more. But from a *private* standpoint, SSO members treat the prospect of paying royalties as a *private* cost akin to any other cost. This in turn implies that SSO members have an incentive to adopt societally-*in*efficient production techniques that avoid patented technology.

As an illustration, consider the following numerical example.[75] Suppose that an SSO is trying to set a standard for the widget-making process. Suppose that there are two alternative ways of making (otherwise identical) widgets, one patented and one unpatented.[76] Suppose that the patented way

---

degree of "dead weight loss" is an unavoidable consequence of the fact that society has chosen to encourage innovation by granting successful inventors the right to exclude others from using their patented innovations, or to charge a fee for such use.

75. This issue is treated in more analytic detail in the Appendix. *See infra* notes 237-40 and accompanying text.

76. In his commentary on our Article, Professor Mark Patterson correctly points out that our focus is on the choice between a patented standard (call it P) and an unpatented standard (U), and that we do not consider a third situation, namely, one in which no standard is adopted at all (N). Mark R. Patterson, *Antitrust and the Costs of Standard-Setting: A Commentary on Teece & Sherry*, 87 MINN. L. REV. 1995, 1997 (2003). He notes that the actions of the SSO in adopting *a* standard generates value. *Id.* at 1999. We agree. His focus is on the *relative* contributions of the SSO and the patent holder when moving from no standard (N) to the patented standard (P). *Id.* at 2011. We agree that such a focus is a worthwhile topic for investigation, but it is not the one we focus on. Our analysis focuses on the choice between P and U.

We agree with Professor Patterson that both the SSO and its members need to be compensated for their efforts in developing the standard. *Id.* at 2000-01. We acknowledge that non-members may be able to "free ride" on the standardization efforts of the SSO and its members. But we are not convinced that such free riding is a significant practical concern, especially in cases when the majority of interested industry participants are members of the relevant SSO. We note that SSOs collect dues from their members, and that SSOs can and do copyright their standards and charge non-members for copies, which reduces (though may not fully eliminate) the "free rider" problem. (See ANSI's prices for copies of its standards at ANSI Electronic Standards Store, *at* http://webstore.ansi.org/ansidocstore/default.asp (last visited Mar. 24, 2003).) We also believe that SSO manufacturer members benefit from the standard by making and selling the standardized product, and that SSO purchaser members benefit from the standard by being able to purchase compatible parts at a lower cost than would be the case if no standard were adopted.

We further believe that the costs of standardization efforts, while clearly non-trivial, are often very small relative to the benefits that SSO member firms receive from standardization. For example, the semiconductor industry is a $155 billion a year industry. *See* DRAM Market Drives Semiconductor Revenue Growth, *at* http://www.computeruser.com/news/02/12/17/news4.html (Dec. 17, 2002). Even if standardization efforts in the industry cost are, say, $750 million per year (which we believe is unrealistically high), that would

consumes $100 in real resources to make a widget, while the unpatented way consumes $130 in real resources. Suppose that in either case widgets can be sold for $150 and that the patent holder demands a royalty of $35 per widget for the use of its patent. If the SSO chooses the unpatented technology as the standard, consumers pay $150 for widgets, firms incur $130 in real resource costs to make the widgets (and thus make a $20 profit), and the patent holder receives nothing. Conversely, if the SSO chooses the patented technology, consumers again pay $150 for their widgets, firms incur $100 in real resource costs and pay $35 in royalties to the patent holder (thus making a $15 profit), and the patent holder receives $35 in royalties.

From a societal standpoint, society is clearly better off if the SSO adopts the patented technology as a standard. Society saves $30 per widget in real resource cost savings. The $35 royalty payment from manufacturers to the patent holder is a "wash" from a societal standpoint; manufacturers are $35 worse off, but the patent holder is $35 better off. But the SSO members clearly prefer to adopt the unpatented technology as a standard, because they make a $20 profit per widget rather than the $15 profit per widget that they would make if they adopted the patented technology as a standard.

---

only amount to less than 0.5% of annual industry revenues.

Moreover, we are aware of examples where the SSO itself developed patented technology in the course of developing its standard, and applied for patents in the name of the SSO. One example is the SynchLink consortium (later known as SLDRAM), which developed an alternative specification for DRAMs. SLDRAM contracted with Mosaid Technologies to develop the technology for SLDRAM chips, and Mosaid assigned its intellectual property rights to the SLDRAM consortium. *See* Press Release, *Mosaid to Design Next Generation Memory Technology*, *available at* http://www.mosaid.com/corporate/press97-04-18.htm (April 18, 1997). Unlike Professor Patterson, we see no need for a sui generis form of intellectual property protection for standards per se.

We disagree with certain aspects of his analysis, and (in particular) with his proposal that the patent holder receives an unwarranted "windfall" when its patented technology is incorporated in the standard, relative to what it would receive under the no-standard (N) case. Patterson, *supra*, at 2009. In our view, there are gains to *both* the users of the (patented) standard and the owner of the patented technology from moving from N to P; we see no reason why *all* of those gains should be attributed to the SSO's actions and should accrue to the users of the standard rather than to the patent holder.

D. Policy Implications

The above analysis suggests that SSOs are likely to be biased toward a societally inefficient attitude towards IP when setting standards, for three reasons: (1) the rules of the SSOs are likely to favor the users of IP rather than the owners of IP, as the former outnumber the latter; (2) SSO participants tend to be engineers, with an engineer's bias against patents; and (3) royalties are treated as a private cost by manufacturers and end-users, despite the fact that from a *societal* standpoint they are best seen as a *transfer payment* rather than a (social) *cost*.[77]

This in turn suggests that both the antitrust authorities and the legislature should tread warily when making public policy in this area. The complaints of those who believe that they are being compelled to "overpay" for the use of others' IP embedded in the standard are frequently and forcefully stated.[78] The more reasoned and quieter countervailing arguments focused on the social benefits of innovation and the need to compensate inventors for their efforts often are downed out by this din. The tension between static and dynamic views of efficiency is nothing new in the context of IP. But it suggests that policies that further burden IP and IP holders will only exacerbate the problem.

III. CHOOSING STANDARDS

A. The Choice Of Standards: Divergent Roles and Divergent Beliefs

In choosing a standard, there is typically a range of alternative candidates that could be adopted as the standard. Setting a standard may mean choosing one alternative and discarding alternatives, but it need not do so. For example, many health and safety standards are "inclusive" in the sense that quite different products (e.g., wood, aluminum, and fiberglass ladders) can all satisfy the standard. In other contexts, the adoption of one standard effectively amounts to

---

77. The major caveat here is that firms may expend real resources in what economists term a "rent-seeking" effort to affect the pattern of transfer payments. For a discussion of the economics of rent-seeking, see Gordon Tullock, *Rent Seeking*, *in* PALGRAVE, *supra* note 65, at 147-49.

78. Examples include the other gasoline refiners in the Unocal case (*see* discussion, *supra* Part I.B), and semiconductor manufacturers in the Rambus case (*see* discussion, *infra* Part IV.D.1).

discarding other alternatives that could have been adopted. For example, assigning a particular function to a particular pin in a new computer chip means that chips that assign that function to a different pin will be incompatible with the standard.

In some contexts, the participants in the standard-setting process will have very similar views as to which standard should be chosen, and the task at hand is simply to facilitate the coordination process by agreeing on a standard. In other cases, however, the participants may have genuine disagreements as to which alternative should be adopted as the standard. In many such cases, the disagreements may arise out of the different beliefs, expertise, or roles that the participants bring to the table.

For example, in designing the next-generation microprocessors, certain functions must be performed for the system as a whole to work, but some of those functions can be implemented either in hardware or in software.[79] The SSO must determine what functions (or portions thereof) will be implemented in hardware and what functions will be implemented in software, in order to coordinate the development efforts. Not surprisingly, different parties may possess different degrees of expertise in the two areas. Software firms may believe that it is better to have certain functions implemented in software, while hardware manufacturers may believe the opposite (or vice versa). SSOs routinely address these sorts of disagreements in the give-and-take process of setting the standard.

From an economic perspective, it is worth noting that the positions that firms take, even on "technical" issues, are likely to reflect the competitive positions and comparative advantages that different firms enjoy. A firm with experience in solving a problem in a particular way will, not surprisingly, favor standards that adopt that familiar approach, as that will give that firm a comparative advantage over other firms that have more experience in alternative solutions.

---

79. For example, the choice of the instruction set for microprocessors (e.g., between Complex Instruction Set Computer (CISC) and Reduced Instruction Set Computer (RISC) architectures) in turn affects the nature of the software that runs on such computers. For a (moderately technical) discussion which emphasizes the hardware/software trade-off, see RISC v. CISC, *at* http://cse.stanford.edu/class/sophomore-college/projects-00/risc/risccisc/ (last visited Mar. 10, 2003).

Some comparative advantage may rest in a firm's accumulated knowledge and expertise, some of which may have arisen from research and development that may also have led to patented inventions. Thus, firms may have a preference for adopting a standard on which they have a patent, not primarily because of the existence of the patent per se, but because the firm has a comparative advantage (relative to other firms) in the technological approach reflected in the patented technology. It may be a difficult task to determine whether a firm supports selecting choice *A* rather than choice *B* as the standard because of its belief that technology *A* is superior, because of its comparative advantage in using *A* rather than *B*, because of its economic legal (patent) position with respect to *A* versus *B*, or because of some combination of the above.

B. The Choice of Standards: Ex Ante Versus Ex Post Assessment

Choosing a standard generally means that some alternative will be adopted and other alternatives discarded. Ex ante, prior to the adoption of the standard, there typically will be a range of feasible alternatives available. The alternative selected as the standard may be significantly superior to the alternatives, and if so, it is likely that the SSO would have been reluctant to adopt a different standard.

But in other contexts, there exists a range of roughly equivalent alternatives, and in such cases the need is to pick one and standardize on it so as to facilitate coordination and avoid fragmentation. In such situations, the chosen alternative may be only slightly superior ex ante to other feasible alternatives, and the SSO could have just as easily chosen another alternative.

One clear historical example involves which side of the road countries require automobile users to drive on. In most countries, cars drive on the right-hand side of the road. In the United Kingdom and certain other countries (notably Japan and some former Commonwealth countries), cars drive on the left.[80] There is little reason to prefer one over the other, and the other alternative could equally well have been chosen.

Clearly, it is a factual question as to the extent that the

80. *See Which Side of the Road Do They Drive On?*, *at* http://www.travel-library.com/general/driving/drive_which_side.html#changing (last updated Aug., 2002).

chosen standard was superior to available alternatives on an ex ante basis. This issue has to be evaluated on a standard-by-standard basis."[81]

However, ex post, after the adoption of the standard, once firms have committed to the standard and have made the requisite investment in complementary assets to manufacture and sell the standardized product, switching to an alternative may be much less feasible, for three reasons. First, the industry may have made investments in implementing the (patented) standard. Products may have been designed to meet the standard, and factories geared up to produce the patented standardized products. While from an economic standpoint those costs are often "sunk costs"[82] (not recoverable), manufacturers clearly do not want to incur the additional costs associated with switching to another alternative.

Second, the need or desire for compatibility (especially backwards compatibility with the existing installed product base) may make it costly to switch to a different standard. Third, and similarly, there is often a significant coordination problem in getting all interested parties to switch to an alternative. For example, computer manufacturers may already have designed their motherboards and computers to work with existing standardized chips, and switching to a different chip design would require changes, not only to the chips themselves, but also to the motherboards and computers. The difficulties associated with coordinating the necessary changes may make it impracticable to switch away from the patented standard.

Here again, the left-hand-versus-right-hand-drive example provides an illustration. Once a country has adopted one alternative, it is important for everyone to comport with the standard so as to avoid collisions. The public and private sectors make substantial investments in automobile design and in road construction that are consistent with the chosen alternative and inconsistent with the other. Changing over is complex, requiring significant social planning and coordination,

---

81. From an economic standpoint, what is important is not the technological feasibility of various alternatives, nor cost considerations (narrowly considered), but rather the overall attractiveness (on a quality/cost-adjusted basis) of the various alternatives.

82. For a discussion of the nature and economic significance of "sunk costs," see F. M. SCHERER, INDUSTRIAL MARKET STRUCTURE AND ECONOMIC PERFORMANCE 414-16 (2d ed. 1980).

as evinced by the Swedish changeover in 1967.[83]

C. STANDARDIZATION, LOCK-IN, AND OPPORTUNISM

The asymmetry between the low ex ante cost of choosing an alternative proposed standard and the higher ex post cost of abandoning an existing standard in favor of a new standard causes concerns about the prospect for "lock-in."[84] In recent years this issue has received particular attention when, after the standard has been adopted and becomes established, someone claims an intellectual property right (whether patent or copyright) over the existing standard.[85]

The argument frequently made is that, had the SSO members known of the IP claim ex ante while they were considering the standard for adoption, the SSO members could have chosen another, non-patented alternative.[86] But once the standard has been chosen and widely adopted, it may be much more difficult to avoid the patent. In other words, the adoption of the standard *may* increase the bargaining position of the IP holder. If so, the patent holder may be able to extract, not only the gains from using its patented technology vis-à-vis other alternatives, but also a portion of the gains from standardization generally.

This prospect for ex post "hold-up" is one of the key rationales for requiring a patent holder to disclose its patents ex ante and to specify ex ante the royalty rates (if any) that it intends to seek for the use of its patents.[87] Armed with such knowledge, the SSO can choose to adopt another alternative if it believes that the patent holder's prospective royalties are too high relative to the benefits of using the patented technology.

Whether the SSO *would* have in fact adopted another alternative had it known of the patent claims raises a complex

---

83. For a discussion, see *Which Side of the Road Do They Drive On?, at* http://www.travel-library.com/general/driving/drive_which_side.html (last updated Aug., 2002).

84. *See* HOVENKAMP ET AL., *supra* note 1, § 35.5b, at 35-37 to 35-46; SHAPIRO & VARIAN, *supra* note 12, 104-05, 116-30.

85. The *Dell* and *Rambus* cases provide examples. *See* HOVENKAMP ET AL., *supra* note 1, § 35.5b, at 35-37 to 35-46.

86. See, *e.g.*, FTC Unocal Compl., *supra* note 27, ¶ 5; Compl. ¶ 65, *In re* Rambus, Inc. (Fed. Trade Comm'n June 19, 2002) (No. 9302) [hereinafter FTC Rambus Compl.], http://www.ftc.gov/os/2002/06/rambuscmp.htm (last visited February 23, 2003).

87. Such disclosure and licensing rules will be discussed in more detail *infra* Parts IV.B and IV.C.

counterfactual question: "What would the SSO have done *if* the world had been different?" The answer is likely to be hotly debated,[88] and depends on the particular facts of the standard at issue. The greater the advantages of the (patented) standard over the alternatives that were considered and rejected at the time the standard was originally set, the less likely it is that an alternative would, in fact, have been chosen.

It is true that some SSOs have an absolute policy against adopting a standard that incorporates a (known) patent.[89] But this must be interpreted carefully. These SSOs retain the option of withdrawing the standard entirely, or of not adopting any standard. There need be no presumption that the SSOs would have adopted different standards than those adopted had they known of the existence of relevant patents.

In the more common case, however, the SSO *is* willing to adopt a standard incorporating patented technology if there are good technical reasons for doing so. In such situations, it seems reasonable to require those who contend that the SSO would have adopted a different standard, *if* it had only known of the existence of the (undisclosed) patent, not only to suggest another non-patented alternative standard and to demonstrate that the alternative was acceptable (on a commercial, not merely a technological, basis), but also to give some basis for believing that the SSO *would* have adopted such an alternative had it known of the existence of the patent.[90]

D. Limitations of the "Hold-Up" Concern

The above analysis suggests that the "hold-up" concern has some natural limitations. First, it is limited to what might be termed "necessary" or "essential" patents—patents that *must* be used in order to practice the standard. Often there are other "desirable" patents that firms wish to use when making and selling products that incorporate the standard. But if those

88. Ex post, once a standard has been adopted and a patent reading on the standard has been asserted, the accused infringers clearly have a strong incentive to claim that the SSO would have adopted some alternative non-patented standard had the SSO only known of the existence of the patent. In our experience, such claims rarely articulate *which* alternative would have been adopted, or demonstrate that the SSO *would* in fact have adopted a different alternative.

89. The best-known example is the World Wide Web Consortium, discussed *supra* note 73.

90. This is likely to require an evaluation of the commercial acceptability of alternative standards on a quality/cost-adjusted basis.

patents involve non-standardized features, or "optional" features that are not required in order to practice the standard, the "hold-up" problem does not arise.

Second, the "hold-up" concern is limited to patents that are "implicated by" the *standard* itself, not necessarily by the *products* made that comport with the standard. As noted above, many standards involve only certain features of the product.[91] For example, an SSO may set a standard for the physical size and pin assignment of a new generation of memory chip.[92] The standard determines those particular characteristics of the chip. But there are many other features of the chip that are not standardized. Some of those features may themselves be patented. Adoption of the standard does not affect such independent patent claims.

For example, a basic patent on the integrated circuit (e.g., Texas Instruments's original Kilby patent) covers *all* integrated circuits, regardless of the physical size and pin assignment of the chip.[93] Where such basic patents exist, the new generation chip infringes on the basic patent regardless of which pin-assignment standard the SSO adopts: The adoption of the standard does not affect the strength of the IP holder's position. In such contexts, the "hold-up" problem again does not arise. Hence, the basic-patent holder should not be precluded from enforcing its patent against those who use the newly standardized pin assignment to make chips merely because the patent holder participated in setting the pin-assignment standard. This in turn implies that the test should be the relationship of the patent to the *standard*, not the relationship of the patent to the *products* made that comport with the standard.[94]

---

91. *See supra* text accompanying notes 5-7.

92. *See, e.g.*, JEDEC SOLID STATE TECHNOLOGY ASSOCIATION, DOUBLE DATA RATE (DDR) SDRAM SPECIFICATION 3-7, http://www.jedec.org/download/search/JESD79R2.pdf (May 2002).

93. For a discussion of the Kilby patent and its significance, see Texas Instruments, *About Jack*, *at* http://www.ti.com/corp/docs/kilbyctr/jackbuilt.shtml (last visited Mar. 10, 2003).

94. It is only when the standard is effectively coextensive with the product that these two inquiries yield the same answer.

Moreover, there may be patents on what might be termed "optional" features of the product—features that, though they may be commercially desirable, are not necessary to practice the standard. Here again, the "lock-in" concern is not an issue, because a firm can avoid the patent while still complying with the standard by eliminating the optional feature.

Third, in many industries in which standards play an important role, the fast pace of technological change drives the continual redesign and reengineering of products. For example, the product life cycle in the semiconductor industry is reported to be as low as ten months.[95] Therefore, even if there may be some "lock-in" of earlier designs, once the existence of the patent is disclosed, the SSO has the opportunity to revise the standards, and manufacturers have the opportunity to redesign their products to avoid incorporating the patented features. In other words, the extent of "lock-in" may be limited by the pace of technological change.

E. "MANIPULATION" OF STANDARDS: ACTIVE AND PASSIVE CONDUCT

The "hold-up" concern is particularly present in two paradigm cases of what might be termed "active" manipulation. In the first case, the patent holder participates in the SSO and steers the standard toward the claims of one of its existing patents or pending patent application, in an effort to "capture" the standard within the claims of its IP. In the second case, the patent holder modifies a pending patent application so that the claims of the pending patent read more closely on the proposed standard (or, more precisely, on future products that comport with the standard).

The pejorative use of the term "manipulation," to some extent begs the question. It *assumes* what needs to be *shown*: namely, that the rationale behind the patent holder's conduct constitutes an improper effort to "capture" the standard. The discussion above[96] of the reasons why different firms might prefer different alternative standards—whether because of divergent technical beliefs, differences in comparative advantage, or intellectual property issues—suggests that distinguishing between intentional "manipulation" and less sinister motives may be difficult.

But in many cases, the concern is not with "active" conduct (as above), but with "passive" conduct: the failure to disclose patent rights that may read on the standard being considered. The issue then becomes whether "active" and "passive" conduct should be treated differently. One obvious difference is that

---

95. *See* Tom Dellecave Jr., *Charged With Change*, Information Week, http://www.informationweek.com/545/50oelec.htm (Sept. 18, 1995).

96. *See supra* Part III.A.

"passive" conduct has a much lower prospect for manipulation of the standards-setting process or its outcome.[97]

IV. SSO IP RULES: MAKING THE CASE FOR CLARITY

A. SSOs and IP Rules: Effect on Participation and Organizational Constraints

Many, if not most, SSOs have policies that address the interplay between the standards they adopt and the IP rights of participants.[98] These policies take several forms. Some policies constrain the SSO itself and the standards that it can adopt (or maintain).[99] Other policies seek to impose duties or obligations on SSO participants.[100]

1. Rules Constraining the SSO Itself

Some (but by no means all) SSOs have policies that prevent the SSO from adopting standards on which some individual or entity has (or claims to have) a patent.[101] Some of these policies are absolute, but most allow exceptions if the patent holder declares that it is willing to license its patent, whether royalty-free or on certain terms, to those making products that comport with the standard.[102] Similarly, some SSOs have policies that call for the SSO to withdraw previously approved standards if it is subsequently discovered that there is a patent that reads on the standard.[103]

These policies—which might be termed "adoption policies"—impose restrictions on the *SSO* itself, *not* on the patent holder. To some extent, this is a consequence of the fact

97. This is because "active" conduct can steer the SSO to adopting a patented standard, while "passive" conduct can at most "capture" a standard that the SSO was willing to adopt in any case.

98. *See* Lemley, *supra* note 17, at 1903-06, 1973-75.

99. *Id.*

100. *Id.*

101. Professor Lemley, in an overview of SSOs, asks, "Can Standard Include IP?" *Id.* at 1973.

102. "[T]wenty-nine of the thirty-six SSOs with [patent licensing] policies required members to license their patent rights on ['reasonable and non-discriminatory'] terms." *Id.* at 1906.

103. According to Professor Lemley's survey, examples include CEN/CENELEC ("RAND to entire world required or standard is withdrawn"); ECMA ("RAND, or the standard will be cancelled"); J Consortium ("RAND, or the standard will be referred back to Committee for consideration"); and JEDEC ("RAND, or possible withdrawal of the standard"). *Id.* at 1974.

that private SSOs are voluntary entities. A voluntary SSO cannot bind the rights of non-participants. If an SSO adopts a standard and subsequently discovers that a non-participant has a patent that reads on the standard, the SSO cannot compel the non-participant to relinquish its patent rights. All the SSO can do is elect to "de-establish" the standard.

Such rules, however, often only address part of the issue. They state what the SSO must do *if* and when it learns of a relevant patent, but this begs the question of how the SSO comes to *know* of the existence of a relevant patent. We turn to that issue next.

2. Rules Affecting Participants

The more significant rules or policies impose certain obligations on SSO participants. These policies take three main forms: *search* rules, *disclosure* rules, and *licensing* rules. Under a search and/or disclosure rule, the question is "to what extent does a participant in an SSO have an obligation to search for, and/or to disclose, the existence of IP (whether issued or pending, whether its own or belonging to others) that may relate to a (proposed or issued) standard?" Under a licensing rule, the question is "to what extent does a participating IP holder have an obligation to license its IP to those practicing the standard, and on what terms?"

The terms of such policies vary widely across different SSOs, as demonstrated by the results of a recent survey by Professor Mark Lemley.[104] His survey results indicate that a "one size fits all" policy approach is inappropriate.

Search and disclosure rules clearly impose burdens on SSO participants. More significantly, they impose *asymmetric* burdens and benefits. The burdens fall primarily, if not exclusively, on IP holders. The benefits accrue primarily to the users of the standard.

Search and disclosure rules rarely pose problems when the participants in the SSO are roughly symmetrically situated. In such circumstances, firms realize that they will incur the burdens sometimes (when their IP is implicated by a proposed standard) but reap the benefits other times. But in contexts where the participants in the SSO are *not* symmetrically situated—and, in particular, where some participants are pure-play technology companies (suppliers of IP) and others are

104. *Id.* at 1904-06, 1973-75.

primarily users of others' IP—then the problem of conflicting interests can become acute.

3. Participation Constraints

Given the consequences of SSO rules and the nature of voluntary participation, SSOs must tread warily. IP holders must believe that their interests will be protected in the standards-setting process, or they may choose not to participate.[105] Indeed, the proliferation of voluntary special-purpose consortia in many technological areas means that a number of different SSOs, to a greater or lesser extent, "compete" with one another to develop standards.[106] Thus, IP holders that believe that a particular SSO does not adequately protect their interests *may* be in a position to leave that SSO and participate in another SSO that provides better protection for their IP rights.

SSO competition, however, is not a panacea for resolving these concerns. Leaving the SSO often entails foregoing any opportunity to affect the SSO's decision. Especially when the IP holder "wears multiple hats" (as both an IP holder and a participant in the product market, whether as a buyer or seller), the holder often faces a significant conflict between these different roles. As an IP holder, the firm may believe that the SSO's policies provide inadequate protection for its IP. But leaving the SSO to avoid being bound by the SSO's policies may prove difficult if the firm believes that it needs to participate in the standards-setting process in its *other* role(s).

Consequently, the (apparent) "voluntariness" of participation in SSOs may be illusory. Even firms with significant IP portfolios may conclude that they have little practical choice but to participate in certain SSOs; the SSO may be "the only game in town." In particular, the SSO may make standards decisions that significantly affect the firm's participation in the product market.[107] As such, the firm may have a business need to participate in the SSO that outweighs the risks that participation imposes on its IP portfolio.

One concern raised by this pragmatic constraint on firms'

105. This factor differentiates voluntary SSOs from regulatory agencies.

106. In 1998, Sun Microsystems participated in eighty-seven different SSOs, and "there are dozens of different SSOs associated with Internet technical standards alone." Lemley, *supra* note 17, at 1907.

107. For example, a standard for next-generation DRAM chips affects both DRAM manufacturers and DRAM purchasers.

ability to leave an SSO is that other SSO participants who are aware of this factor may take advantage of the firm's predicament by setting onerous rules that adversely affect the IP holders' interests. For instance, the SSO may adopt rules that impose a burden that almost, but not quite, causes the IP holder to elect to quit.

B. Search and Disclosure Rules: Representation and Burden

Firms are represented at SSOs by individuals.[108] The representatives are typically chosen for their technological knowledge of the issue under discussion.[109] A large firm may have hundreds of such representatives attending different SSO meetings.[110] The representatives are not likely to be aware of the scope of their firms' patent portfolios.[111] They are even less likely to be aware of their firms' pending patent applications. Moreover, determining whether a particular patent reads on a proposed standard is often a complex task in claims construction and patent interpretation.[112] It is rare for representatives to have both the requisite knowledge and the relevant expertise to evaluate patent claims.

1. Search

The fact that firms' representatives to the SSO often have limited knowledge of their firms' patent portfolios implies that the *scope* of the obligation (if any) to search for potentially relevant patents is extremely important. As a conceptual matter, there clearly is a continuum of possible search and disclosure rules that might be adopted. At one extreme of the continuum, the representatives may have no search or disclosure obligations whatsoever. Or the representatives might merely have an obligation to disclose patents (and/or patent applications) of which they personally are aware and which they personally believe may implicate the standard. At the other extreme, one might impose on the firm a "disclose it

108. This and the following points are based on our discussions with senior management at major high-tech firms, including Texas Instruments and Hewlett-Packard. *See also* Lemley, *supra* note 17, at 1907.

109. *Id.*

110. *Id.*

111. *Id.*

112. *See* Markman v. Westview Instruments, Inc., 517 U.S. 370, 388-89 (1996).

or lose it" obligation to search for and disclose all potentially relevant patents (or pending applications). Under such a proposal, a firm might be required to certify that it had searched for and disclosed all of its relevant IP. The certification would preclude the firm from subsequently asserting any undisclosed IP.

An obligation to search for "implicated" IP can be extremely onerous. It is a major task to search a patent database and to compare it against the proposed standard. Patent searching is especially problematic when the standard evolves over time. Further, it is often difficult to know whether a patent "reads on" a proposed standard, as that may entail a major effort at claims construction and interpretation.[113]

A search requirement is especially onerous for IP owners who have substantial numbers of patents. Many firms in high-tech industries have thousands of patents, hundreds of which may be potentially relevant to a proposed standard.[114]

Moreover, with the advent of computerized patent databases, there is no particular reason why the patent holder has any comparative advantage over other SSO participants in searching for its potentially relevant patents.[115] As such, one would expect that it would be more cost-effective for the SSO to take on the responsibility of searching for potentially relevant patents, raising the necessary funds to pay for such searches by charging a fee to SSO participants.[116] One key virtue of such

---

113. *Id.* In the closely-analogous licensing context, determining whether a patent holder's patent portfolio reads on a (prospective) licensee's products can be extremely expensive and time consuming, taking a year of effort and hundreds of engineering man-hours. *See* Grindley & Teece, *supra* note 68, at 19, n.42.

114. For example, IBM received over 22,000 patents during the 1993-2002 period, and has an active portfolio of 21,000 U.S. patents and 38,000 patents worldwide. IBM, *State-of-the-Art Patent Portfolio*, http://www.ibm.com/ibm/licensing/patents/portfolio.shtml (last visited Mar. 10, 2003).

115. The main caveat here is pending patent applications. Under current U.S. law, most (but not all) pending applications are "thrown open" for public inspection eighteen months after filing. *See* 35 U.S.C. § 122(b)(1)(A) (2000). Until that date, the application is confidential, and others are not in a position to know whether such an application has been filed. *See id.* §122(a).

116. The economic cost of performing searches is not eliminated when the obligation to search is imposed on SSO members; it just means that the costs are borne by patent holders rather than being spread across all SSO participants. As such, an SSO-based search coupled with a participation fee to finance the search has the potential for allocating the cost of conducting the search more equitably across the various SSO participants than a policy that requires each SSO participant to search its own patents.

an approach (rather than relying on SSO members to search their own patent portfolios[117]) is that an SSO-conducted search would have a much better chance of finding potentially relevant patents that did *not* belong to SSO participants.

In addition, the scope of any duty to search is difficult to specify in advance. Is a "good faith" effort to discover potentially relevant patents sufficient? How does one determine whether a particular search was adequate? At what stage during the standards-setting process must a search be made? On the one hand, earlier disclosure is (other things equal) preferable.[118] On the other hand, the proposed standard itself is likely to evolve over time, meaning the target of the search—patents within the proposed standard's scope—will also change over time and trying to conduct a patent search against such a "moving target" is difficult.[119]

Professor Lemley's survey of SSO rules demonstrates that, of the twenty-one SSOs that he studied that addressed the question whether there was a duty to search, seventeen imposed *no* duty to search, and one had what Professor Lemley called a "*de facto*" duty (in the sense that the patent holder agreed not to sue users of the standard for any patents not disclosed to the SSO).[120] The European Telecommunications Standards Institute follows the European Community policy which (as Professor Lemley puts it) "requires that the patent owner conduct a search unless the standard-setting body commits to do the search itself."[121] The Frame Relay Forum required a "reasonable" search.[122]

According to Professor Lemley, only *one* U.S.-based SSO, the National Institute of Standards and Technology (NIST),

---

117. Another alternative is to impose on each SSO participant the obligation to search, not only its own patent portfolio, but also the publicly available information (such as patent databases), for potentially relevant patents belonging to others. The key advantage of such a proposal is that multiple searches, conducted by different individuals or firms, are more likely to discover a greater number of potentially relevant patents, especially since the issue of whether a particular patent is or is not relevant to the proposed standard is likely to be one on which parties can differ. The obvious disadvantage is the likely resulting duplication of effort.

118. Early disclosure allows the SSO to explore the relative merits of patented and non-patented alternatives in a timely fashion.

119. This last issue is especially significant with respect to pending patent applications, which themselves are "moving targets."

120. Lemley, *supra* note 17, at 1905, 1973-75, 1978 n.387.

121. *Id.* at 1977 n.373.

122. *Id.* at 1905.

formerly the National Bureau of Standards, imposes a more extensive duty to search.[123] And NIST appears to be the exception that proves the rule, since NIST appears to have virtually *no* role in actually *setting* the sorts of standards that are at issue here.[124]

2. Disclosure

Like search rules, the scope of any disclosure obligation is also difficult to pin down in advance.[125] There are two difficulties with the scope of a disclosure obligation. First, *over*disclosure can be as problematic as *under*disclosure. To take an extreme example, a firm could simply list *all* of its issued patents and tell the SSO that it is up to the *SSO* to determine whether any of those patents are potentially relevant to the proposed standard. Such a "core dump" type of disclosure is not very helpful. (Nevertheless, if the SSO has a stringent disclosure policy and there are significant sanctions for failure to disclose, the only safe course for a firm with valuable IP may be to make such an overdisclosure.)

Second, the question whether a particular patent does or does not read on a proposed standard is often unclear, involving claim construction and interpretation.[126] Is disclosure of the

123. *Id.*

124. The NIST website indicates that "[o]ur laboratories have the job of developing ever more accurate ways to measure length, time, mass, temperature, and the other physical quantities that are fundamental to all types of technology." *See* NIST, *Standards*, *at* http://www.nist.gov/public_affairs/standards.htm#Documentary (last updated Mar. 18, 2002). The NIST website also indicates, "In the United States, responsibility for setting product and process standards rests almost exclusively with the private sector. . . . Because of its measurement expertise and its impartiality, NIST is called upon to provide technical support that contributes to the development of high-quality domestic and international standards." *Id.*

According to the 2000 "Memorandum of Understanding" between NIST and ANSI, "NIST's role . . . is to coordinate Federal activities in voluntary standards and to ensure adequate representation of U.S. interests in all relevant international standards organizations." *See* NIST/ANSI Memorandum of Understanding, http://ts.nist.gov/ts/htdocs/210/ansimou.htm (last visited Mar. 10, 2003). Also, "NIST facilitates information exchange between Federal agencies and the private sector on voluntary standards activities . . . . It works with ANSI, Federal agencies, and the private sector to ensure that U.S. interests can participate appropriately in international standards activities to enhance U.S. international competitiveness." *Id.*

125. See discussion *infra* Part IV.E for why it is important that the scope of any disclosure obligation be clearly specified.

126. *See* Markman v. Westview Instruments, Inc., 517 U.S. 370, 388-89 (1996).

existence of the patent enough, or does the patent holder have an obligation to explain why (or how) it believes that the patent reads on the standard?

Many SSOs explicitly disclaim any effort to interpret the patent or to determine whether or not a patent reads on a proposed standard.[127] This unwillingness is not surprising, given that SSOs themselves are generally not comprised of patent lawyers and rendering opinions would increase the SSO's potential liability exposure should the SSO get it wrong.[128]

But once a patent has been disclosed, the SSO must determine whether or not to adopt the standard. If the SSO really is concerned about the prospect that a patent reads on a standard, there would appear to be little choice but to try to interpret the patent. The SSO *itself* (as an entity) may choose not to do so, instead leaving it up to the members of the SSO committee considering the standard to determine whether or not to adopt the proposed standard despite the knowledge of the potentially relevant patent.

A "disclose it or lose it" policy is likely (at a minimum) to delay the adoption of a proposed standard. Rational corporate and/or patent counsel would likely require that participation and subsequent disclosures would have to be "vetted" by patent counsel, because the risk of inadvertently losing the ability to enforce an important patent would otherwise be significant. Economics suggests that firms will rationally err on the side of *over*disclosure,[129] which will force the SSO to examine large

---

127. *See, e.g.*, IEEE STANDARDS ASS'N, COMMENTS REGARDING COMPETITION AND INTELLECTUAL PROPERTY, *at* http://www.ftc.gov/os/comments/intelpropertycomments/ieee.pdf (Apr. 17, 2002) (noting that "the [Institute of Electrical and Electronics Engineers (IEEE)] working group [developing the proposed standard] does not attempt to determine whether" the patent applies "to the proposed standard"). The IEEE provides the following "public notice" in connection with its standards: "The IEEE shall not be responsible for identifying patents for which a license may be required by an IEEE standard or for conducting inquiries into the legal validity or scope of those patents that are brought to its attention." *See* IEEE, *IEEE-SA Standards Board Operations Manual*, § 6.3.1, *at* http://standards.ieee.org/guides/opman/sect6.html (last modified Jan. 18, 2002).

128. The concern is that, if the SSO undertakes such an investigation and concludes that the patent does not read on the standard, users of the standard who are accused of infringing a patent could sue the SSO for negligence should the SSO's evaluation later prove incorrect. Defending against such negligence claims can be costly, even if the SSO were to ultimately prevail on the merits.

129. The economic logic here is that, while both under- and over-disclosure impose costs on the firm, the private costs of underdisclosure (in the form of

numbers of marginally relevant patents in order to determine the patent's significance.

More significantly, a "disclose it or lose it" policy is likely to deter firms with significant IP portfolios from participating in an SSO that has such a policy. In many cases, the risk of losing the ability to enforce patents is likely to outweigh the gains the firm receives from participating in the standards-setting process. This is especially true since firms can get most of the benefits from standardization by adhering to the standard once set, even if they do not participate in setting the standard. However, such firms (which presumably have large IP portfolios because they have successfully engaged in significant R&D may be precisely the firms whose participation in the standards-setting process is the most significant and whose absence would most likely be detrimental.[130]

Again, the results of the Lemley survey are enlightening. Of the thirty-six SSOs surveyed that currently have some policy regarding IP, only twenty-four "imposed either an express or implied obligation that members disclose IP rights of which they are aware," though others "generally imposed other conditions that obviated the need for disclosure."[131] Of those

---

the prospect that the firm will be found to have waived its right to assert its patents) are likely to outweigh the private costs of overdisclosure.

130. Professor Janice Mueller argues, "An expansive disclosure requirement is not likely to chill industry participation in standards-setting, because 'participation in standards-setting is motivated by commercial self-interest and is not a form of community service.'" Mueller, *supra* note 27, at 933 (internal citation omitted). This is an unsubstantiated *empirical* claim, for which Professor Mueller provides no basis. Moreover, it is precisely *because* participation *is* "motivated by commercial self interest" that the SSO has to be concerned that onerous search and disclosure rules can tip the balance of the participate-versus-do-not-participate cost-benefit calculation and will thereby cause some participants to elect not to participate.

Yet Professor Mueller's argument contains a grain of truth. It may well be, as discussed above, that some firms believe that they have little choice but to participate in certain "voluntary" SSOs, despite the risk to their valuable IP, because the SSO is "the only game in town." However, we draw a very different conclusion than Professor Mueller does. Her conclusion is that SSOs can and should impose stringent search and disclosure rules, and stringent sanctions for non-disclosure, in order to ensure that the users of the standard can rest assured that their use of the standard will not infringe other participants' patents. *Id.* at 945-46. By contrast, we are concerned that such rules will inappropriately favor the users of IP and harm the owners of IP, who may have little realistic choice but to participate. As discussed in more detail in the Appendix *infra*, we believe that a bias in favor of the users of a standard can be societally inefficient.

131. Lemley, *supra* note 17, at 1904.

that imposed a duty, six explicitly addressed the question whether the duty to disclose also included pending patent applications.[132] The rules of those six were mixed, with four requiring disclosure of all pending applications, one requiring disclosure only of already-published applications, and one requiring disclosure only from the proponent of the standard.[133]

The combination of search and disclosure obligations in Professor Lemley's survey is particularly striking. While most SSOs imposed a duty to disclose known relevant patents, only *four* required a search.[134] In other words, in the vast majority of cases the participant has a duty to disclose patents of which it *is* aware, but has *no* affirmative duty to search its firm's patent portfolio for potentially relevant patents.

It can be argued that imposing disclosure duties without requiring a patent search may do little or nothing to protect other SSO participants, or users of the standard, from future patent infringement claims. Similarly, such a policy may also encourage "willful ignorance" on behalf of patent holders. If a firm's representative to the SSO is aware of relevant patents (and/or patent applications), the representative must disclose them, so the obvious course of action is either to choose not to look for potentially relevant patents and thus remain ignorant of possible patent claims, or to send a representative who has not been informed of such patents to the SSO meetings.

These arguments, while in some sense clearly correct, ignore three considerations. First, imposing a duty to disclose *known* patents is relatively costless, while imposing a duty to *search* for potentially relevant patents can be quite costly to firms with significant patent portfolios. Second, as noted above,[135] the individual representatives to an SSO are typically chosen for their engineering expertise, not for their knowledge of their firms' patent portfolios. Third, if the firm's representative is not aware of its firm's patents that potentially read on the proposed standard, then the representative logically *cannot* be in a position to try to "actively" manipulate the standards-setting process in ways that benefit its firm.[136]

---

132. *Id.* at 1905.

133. *Id.*

134. *Id.* at 1905, 1961 n.298 ("[O]nly four out of twenty-four SSOs that require disclosure impose a search obligation.").

135. *See supra* note 77 and accompanying text.

136. As such, this situation is in many ways akin to the situation in which a *non*-participant has a potentially relevant patent. It is widely acknowledged

It is worth noting that, based on the results of Professor Lemley's survey, it appears that the search and disclosure rules of the two SSOs involved in two controversies that have received a significant amount of attention—the Video Electronics Standards Association (VESA) in *In Re Dell*[137] and the Joint Electron Device Engineering Council (JEDEC) in the Rambus matter[138]—are significantly more stringent (in the sense of purportedly imposing strong duties on patent holders) than appears to be the norm.[139] VESA rules required the

---

that SSO rules *cannot* bind non-participants. *See, e.g.*, Lemley, *supra* note 17, at 1945 ("Because SSO rules necessarily bind only members of the SSO, exit is always an option.").

137. *See In re* Dell Computer Corp., 121 F.T.C. 616 (1996).

138. *See* FTC Rambus Compl., *supra* note 86.

139. We are aware that the question as to what the scope of the search and disclosure rules of VESA and JEDEC actually were, both on paper and as applied in practice, is a hotly disputed topic. For example, Infineon and the FTC argued that JEDEC rules required disclosure of all pending patent applications that "might be involved in" a proposed standard. See *id.* ¶ 21.

Rambus argued (1) that prior to 1993 the JEDEC rules did not require disclosure of pending patent applications (but only sought disclosure of issued patents), (2) that the October 1993 "Manual of Practice and Procedure" that indicated that disclosure of pending patent applications was required was an internal manual for the use of JEDEC personnel and committee chairmen and was not disseminated to rank-and-file members such as Rambus, (3) that the 1995 Member's Manual, which (unlike the 1993 Manual) was disseminated to JEDEC members, required the disclosure of patent applications by "first presenters" of technology for standardization, thus supporting the conclusion that other JEDEC members not making such a presentation had no similar obligation, (4) that Rambus was never informed that the policy had been changed to require that all members disclose pending patent applications, (5) that the chairman of the JEDEC committee in which Rambus participated consistently displayed the older policy (which sought disclosure only of issued patents, *not* of pending patent applications) at all of the meetings that Rambus attended, so that Rambus had a good-faith belief that the policy only sought disclosure of issued patents, and (6) that, throughout the relevant period, JEDEC rules only sought disclosure of patents or pending patent applications the use of which was "required" in order to practice the standard. *See* Answer of Respondent Rambus, Inc., *In re* Rambus, Inc., No. 9302 (July 29, 2002), *at* http://www.ftc.gov/os/adjpro/d9302/020729arri.pdf (last visited Mar. 26, 2003).

We note that the Federal Circuit has recently ruled, in connection with Infineon's claim that Rambus committed fraud on JEDEC by failing to disclose its patent applications, that Rambus did *not* commit fraud, because the technology claimed in its patent applications while it was a member of JEDEC was not required in order to practice the JEDEC SDRAM and DDR SDRAM standards. *See* Rambus, Inc. v. Infineon Technologies AG, 2003 U.S. App. LEXIS 1421, at *39-*66 (Fed. Cir. 2003).

Resolving this disputed factual issue would take us too far afield. But the example illustrates that one key factor is the clarity of the search and

representative to certify that its firm did not have any relevant patents;[140] the 1993 JEDEC Manual of Procedure required disclosure of pending patents.[141]

C. LICENSING POLICIES

In addition to search and disclosure policies, many SSOs have what might be termed "licensing policies." Such policies most commonly require participating IP holders whose patents have been identified as being relevant to the proposed standard to agree to license whichever patents are "necessary" to make products that comply with the standard to anyone seeking a license.[142] The IP holder must typically also agree to license its patents either royalty-free or on "reasonable and non-discriminatory terms." If the patent holder is not willing to agree to abide by these terms, the SSO will often refuse to adopt the proposed standard or will withdraw the standard if it has already been promulgated.[143]

1. The "One Patent, One Standard" Problem

The implicit paradigm seems to be that a firm has a single patent that reads on the proposed standard, and that the task is to determine the appropriate terms for licensing that *particular* patent. The obvious complication is that, in many high-tech industries, much licensing does not fit this "one patent, one product" paradigm. Instead, it is common to see broad portfolio-wide cross-licenses, often with "balancing payments" reflecting the difference in value between the firms' patent portfolios.[144] As discussed in more detail below, such a practice makes it difficult to determine whether proposed terms

---

disclosure rules adopted by the SSO, and the extent to which those rules are clearly conveyed to all participants. We discuss this issue in more detail in Part IV.E *infra*.

140. *See* HOVENKAMP ET AL., *supra* note 1, § 35.5, at 35-37 to 35-38 (Supp. 2003).

141. *Id.* at 35-38.

142. For example, the IEEE patent policy refers to patents "whose use would be *required* to implement the proposed IEEE standard." IEEE, *IEEE-JA Standards Board Bylaws*, *at* http://standards.ieee.org/guides/bylaws/sect6-7.html (last visited Mar. 10, 2003) (emphasis added). There may be other patents—for example, patents on product features that are seen as desirable, but which are not strictly necessary to practice the standard—which relate to standardized products, but which do not fall within the scope of the licensing requirement.

143. *See* examples cited *supra* note 103.

144. *See* Grindley & Teece, *supra* note 68, at 9, 33.

are "reasonable and non-discriminatory."

2. "Royalty-Free" and "Reasonable Terms"

In some contexts, the patent holder may be willing to license its patents royalty-free to all interested parties. From an economic standpoint, this is most likely to occur under one of two conditions: (1) when the patented technology represents only a minor improvement over the next-best non-patented alternative (so that the value of the patent is small), or (2) when the patent holder will benefit from others' adoption of its patented technology as a standard because the patent holder has other complementary capabilities that will enable it to profit from its innovation in a manner other than collecting royalties.

The latter is most common when the patent holder is also a seller in the product market and wants the SSO to adopt the firm's patented technology as a standard to encourage market acceptance of its technology and thus of the products that it sells.

Setting royalty payments at "zero" provides two obvious economic advantages. First, it has what economists and game theorists term a "focal point" clarity[145] that other royalty rates do not have. Once one moves away from zero, no other royalty rate has any particular reason to commend it in all cases. There is nothing special about 1%, 3%, or 7%. Instead, choosing a particular *non*-zero rate must consider the particular advantages of the patented technology relative to non-patented alternatives, which is inherently a fact-specific investigation.

Second, a royalty-free license is administratively simple (though determining the *scope* of the license may not be). By contrast, a royalty-bearing license requires mechanisms to monitor sales volume, to pay and collect royalties, and (if necessary) to audit royalty payments to make sure that the patent holder has been properly compensated. For both of these reasons, it is not surprising that a number of SSOs select "zero" (royalty-free) as the *only* pre-specified numerical royalty rate (as opposed to the *un*specified "reasonable" alternative, which does not select a particular number) when setting their

---

145. For a discussion of "focal points" and their economic significance, see THOMAS C. SCHELLING, THE STRATEGY OF CONFLICT 111-14 (1960).

licensing policies.[146]

Nevertheless, the obvious *dis*advantage of a "zero" (royalty-free) license is that it fails to compensate the patent holder for others' use of its patented technology. From an economic standpoint, there is *no* necessary reason why the appropriate "price" (royalty rate) for use of the patented technology should be *zero*, as it is in a royalty-free license, merely because that technology was incorporated into a standard. Presumably, the SSO adopted the patented technology as a standard, rather than some other alternative, because it believed that using the patented technology was superior to the alternatives. As such, the general economic presumption is that the appropriate price would be *greater* than zero. Thus, though SSOs list royalty-free licenses as one option (indeed, the only numerically pre-specified option), there is and should be *no* presumption that "zero" (royalty-free) is the appropriate price for patented technology that reads on a standard.

3. "Reasonable Terms": When Announced?

One key issue concerning patents is whether the patent holder must announce the terms for a patent license in advance. If so, there are potential antitrust concerns. Typically, the other participants in the SSO are the most likely potential licensees for the patent.[147] This raises the potential for collusive, oligopolistic "price fixing" in the technology market. The SSO members would, in effect, say to the patent holder, "We will collectively reject a standard that incorporates your patented technology unless you agree to license it to us at pre-specified rates that we collectively find acceptable." In other contexts, this clearly would amount to a group boycott.[148]

146. Professor Lemley's survey of SSO licensing practices is misleading in this regard. For example, he lists W3C as having a "[r]oyalty-free license requested but not required; RAND as alternative" licensing policy. Lemley, *supra* note 17, at 1973. But he lists JEDEC as an SSO with a "RAND" policy. *See id.* at 1974. The current JEDEC policy in fact specifies that the patent holder must agree to license either royalty-free *or* on RAND terms. *See* JEDEC, *JEDEC Manual of Organization and Procedure*, at 20, Annex A.1, *at* http://www.jedec.org/Home/manuals/JM21L.pdf (July 2002). Our review of the licensing policies of other SSOs confirms that many other SSOs have similar policies.

147. The prospective licensees are the firms that intend to manufacture and sell the standardized products. As noted above, such firms are likely to participate in the standards-setting process.

148. *See* HOVENKAMP ET AL., *supra* note 1, § 35.6, at 35-46 to 35-54.3 (Supp. 2003).

If advance notice of the patent license's terms is not given, the requirement that the licensing terms be "reasonable" may have little "teeth." What the patent holder believes is "reasonable" may strike potential licensees as excessive.[149] One difficulty is that circumstances (and thus whether a particular rate is "reasonable") vary, not only across different patents and different technology but over time. No single rate is "reasonable."

The more significant difficulty is that the adoption of a standard can affect the bargaining position of the parties. Ex

149. Professor Lemley discusses the fact that most SSO's "RAND" policies provide little guidance as to "what those obligations mean in practice." Lemley, *supra* note 17, at 1906. He proposes that SSOs should "give content to the reasonable and nondiscriminatory licensing requirement." *Id.* at 1964, 1964-67.

We sympathize with Professor Lemley's concern that having a "RAND" policy without more may do little to provide guidance as to the *rates* that the patent holder can charge. However, a RAND commitment *does* constrain the patent holder in other significant ways.

If a standard is adopted (and especially if that standard becomes the industry norm), manufacturers want the ability to make products that comport with that norm/standard. As such, they would be concerned if patent holders could "pick and choose" by licensing some firms and refusing to license others. (This concern is especially significant if the patent holder itself competes in the product market; in that event, the patent holder could assert its patent to exclude others from competing, thereby reserving the market to itself.) And firms are concerned, not merely about obtaining a license, but about obtaining a license on terms that enable them to compete on a "level playing field."

A patent holder's commitment to license any interested party on (unspecified) RAND terms ensures (a) that all potential licensees can obtain a license (*if* they are willing to pay the royalty rates sought), and (b) that all will be offered the same terms and will be assured the opportunity to obtain a license without being put at a comparative disadvantage. Such a commitment to license on RAND terms limits the patent holder's ability to play potential licensees against one another.

But once such a "level playing field" is attained, then much of the concern that the prospective licensees have about their potential competitive position disappears. To be sure, firms would prefer not have to pay royalties, just as they would prefer not to have to pay their rent or their income taxes. But so long as every firm must pay, then the cost of the royalties can be built into the price of the product being sold, just as the cost of the raw materials and labor needed to make and sell the product is likewise built into the price. That is, prospective licensees may rationally be far more concerned about the "non-discriminatory" aspect of the RAND requirement than they are about the "reasonable" aspect.

This, in turn, implies that from an economic and organizational behavior perspective, it is quite rational for SSOs to pay much more attention to the requirement that licenses be available on (unspecified) RAND terms than they pay to the question of what the "reasonable" royalty rates should be.

ante, before an SSO adopts a standard, the patented technology may be only slightly better than the alternatives. Suppose, for instance, that there are two alternatives: an unpatented public-domain technology, and a patented technology that, if adopted, would result in a 5% cost savings. If the parties were to negotiate ex ante over royalty rates, one would expect that the negotiated rate would be no more than 5%, and likely less.

Once the patented technology is adopted as a standard, however, firms may commit to the standard and invest in complementary assets needed to make and produce the newly standardized products. Ex post, the cost of switching to the unpatented alternative may now be much greater, as the industry is "locked in" to the patented standard. Suppose, for example, that firms in the industry would find it worthwhile to switch only if the patent holder demanded greater than a 20% royalty. Then the patent holder may be able to extract (say) a 15% royalty ex post, despite the fact that the ex ante value of the patented technology is no more than 5%.

The above discussion implies that a royalty rate may be reasonable ex post (in the sense that it could be agreed to in an arm's length negotiation between a willing licensor and a willing, albeit reluctant, licensee) but *not* reasonable ex ante.

4. "Reasonable" Terms: How Determined?

In any case, many if not most SSOs do not get involved in determining whether proposed license terms *are* "reasonable."[150] Indeed, most SSOs do not require the patent holder to announce its proposed license terms in advance of

150. To our knowledge, most SSOs do *not* try to determine whether proposed terms *are* "reasonable," whether ex ante or ex post. For example, the IEEE *Public Notice* explicitly says, "The IEEE makes no representation as to the reasonableness of rates and/or terms and conditions of the license agreements offered by patent holders." IEEE, *IEEE-SA Standards Board Operations Manual cl. 6.3.1, at* http://standards.ieee.org/guides/opman/sect6.html (last visited Mar. 10, 2003).

ANSI does not require that terms be specified in advance, but the ANSI Board of Standards Review will review claims that the royalties being sought are not "reasonable." *See* AM. NAT'L STANDARDS INST., GUIDELINES FOR IMPLEMENTATION OF THE ANSI PATENT POLICY, http://public.ansi.org/ansionline/Documents/Standards%20Activities/American%20National%20Standards/Procedures,%20Guides,%20and%20Forms/ (last updated Mar., 2003); *see also* Lemley, *supra* note 17, at 1906 ("While 'reasonable and nondiscriminatory licensing' thus appears to be the majority rule among SSOs with a patent policy, relatively few SSOs gave much explanation of what those terms mean or how licensing disputes would be resolved.").

adopting the standard, but merely require a statement that the patent holder is willing to license on "reasonable and non-discriminatory" terms.[151] Instead, royalty rates are left to the parties for negotiation.

There are four likely reasons that SSOs avoid addressing the reasonableness of licensing terms. First, determining whether a particular rate is "reasonable" often goes beyond the competence and/or expertise of the SSO or its participants. The "reasonableness" of royalty rates is an economic or business issue, not a technical one. The SSO representatives, generally chosen for their technical knowledge of the technology being standardized, frequently have little or no experience or expertise in negotiating royalty rates or determining what an appropriate rate should be.[152]

Second, trying to determine an appropriate royalty rate is often difficult when technology is changing rapidly, as is often the case in the context of standards for the next-generation of products. From an economic standpoint, patent holders are naturally reluctant to quote a royalty rate for their patents in advance.[153] This is especially true because of the asymmetric (one-way) nature of the commitment. Asking the patent holder to commit to a royalty rate prior to the standard's adoption would, presumably, be binding on the patent holder, in the sense that the patent holder could not *increase* the rate, though it could always agree to accept a *lower* royalty. But the rate, would *not* be binding on the prospective licensees. They would have made no commitment to *take* a license, to agree to *pay* royalties. Not surprisingly, patent holders are reluctant to constrain their future negotiating position in such an asymmetric fashion.

Third, there are potential antitrust issues that might arise if the SSO (or its members, as potential licensees of the patent) were to try to determine whether a proposed rate was "reasonable."[154] Antitrust issues clearly would arise if the SSO

151. For a survey of licensing policies, see Lemley, *supra* note 17, at 1906, 1973-75.

152. We obtained this information in conversations with senior management at several major semiconductor companies.

153. In large part, this is because they cannot be aware of the scope of the ultimately issued patents, and thus of what it is that they will be licensing.

154. The concern is that the SSO manufacturer members are often the most likely licensees of the patented technology, and collective action by manufacturers could amount to collusive buyer-side price fixing in the technology market.

explicitly conditioned its acceptance (or rejection) of a proposed standard implicating a patent on the asked-for royalty rates. Indeed, even asking the patent holder to announce its proposed rates in advance, and then having the SSO determine whether or not to adopt the standard in light of the announced rates, may be problematic from an antitrust standpoint.

Fourth, SSOs are aware that the issue of royalty rates pits the interests of some participants (the patent holders) against those of other participants (the prospective licensees), and the SSOs do not want to "take sides" in such matters.

Unfortunately, the reluctance of SSOs to get involved in determining (ex ante or ex post) whether proposed licensing terms are "reasonable" can lead to an ex post situation in which: (1) a standard has been adopted; (2) a participant has a patent that reads on the standard; (3) the patent holder contends that it is willing to license its patent on "reasonable" terms (albeit at rates that were not announced in advance); and (4) the users of the standard, as prospective licensees, protest that the royalty rates are unacceptably high and are thus anything *but* "reasonable."

5. Antitrust Remedies

In our view, the antitrust authorities have shown what we believe to be an unfortunate tendency to propose royalty-free as a remedy in cases where the antitrust authorities have contended that the patent holder violated the antitrust laws by failing to disclose its patents (or pending patent applications). For example, the *In Re Dell* consent decree called for Dell to allow others to use its VESA patent on a royalty-free basis.[155] The recent Federal Trade Commission (FTC) complaint against Rambus effectively seeks to compel Rambus not to enforce its patents against users of the JEDEC synchronous dynamic random access memory (SDRAM) and double data rate (DDR) SDRAM standards; if successful, the action amounts to forcing

155. *See In re* Dell Computer Corp., 121 F.T.C. 616, 619-23 (1996). Our understanding (based on discussions the authors had with parties familiar with the case) is that one likely explanation for Dell's willingness to agree to this remedy is that, by the time of the consent decree, the VESA bus standard at issue was obsolescent and rapidly becoming obsolete. We also understand (on the same basis) that others had challenged the validity of Dell's patents. Thus, Dell may not have been giving up much in the way of potential revenue by agreeing to license its patents on a royalty-free basis in order to settle the matter.

a royalty-free license on Rambus.[156]

Such a remedy typically goes far beyond the SSO's requirements, which *allow* a patent holder to agree to license its patent royalty-free, but *also* generally allow a patent holder to agree to license on "reasonable and non-discriminatory" royalty-bearing terms.[157] The above analysis suggests that a more appropriate remedy would be to require a patent holder to license its patent to others on terms that are "reasonable" in the ex ante sense.

To be sure, it can be difficult to determine ex post what the ex ante "reasonable" terms would be. It is far easier administratively to propose a royalty-free license as a remedy. The courts are routinely asked, however, to determine ex ante reasonable royalty rates in the context of patent infringement damages.[158] We know of no reason why the analysis that courts use in patent damages cases, such as the well-known *Georgia Pacific* factors,[159] could not be applied equally well in antitrust cases alleging wrongful non-disclosure. The harsh nature of the proposed royalty-free license as a remedy for (allegedly) wrongful non-disclosure strikes us as out of line with the nature of the offense. In particular, the royalty-free license remedy does not set damages equal to the difference between the actual world and the but-for world.

6. "Non-Discriminatory" Terms

The common requirement that the patent holder agree to license its patent on "non-discriminatory" terms is not likely to be controversial in the context of cash-only single-patent licenses. That is, if the patent holder seeks, for example, a 1% royalty for the one patent that is implicated by the standard, its terms are "non-discriminatory" if every licensee is asked to pay the same 1% royalty.

But even here there may be problems. Suppose, for example, that both the patent and the standard read on a

---

156. *See* FTC Rambus Compl., *supra* note 86.

157. In essence, it amounts to a confiscation of the patent holder's rights, in favor of the users of the standard.

158. Indeed, the patent statute requires that damages be set at a level "adequate to compensate for the infringement, but in no event less than a reasonable royalty for the use made of the invention by the infringer." 35 U.S.C. § 284 (2001).

159. *See* Georgia-Pacific Corp. v. United States Plywood Corp., 318 F. Supp. 1116, 1120 (S.D.N.Y. 1970).

variety of different products (e.g., multiple generations of computer chips). The patent holder may elect to seek different royalties for different products: For example, it may seek a 1% royalty on older-generation chips and a 2% royalty on newer-generation chips. If it offers these same terms to all firms, knowing that they will apply differently to different firms depending on their product mix, this may be discriminatory.

On the one hand, firms that predominantly sell newer-generation chips will complain that they are being discriminated against, because they must pay 2% while firms selling older-generation chips only pay 1%. The terms change the competitive position of the newer chips vis-à-vis older chips, making newer chips relatively more expensive than older chips and shifting demand away from newer chips toward older chips. On the other hand, the patent holder can truthfully point out (1) that its patent is more valuable to makers of newer-generation chips, so that the royalty terms should reflect that difference in value, and (2) by offering the *same terms* to all potential licensees, it did not "discriminate" in favor of or against any particular firm.

More significantly, the "non-discriminatory" requirement is likely to be controversial in the context of the practice of *portfolio* licensing, whereby the firm does not license its patents individually, but instead grants a license for its entire portfolio of patents.[160] It is also likely to be controversial where some users of the standard already *have* portfolio licenses from the patent holder.

Suppose, for example, that a firm like IBM, which has tens of thousands of patents,[161] has a patent that is implicated by a newly-adopted standard. IBM already has many preexisting licenses with many firms.[162] Many of those licenses are portfolio licenses and thus already include a license to use the patent at issue. Those licensees would naturally contend that they do not need to take *another* license for the same patent, merely because the patent has now been implicated by a standard.

But other firms that do not yet have licenses may argue that they are being asked to pay a disproportionately high price

---

160. *See* Grindley & Teece, *supra* note 68, at 9, 33.

161. *See supra* note 114.

162. We know this because of our discussions with IBM licensing executives.

(relative to the price paid by IBM's other licensees) for access to the patent. For example, suppose that IBM offers a portfolio-wide license to its thousands of patents for a 5% royalty, but asks a 1% royalty for a single-patent license.[163] Users of the standard, who need the single patent implicating the standard but who do not need access to IBM's thousands of other patents, will, not surprisingly, argue that they are being charged a price that is out of proportion to the price paid by others—and thus that they are being discriminated against.

This issue has already arisen in connection with various patent pools.[164] For example, patent pools may seek to license the pooled patents on a "bundled" basis, with a single royalty rate for the use of all of the patents in the pool. For a number of legitimate reasons, the patent pools administrator may prefer not to offer the pooled patents on an a la carte basis, that is, setting separate royalty rates for each patent and allowing licensees to take licenses only for those patents that they elect to license.[165] But firms that (because of their existing cross-

---

163. IBM's current policy reads,

> The following practice applies *to the extent* IBM has committed to provide patent licenses as required by a Standards Organization: Upon written request, IBM is willing to grant nonexclusive licenses under its patents (*to the extent required* for the practice or implementation of standards), on reasonable and nondiscriminatory terms and conditions, *to those who respect IBM's intellectual property rights*.

IBM, *IBM Standards, Practices, available at* http://www.ibm.com/ibm/licensing/standards/practices.shtml (last visited Mar. 10, 2003) (emphases added). It further states, "The royalty rates are 1% of the selling price for each patent used, up to a maximum of 5% of the selling price, for each licensed product." *See id.*

164. For an extensive discussion of patent pools and the antitrust issues they raise, see HOVENKAMP ET AL., *supra* note 1, § 34, at 34-1 to 34-34 (Supp. 2003).

165. From an economic and organizational behavior perspective, one concern is that, if the managers of the patent pool were to grant such licenses, they would have to monitor the subsequent conduct of the licensee to make sure that the licensee was not using the unlicensed patents as well as the licensed ones. Such monitoring can be costly, because it may involve a detailed patent infringement-style evaluation of the licensee's product.

One of the reasons that the patent pool is willing to license is to avoid incurring the ongoing costs of such investigations.

Another concern is that it may be more difficult for the various owners of the pooled patents to agree on the appropriate royalty rates for such a la carte licenses than it is to agree on both (a) an overall rate for the pooled patents, and (b) a division of the royalty income among the patent holders.

Moreover, if all of the patents in the pool are "blocking," that is, it is necessary to have a license to each of the patents in the pool in order to make

licenses) already have a license to some, but not all, of the patents in the pool contend that they should not have to pay the same royalties as others who do not have such cross-licenses.[166] This raises the difficult question of how much "credit" against the "pooled" rate should be allowed for the existing in-licenses.

The "non-discriminatory" issue is also likely to be controversial in the context of *cross*-licenses, in which one firm out-licenses its patents to another firm in exchange for in-licenses to the other firm's patents.[167] As one of us has discussed extensively elsewhere, cross-licenses can be royalty-free, or they can involve some "balancing payment" that reflects the difference in the value of the firms' patent portfolios.[168] Firms that are being asked to pay a higher *cash* royalty than other firms (especially when the other firms have a royalty-free license) may well disagree as to whether their out-licensed patent portfolio is being valued in a "non-discriminatory" fashion vis-à-vis other firms' portfolios.

Both the portfolio license and the cross-license issue discussed above reflect the complications that arise when one moves away from the simple "one patent, one standard, one license" paradigm, into the more realistic world in which firms have numerous patents and licenses cover more than one patent.

Another concern involves the question of how one determines whether or not license terms are "non-discriminatory." License terms are often extremely confidential business information (for both the licensor and the licensee). Establishing whether license terms are "non-discriminatory" inherently involves comparing the rates charged to one licensee to the rates charged to other licensees. This leads to an

---

and sell a commercially viable product, then a la carte licensing makes no economic sense, since a firm that took a license to some but not all of the patents in the pool would be unable to make a commercially viable product without licenses to the other blocking patents. (There are two main caveats here. First, some firms may already have a license to some, but not all, of the necessary patents. Second, firms may disagree about whether particular patents in the pool are or are not "blocking.")

166. *See, e.g.*, *Hewlett-Packard Company for FTC/DOJ Hearings on Competition and Intellectual Property Law and Policy in the Knowledge-Based Economy*, 4-6, *at* http://www.ftc.gov/opp/intellect/020417jeffretfromm.pdf (last visited Mar. 10, 2003) (statement of Jeffrey Fromm).

167. For a general discussion of cross-licensing in high-tech industries, see Grindley & Teece, *supra* note 68.

168. *Id.* at 9, 18-19, 33.

inherent conflict between maintaining confidentiality and demonstrating that the terms are "non-discriminatory." This conflict is even more pronounced when the other license terms (e.g., the scope of the license or the royalty base) are different across different licensees.

D. Pending Patent Applications

Pending patent applications present a number of complications not present in the context of issued patents. First, as noted above, until recently the PTO kept pending patent applications confidential.[169] The public policy rationales for that confidentiality are well recognized.[170] Research projects are often multi-faceted, and the results of that research are developed over time. Some inventions may be fully developed and ready to patent before other inventions that flow from the same research. Other things being equal, a firm with an invention wants to apply for a patent as quickly as possible in order to preserve its priority date.[171] But if its application is immediately made public, others can see the disclosure and rush in to file patent applications that "flank" the initial application. This sort of "patent flooding" can deprive the initial inventor of the fruits of its invention. Allowing the patent applicant a "grace period" of confidentiality before the patent application is thrown open gives the inventor the opportunity to finish any remaining research and prepare applications in an orderly fashion.

Moreover, the quid pro quo of patent law is that the government grants the inventor the right to exclude others from using the patented invention for a period of years in exchange for the inventor's public *disclosure* of the invention.[172]

---

169. This was changed by 35 U.S.C. § 122(b)(1) (2000).

170. For an economic analysis of the pros and cons of confidentiality and disclosure, see REIKO AOKI & YOSSI SPIEGEL, PRE-GRANT PATENT PUBLICATION, R&D, AND WELFARE, (May 16, 2001) (unpublished manuscript, *at* http://www.tau.ac.il/~spiegel/papers/reiko.pdf); DANIEL K.N. JOHNSON & DAVID POPP, FORCED OUT OF THE CLOSET: THE IMPACT OF THE AMERICAN INVENTORS PROTECTION ACT ON THE TIMING OF PATENT DISCLOSURE, (July 2001) (Nat'l Bureau of Econ. Research, Working Paper No. 8374, 2001, *at* http://papers.nber.org/papers/w8374.pdf).

171. Even under a first-to-invent system, the priority date for evaluating prior art is the filing date. As such, the innovating firm has an incentive to file as early as possible, so as to prevent later publications from being considered as "prior art."

172. *See* 35 U.S.C. § 122(b)(1) (2000).

If the PTO chooses not to grant the patent, the inventor does not get the right to exclude others, but retains the opportunity to try to keep its invention a trade secret. Mandatory disclosure to the SSO precludes this opportunity.

Though the confidentiality issue is now less significant than it used to be (because the change in the law now means that many, but not all, pending applications are "thrown open" for public inspection eighteen months after filing), the issue is still significant (1) for those applications which are not thrown open and (2) during the eighteen-month period for those applications that are thrown open.

Second, any interested party (or the SSO itself) can search for issued patents that are potentially relevant to a proposed standard.[173] But only the applicant and the PTO are aware of not-yet-disclosed, thus confidential, pending patent applications. Therefore, to the extent that the SSO believes that early disclosure (during the eighteen month confidentiality period) is needed, it inherently has to rely on the patent applicant to disclose the existence of the application.

Third, and most significantly, while the application is pending, the applicant cannot know (1) *whether* any patent will ultimately issue from a given application or (2) what the *scope* of any (ultimately) granted patent may be.[174] It is widely recognized that the scope of patent claims is often narrowed (sometimes drastically) during the patent prosecution process. As such, it is often very difficult to reach a conclusion about whether a patent application will ultimately mature into a patent that reads on the proposed standard. And it is virtually impossible to set "reasonable" ex ante licensing terms for the patent *if and when* it issues.

As with issued patents, there is a continuum of possible disclosure rules, ranging from no disclosure obligation, to an obligation to disclose the *existence* (but *not* the content) of the application,[175] to an obligation to disclose the contents of the

---

173. The advent of computerized patent databases, such as the official United States Patent and Trademark Office (PTO) database, *at* http://www.uspto.gov/patft/ (last visited Mar. 10, 2003) and the Thompson/Delphion database, *at* http://www.delphion.com (last visited Mar. 10, 2003), among many others, has made it significantly easier to conduct such searches.

174. By definition, these issues have already been resolved for issued patents.

175. At least one major SSO, the IEEE, seems to propose such a policy: "Given the confidentiality of the information [in pending applications], it is

initial application, to a continuing obligation to inform the SSO of any change in the status of the pending application. As noted above, of the SSOs surveyed by Professor Lemley, only four required disclosure of all pending patent applications, while one required disclosure of published applications, and one required disclosure only from the proponent of the standard.[176] Again, given the variety of policies that have been adopted by different SSOs, there is no reason why a "one size fits all" policy is appropriate.

1. Non-Disclosure Does Not Equal Lack of Knowledge

Even if the patent holder (or applicant) does not disclose the existence of its patent (or application) to the SSO, some or all of the other SSO participants may be aware of the existence of the patent (or application), whether as a result of their own investigations or as a consequence of disclosures in other contexts. If so, the patent holder's "failure" to disclose would appear to be relatively innocuous.

For example, if the other SSO members are aware that a firm has a patent that it has not disclosed to the SSO, and the SSO elects to adopt the standard anyway, that action implies that they are not concerned about the patent issue. The SSO may adopt the standard because they believe that the patent either does not read on the standard or that the patent will not be enforced (possibly because of validity concerns). Similarly, if aware that a firm has a pending patent application, but the SSO elects to adopt the standard anyway, this again implies that they are not concerned about the patent issue, possibly because they believe that no patent will ultimately issue (or that the scope of any issued patent will not read on the standard).

For example, in the Unocal case discussed in Part I.B above, by disclosing the results of its research to Auto/Oil, Unocal presumably put other firms on notice that Unocal had conducted its own research, and those firms should have reasonably inferred that Unocal would have applied for a

recognized that only a very limited amount of information can be expected to be disclosed: i.e., that a patent application has been filed in a subject area." Public Comments filed by the IEEE Standards Association in connection with the FTC/DOJ Hearings on Competition and Intellectual Property Law and Policy in the Knowledge-Based Economy, *at* http://www.ftc.gov/os/comments/intelpropertycomments/index.htm (last visited Apr. 22, 2003).

176. *See* Lemley, *supra* note 17, at 1905.

patent on its innovation. Indeed, several other oil companies filed for patents for their own formulations of reformulated gasoline and diesel fuel that complied with the CARB regulations and did not disclose those patent applications to others.[177] If the existence of a patent application had been an issue, the obvious solution would have been for CARB to ask Unocal whether it had any potentially relevant patents or pending patent applications. CARB did not do so.[178]

The Rambus case provides another example.[179] Rambus did not disclose to the standards-setting organization, a JEDEC committee, that it had pending U.S. patent applications.[180] But Rambus had filed European counterparts to its U.S. applications, and under European patent law the European applications had been thrown open eighteen months after filing. One European-based JEDEC member explicitly informed other JEDEC members of the existence of Rambus's European application, which contained a cross-reference to the existence of the counterpart (and confidential) U.S. application.[181] Since the European application was open for public inspection, one presumes that, had the other JEDEC members been concerned about the issue, they could have obtained and examined the European counterpart.

Moreover, Rambus had entered into limited-field-of-use licenses with many of the JEDEC members for its proprietary

---

177. *See* Unocal, *Unocal Patent Issues Timeline*, *at* http://www.unocal.com/rfgpatent/rfgtime.htm (last visited Apr. 22, 2003).

178. Deposition testimony in Unocal case (on file with authors).

179. David Teece testified as Rambus's damages expert in the patent infringement case brought by Rambus against Infineon. Rambus, Inc. v. Infineon Technologies AG, 155 F. Supp. 2d 668 (E.D. Va. 2001), *rev'd and remanded*, 318 F.3d 1081 (Fed. Cir. 2003).

The discussion in this Article is based on publicly available information (including the transcript of that trial), and does not reflect any confidential or proprietary information. Some of the relevant factual background can be found in the FTC's administrative complaint against Rambus. FTC Rambus Compl., *supra* note 86. The views expressed in this Article are those of the authors, and do not necessarily reflect Rambus's position.

180. We do not address here the question whether Rambus had any *duty* to disclose its pending applications to the JEDEC committee, or the related question whether the (acknowledged) "failure" to disclose was wrongful. That is a hotly-contested topic, and to address the issue fully would take more space than is appropriate here. We note, however, that the Federal Circuit has ruled that Rambus did *not* commit fraud on JEDEC by not disclosing its pending patent applications. Rambus, Inc. v. Infineon Technologies AG, 318 F.3d 1081, 1106 (Fed. Cir. 2003).

181. *See* FTC Rambus Compl., *supra* note 86, ¶ 77.

technology, and thus the Rambus licensees were aware of the likelihood that Rambus had pending patent applications.[182] These facts suggest that at least some other JEDEC members were aware of the existence of Rambus's U.S. patent applications and of the possibility that Rambus's patent applications might read on the proposed standard.[183] Furthermore, JEDEC meeting minutes indicate that at least some other firms apparently believed at the time that Rambus's patent applications were "stuck" in the PTO and would not ultimately issue (or would be limited by prior art).[184] The fact that the JEDEC committee adopted the standard despite this knowledge suggests that the other JEDEC members were not concerned about Rambus's pending patent applications either because they believed that the patents would not issue, because they felt that the Rambus technology was sufficiently superior to the available alternatives, or because they believed that they would be able to obtain licenses from Rambus.

Obviously, there are intermediate cases, in which *some* but

---

182. We are aware, from discussions with Rambus executives, that, in at least one Rambus license signed in 1995, the licensee (Hynix) agreed to pay Rambus royalties on a relatively broad range of DRAM types, not merely Rambus's proprietary RDRAMs.

183. In its Complaint, the FTC alleged that it was only after attending JEDEC meetings that Rambus submitted additional patent claims to the PTO that (allegedly) expanded the scope of the patent to cover the proposed standard. FTC Rambus Compl., *supra* note 86, ¶¶ 48, 54, 61, 64, 68.

Under U.S. patent law, a patent applicant is allowed to amend the patent claims and to add additional claims while still retaining the priority date of the original patent application, so long as the new claims do not introduce any "new matter" into the application. 35 U.S.C. § 132(a) (2000); Kingsdown Med. Consultants, Ltd. v. Hollister, Inc., 863 F.2d 867, 874 (Fed. Cir. 1988).

Rambus contended, and the PTO examiner (by issuing the patent) apparently agreed, that the ultimately-issued patent claims involved nothing more than the invention that was disclosed in its initial patent filing, which Rambus submitted over a year before it joined JEDEC and several years before the meetings at issue. *See* FTC Rambus Compl., *supra* note 86, ¶¶ 34, 37.g, 37.i, 38.

Even if Rambus *had* disclosed the existence of its U.S. application, it is by no means clear that the other JEDEC members would have believed that the U.S. application read on the proposed standards. Other JEDEC members were aware of the European counterpart application, and presumably examined it. *Id.* ¶ 77. If they believed that the Rambus application would *not* apply to the products being standardized by JEDEC, then presumably they would not have acted differently than they in fact did.

184. *See* Committee Meeting Minutes of JEDEC Solid State Technology Association (Dec. 6, 1995) (on file with author).

not all SSO participants are aware of the existence of a potentially relevant patent or patent application. Our main point here is that it is a mistake to confuse the lack of disclosure with the proposition that other SSO participants were unaware of the existence of the patent or patent application. What should matter is what is known by the SSO, not how the knowledge was acquired.

2. Public Policy Regarding Disclosure of Pending Patent Applications

It might be argued that Congress's recent revision of the patent statute, that throws open most pending patent applications eighteen months after initial filing (which we will refer to below as "early disclosure"), is likely to render the disclosure issue largely moot in the context of standards-setting. Estimates suggest that 75-85% of applications will be thrown open, putting participants in the SSO in a position to search pending applications and ending reliance on the applicant's disclosure.[185] Alternatively, it might be argued that Congress's decision represents a preference for openness, so that there should be a presumption that SSO rules should require disclosure. A closer examination casts doubt on both suggestions.

Addressing the second suggestion first, it appears that Congress amended the statute to bring U.S. patent law into closer conformity with the patent laws of other countries, notably European countries and Japan, which have long had a policy of early disclosure.[186] In the debates leading up to the adoption of the new statute, a number of scholars investigated whether the change was desirable from a public policy perspective. The general conclusion was that early disclosure, by reducing the competitive advantage that innovators could gain from their innovations, ran substantial risks of reducing the returns to innovation, reducing the incentive to innovate and thereby reducing the rate of innovation and the resulting economic growth. In other words, as a stand-alone policy, mandatory early disclosure does not appear to be desirable from an overall social welfare standpoint.

We believe that Congress's enactment of early disclosure legislation, in light of the contemporary evidence of the

185. *See* Johnson & Popp, *supra* note 170, at 33.
186. *See* the articles cited in note 170 *supra*, and sources cited therein.

detrimental effects of early disclosure, can best be interpreted as Congress's recognition of the U.S.'s obligation (under international treaties) to modify U.S. patent law to harmonize it with the patent law of other countries, and of the overall benefits of such harmonization, rather than as an endorsement of early disclosure per se. In fact, the form of the statute reflects Congress's skepticism of early disclosure. Early disclosure does not apply to certain types of patent applications; in particular, the application will *not* be disclosed if the applicant commits not to file foreign patent applications for the invention.[187] This exception suggests that Congress made the minimum disclosure change necessary to harmonize U.S. patent law with foreign patent law. Where the issue of harmonization did not arise—as it does not with applications filed only in the United States—Congress chose *not* to adopt early disclosure, but to retain the earlier no-disclosure policy instead.[188] This clearly casts doubt on any suggestion that Congress favors early disclosure as a general policy matter. On the contrary, the "carve-outs" in the statute suggest that Congress preferred a no-disclosure regime, but felt constrained by the United States' treaty obligations.

The fact that some applicants can opt out of early disclosure (albeit at some economic cost, such as foregoing the opportunity to file for foreign patent protection) in turn suggests that non-disclosure of at least some pending applications is likely to persist into the foreseeable future. Hence, SSOs will still have to come to terms with the issue of disclosure rules for pending patent applications. We believe, however, that Congress's reluctance to adopt a widespread early disclosure regime suggests that the public policy rationales favoring secrecy of pending applications are still significant.

E. The Need for Clarity

Regardless of what search, disclosure, and licensing rules the SSO adopts, from a public policy standpoint it is important that the rules be (1) clearly stated, (2) clearly communicated to participants, and (3) clearly enforced in an even-handed manner. It is also important (4) that the consequences of a participant's failure to comply with the rules be explicitly laid

---

187. *See* 35 U.S.C. § 122(b)(2)(B)(i) (2000).
188. *Id.*

out.

Some SSOs have no explicit policies, believing (falsely) that "everyone knows" the appropriate policies.[189] The wide divergence in actual policies noted in Professor Lemley's survey undercuts the credibility of this claim.[190]

Other SSOs have ambiguous policies. For example, they may not make it clear whether the disclosure policy applies to pending patent applications as well as issued patents.[191] More commonly, policies can fail to indicate whether a duty to search for potentially relevant patents exists; or, if such a duty exists, the extent of the duty's scope. For example, is a "good faith" investigation enough, or is something more required? Such policies also fail to provide whether a search and/or disclosure obligation applies to the firm's representative to the SSO or instead reaches to the entire firm.

Moreover, regardless of what the rules say on paper, there is a significant question whether the SSO applies those rules in practice in an even-handed manner that is consistent with the written rules. For example, suppose that on paper the SSO has a strong policy requiring participants to search for and disclose all potentially relevant patents and patent applications. Given the sheer number of existing patents and new patent applications in many technological fields, if such a policy were interpreted literally, one would expect that the SSO would be deluged with hundreds or thousands of disclosures.

The limited amount of available evidence suggests that, in practice, disclosure is in fact much more limited. SSO representatives are apparently disclosing only a tiny fraction of their firms' potentially relevant patents.[192] To be sure, the representatives may be doing so on technological grounds, disclosing only the patents that they believe are clearly

189. Professor Lemley reports that, of the forty-three SSOs he studied, "four had no policy at all" and "two had statements on their website about IP rights but no official policy." Lemley, *supra* note 17, at 1904.

190. Professor Lemley notes that "[t]he fact that different SSOs have different rules governing IP rights (or no rules at all) means that it is very difficult for IP owners to know ex ante what rules will govern their rights." Lemley, *supra* note 17, at 1906-07.

191. *See* Lemley, *supra* note 17, at 1904 ("There was rarely discussion of the problem of pending patent applications.").

192. *See* Committee Meeting Minutes of JEDEC Solid State Technology Association, RAM Devices Subcommittee (May 24, 1995) (on file with author). We have also had several discussions with various SSO representatives to obtain this information.

relevant to the standard under consideration. But this limited disclosure, provides no assurance that the patent holder will not subsequently assert an undisclosed patent against those practicing the standard. A more realistic interpretation is that the representatives simply are not aware of the full scope of their firms' patent portfolios.

Similarly, even when the SSO's written policies require that patent holders publicly announce the royalty rates that they will seek on their patents that are implicated by proposed standards before the standard's adoption, such policies appear to be frequently ignored.[193] If a written rule is "honored in the breach" by many, if not most, SSO participants, then it is clearly objectionable to argue that the rule should nonetheless be interpreted at face value when subsequent legal disputes arise.

Lack of clarity in the rules—both as written and as applied—clearly is something that SSOs can and should address. As the Federal Circuit noted in the *Rambus* case, "Just as lack of compliance with a well-defined patent policy would chill participation in open standard-setting bodies, after-the-fact morphing of a vague, loosely defined policy to capture actions not within the actual scope of that policy likewise would chill participation in open standard-setting bodies."[194] But again there should be no presumption that, because the rules are cloudy, the appropriate solution is to err on the side of requiring disclosure and penalizing non-disclosure. On the contrary, the fact that many, if not most, SSOs impose *no* duty to search, and merely expect the individual representative to disclose any potentially relevant patents of which she

---

193. For example, Professor Mueller cites a policy of the Internet Engineering Task Force (IETF) to the effect that the IETF requires patent holders to agree to license their implicated patents on "openly-specified . . . terms." Mueller, *supra* note 27, at 906 n.55. Yet an examination of the letters from various patent holders who have agreed to license their necessary implicated patents indicates that the claimed obligation that the terms be "openly specified" is almost entirely ignored in practice. See the letters posted at the IETF Page of Intellectual Property Rights Notices, *at* http://www.ietf.org/ipr.html (last visited Mar. 10, 2003). Virtually all of the letters merely indicate that the patent holder agrees to license on "reasonable, non-discriminatory terms," *without* specifying what the royalty rate will be. The "exception that proves the rule" is when the patent holder is willing to license on a royalty-free basis. Mueller, *supra* note 27, at 906 n.55; *see also id.* (listing examples of patent letters).

194. Rambus, Inc. v. Infineon Technologies AG, 318 F.3d 1081, 1102 n.10 (Fed. Cir. 2003).

individually is aware, suggests that the "default" rule should err in the other direction.

F. SANCTIONS FOR NON-COMPLIANCE

A significant question that appears to have received almost no attention is: what remedies and/or sanctions (if any) does the SSO have when its members fail to comply with the SSO's search, disclosure, and licensing rules?

Voluntary SSOs have to be concerned that, if they impose onerous rules, firms may elect not to participate.[195] This "participation constraint" limits the sanctions that SSOs can impose on firms for failure to comply with the SSO's rules. The typical remedy for non-disclosure imposed by the SSO is quite limited: the SSO will elect not to issue the proposed standard, and will "decertify" or withdraw the standard if it is subsequently discovered that there is a patent that reads on the standard, *and* the patent holder is *not* willing to agree to license its patent on "reasonable and non-discriminatory" terms.[196]

Taking the latter point first, this raises the obvious question: does a willingness to license on "reasonable and non-discriminatory" terms to all interested parties effectively "purge" any failure to disclose earlier? On the face of it, the answer would appear to be "yes." If the patent holder is willing to license on reasonable terms, there is no reason to believe that the SSO will withdraw the standard, and no reason to believe that the SSO would have acted any differently had it known of the existence of the patent before the standard's adoption. The major caveat here involves the distinction drawn above between ex post and ex ante "reasonableness."[197]

The power of the "threat" to withdraw the standard as a sanction for non-disclosure varies across different standards, depending on how easy it is for industry participants to switch to a different standard. In other words, the threat's power depends on the degree of "lock-in" involved. In some cases, once firms in the industry gear up to make and sell a standardized product, it may be extremely difficult to change to

195. *See supra* discussion Part IV.A.3.

196. As noted, a relatively small number of SSOs have policies preventing the adoption, or maintenance, of any standard on which there is a patent unless the patent holder is willing to license its patent royalty-free. *See supra* Parts IV.A and IV.C.

197. *See supra* Part IV.C.3.

another, unpatented alternative.[198] Merely "withdrawing" a standard *as* a formal standard may have little effect on industry behavior or the ability of the patent holder to induce others to pay royalties. Moreover, withdrawing a standard on a going-forward basis will not undo the fact that industry participants previously used the standard, thereby infringing the patent and exposing themselves to liability for damages for past infringement.

On a more conceptual level, withdrawing the standard imposes costs on other firms, not just on the firm that failed to comply with the SSO's rules. Ideally, one would want to find a sanction for non-compliance that penalized only the offending firm, not others.

On the face of it, the fact that SSOs may have relatively weak sanctions for non-compliance might be thought to justify the antitrust authorities in imposing stronger sanctions under the antitrust laws. Such a policy is seriously questionable, for several reasons.

First, if sanctions for non-compliance are relatively weak, the participants in the SSO, and other non-participants who make and use the standardized products, presumably are, or should be, *aware* of that fact. They presumably are thus aware that their protection against non-compliance is weak, and they can make their plans accordingly. Second, as noted above,[199] imposing strict sanctions for non-compliance would run the risk that firms would choose not to participate in the SSO in the first place.

In other words, firms "know what they are getting into" when the sanctions for non-compliance are clearly specified in advance. In such situations, intervention by the antitrust authorities to impose *additional* sanctions for non-compliance, beyond the ones imposed by the SSO itself, essentially amounts to changing the "ground rules" ex post. We see *no* reason why, if an SSO chooses to impose weak sanctions for non-compliance, and if those sanctions are clearly laid out in advance, the antitrust authorities should "fix" the situation by imposing additional sanctions beyond those imposed by the SSO.

---

198. The FTC has argued that the *Rambus* case, involving semiconductor manufacturers and JEDEC SDRAM standards, involves such a "lock-in" to the standard. *See* FTC Rambus Compl., *supra* note 86, ¶¶ 105-09. The extent to which any such "lock-in" exists, or how significant it is in a dynamic industry such as DRAMs, is a disputed factual question.

199. *See supra* Part IV.A.3.

V. CAPTURING THE BENEFITS OF STANDARDIZATION THROUGH SSO-RULE CLARITY

A. THE BENEFITS OF STANDARDIZATION AND THE NEED FOR SPEED

It is widely recognized that standardization can yield significant social benefits.[200] It is often less widely recognized that factors that delay the standards-setting process can impose significant adverse social costs. To generate the greatest social benefits, the standards-setting process must be both timely and efficient. This is especially true in fast-moving, high-tech industries. It does little good to take twenty-four months to set a standard when product life-cycles are eighteen months. A delay in setting a standard often means that new products get delayed, or markets do not develop.[201]

1. The Social Costs of Delay

Public policies that slow the adoption of standards can have very detrimental economic effects. By way of illustration, suppose that the product life-cycle in the industry is three years, and that some factor delays the adoption of a standard for the next-generation product by six months. This delay causes the loss of *one-sixth* (the three-year product life-cycle, divided by the one-half-year delay) of the overall social gains realized by moving to the next generation product. It is not uncommon in many high-tech industries for new generation products to represent a 50-100% improvement or more (in terms of net consumer satisfaction) over the earlier product, as anyone who recalls the early days of cellular telephones or personal computers can appreciate.[202] If so, a six-month delay

200. *See, e.g.*, Lemley, *supra* note 17, at 1896-98.

201. Examples include digital audio tape (DAT), high-definition television, and cellular telephone technology. For an extended discussion of these and other examples, see GRINDLEY, *supra* note 5, at 99-130, 195-234, 235-72.

202. In the semiconductor industry, "Moore's Law" asserts that the number of transistors per square inch of silicon in integrated circuits doubles every twelve to eighteen months. For an extended discussion, see Ilkka Tuomi *The Lives and Death of Moore's Law*, *at* http://www.firstmonday.dk issues/issue7_11/tuomi/ (last visited Apr. 23, 2003). A recent study suggests that, on a quality-adjusted basis, prices of desktop personal computers fell at a compound annual rate of 27% over the 1976-99 period. Ernest R. Berndt et al., *Price and Quality of Desktop and Mobile Personal Computers: A Quarter Century of History*, 2, *at* http://www.nber.org/~confer/2000/si2000/berndt.pdf (July 17, 2000).

can result in losing 8-16% of the overall social value of the product line as a whole.

2. Consortia, Organizational Structure, and Efficiency

Special-purpose SSOs, sometimes called consortia, are often faster, and can be more focused and nimble, than more "formal" SSOs in setting standards.[203] Participation in special-purpose SSOs is often limited to key players whose agreement is crucial to developing the standard.[204] With fewer participants, it is often faster and easier to reach consensus. Moreover, consortia are often less constrained by "due process" considerations than formal SSO counterparts that have formalized agendas and procedures. This in turn has led to a major shift away from the use of "formal" SSOs, and the proliferation of informal/voluntary SSOs and special-purpose consortia.

The difference is especially apparent in the distinction (discussed above[205]) between voluntary standards and regulations. Regulatory bodies are often constrained by due-process considerations in ways that voluntary SSOs are not.[206] As such, the regulatory process can often be much more time-consuming than private standards-setting.

The shift from more-formal SSOs to less-formal consortia can potentially have an adverse effect on the openness and transparency of decision making if consortia lack the types of due process procedural safeguards that more-formal SSOs have. In our view, however, the absence of complex procedural rules should be seen as an advantage, not a disadvantage. So long as the SSO's rules are clearly specified in advance, firms can "know what they are getting into" and can decide whether to participate and whether to rely on (or adopt) the SSO's standard accordingly.

---

203. The authors have discussed this with participants at the FTC/DOJ panel on standards setting and antitrust, and, in particular, with Carl Cargill, Director of Corporate Standards, Sun Microsystems, Inc. (Apr. 18, 2002).

204. For a discussion of the pros and cons of limited-membership SSOs, see HOVENKAMP ET AL., *supra* note 1, § 35.3, at 35-21.

205. *See supra* Part I.A.2.

206. *See supra* note 24 and accompanying text.

B. EFFORTS TO BLOCK OR DELAY THE ADOPTION OF A STANDARD

This Article's premise so far has been that the patent holder wants the SSO to adopt a standard that incorporates its patent. But in some cases, a patent holder may want to *prevent*—or if that is not possible, delay—the adoption of a standard. For example, an incumbent firm may be concerned that an SSO is developing a standard that will enable other firms to compete more effectively with the incumbent.[207] If the incumbent can prevent or delay the adoption of such a competing standard, it may be able to protect its current market position.

Further, if the incumbent firm holds a potentially relevant patent, it may be able to manipulate the SSO's rules to prevent or delay the adoption of a competing standard. Suppose, for example, that the SSO rules prevent the SSO from adopting a standard once a patent has been identified that potentially reads on the standard, unless the patent holder is willing to agree to license its patent on reasonable and non-discriminatory (RAND) terms. The incumbent firm need only assert that it has a potentially relevant patent and refuse to agree to license on RAND terms to block the standard's adoption. By blocking the SSO standard, the incumbent protects its market position.

Note that this tactic requires only that the incumbent *assert* that it has a potentially relevant patent. It does *not* require that the patent *actually* read on the proposed standard, at least if the SSO is not willing to take a position on that issue. Many SSOs are unwilling to do so, instead taking at face value the assertions of the patent holder.

C. RESTRICTIONS ON PARTICIPATION

Above we discussed one sort of "participation constraint": the concern that onerous rules will discourage participation in the SSO by patent holders. But there is another, unrelated "participation" concern: should any interested party be allowed

207. For example, Microsoft has been accused of trying to prevent or delay the development of operating systems or middleware such as Java that would undercut its market position. *See* United States v. Microsoft Corp., 84 F. Supp. 2d 9, 22-24, 32-37 (D.D.C. 1999); Franklin M. Fisher & Daniel L. Rubinfeld, U.S. v. Microsoft—*An Economic Analysis*, 46 ANTITRUST BULL. 1, 19-23 (2001).

to participate in the SSO, or are "closed" SSOs acceptable?[208]

In this regard, it is important to draw a distinction between *participating* in *setting* a standard and *practicing* that standard. Unless the standard implicates a patent, *anyone* can practice the standard, even those who did not participate in setting it. Indeed, one concern that many SSOs have is that non-participants can "free-ride" on the standardization efforts of others. Operating an SSO involves costs; those costs have to be recouped somehow,[209] and the SSO is often concerned about how to do so. One way is to charge a significant fee, much larger than the cost of reproduction, for copies of the standards adopted by the SSO.[210]

Concerns about firms' ability to participate in setting the standard are fundamentally different from concerns that firms need to produce products that comport with the standard in order to participate in the product market, i.e., in that the ability to practice the standard is (in antitrust terms) an "essential facility." As a general matter, unless the standard involves patented technology that the SSO-member patent holder is not willing to license to non-members on RAND terms,[211] non-participants in the standards-setting process can still *practice* the standard.

Instead, the inquiry turns to whether those excluded from participating are disadvantaged either (1) in terms of setting the standard or (2) in terms of access to the patented technology incorporated in the standard. With respect to the former issue, the non-participants' concern is that they may disagree with the standard ultimately chosen. But in a world where firms disagree about the best standard, that is likely to be the case for some participants as well. The relevant issue is whether the standards-setting process is systematically biased

---

208. For an extensive discussion of this issue, see HOVENKAMP ET AL., *supra* note 1, § 35.3.

209. *See* Patterson, *supra* note 76, at 2003-04.

210. Some SSOs charge significant fees (especially to non-members) for copies of their standards. For example, ANSI promotes membership by indicating that members "achieve up to three-times financial payback on member dues via site licensing of electronic standards collections." American National Standards Institute, *The Benefits of ANSI Membership*, *at* http://public.ansi.org/ansionline/Documents/Membership/why_buy_ansi.pdf (last visited Mar. 10, 2003).

211. Issues arise if licenses are only available to SSO members, or if non-members are charged higher royalties than members are charged. *See* HOVENKAMP ET AL., *supra* note 1, § 5.7, at 35-54.2. Even here, there is a countervailing "free rider" issue.

by excluding non-participants.

With respect to the latter (access to patented technology) issue, if the patent holder agrees to license all interested parties (whether SSO members or non-members) on RAND terms, then non-members can obtain access to the standardized technology on non-discriminatory terms, and the issue does not arise. If, however, the SSO members agree only to license their IP to each other (and not to non-members), or agree to charge higher royalty rates to non-members, then the issue of discrimination *may* become significant.[212]

The example in the previous section suggests that SSOs may legitimately want to limit participation in the standards-setting process by those whose interests would be served by preventing or delaying the adoption of such a standard. Unfortunately, it may be very difficult to determine whether this is in fact the case. Ascertaining true motives can be difficult, especially since large incumbent firms are likely to be affected by proposed changes and, thus, presumably have a strong interest in having their views considered. And accusations of misconduct are easy to make but difficult to resolve. What one firm may see as "delay" or "obstructionist behavior" on the part of another firm can often be defended as "deliberate consideration of relevant factors."

Moreover, considerations of organizational efficiency suggest that the standards-setting process is likely to work more efficiently and more quickly with fewer participants, especially if those participants have interests that are reasonably well aligned with one another. One suspects that this is the reason for the proliferation of special-purpose standards consortia; they can limit participation to a small number of key players, and are not constrained by the "due process" and other procedural considerations that more formal

---

212. We stress "may" here because, to the extent that the SSO depends on member contributions (e.g., membership dues) to fund its operations, it is perfectly reasonable for the SSO to charge non-members (who do not pay dues) a higher royalty rate for the SSO's IP than it charges to non-members.

Similarly, to the extent that SSO members have contractually agreed to a mutual licensing policy (so that each member agrees, as part of the quid pro quo membership agreement, that it will agree to out-license its patents to other members in exchange for their agreement to do the same), there is no reason why *non*-members (who have *not* agreed to be bound by such a commitment) should get the benefit of the lower (member) royalty rates.

As such, it can often be a difficult factual issue to determine whether any differential treatment of members and non-members is legitimate.

SSOs must follow.

Efficiency considerations also suggest that there is no necessary reason for a "one firm, one vote" rule in setting standards. Some firms are much more likely to be affected by which standard is selected than other firms are. Absent some procedural safeguards (such as a supermajority voting requirement), some firms may well elect not to participate. Nonetheless, the choice of a standard clearly can affect non-participants. Affected individuals or firms naturally want to "have a voice" in decisions that affect them.

Different SSOs can and do "compete" with one another via the participation rules and procedures they adopt. These include such factors as membership qualifications, voting rules (e.g., one-firm-one-vote, one-attendee-one-vote, votes proportional to market share, and supermajority rules), procedures (e.g., agenda setting, the role of committees, and opportunities for comments), and the overall cost and speed of the standard setting process. Our point here is that, again, there is no reason why a "one size fits all" approach is appropriate.

However, we believe that one comment is in order. There is a fundamental asymmetry between participants and non-participants. Non-participants cannot be bound by the rules of a voluntary SSO. In particular, a non-participant can assert its patents against those practicing the standard, without being subject to search, disclosure, or licensing obligations.

If participants are bound by the SSO's rules, and (in particular) to extend the benefits of such rules (in particular, rules requiring licensing on RAND terms) to non-members, then non-participants can remain free to assert their own patents against participants as they see fit, while retaining the ability to compel participants to grant licenses for *their* patents on "reasonable and non-discriminatory" terms. Such an asymmetric situation provides an incentive for firms *not* to participate in the standards-setting process.[213] This in turn implies that it may well be rational for the SSO to implement asymmetric rules aimed at counteracting this incentive.

One possible resolution to this asymmetry would involve having a "two-tiered" membership in the SSO. "Full" members would be able to fully participate in the standards-setting

213. This is especially likely if non-participants can argue that they should be able to claim the benefits, under the legal theories of equitable estoppel or third-party beneficiary, of disclosure policies for SSOs for which they are not members.

process, including proposing new standards, proposing changes, and voting on proposed standards. "Observer" members would be able to observe, and be informed of, the standard being set, but would not otherwise be able to participate. The quid pro quo would be that "full" members would have broader disclosure and/or licensing obligations than "observer" members would. The key idea here is that "observer" members would have little or no ability to actively "manipulate" the standards-setting process in their favor.

D. Adoption of Implicated Standards

The analysis thus far has been based on the premise that the SSO wants to *avoid* adopting a standard if there are patents that read on the standard. It is worth considering the *converse* situation, in which the SSO adopts a standard when it knows (or has good reason to know) that there are potentially relevant patent claims that may read on the products that comply with the standard.

Such conduct need not be entirely intentional. An SSO may be aware of the patent claims (or the existence of the pending patent application), but may conclude that the patent does not read on the proposed standard, or that the patent is invalid, or that the patent application will not issue. The SSO may, thus, adopt the standard despite its knowledge of the patent or pending application. The patent holder (applicant) clearly may disagree with the SSO's assessment of the prospect that a patent will ultimately issue; such a discrepancy on the merits is unsurprising.

Many SSOs have policies that, at least on paper, prevent the SSO from adopting a standard when there is any doubt as to whether a known patent reads on the proposed standard, unless the patent holder has agreed to license its patent to all potential users of the standard, whether royalty-free or on a "reasonable and non-discriminatory" basis. However, the extent to which such policies are honored in practice is an open question.

The concern here is that ex post, once (1) the standard is adopted, (2) firms are making products comporting with the standard, and (3) the patent holder seeks payment for the use of its patent, the accused infringers may argue that it is inappropriate under the antitrust laws for a participating patent holder to assert its patents against the infringing standardized products. If this argument is accepted, the users of the standard can use the leverage of the antitrust law to compel the patent holder to allow them to use the patent. If the antitrust remedy is a compulsory license (whether royalty-free or a license at a court-determined rate), then the other SSO members have in effect been able to obtain a compulsory license at a court-ordered rate (or for free), rather than having to negotiate and pay for a license.

As noted above, we believe that SSOs tend to be biased in favor of the users of standards and against the interests of patent holders.[214] The prospect that SSOs may be able to use the antitrust laws to compel compulsory licenses (especially if such licenses are royalty-free) is likely to exacerbate this bias. In particular, we believe that it makes no sense on public policy grounds to allow SSO members effectively to "expropriate" others' patented technology merely by adopting a standard when they know (or have reason to know) of potential patent issues.

The Rambus case provides a possible example.[215] Rambus withdrew from JEDEC in June 1996, some three years prior to JEDEC's adoption of standards for the next-generation of dynamic random access memory (DRAM) (known as DDR SDRAMs) in August 1999.[216] In its withdrawal letter, Rambus explicitly stated that it had both issued patents and pending patent applications, and informed JEDEC that it intended to enforce its patent rights against other firms.[217] To the extent that other JEDEC members were aware of Rambus's patent specification and/or claims (whether in its issued patents or its pending patent applications), but chose to adopt a DDR SDRAM standard (or later standards) incorporating Rambus's technology despite that awareness, this clearly raises concerns.

E. ANTITRUST INTERVENTION AND CLARITY

As noted above,[218] we believe that clarity of the SSO's rules is a key *desideratum*. Unfortunately, in our opinion, ex post antitrust enforcement efforts are often likely to *reduce* clarity and predictability, rather than enhance it.

Up to this point, this Article has tacitly assumed that

---

214. *See supra* Part II.B.
215. *See* discussion *supra* Part IV.D.1.
216. FTC Rambus Compl., *supra* note 86, ¶¶ 27, 82-83.
217. *Id.* ¶¶ 48, 56, 63-64, 67, 72, 82-85.
218. *See supra* Part IV.E.

standards-setting activities can potentially raise antitrust concerns. But in our experience the nature of those concerns, and the legal basis for intervention, has rarely been articulated clearly.[219]

We believe that the typical context[220] involves the claim that, by manipulating the standards-setting process (whether "actively" in an effort to "capture" a standard, or "passively" by improperly failing to disclose a relevant patent), the patent holder has gained improper market power in the technology market. Absent the need to comport with the standard (i.e., absent the "lock-in"), firms might (if feasible) find a way to avoid infringing the patent, by adopting an alternative technology.[221] But given that firms have a strong economic incentive to comport with the standard, the patent holder may be able ex post to extract a much higher price for the use of its patented technology than it would have been able to do absent the standard.

The antitrust concern here is *not* the proposition that the standard enhances the patent holder's market power per se. This is most readily seen in connection with patents held by *non*-participants in the standards-setting process. Adoption of a standard can confer a substantial windfall gain on non-participant patent holders, who (just like participant patent holders) may be able to extract higher royalties for the use of their patents than they would have been able to do absent the standard. But we know of no one who suggests that such conduct is an antitrust violation. Consequently, the "evil" that the antitrust law seeks to address in these contexts is the

---

219. Is the concern one of a conspiratorial agreement under section 1 of the Sherman Act, or monopolization or attempted monopolization under section 2 of the Sherman Act? If so, presumably the challenger must establish the other elements of any such claims.

The distinction between an enforcement action by the FTC or DOJ and a private antitrust action can be significant here. In particular, section 5 of the Federal Trade Commission Act gives the FTC the authority to intervene in the context of "unfair methods of competition," but does not create a private right of action. *See* 15 U.S.C. § 45 (1994); 1 AM. BAR ASS'N, SECTION OF ANTITRUST LAW, ANTITRUST LAW DEVELOPMENTS (FOURTH) 759-60 n.22 (1997). Despite this, private litigants (such as Infineon in *Rambus v. Infineon*) have argued that the FTC's *In re Dell* consent decree reflects a more general antitrust policy regarding standards-setting activities.

220. Obviously, there are other antitrust claim contexts, such as the "stuffing the ballot box" conduct in *Allied Tube & Conduit Corp. v. Indian Head, Inc.*, 486 U.S. 492, 496-97 (1988).

221. *Cf.* HOVENKAMP ET AL., *supra* note 1, § 35.5b, at 35-41.

manipulation that led to the enhanced value of the patent, not the fact that a patent reads on a standard or the enhanced value per se.

It is one thing for the antitrust authorities to adopt clearly specified rules on an ex ante basis governing standards-setting organizations and practices. For example, if the antitrust authorities believe that the public interest will be best served by requiring fully open participation, then they could announce an ex ante rule (or ask Congress to pass a statute) requiring all SSOs to be fully open to all interested parties. If the antitrust authorities believe that the public interest would be best served if all standards were "open," in the sense that they did not implicate patent rights, then the antitrust authorities could announce an ex ante rule (or ask Congress to pass a statute) requiring that standards be "open" in that sense.[222]

But it is something quite different for the antitrust authorities to use enforcement actions applying general antitrust principles to penalize conduct on an ex post basis in contexts where the rules are not clear, or, indeed, where the rules are explicitly to the contrary. For example, many of the public comments on the *In Re Dell* consent decree expressed concern that the scope of that ruling was unclear.[223] Was it intended to apply solely to cases (such as the situation described by the FTC majority in *Dell*[224]) where the SSO's policies required the firm's representative to certify in writing that his or her firm had no patents that read on the proposed standard? Or did the prohibition extend to other cases? For example, what is the appropriate antitrust rule (or policy)

---

222. We stress that we do not believe that either of these rules is desirable as a general policy.

223. The *Dell* matter involved a consent decree, and, in theory, consent decrees have no precedential effect. *See* 2 AREEDA & HOVENKAMP, *supra* note 24, ¶ 327c. This does not mean, however, that interested parties were not concerned about the general policy that underlay the FTC's reasoning in *Dell*, as the number of public comments demonstrates. *See In re* Dell Computer Corp., 121 F.T.C. 616, 623 (1996).

224. *See In re* Dell Computer Corp., 121 F.T.C. 616, 623 (1996). It is our understanding, based on conversations with knowledgeable individuals, that Dell contended that the scope of the VESA certification obligation was limited to the representatives' personal knowledge. However, Dell was willing to enter into the consent decree rather than litigate this disputed issue. Then-Commissioner Azcuenaga's dissent in *In Re Dell* took the Commission majority to task (in our opinion, correctly) for going beyond what the parties had stipulated to in the consent decree when writing the majority opinion (in particular, whether Dell's conduct was "intentional"). *Id.* at 639 (Azcuenaga, Comm'r., dissenting).

toward disclosure when the SSO's policies make it clear that it imposes *no* obligation to search for potentially relevant patents and that any disclosure obligation is limited to the personal knowledge of the individual representative?

F. PROBLEMS WITH "ONE SIZE FITS ALL" POLICIES

It is common for commentators to suggest that the rules "should" or "must" be one way or another. For example, Mueller recently proposed that "[a]ny firm that participates in creating an industry standard and thereafter obtains patent rights in some aspect of the standard *must*, at a minimum, disclose the existence of any patents or pending patent applications that may be relevant to the standard."[225]

Such a proposal can be understood in one of two ways. The first is as a *mandatory* rule, specifying what the rules should be—whether as a general matter of public policy or as a consequence of application of antitrust principles—allowing for no deviation. The second is what is often termed a *default* rule, to be thought of as the general proposition to be applied in the absence of evidence to the contrary, but one that can be changed by the SSO if it chooses to do so.[226]

These two interpretations have fundamentally different bases and policy implications. In our opinion, it is simply unnecessary to adopt mandatory rules in this area. SSOs are perfectly capable of adopting their own search, disclosure, and licensing rules, and of adapting those rules to the needs of the SSO participants. The results of Professor Lemley's survey indicate that SSOs have a variety of different rules.[227] There is no reason why a "one size fits all" mandatory-type approach is appropriate.[228]

---

225. Mueller, *supra* note 27, at 929 (emphasis added) (internal citations omitted).

226. For a discussion of the law and economics of default rules, see generally, Ian Ayres, *Default Rules for Incomplete Contracts*, *in* 1 PALGRAVE, *supra* note 65, at 585-90.

227. *See* Lemley, *supra* note 17, at 1904-07.

228. Professor Lemley has expressed concern that the variety across SSO rules does not appear to be indicative of any intentionally adopted policies. He sees no indication that the SSOs explicitly considered the range of alternative policies and intentionally chose the particular alternative that it did based on any kind of cost-benefit analysis. As such, he suggests that the antitrust authorities need not give deference to SSO policies in this area. *See* Lemley, *supra* note 17, at 1954-57.

We agree that the existing policies often appear to be a matter of

We find it is extremely telling that, at the recent FTC and Department of Justice (DOJ) hearings on the intersection between antitrust and intellectual property, *both* of the comments from SSOs expressed the belief that the current system worked reasonably well, and expressed concern that the antitrust authorities might adopt a "one size fits all" interventionist approach to standards issues.[229] We believe that those comments, coupled with the results of Professor Lemley's survey showing the wide diversity of policies across SSOs,[230] strongly suggest that the antitrust authorities should proceed cautiously in this area.

In particular, we are concerned that antitrust intervention may reduce the clarity of the rules, thereby making participation in SSOs more risky and reducing the willingness of firms with valuable IP (and which therefore presumably have much to contribute to selecting the appropriate standard) to participate. If the SSO's rules are unclear, the obvious public policy solution is to encourage SSOs to adopt clearer rules on a going-forward basis.

Most significantly, we believe that intervention runs a significant risk of slowing down the standards-setting process, thus delaying the adoption of new standards and new products made in accordance with those standards, to the detriment of consumers and of society generally.

This is not, of course, to suggest that there will never be an appropriate role for antitrust scrutiny of the standards-setting actions of SSOs or their participants. There is no question but that the activities of SSOs can affect non-participants, and one

---

historical accident and are not likely to reflect a well-thought-out or "optimal" degree of diversity. But in our opinion this is all the more reason for the antitrust authorities to tread warily in this area. This implies either that the SSOs *themselves* did not believe that the issue was important enough to carefully consider the range of alternative policies and to select the "best" policy, or that the choice of search, disclosure, and licensing rules reflects a degree of political compromise within the SSO. In either case, it is not likely that the antitrust authorities are in a better position than the SSO itself to strike a better balance of the competing interests of the various SSO participants.

229. *See* IEEE Standards Association, *Comments Regarding Competition and Intellectual Property*, *available at* http://www.ftc.gov/os/comments/intelpropertycomments/ieee.pdf (Apr. 17, 2002); Amy A. Marasco, *Standards-Setting Practices: Competition, Innovation and Consumer Welfare*, *available at* http://www.ftc.gov/opp/intellect/020418marasco.pdf (Apr. 18, 2002) (testimony by Amy A. Marasco, Vice President and General Counsel, American National Standards Institute).

230. *See* Lemley, *supra* note 17, at 1904-07.

rationale for antitrust intervention is to protect the interests of such non-participants from being adversely affected by decisions in which they did not participate or could not exert influence. And there are obvious examples of manipulation of SSO rules/policies, such as the "stuffing the ballot box" example of *Allied Tube*,[231] in which antitrust intervention may be the only solution.

But we believe that the antitrust authorities are likely to give too little weight to the fact that SSOs, as voluntary organizations, must often walk a fine line between competing interests. In our view, ex post intervention runs the serious risk of failing to recognize the ex ante balancing of competing interests.

CONCLUSION

Standards have come to play an increasingly significant economic role in fostering the development and adoption of new technology. Of particular importance in many fields are formal standards adopted by voluntary SSOs. Since the source of new technology is new innovation, and since Western societies have chosen to use the patent system (along with trade secrecy and copyright) as a major social incentive mechanism to encourage innovation, SSOs must ipso facto come to grips with the issue of how to deal with standards that involve patented innovations. A recent survey by Professor Mark Lemley indicates that SSOs have a variety of types of rules—notably search, disclosure, and licensing rules—to address these issues, and that SSOs vary widely in the stringency of the rules that they adopt.[232]

Private standards-setting by SSOs represents a classic example of what organizational theorists term a "private ordering." Such private orderings can be both a possible substitute for, and a strong complement to, "public" orderings such as those reflected in the antitrust laws and other statutes. As such, private SSOs have the flexibility to experiment with alternative procedures and rules, and to adapt those rules and procedures to particular circumstances in a way that a "one size fits all" approach can never achieve.

Consequently, we believe that though there clearly is a legitimate role for antitrust policy in connection with standards

231. Allied Tube & Conduit Corp. v. Indian Head, Inc., 486 U.S. 492 (1988).
232. *See* Lemley, *supra* note 17, at 1904-07.

setting, the antitrust authorities should tread warily in intervening in this area. Intervention runs the risk of reducing the clarity and predictability of rules, thereby increasing uncertainty and making participation in SSOs by firms with valuable intellectual property more risky and, therefore, less likely. As such, intervention runs the risk of delaying the adoption of standards, thereby reducing the economic gains from standardization and reducing social welfare generally.

APPENDIX

ON THE PRIVATE AND SOCIAL EFFICIENCY OF STANDARDS INVOLVING PATENTS

For simplicity and analytic convenience, we consider the case in which the characteristics of the *product* are fixed, and the issue is whether a standards-setting organization (SSO) should adopt (as a standard) either a patented *process* (denoted P) of making the product or an alternative unpatented process (denoted U).

We assume throughout that the SSO must select either P or U as a standard, and that the adoption of a standard makes it much more expensive for firms to subsequently switch to the alternative.

Obviously, if the patented alternative P costs *more* than the unpatented alternative U, it would make no sense for the SSO to adopt P as the standard.[233] As such, we restrict our

---

233. In practice, the SSO might not know whether P is (or will turn out to be) lower cost than U at the time that the SSO must adopt the standard. For example, U might be the existing technology, while P might be a new, unproven but promising technology that has the prospect (but not the certainty) of lowering costs.

Analytically, if the SSO must make a once-and-for-all irrevocable decision to adopt either P or U, the appropriate approach is straightforward: the SSO should calculate the ex ante expected costs of using P and U (given the best available information at the time the decision is made), and choose the alternative with the lower expected cost. However, if the SSO can delay adoption of the standard until more information about cost is known, or if adopting a standard does not rule out future use of the rejected alternative, the analytic framework is more complicated, involving a "real options" approach. For a discussion of the concept of "real options" and its economic significance, see TOM COPELAND & VLADIMIR ANTIKAROV, REAL OPTIONS: A PRACTITIONER'S GUIDE (2001).

Another, related but conceptually distinct issue involves the appropriate choice of standards in contexts where technology is changing over time. The concern is that, by adopting a particular technology or feature as the standard, it may be more difficult to switch to another technology/feature that subsequently becomes feasible, less costly or more desirable. This "lock-in" feature of standards is well recognized.

It might be argued that that standardization is undesirable, on the grounds that adopting any given technology as a standard makes it more costly and/or difficult for the industry to adapt to new technologies as they develop. In our opinion, such a conclusion is unwarranted. While it is clearly true that premature standardization can impose significant social costs (as the history of efforts to develop a high-definition television standard demonstrate), the difficulty is knowing in advance when standardization is "premature."

attention to the case where P is a lower cost alternative than U, ignoring for now the issue of royalties. By "lower cost" we mean that fewer real economic resources (e.g., labor, materials, capital, etc.) are used in producing the good using P than using U. Thus, choosing P rather than U "frees up" those resources for use elsewhere in the economy.

We denote by C_P the cost (per unit) of making the product using the patented alternative P, and by C_U the cost (per unit) of making the product using the unpatented alternative U. (For simplicity, we ignore fixed costs, or assume that the fixed costs are the same regardless of which alternative is chosen.) Then the per-unit cost savings from adopting the patented alternative P are $S = C_U - C_P$. In light of the above discussion, we assume $S > 0$.

Assume for now that, prior to the adoption of the standard, the patent holder commits itself to demanding per-unit royalties[234] of R_P for the use of its patented technology.

Suppose further that, once the standard is adopted, competition among users of the standard (i.e., manufacturers of the product) will drive the price of the product down to the marginal cost of production.[235] Thus if the unpatented alternative is adopted, the price will equal C_U, while if the patented alternative is adopted, the price will equal the sum of the manufacturing cost-per-unit C_P and the per-unit royalty R_P.

With a downward-sloping demand curve D with $Q = D(P)$, the market-clearing price if the unpatented alternative is adopted will be $Q_U = D(C_U)$, while the market clearing price if the patented alternative is adopted will be $Q_P = D(C_P + R_P)$.

Using 20/20 hindsight provides little useful policy guidance. And the alternative is often to forego the benefits of standardization altogether, a "cure" that is likely to be worse than the "disease."

234. The case where the patent holder asks for running royalties expressed as a percentage of the selling price (rather than on a flat per-unit basis) is similar. If the patent holder seeks a fully-paid-up-lump-sum (FPULS) license rather than running royalties, the economic analysis is significantly different. Fundamentally, with running royalties, the royalty payments impose a marginal cost on the licensee and thus affect the licensee's profit-maximizing output choice, while under a FPULS license the marginal royalty cost is zero.

235. If there are positive recurring "fixed" costs, in a free entry/exit equilibrium, economic theory implies that price will be driven down to average total cost. We ignore this complication in the text, as it does not fundamentally alter the analysis.

If entry/exit is otherwise difficult, so that the market is oligopolistic, the basic insight of the text is unchanged, but the mathematical derivation is more complex.

There are three cases, which we consider in turn:

(1) the royalties demanded may equal the cost savings ($R_P = S$);

(2) the royalties demanded may be less than the cost savings ($R_P < S$); or

(3) the royalties demanded may exceed the cost savings ($R_P > S$).

In the first case, with $S = C_U - C_P$ and $R_P = S$, the market clearing price if the patented alternative is adopted, ($C_P + R_P$), will just *equal* the market-clearing price C_U if the unpatented alternative is adopted, and so the market-clearing quantities in the two cases will be the same. Consumers will be indifferent between the two alternatives. Manufacturing firms will likewise be indifferent between the two alternatives; in either case, their total cost per unit is $C_U = C_P + S = C_P + R_P$.

From a *societal* perspective, however, the patented alternative is clearly *superior* to the unpatented alternative. If the unpatented alternative is adopted, society consumes more real resources (costing C_U) to make the product. If the patented alternative is adopted, the cost of production is only C_P, with a per-unit real-resource savings of S. This cost savings all accrues to the patent holder in the form of a royalty payment R_P. In other words, consumers and manufacturing firms are indifferent between the two alternatives, but the patent holder is strictly better off, and thus, society is strictly better off, if the patented alternative is adopted.

Turning to the second case, where the royalties demanded are *less* than the cost savings ($R_P < S$), if the patented alternative is adopted as the standard, competition among manufacturing firms drives prices down to ($C_P + R_P$) which is *less* than C_U. Quantity $Q_P = D(C_P + R_P)$ will be greater under the patented alternative than under the unpatented alternative $Q_U = D(C_U)$.

Consumers benefit from adopting the patented alternative as a standard both because they pay lower prices [($C_P + R_P$) is less than C_U] and because, at the lower prices, more consumers are willing and able to purchase the good ($Q_P > Q_U$), thereby receiving additional consumer surplus. Manufacturing firms are indifferent, as they still earn zero profits in equilibrium. And society is better off because it avoids the higher real resource costs associated with using the unpatented alternative. In essence, some of those cost savings accrue to the patent holder in the form of royalties, and some of them accrue to consumers in the form of lower prices.

This leaves only the third case, in which the royalties demanded exceed the cost savings ($R_P > S$). This case is illustrated in Figure 1. Let us compare what would happen if the SSO adopted the patented alternative with what would happen if the SSO adopted the unpatented alternative.

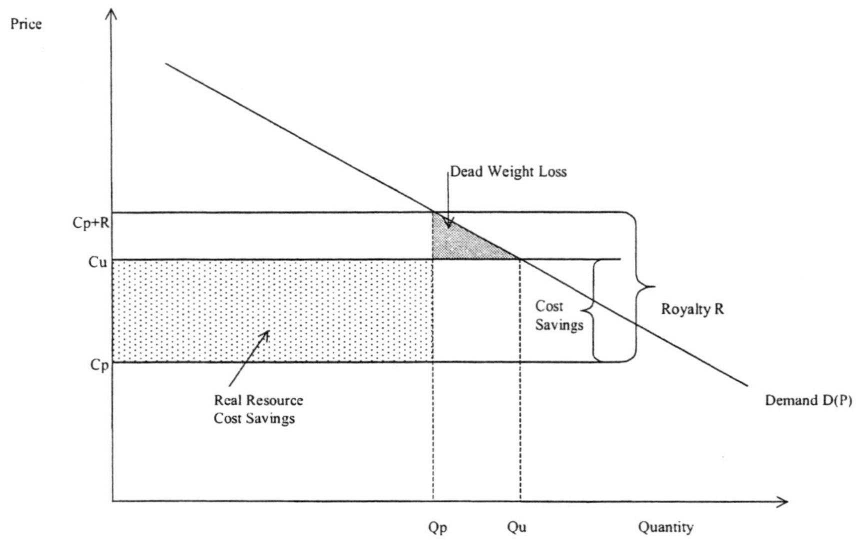

Figure 1

Because the market-clearing price (of C_P+ R_P) is higher with the patented standard than with the non-patented alternative (at C_U), the quantity demanded is lower if the patented standard is adopted [at $Q_P = D(C_P + R_P)$] than if the non-patented standard is adopted [at $Q_U = D(C_U)$]. This in turn implies that there is what economists term a "dead weight loss" associated with adopting the patented standard, in the form of the shaded triangle shown in Figure 1. This represents the lost consumer surplus from consumers who would have purchased at a price above C_U but below (C_P+ R_P) but who choose not to purchase given that the price is (C_P+ R_P).

Set against this dead weight loss, however, is the fact that, if the *unpatented* standard is adopted, society uses *more real resources* to produce the goods that are produced. The manufacturing firms incur the real resource cost C_U rather than the lower real resource cost C_P on each of the $Q_U = D(C_U)$ units produced. This is shown by the shaded rectangle in Figure 1. In essence, when the SSO adopts the unpatented standard, society foregoes the savings $S = C_U - C_P$ on each unit produced.

Note that, from a *private* standpoint, the SSO-member manufacturing firms would rather incur the higher cost C_U (and forego the cost savings S) rather than incur the lower costs C_P associated with the patented standard, because by doing so they can avoid paying royalties R_P which are larger than the savings S. That is, from a bargaining theory standpoint, if the patent holder announces its royalty terms R_P ex ante (i.e., *prior* to the adoption of the standard), the SSO-member manufacturing firms will choose to adopt the *un*patented alternative as the standard because, from a private-cost standpoint, they strictly prefer the unpatented alternative (with its cost of C_U) to the patented alternative (with its higher private cost of $C_P + R_P$).

But from a *societal* perspective, the net result is that the reduction in dead weight loss (the shaded triangle) can be outweighed by the extra real resources (the shaded rectangle). It is an *empirical* question which of these two offsetting factors is larger, and, thus, it is an empirical question whether society as a whole is better off adopting the patented alternative as the standard. But the key point here is that, in making that determination, the amount of the royalty payments is almost entirely *irrelevant.* From a *societal* standpoint, the royalty payments are a "wash."[236]

The above discussion of case (1) assumed that the patent holder was able to charge a royalty R_P that exceeded the cost savings S. In practice, however, faced with such a royalty demand, the SSO members would likely tell the patent holder that they would choose the non-patented standard instead so that the patent holder would receive no royalties. Faced with a choice between maintaining its high royalty demand (and receiving no royalties) or dropping its royalty demand to a

236. This needs to be interpreted with some care, because of the phenomenon known as "rent-seeking." *See supra* note 77. Firms can and do expend real resources in an effort to appropriate larger fractions of the overall economic "rents" (surplus) for themselves. The real resources consumed in such rent-seeking behavior themselves are a social cost that must be considered.

lower level (so that the SSO would adopt the patented standard and the patent holder would receive royalty revenues), and given that a rational patent holder prefers revenues to no revenues, one would expect that, at least when the royalty rates are negotiated ex ante (prior to the adoption of the standard), the patent holder would moderate its royalty demands so that the situation fell within case (1) or case (2) above rather than case (3).

This in turn implies that adopting the patented alternative is a win-win-win situation. Society is better off because fewer real resources are used in producing the product. Manufacturers are better off (or at least no worse off) because the cost of the royalties is no more than the cost savings from using the patented technology. And, the patent holder is better off because it receives income from its invention.

However, when the royalty rates are negotiated ex post (*after* the adoption of the standard), things are somewhat more complex. It is still the case that society is better off using the (lower-cost) patented standard than the (higher-cost) unpatented alternative. And, the patent holder is better off. But when royalties are negotiated ex post, the patent holder may be able to extract royalties that exceed the cost savings. In such situations, the patent holder may be able to extract royalties that exceed the cost savings. Hence, manufacturers will be worse off having adopted the patented technology than using the unpatented (but higher cost) alternative.

Analytically, much the same economic issues arise in the more realistic case where the standard in question involves some feature of the product, and the issue is whether to adopt a patented version of the feature or a non-equivalent but unpatented version as a standard. The advantage of the current approach is that, because the features of the product are fixed, we can assume that the demand curve for the product is the same regardless of which standard is adopted.

In the more realistic case, it is necessary to control for the fact that consumers will have different willingness-to-pay for products with different features. It can be shown analytically that, when consumers uniformly value the patented product more than the unpatented alternative, and the extent of that preference is the same across all consumers, the same analysis developed above for cost-saving innovations applies to such quality-enhancing innovations, and the same conclusions apply.

A Primer on Trademarks and Trademark Valuation[1]

MICHAELYN CORBETT
LECG, LLC

MOHAN RAO
LECG, LLC

DAVID TEECE
LECG, LLC

A trademark is a distinctive word, phrase, name, or symbol that is used in commerce to indicate the source of a good or service and to distinguish it from the goods or services of others. Although trademarks have existed for centuries, they have become of great importance in recent years. In fact, it is hard to imagine key elements of today's global commerce proceeding without trademark protection. Sought-after products around the globe have widely recognized trademarks: Coca-Cola, McDonald's, Sony, Disney, Microsoft, and Apple, to name just a few. Like patents, trademarks can constitute a significant portion of a firm's asset value and, therefore, need to be strategically developed and protected. As recognition of their increasing importance in commerce, damage awards in trademark infringement litigation have reached tens of millions of dollars, with some awards exceeding $100 million. In this chapter, we provide a primer on trademarks and trademark valuation. We discuss the economic principles of licensing and describe some of the commonly used approaches to trademark valuation, particularly in the context of licensing trademarks.

WHAT IS A TRADEMARK?

A trademark is a distinctive word, phrase, name, or symbol that is used in commerce to indicate the source of a good or service and to distinguish it from the goods or services

Economic Damages in Intellectual Property: A Hands-on Guide to Litigation, pp. 281–296, edited by Daniel Slottje.

of others. In addition, domain names on the Internet are also subject to trademark laws.[2] Some examples of trademarks include Rolex for timepieces, the Pillsbury doughboy symbol for baked goods, and the "Got milk?" slogan of America's milk processors.

What Constitutes an Acceptable Trademark?

To be protectable, trademarks must be inherently distinctive or possess "secondary meaning" (i.e., the public associates a particular symbol with a single entity). Furthermore, the trademark must be nonfunctional. Commonly used words for objects, such as *belt, glass,* or *chair,* generally cannot be registered as trademarks. For instance, a company cannot register the trademark "apple" to describe a product of fruit that grows on a tree because that merely describes what it is. However, a company might be able to register the trademark "Apple" to identify a particular brand of computer product because it is not descriptive in that context at all. Instead it is a highly arbitrary identifier and subject to trademark protection. Geographic names and surnames are generally off limits to trademark protection. In addition, trademarks cannot be lewd or immoral. However, the criteria for trademark registration can be a challenge to understand. While companies and individuals cannot simply register ownership of common words or phrases, they can turn a *distinctive* combination of words into a trademark. For example, the Wheaties cereal (owned by General Mills) is associated with the well-recognized trademarked slogan "The Breakfast of Champions."

A trademark is likely to fall under one of three categories:[3]

1. *Coined words (or "fanciful" words)* are invented words without any real meaning in any language (e.g., Prozac or Lexus). Coined words often have the advantage of being easy to protect, as they are more likely to be considered distinct. The downside is that some coined words may be more difficult for consumers to remember. They are neither descriptive of the products they represent nor easy to associate with other things more familiar to a consumer.
2. *Arbitrary marks* are trademarks that consist of words that have a real meaning in a given language. The meaning of such words, however, has no relation to the product itself or to any of its qualities (e.g., Dove for soap or Polo for clothing). While the level and ease of protection is generally high for arbitrary marks, as with coined words, greater effort may be required to create a direct association between the mark and the product in consumers' minds.
3. *Suggestive marks* are marks that hint at one or some of the attributes of a particular product. For example, the trademark Coppertone attempts to characterize the results of using a particular suntan lotion, intending to capture one of the most appealing aspects of such products (i.e., a deep, dark tan). Another example is the trademark "Lucent" used by Lucent Technologies to associate "clear" with telecommunication products. The appeal of suggestive marks lies in their ability to act as an advertising mechanism to create a direct association in the mind of consumers between the trademark, certain desired qualities, and the product. The perceived risk of suggestive

marks, however, is that they may be considered not sufficiently distinctive to meet the criteria for trademark protection.

Different Types of Trademarks

There are four types of trademarks:

1. *Service marks.* A service mark is a type of trademark that identifies and distinguishes the source of a service rather than a product. Examples of service marks include Kinko's for photocopying services and FedEx for express delivery.
2. *Collective marks.* Collective marks are used in indicating membership in a group or an association where only members of the group or organization can use the collective mark. Examples of collective marks include the AFL-CIO, Rotary International, or the CPA designation to indicate members of the Society of Certified Public Accountants.
3. *Certification marks.* Certification marks are used in certifying the quality, regional origin, or other origin of a product or service. A certification mark certifies that a product or service meets a certain standard of quality or is of a regional origin. An owner's function is to exercise control over use of the mark by others, ensuring that their products and services meet specified quality standards. Incidentally, use of a certification mark by others may not be refused as long as the required standard compliance is met.[4] Examples of certification marks include wine from the Champagne region of France, the Good Housekeeping Seal of Approval, and the Certified Vegan symbol on food products administered by Vegan Action.
4. *Trade dress.* In addition to a label, logo, or other identifying symbol, a product may come to be known by its distinctive packaging or manner of presentation. *Trade dress* refers to the way in which a product—or place of business—is "dressed up" to go to market. Examples of trade dress include the yellow packaging of Kodak film, the Golden Arches of a McDonald's restaurant, and the distinctive red label on a Campbell's Soup can. The particular style and ambience of a retail outlet can also, in some circumstances, constitute trade dress. Exhibit 18.1 presents examples of the different types of trademarks.

A Trademark Is Not the Same as a Trade Name or Brand

Trademarks are closely related to trade names and brands. A trade name is used to identify a company or a business and serves as the name of the company or a business. In contrast, a trademark or service mark is used to identify the source of the products or services that the company or business sells or provides. In practice, however, the distinctions between trademarks and trade names and the laws that govern them are not always clear. For instance, only trademarks can be registered in the federal or state trademark registries. Trade names are registered at the state level, often with the secretary of state. Trademarks are

EXHIBIT 18.1 TYPES OF TRADEMARKS

Service Marks	Burger King Blockbuster Roto Rooter
Collective Marks	AAA (American Automobile Association) Realtor CFE (Certified Fraud Examiner)
Certification Marks	Harris Tweeds (a special weave from a specific area in Scotland) Stilton cheese (a product from the Stilton locale in England) Star-K Kosher
Trade Dress	Coca-Cola bottle Distinctive decor of Two Pesos Mexican restaurant Red border of *Time* magazine cover

afforded a higher level of protection than trade names. For instance, a trademark holder may face a legal challenge if the trademark is not inherently distinctive; the trade name holder bears no burden to prove the distinctive nature of the underlying name. Moreover, while both concern themselves with "similarity," trade name law is less concerned with confusion of marks and more with deception and the resulting harm to business reputation.[5] For instance, Philip Morris, a company name, is a trade name whereas Marlboro is a trademark for a specific brand of cigarettes from Philip Morris.

Brands and trademarks also are closely linked, although there are subtle distinctions between the two concepts. A brand can be loosely thought of as an accumulation of assets that may or may not include trademarks. For instance, the McDonald's "brand" includes trademarks such as Big Mac, the Golden Arches symbol, and the Ronald McDonald name and logo. In contrast, Linoleum is also a brand; however, there is no longer a trademark associated with the "linoleum" product. The close relationship among trademarks, trade names, and brands means that it can often be difficult to separate the incremental financial contributions by each of these intangible assets for valuation purposes.

A Trademark Is Not the Same as Goodwill

An enterprise typically contains a portfolio of intangible assets, which may include various forms of intellectual property. For many years the bundle of intangible assets was simply referred to as *goodwill*. Trademarks were often simply considered goodwill or patronage—the ability to attract and retain customers who will recommend the business to others. As valuation experts have begun to focus on intangibles, efforts have been made to break down the "black box" known as goodwill and identify and analyze the individual components of a firm's intangible assets. Thus, for the purpose of trademark valuation, it is now well recognized that it is no longer necessary to fall back on the use of the term *goodwill*.

Why Do We Care about Trademarks and Why Are They Needed?

Trademarks serve a variety of economic and business purposes. The traditional rationale for trademark protection is to allow an enterprise's customers to distinguish its products or services from those of its competitors. In that regard, a trademark is in some ways the "face" of a business. Thus, a fundamental principle underlying trademarks is protection of the reputation of particular products or services from "copycat" producers that could confuse consumers regarding which product or service they are purchasing. Alternatively, actual competitors may begin to mimic certain characteristics of a successful trademark in ways that could lead to consumer confusion. For example, in 1991, Merriam-Webster, one of the market leaders in college dictionaries, sued a competitor, Random House, after the latter changed the name of its dictionary from *RH College Dictionary* to *Random House Webster's College Dictionary*.[6] In addition, Random House adopted a trade dress similar to *Merriam-Webster's Ninth New Collegiate Dictionary*. Among the similarities between the two competing works were the bright red dust jackets, the use of the generic word *Webster's* in combination with the descriptive term *College* or *Collegiate* on the cover, and the name *Webster's* in large white vertical letters on the spines. An important question in this litigation was to determine the likelihood of customer confusion due to Random House's new dress cover on its dictionaries.

Second, a successful trademark is seen as a mark of quality assurance. A customer who is pleased with the quality of a product or service will continue to associate high quality with that trademark. Overtime, a consistently high-quality product is likely to result in a trademark with a strong reputation, which, in turn, could translate into significant value for the owner. For example, Cadillac is often associated with class and luxury while Toyota is frequently associated with dependability.

Third, trademarks provide a mechanism by which to protect an owner's investment in advertising and promotion. Firms often spend significant resources in advertising and promoting their trademarked products. General Motors, for instance, spent almost $4 billion on advertising in 2004. By comparison, the U.S. government spent over $1.2 billion on advertising in the same year. These investments, however, may not generate rewards until many years into the future. Without trademark protection, owners might not invest in developing distinctive and widely recognized brands and trademarks, to the detriment of consumers. Consequently, trademarks are important from a commercial perspective and should be properly viewed as business assets.

Exhibit 18.2 lists the 10 leading national advertisers as measured by 2004 advertisement spending.

In its annual special report, *BusinessWeek* magazine presents the world's most valuable brands as determined by Interbrand Corporation. Interbrand values brand assets based on discounted future earnings. As discussed earlier, brand value reflects, in part, the intellectual property assets, including trademarks, of the owner. Without endorsing its methodology, Exhibit 18.3 presents Interbrand's list of the 10 most valuable brands in 2005.[7]

EXHIBIT 18.2 10 LEADING NATIONAL ADVERTISERS BY 2004 ADVERTISING SPENDING

Rank	Advertiser	Advertising Expenditures (in billions US$)
1	General Motors	4.0
2	Procter & Gamble	4.0
3	Time Warner	3.3
4	Pfizer	3.0
5	SBC Communications	2.7
6	DaimlerChrysler	2.5
7	Ford Motor	2.5
8	Walt Disney	2.2
9	Verizon Communications	2.2
10	Johnson & Johnson	2.2

Source: Advertising Age, June 27, 2005, p. 6.

How Does One Obtain a Trademark?

There are a number of ways in which trademark status can be obtained. In the United States, individual state statutes govern the registration and protection of state trademarks and service marks. Registration of state trademarks is typically accomplished through the secretary of state. There also are federal trademarks for goods and services in interstate commerce, governed by federal law at 15 U.S.C. § 1125 (the Lanham Act). Federal trademarks are registered through the U.S. Patent and Trademark Office (USPTO). Finally, rights in trademarks and service marks may arise solely from use, even without state or federal registration. These are common law rights.[8]

EXHIBIT 18.3 10 MOST VALUABLE BRANDS OF 2005

Rank	Advertiser
1	Coca-Cola
2	Microsoft
3	IBM
4	GE
5	Intel
6	Nokia
7	Disney
8	McDonald's
9	Toyota
10	Marlboro

Source: BusinessWeek, August 1, 2005, pp. 90–94.

At the global level, trademarks are covered by a number of multilateral treaties and agreements such as the Paris Convention, the Madrid Agreement, the Trade Related Intellectual Property Rights (TRIPS) agreement, the Trademark Law Treaty, the Nice Agreement, and the Vienna Agreement.[9]

Currently, registration of a trademark is effective for 10 years, but registration can be renewed indefinitely so long as the mark continues to be in active use and has not become generic. Unlike patents, active trademarks can last in perpetuity.

How Are Trademarks Protected and How Can They Be Damaged?

The Trademark Act of 1946 (also known as the Lanham Act) is the U.S. law governing trademarks and is administered by the USPTO. It protects products and services that have trademarks from use that is "likely to cause confusion, or to cause mistake, or to deceive."[10] The Lanham Act was amended in 1988 with the passage of the Trademark Law Revision Act (TLRA). The TLRA remedied the problem by which a business or individual seeking to register a trademark in the United States must be using the mark in interstate commerce before applying for registration, a practice not required in most other countries. The TLRA gave all applicants the choice of applying to register marks on the basis of preapplication use in commerce or on the basis of a bona fide intention to use the mark in future commerce (intent-to-use).[11] The TLRA also contained provisions to help deal with the high volume of inactive, but registered, trademarks, including a reduction in the renewal period from 20 to 10 years.

In January 1996, the Federal Trademark Dilution Act (FTDA) was signed in the United States, providing owners of famous trademarks with a federal cause of action against those that lessen the distinctiveness of such marks by the use of the same or similar trademarks on similar or dissimilar products or services. Thus, the antidilution statutes can apply even if it were unlikely that customers would confuse the source of the goods or services with those sold by the owner of the famous mark. The antidilution laws essentially protect trademarks from two forms of injury: blurring and tarnishment.[12]

Blurring occurs when the distinctiveness of a famous mark is lessened through overuse by others in different contexts, thereby reducing its value even though a consumer may not be confused as to the source of the goods. For instance, a ball point pen manufacturer using a mark similar to "Rolls-Royce" on its products may serve to diminish the mental association made between "Rolls-Royce" and the line of goods it is normally associated with—ultra-luxury cars. Therefore, the strength of Rolls-Royce's trademark could be weakened if the public now associates the mark with a different source or a "lesser" product.

Tarnishment damages the goodwill or reputation of the mark by associating it with an inferior product or showing it in a distasteful light. Tarnishment occurs when a party uses a famous mark in association with unwholesome or defective goods or services, particularly when the unauthorized use involves illegal activity or crude behavior. For example, if an X-rated movie portrays actresses in Chicago Bears' cheerleading uniforms, the Chicago Bears' reputation could be tarnished. Tarnishment may also occur when an individual attempts to parody a famous trademark in a derogatory fashion.

The right to exclude others from using a trademark is not absolute, however. Rather, trademark owners only have the right to protect their goodwill by prohibiting others from misrepresenting the marks in commerce or otherwise using the marks in a damaging manner to the trademark holder. A trademark owner's ability to stop others from using its trademark, or a confusingly similar one, depends on such factors as whether:

- The trademark is being used on competing goods or services (goods or services compete if the sale of one is likely to affect the sale of the other)
- Consumers would likely be confused by the dual use of the trademark
- The trademark is being used in the same vicinity or is being used on related goods (such as goods that will likely be noticed by the same customers, even if they do not compete with each other)

Improper use of a trademark, or allowing others to use a trademark improperly, can result in the trademark becoming generic and open to use by others in the public domain. Once a trademark ceases to distinguish products or services as produced from a single source, the trademark's legal protection may be compromised. Examples of products that have become generic terms in the United States but were once protectable trademarks include "aspirin," "nylon," and "corn flakes." In fact, herein lies a dilemma for trademark owners: A widely recognized trademark reflects success in getting consumers to associate a specific product with that trademark or brand. However, using the trademark to describe a class of products (such as Kleenex for tissuess) puts the trademark at risk. Thus, trademark owners have a strong incentive to vigilantly guard the use of their trademarks. Gordon Smith cites the rather imaginative efforts by Xerox Corporation to protect its trademark from becoming generic through advertising campaigns in the media that plead: "When you use 'Xerox' the way you use 'aspirin,' we get a headache."[13]

The rights to a trademark can also be lost through abandonment if its use is discontinued with an intent not to resume. The basic premise is that trademark law only protects marks that are being actively used and parties are not entitled to "warehouse" potentially useful marks. For instance, in a prominent case, a federal court in New York found that the Los Angeles Dodgers had abandoned rights to the Brooklyn Dodgers trademark.[14] In a more dramatic example, Procter & Gamble (P&G) allowed the trademark on its White Cloud toilet paper to lapse in 1994 when its attention was focused on the more expensive Charmin brand toilet paper. The White Cloud trademark was promptly picked up by a private entrepreneur and sold to Wal-Mart. Wal-Mart, in turn, started using the White Cloud trademark in competition with P&G's Charmin. As P&G executive Tom Muccio later commented: "The beauty about White Cloud for Wal-Mart was that we had built that Mercedes image and they brought it in at Chevy prices."[15]

Trademarks and the Internet

The emergence of the Internet and its unprecedented expansion led to new complications for the ownership and protection of trademarks. While trademarks are used in a variety of ways on the Internet, mirroring their use in non-Internet commerce, a primary area of concern was related to domain name registration.[16] A private company, Network Solutions Incorporated (NSI), began registering domain names in the United States under an exclusive contract with the National Science Foundation in 1993.[17] Names were registered on a first-come, first-served basis with no checking of trademark status or violations. In fact, NSI took the position that it was not responsible for trademark disputes, which it expected would be resolved between the parties through normal legal channels.

However, there are several significant problems associated with the Internet and trademarks. A trademark such as "Apple" can be famous in a number of different lines of commerce. For instance, it is the trademark of Apple Computer but also of Apple Records, the Apple Club, the Apple Bank, Fiona Apple, and the American Professional Partnership for Lithuanian Education (APPLE), to name just a few. The obvious question is who should have right to the domain name www.apple.com? There is no easy way to map a similar trademark associated with different lines of commerce with a single domain name reflecting that trademark. This technical limitation led to a second problem known as cybersquatting.

Cybersquatting is the registering, selling, or using of a domain name with the intent of profiting from the goodwill of someone else's trademark. For a while, the practice was rampant.[18] Early disputes related to trademarks used as domain names were typically settled privately or through arbitration and often involved the payment of "ransom" to obtain the domain name. One of the first disputes resolved through arbitration involved the Princeton Review Management Corporation and Stanley H. Kaplan Educational Center, Ltd., two well-known test preparation firms and fierce competitors. Princeton Review registered the domain name kaplan.com. Potential customers visiting kaplan.com would be welcomed to the Princeton Review. Web users would be asked to contribute to a list of complaints against Kaplan, evidently for the purpose of taunting and belittling Kaplan.[19] Kaplan sued Princeton Review for trademark infringement and eventually gained rights to the kaplan.com domain name. Under the provisions of the 1999 Anti-cybersquatting Consumer Protection Act (ACPA), a victim of cybersquatting in the United States can now sue or fight a cybersquatter using an international arbitration system.[20]

How Are Trademarks Valued?

Trademarks are typically valued for one of three purposes: (1) in the context of a licensing transaction or acquisition, including as part of a business acquisition; (2) for regulatory compliance, such as in transfer pricing; and (3) in the context of litigation. The economic principles behind the valuation remain essentially the same in each context. Next we discuss some of the commonly used valuation approaches, particularly in the context of trademark licensing.

Principles of Licensing

A license represents a transfer of certain rights to an asset from the licensor to the licensee. A licensing transaction occurs because both the licensor and the licensee expect future economic benefits from the transfer of rights. There are typically two primary channels for a trademark owner to exploit its intellectual property: (1) direct exploitation and (2) indirect exploitation. Direct exploitation allows the owner to retain the full bundle of intellectual property rights to an asset but also requires that the owner have the necessary resources to fully exploit the asset. For example, a regional manufacturer may be unable to meet the challenges of realizing expanded commercialization of its trademarked product. A university might have valuable trademarks that could be exploited by developing logo apparel, but it does not have the capability or the desire to get into apparel manufacturing. Faced with this issue, an alternative channel for an owner to exploit its property without putting additional capital at risk is to license it, that is, indirect exploitation. But licensing the technology may also come with risks. For example, the licensor faces the risk that the licensee may not adequately market a product or may not be able to fulfill its financial obligations to the licensor. Therefore, from a licensor's perspective, the decision to exploit the technology on its own or through licensing requires a careful assessment of the relative expected benefits and expected costs associated with each option.

A potential licensee would be willing to license a technology if it generates *incremental* economic benefits over alternative business opportunities available to the licensee. For instance, an apparel manufacturer might license a trademark if it allows the manufacturer to enter a new market, serve a larger market, or earn a higher return on investment. Like the licensor, the licensee also faces a number of risks associated with the licensing transaction. First and foremost, the licensee undertakes the risk of deploying capital to manufacture, distribute, and sell the licensed products. While the licensee's capital investment and fixed costs may be known, the success of the licensed product is not guaranteed. For instance, changes in the state of the economy or in industry circumstances might make the business model unsustainable.

In a licensing situation, a trademark holder (the licensor) effectively "rents" certain rights to the trademarked property to another party (the licensee). The rental payment in return for an agreed upon use of the intellectual property is referred to as a licensing or royalty payment. Generally, the licensor will not rent out all rights to a trademark but rather set out specific rights of use for the licensee. Some elements of the rights include the term of use, geographic territory, termination provisions, and trademark quality control. In addition, a trademark license may be bundled in a licensing agreement with other rights to intellectual property.

Royalty payments are generally determined through negotiation between the licensor and licensee and, as such, should represent a market-based outcome. The negotiated fee and royalty rate should reflect a *fair market value* for the asset to be transferred and should make both the licensor and the licensee better off. Fair market value is commonly defined as the amount at which an asset would change hands between a willing licensor and a willing licensee, neither being under any compulsion to buy or sell and both having reasonable knowledge of the facts.

Division of the economic benefits of an asset between a licensor and a licensee often entails a valuation analysis. There are three widely used approaches to valuing intellectual property, including trademarks: the *cost approach, market approach,* and *income approach.*[21] We consider each of these approaches next.

Cost Approach

The cost approach values assets based on the cost to create and develop the assets. The premise behind the cost approach is that no party involved in an arm's-length transaction would be willing to pay more to use the property than the cost to replace the property. In the context of patents, for instance, a potential licensee would not pay more to license a patent than the cost to design around the technology contributed by that patent. An alternative to designing around the technology would be to purchase the technology. Accordingly, a potential licensee would not pay more to license the technology than it would have to pay to purchase or create the technology.

Unlike patents, it may not be feasible to "design around" famous trademarks, that is, create comparable trademarks in any meaningful way. For instance, one cannot easily build characters comparable to Mickey Mouse, Elmo, or Dora by incurring the same costs as those required to build these trademarks. But the cost approach may be of value when the assets under consideration involve less famous trademarks where it is possible to develop comparable trademarks.

One possible starting point in applying the cost approach may be to collect information to estimate the cost to "rebuild" the trademark. This is often done by calculating the present value of the past costs incurred to build a comparable trademark. Examples of expenditures that might go into creating a trademark include concept development, trademark searches, consumer testing, package or logo design, and advertising and promotional campaigns.

In addition to the difficulty in "designing around" a famous trademark, the cost approach can have serious additional limitations. The backward-looking nature of the cost approach does not fully measure the economic benefit that may be derived from the trademark. For instance, two trademarks that cost approximately the same to create may have very different income streams associated with them. Nonetheless, the cost approach may still be useful in instances where a comparable trademark can be created and where the costs for such creation are known or can be estimated.

Market Approach

The market approach references a market with comparable transactions to determine the fair market value of an asset. The degree of reliance on comparable transactions depends on an assessment of the transactions to determine if they are sufficiently similar to provide an indication of the fair market value for the assets in question. Factors to consider include the nature of the assets being transferred, the industry and products involved, agreement terms, and other factors that may affect the agreed-on compensation

The market approach is often helpful in determining the running royalty rates in specific licensing transactions based on similar transactions in the marketplace. As an example, suppose that you are an executive in the licensing department of a university and are considering a potential transaction with an apparel manufacturer to produce university logo apparel under license.[22] One approach would be to look for comparable transactions between other similarly situated universities and evaluate their licensing terms. Exhibit 18.4 presents publicly available data on comparable licensing terms for a selected sample of universities. From the table, one can determine that the average comparable royalty rates range from a low of 7.5 percent for Harvard University to a high of 15 percent for the University of Oregon. These royalty rates for comparable transactions provide both the licensor (the university) and the licensee (the apparel manufacturer) information of the "market" for university trademarks. The university can pick a rate in this range after further refining the set of comparables and use that rate as a basis for its much broader licensing program.

While the market approach can be extremely useful in valuing intellectual property for licensing purposes, finding comparable transactions can be tricky. First, by its very nature, intellectual property is somewhat unique. Any market approach analysis will likely require reasonable adjustments. Second, the details of many licensing transactions involving trademarks are not publicly available. However, the market approach can be quite useful in valuing trademarks in instances where such information is available and comparable transactions can be found.

EXHIBIT 18.4 MARKET APPROACH—SELECTED UNIVERSITY TRADEMARKS

Licensor	Low Rate (%)	High Rate (%)	Summary of Terms
University of Oregon	8.0	15.0	Limited nonexclusive right to use selected university marks on licensed articles for a percent of net sales; guaranteed minimum royalty payment of $1,000 for apparel and $500 for other products
University of Arkansas	8.0	10.0	Limited nonexclusive right to use selected university marks on licensed articles for a percent of the wholesale cost; annual advance royalty guarantee of $200
Michigan State University	8.0	8.0	Limited nonexclusive right to use selected university marks on licensed articles for a percent of net sales, unspecified advance royalty fee required
Harvard University	7.5	7.5	Limited nonexclusive right to use selected university marks on licensed articles for a percent of the net sales price; annual advance royalty guarantee of $1,000

Source: Office of Trademark Management, University of Oregon; University of Arkansas, Business Affairs; University Licensing Programs, Michigan State University; Office for Technology and Trademark Licensing, Harvard University.

Income Approach

The income approach is a method used to value intellectual property assets based on the present value of the future income stream generated by an asset. There are three major inputs to the income approach: (1) expected future cash flows from the asset; (2) economic life of the asset; and (3) business risk associated with the realization of the cash flow stream. The key goal is to estimate the present value of incremental profits generated by the asset over its economic life, taking into account the risk associated with generating those profits. Once the present value of the incremental profits is determined, these profits are split in some manner between the licensor and licensee, typically in the form of a royalty.

Returning to the example we used earlier while discussing the market approach, consider how the university and the apparel manufacturer might negotiate the terms of a licensing transaction under the income approach. Exhibit 18.5 presents data on potential sales by the apparel manufacturer. Suppose that the terms of a potential license to produce logo apparel is for four years. The apparel manufacturer can produce either the licensed logo apparel or its own "house brand" apparel (its next best alternative). The logo apparel is likely to command a price premium over the house brand apparel (e.g., say, $18.20 for the logo apparel versus $14 per unit for the house brand). Furthermore, the logo apparel is likely to result in greater sales as well. The exhibit provides data on unit sales, prices, sales revenue, and costs to the apparel manufacturer under each option.

Based on the data in Exhibit 18.5, we can calculate the *incremental* profits attributable to the university trademark for each year. Incremental profits are simply the additional profits generated by the logo apparel versus the house brand. Since these profits are estimated for future years, we need to adjust for the time value of money. In addition, there is some risk associated with the revenue streams for future years, so we need to discount those future cash flows as well. A discount rate is usually used to account for both the time value of money and the risk associated with cash flows. For the purposes of our example, we use a discount rate of 10 percent. Using the discount rate, we calculate the net present value (NPV) of sales and NPV of the incremental profits. The NPV of incremental profits as a percent of the NPV of sales provides the *maximum* royalty (21 percent in this case) that a potential licensee would be willing to pay to license the trademark. In all likelihood, however, a licensee would expect to pay something less than the expected NPV, so as not to remit all of its expected profits to the licensor. Thus, in actual licensing transactions, the licensor and licensee would negotiate within this range of 0 to 21 percent. Alternatively, the licensee could seek to evaluate whether the incremental profit *after* paying a royalty to the licensor provides sufficient return to justify licensing. Thus in the current example, even if the university sets a royalty rate of 10 percent to license its logo, the licensee would earn 11 percent in incremental profit. The licensee must decide whether this option makes sense in light of other opportunities it might have.

Conclusion

A trademark can be defined as a distinctive word, phrase, name, or symbol that is used in commerce to indicate the source of a good or service and to distinguish it from the goods

EXHIBIT 18.5 DISCOUNTED CASH FLOW MODEL

Sales of University Logo Apparel				
	2006	**2007**	**2008**	**2009**
Unit Sales	6,710	6,820	6,963	7,095
Price per Unit	$18.20	$18.20	$18.20	$18.20
Sales Revenue	$122,122	$124,124	$126,727	$129,129
Costs per Unit	$9.80	$9.80	$9.80	$9.80
Total Costs	$65,758	$66,836	$68,237	$69,531
Operating Profits (A)	$56,364	$57,288	$58,489	$59,598
Sales of House Brand Apparel				
	2006	**2007**	**2008**	**2009**
Unit Sales	6,100	6,200	6,330	6,450
Price per Unit	$14.00	$14.00	$14.00	$14.00
Sales Revenue	$85,400	$86,800	$88,620	$90,300
Costs per Unit	$9.00	$9.00	$9.00	$9.00
Total Costs	$54,900	$55,800	$56,970	$58,050
Operating Profits (B)	$30,500	$31,000	$31,650	$32,250
Incremental Profit Attributable to Trademark (A − B):	$25,864	$26,288	$26,839	$27,348
NPV of Sales:	$116,439	$107,589	$99,859	$92,502
NPV of Incremental Profit:	$24,660	$22,786	$21,149	$19,591
NPV of Incremental Profits as Percent of NPV of Sales:	21%			

or services of others. Trademarks serve important business and economic purposes and are often valuable business assets whose worth merits protection. A trademark's worth may be evaluated for a number of purposes, including in the context of licensing, acquisition, regulation, or litigation. The three most common valuation techniques for trademarks are the income approach, the market approach, and the cost approach. While the income and market approaches are generally the most appropriate in the context of valuing trademarks, use of any one of the techniques may depend on the information available and the particular circumstances of the valuation.

Notes

1. The opinions expressed in this chapter are those of the authors. The information contained in this chapter does not constitute legal advice, and you should consult with your own legal counsel regarding trademark and other intellectual property law.

2. Lechter (2002), p. 152.
3. *See, e.g.*, Smith and Parr (2005), p. 38.
4. Belson (1999), § 13.1–13.2.
5. For a discussion of trade names in the State of Florida, *see* Beckman (2003).
6. *Merriam-Webster, Inc. v. Random House, Inc.*, 35 F.3d 65 (2nd Cir. 1994).
7. It is not clear to us that Interbrand Corporation's methodology properly disaggregates all of the relevant intangibles.
8. Daniel (1999), §§ 15.2–15.3.
9. "International Treaties," International Trademark Association.
10. 15 U.S.C § 1114(1)(a), (1995).
11. Daniel (1999), §§ 15.2–15.3.
12. Small and McKay (2002), pp. 173–176.
13. Smith (1997), p. 49.
14. *Major League Baseball Properties, Inc. v. Sed Non Olet Denarius, Ltd.*, 817 F. Supp. 1103 (S.D.N.Y. 1993).
15. Ellison, Zimmerman, and Forelle (2005), p. A1.
16. A domain name is a computer "address" for a reserved site where one's presence located on a computer linked to the Internet can be found.
17. Meadows (1997), p. 6.
18. Small and McKay (2002), pp. 173–176.
19. Brunel (1996), Chapter 3.
20. The Internet Corporation for Assigned Names and Numbers (ICANN) was established in 1998 by the U.S. government to act as the new technical coordinating body for the Internet. ICANN was designed to alleviate many of the problems confronting NCI with respect to trademark disputes. *See, e.g.*, Girasa and Girasa (2002), pp. 224–225.
21. Smith and Parr (2000), pp. 163–173.
22. Trademark licensing and merchandising is big business for universities. University-related logo retail market was estimated at about $3 billion in 2003. *See* Zaslow (2003).

References

"National Advertisers Ranked 1 to 50." *Advertising Age*, June 27, 2005.

Belson, Jeffrey. "Certification Marks in the Competitive Commercial Environment." In *Intellectual Property in the Global Marketplace: Volume 1, Electronic Commerce, Valuation, and Protection*, ed. Melvin Simensky, Lanning G. Bryer, and Neil J. Wiklof, 2nd ed. (New York: John Wiley & Sons, 1999).

Brunel, André. "Trademark Protection for Internet Domain Names." In *The Internet and Business: A Lawyer's Guide to the Emerging Legal Issues* (Wakefield, MA : Computer Law Association, 1996).

Daniel, Al J. Jr. "Imperative Strategies for Protecting Intangible Assets: The U.S. Market." In *Intellectual Property in the Global Marketplace: Volume 1, Electronic Commerce, Valuation, and Protection*, ed. Melvin Simensky, Lanning G. Bryer, and Neil J. Wiklof, 2nd ed. (New York: John Wiley & Sons, 1999).

Ellison, Sarah, Ann Zimmerman, and Charles Forelle. "P&G's Gillette Edge: The Playbook It Honed at Wal-Mart." *Wall Street Journal Online,* January 31, 2005.

Girasa, Rosaria J., and Roy J. Girasa. *Cyberlaw: National and International Perspectives* (Upper Saddle River, NJ: Prentice Hall, 2002).

"The 100 Top Brands." *BusinessWeek,* August 1, 2005.

"International Treaties." International Trademark Association, updated April 2005.

Lechter, Michael A. "Copyright, Software, and Web Site Issues in the Internet World." In *The Licensing Best Practices: The LESI Guide to Strategic Issues and Contemporary Realities,* ed. Robert Goldscheider (Hoboken, NJ: John Wiley & Sons, 2002).

Meadows, Joan. "Comment: Trademark Protection for Trademarks Used as Internet Domain Names," *University of Cincinnati Law Review* 65, no. 1323 (Summer 1997).

Small, Thomas M., and Kenneth D. McKay. "Trademarks, Trade Names, and Trade Dress." In *The Licensing Best Practices: The LESI Guide to Strategic Issues and Contemporary Realities,* ed. Robert Goldscheider (Hoboken, NJ: John Wiley & Sons, 2002).

Smith, Gordon V. *Trademark Valuation* (New York: John Wiley & Sons, 1997).

Smith, Gordon V., and Russell L. Parr. *Intellectual Property: Valuation, Exploitation, and Infringement Damages* (Hoboken, NJ: John Wiley & Sons, 2005).

Smith, Gordon V., and Russell L. Parr. *Valuation of Intellectual Property and Intangible Assets,* 3rd ed. (Hoboken, NJ: John Wiley & Sons, 2000).

Zaslow, Jeffrey. "Sports Fans Snap Up Souvenirs of Winners Beating Losers: Mascots Boiled or Grilled?" *Wall Street Journal,* November 12, 2003.

Part III

ORGANIZING RESEARCH AND DEVELOPMENT

ELSEVIER

Journal of Economic Behavior & Organization
Vol. 31 (1996) 193–224

JOURNAL OF
Economic Behavior
& Organization

Firm organization, industrial structure, and technological innovation[1]

David J. Teece

Mitsubishi Bank Professor, Walter A. Haas School of Business, University of California, Berkeley, CA 94720-1930, USA

Received 29 July 1994

Abstract

The formal and informal structures of firms and their external linkages have an important bearing on the rate and direction of innovation. This paper explores the properties of different types of firms with respect to the generation of new technology. Various archetypes are recognized and an effort is made to match organization structure to the type of innovation. The framework is relevant to technology and competition policy as it broadens the framework economists use to identify environments that assist innovation.

Keywords: Organizational structure; Innovation; Market structure-innovation relationships; Proprietary technology; Vertical Integration; Flexible hierarchy

1. Introduction

It is increasingly recognized that the dynamism of a competitive private enterprise system flows from the development and application of new technology and the adoption of new organizational forms. As a result, attention is being focused on trying to develop a better understanding of the institutional environment in which these activities take place. In market economies, the business firm is clearly the leading player in the development and commercialization of new products and processes.[2] However, much of the literature

[1]I am grateful for helpful comments and conversations with Glenn Carroll, Hank Chesbrough, Niel Kay, Ralph Landau, Richard Nelson, Nathan Rosenberg, Oliver Williamson, and two anonymous referees

[2]In fact, there are governments, universities, and also professional societies in the system, and certain activities that firms cannot be expected to do on their own because the returns are so low, are picked up by other institutions, or receive public monies, or both.

0167-2681/96/$15.00
PII S0167-2681(96)00895-5

in economics proceeds as if the identity of the firm in which innovation is taking place is of little moment. Moreover, the links between firm structure and strategy and the innovation process are poorly understood.[3]

In this paper, it is suggested that the formal and informal structures of the firm, as well as the network of external linkages that they possess, has an important bearing on the strength as well as the kind of innovative activity conducted by private enterprise economies.[4] Frameworks are presented to indicate how firm structure and the nature of innovation are linked. The approach adopted eschews optimality and embraces comparative analysis, in the spirit of Williamson[5] (1975, 1985), whereby alternatives are compared to each other rather than to hypothetical ideals.The institutional context is also considered. In particular, the role of capital markets is at least addressed, and the legal infrastructure is not assumed away completely. Indeed, various aspects of the legal system, and in particular intellectual property law, are explicitly considered.

The general approach adopted involves (1) identifying the fundamental characteristics of technological development, (2) determining the factors that affect innovation at the level of the firm, (3) identifying distinctive archetypes or governance modes for firms, and (4) choosing from available alternatives the organizational forms better suited to deal with various types of innovation. It is hoped that analyzing innovation in this manner will help broaden the agenda for industrial organization economists and organization theorists as they begin to grapple with understanding one of the most distinctive features of modern capitalism.

2. Fundamental characteristics of technological development

It is impossible to identify the organizational requirements of the innovation process without first specifying underlying properties of technological innovation. Fortunately, there appears to be on emerging consensus among scholars who study the innovation process with respect to the stylized facts. In the main, these appear to characterize innovation independently of the organizational context in which it takes place.

2.1. Uncertainty

Innovation is a quest into the unknown. It involves searching and the probing and reprobing of technological as well as market opportunities. With hindsight, much effort is spent traveling down blind alleys. Serendipity and luck play an important role. There are various types of uncertainty. Tjalling Koopmans (1957) has made a useful distinction between primary and secondary uncertainty. Both are critical in the context of innovation.

[3]For a review, see Dosi et al. (1988).

[4]The following statement by Little (1985, p.14) is representative of accepted views: "Our work among innovative companies indicates that the management decision on how to organize for innovation is critical."

[5]The approach rejects assumptions of temporal equilibrium. The framework does not assume that the selection process immediately weeds out all organizations that do not match the business environment at a particular point in time. While the organizational system is seen as gravitating toward an end point or equilibrium, it takes so long to reach it that the environment is likely to change again in the interim, leading to a state of perpetual disequilibrium.

Secondary uncertainty arises "from lack of communication, that is, from one decision-maker having no way of finding out the concurrent decisions and plans made by others." Primary uncertainty arises from "random acts of nature and unpredictable changes in concurrent preferences" (1957, pp.162,163). Williamson recognizes a third kind of uncertainty, which he calls behavioral uncertainty, which is attributable to opportunism. Such uncertainty can lead to ex post surprises.[6] It is important to note that secondary uncertainty can be affected by changing the boundaries of the organization. As Richardson (1990) and Williamson (1975) have explained, vertical integration can facilitate the coordination of complementary investments through the sharing of investment plans. Secondary uncertainty is thus a function of organizational form.

2.2. *Path dependency*

Technology often evolves in certain path dependent ways, contoured and channeled by what might be thought of as technological paradigms (Dosi, 1982, 1982). A technological paradigm is a pattern of solutions to selected technical problems which derives from certain engineering relationships. A paradigm identifies the problems that have to be solved and the way to inquire about them; within a paradigm, research efforts become channeled along certain trajectories.[7] Relatedly, new product and process developments for a particular organization are likely to lie in the technological neighborhood of previous successes.

2.3. *Cumulative nature*

Technology development, particularly inside a particular paradigm, proceeds cumulatively along the path defined by the paradigm. The fact that technological progress builds on what went before, and that much of it is tacit and proprietary, means that it usually has significant organization-specific dimensions. Moreover, an organization's technical capabilities are likely to be "close in" to previous technological accomplishments.[8]

---

[6]Uncertainty also makes information a valuable commodity. Information about which outcomes will occur, or are more likely to occur, will obviously have great value. Information, of course, itself has very special characteristics. It is not only an indivisible commodity, in which case the classic problems of allocation in the presence of indivisibilities will be present, but it is also highly tacit, as discussed below. Often it cannot be readily articulated and codified in language. Combined with the absence of legal protection, these features make it difficult to trade.

[7]Examples of technological paradigms include the internal combustion engine, biotechnology, and tungsten filament lighting. Technological discontinuities occur when new paradigms emerge. Thus new technologies are more threatening to existing skills and capabilities if they embody a new paradigm. The emergence of microelectronics, which carried with it a new paradigm, was far more threatening to the skills of incumbents than the emergence of the facsimile, which fused the technology of the telephone and the copier.

[8]Specific technological skills in one field (e.g. pharmaceuticals) may be applicable in closely-related fields (e.g. pesticides) but they are unlikely to be of use in distant fields (e.g. aircraft). See Teece (1988), Teece et al. (1994).

2.4. *Irreversibilities*

Technological progress exhibits strong irreversibilities. This follows not just because innovation typically requires specialized investments, but because the evolution of technologies along certain trajectories eliminates the possibility of competition from older technologies, even if relative prices change significantly. Thus mechanical calculators are unlikely to ever replace electronic ones, even if the relative prices of silicon and steel were to switch by a factor of 1 000[9] in favor of steel.

2.5. *Technological interrelatedness*

Innovation is characterized by technological interrelatedness between various subsystems. Linkages to other technologies, to complementary assets, and to users must be maintained if innovation is to be successful. If recognizable organizational subunits such as R&D, manufacturing, and marketing exist, they must be in close and continuous communication and engage in mutual adaptation if innovation in commercially relevant products and processes is to have a chance of succeeding. Moreover, successful commercial innovation usually requires quick decision making and close coupling and coordination among research, development, manufacturing, sales and service. Put differently, organizational capacities must exist to enable these activities to be closely coordinated, and to occur with dispatch.

2.6. *Tacitness*

The knowledge developed by organizations is often highly tacit. That is, it is difficult if not impossible to articulate and codify (Polanyi, 1962, Winter, 1987). A corollary is that technology transfer is often difficult without the transfer of key individuals. This simultaneously explains why imitation is often costly, and why the diffusion of new technology often depends on the mobility of engineers and scientists (Teece, 1977, Nelson and Winter, 1977). Relatedly, an organization's technology ought not to be thought of as residing in some hypothetical book of blueprints, or with some hypothetical chief engineer, but in an organization's system and habits of coordinating and managing tasks. These systems and habits have been referred to as organizational routines (Nelson and Winter, 1982). It is the performance of these routines that is at the essence of an organization's technological capacity.

2.7. *Inappropriability*

Under many legal systems, the ownership rights associated with technical know-how are often ambiguous, do not always permit rewards that match contribution,[10] vary in the degree of exclusion they permit (often according to the innate patentability or

---

[9] If sailing ships ever replace propeller-driven ships, it will be with such a different sailing technology as to be almost unrecognizable from nineteenth century counterparts. And if the prop-fan recaptures markets from the fan-jet, it will also be with a markedly different prop and engine.

[10] For instance, it is possible to receive a patent which is arguably too narrow or too broad in relation to the patent holder's contribution to economic welfare. Moreover, in many cases legal protection for technical contributions may simply not be available, or if available may be difficult to enforce.

copyrightability of the object or subject matter) and are temporary. Or as Arrow (1996) put it, technical information is a "fugitive resource, with limited property rights." Accordingly, investment in innovative activity may not necessarily yield property which can be reserved for the exclusive use of the innovator. But the activity may nevertheless still be valuable enough to attract some investment, depending in part on other institutional arrangements to be examined later. The degree to which new products and processes are protectable under intellectual property law will henceforth be referred to as the intellectual property regime. For expositional simplicity, regimes will be classified as strong if patents and copyrights are effective, and weak otherwise. Clearly, the industrial world does not readily bifurcate, and there exists a continuum of appropriability regimes, as data assembled by Levin et al. (1987) make apparent.

The market for know-how is further confounded because in order to provide full information to the buyer, the seller of know-how may have to disclose the object of the exchange, but in so doing the basis for the exchange evaporates, or at least erodes, as the potential buyer might now have in its possession that which he was seeking to acquire. Hence, transactions in the market for know-how must proceed under conditions of ignorance. Accordingly, at least until reputations become established, exchange is likely to be exposed to hazards. Optimal resource allocation is unlikely to result.

3. Organizational and market determinants of the rate and direction of innovation

While our understanding of innovation has been enriched in recent years, the basic framework employed in policy debates about innovation, technology policy, and competition policy are often remarkably naive and highly incomplete. Even elementary considerations such as those identified in Section 2 are frequently neglected. In economics, for instance, it is not uncommon to find debate about innovation policy collapsing to a rather outmoded discussion of the relative virtues of competition and monopoly, as if they were the key determinants of innovation. Clearly there is much more at work. In this section, various classes of variables – some economic some organizational – are identified that impact the rate and direction of innovation. Subsequent sections will identify distinct types of organizations based on various organizational attributes. A final section will then endeavor to match these organizations to different types and levels of innovation.

3.1. Monopoly power

One reason why our understanding of innovation has not proceeded faster in the last half century is that many researchers, particularly industrial organization economists, have overly focused on just one variable: the degree of market power that a firm or firms may have. The evidence is unequivocal that competition and rivalry are important for innovation; but few believe that the world of perfect competition in which firms compete in highly fragmented markets using identical nonproprietary technologies is an organizational arrangement that any advanced economy would aspire to achieve.

Nevertheless, many policy debates proceed on the assumption that fragmented markets assist innovation.[11]

Schumpeter was among the first to declare that perfect competition was incompatible with innovation. He noted, "The introduction of new methods of production and new commodities is hardly conceivable with perfect – and perfectly prompt – competition from the start. And this means that the bulk of what we call economic progress is incompatible with it."[12] However, the Schumpeterian notion that small entrepreneurial firms lack financial resources seems archaic, at least in countries with a vigorous venture capital market. In any case, the Schumpeterian debate seems a little beside the point, as there is an enormous number of variables that can potentially intervene between the generation of monopolistic rents and the allocation of resources to the development of new products and processes. Consider, first, single product firms. The notion that innovation requires the cash flow generated by the exercise of monopoly power assumes both that (1) capital markets are inefficient, and (2) that monopolistic levels of internal cash flows are adequate to fund the requisite R&D programs. If capital markets are operating according to what Fama (1970) has called *strong form efficiency*, then cash flow is unimportant because firms with high yield projects will be able to signal their profit opportunities to the capital market and the requisite financing will come forth on competitive terms. Thus if there is strong form efficiency and zero transaction costs (its corollary), cash will get matched to projects whether or not the cash is internally generated.

In fact, the world is not properly characterized by zero transaction costs, but that does not mean that the availability of internal cash flows from monopoly (as compared to competitive) product market positions is what makes the difference between being able to fund a project and not being able to fund it. Significant innovative efforts involve expenditures in a particular year which may be many times the available cash flows. So the availability of marginally higher cash flows occasioned by monopoly power are unlikely to grossly change the financial picture, except in unusual circumstances.

Furthermore, even in the absence of adequate internal cash flow, firms need not go to the capital market to find the requisite financing. The "Schumpeterian" view of the innovation processes appears to be one that involves full integration, from research, development, manufacturing and marketing. But the financial requirements associated with developing and commercializing new products and processes can be accomplished with a myriad organizational arrangements including research joint ventures, co-production, and co-marketing arrangements. With such arrangements, there is the possibility that the capital requirements associated with a new project could be drastically reduced for the innovator. Economies of scale and scope can often be captured through interfirm arrangements. In some instances they cannot.[13]

---

[11]Clearly rivalry and competition are important to innovation, but belief in the virtues of perfectly competitive systems is lore, reflecting casual empiricism and prejudice and not careful theorizing and empirical study. The same is true for monopoly.

[12]Schumpeter (1934, p.105).

[13]For a managerially oriented analysis of the limits of outsourcing in the context of innovation, see Chesbrough and Teece (1996).

The link between market power and innovation in specific markets is further undone if the multidivisional multiproduct firm is admitted into the scene.[14] The basic function and purpose of the multiproduct structure is to allocate cash generated everywhere to high-yield purposes anywhere. If a multidivisional multiproduct firm does operate this way, and there is plenty of evidence to suggest that they can and do, then the link between market power in a particular market and the funding of innovation in that market is undone. In a multiproduct firm selling products in markets A through Z, the cash generated by virtue of power in market A can indeed fund innovation relevant to market A, but it can equally well fund innovative activity in market Z. The capital market inside the multiproduct firm thus unlocks the relationship between market structure and innovation proposed by Schumpeter.

When firms do go into the capital market they generally have multiple sources of funding available. Generically, these can be split into debt and equity. The various types of debt and equity can on the one hand be thought of as financial instruments or, as Williamson suggests, as different "governance structures" (Williamson, 1988, 1996). Williamson explains that the decision by firms to use debt or equity to support individual investment projects is likely to be linked to the redeployability of the underlying investment. Since new product development programs commonly involve investment in assets that are substantially irreversible (like R&D) and/or non-redeployable (like specialized equipment), debt is only of limited value in financing innovation, unless a firm has collateral and is under-leveraged to begin with. Accordingly, the fund sources generally available to support new product development are internal cash flow and new equity. In instances when a firm does not already have substantial cash flows, then equity is the major source of new funds. The role of equity is made distinct if it is considered in the context of "start-up" firms which do not already have free cash flows. Investors have obvious problems in evaluating the prospects for new products and processes, and the best investees have problems, though less serious, in identifying the best investors.[15]

Now consider internally generated cash flow. Even in the United States where there is a vibrant venture capital market, internal "free" cash flow is the major source of private financing for innovation. A firm's cash flow is not just a function of price-cost margins in the product markets (sometimes suggested as a proxy for monopoly power) but its existing asset structures and the need for new investment in existing businesses. A business can clearly be a "cash cow" even if it is earning only competitive returns. This would be true if the firm was gradually divesting itself, or simply harvesting its position

[14]See Kay et al. (1990), undated working paper.

[15]The investors' problems are rather obvious. The investor has the difficult challenge of calibrating investment prospects in an environment where there is usually high market uncertainty, high technical uncertainty, and bountiful opportunism and optimism. Several kinds of opportunism are possible. One is simply that the technology can be misrepresented. This tendency, however, can be checked if the investor hires technical consultants to validate the entrepreneurs' claims. Another is that the tenacity and veracity of the entrepreneur are difficult to calibrate, with consequences much more unfortunate for the investor than for the entrepreneur. Ascertaining whether the entrepreneurs' optimism is honest yet misplaced is perhaps even more difficult. There is "much evidence that in the context of planning and action most people are prone to extreme optimism in their forecasts of outcomes, and often fail to appreciate the chances of an unfavorable outcome" (Kahneman and Lovallo, 1995, p.2). Decision makers often take risks because they deny their existence or underestimate their extent (March and Shapira, 1987).

in a particular market. Internally generated cash can be readily allocated by management and is not typically constrained by covenants.

Over the last half decade, a controversial body of literature has emerged which, in essence, argues that free cash flows must be distributed to shareholders, rather than being invested internally in discretionary projects, if firms are to operate efficiently (Jensen, 1989). The basic idea is that the discipline of debt is needed to cause capital to be channeled to high-yield uses in the economy, as well as in the firm. There are severe problems with this thesis, not least of which is that debt holders are loss averse and not at all business-opportunity driven. While it may indeed be the case that free cash flows do sometimes get misallocated by managers, to delimit them in the manner proposed by advocates of the free cash flow hypothesis is to force the firm into equity markets to finance innovation. For reasons explained earlier, this is not always desirable because the new issues markets, both public and private, have disabilities with respect to recognizing and funding new opportunities.

To summarize, innovation clearly requires access to capital. The necessary capital can come from cash flows or from equity. At least with respect to early stage activity, debt financing is unlikely to be viable, unless the firm has other assets to pledge. However, certain downstream investments needed to commercialize innovation can be debt financed if they are redeployable. Alternatively, alliances can be entered which reduce the need for new investment in complementary assets. The point, however, is that there are many factors besides firm size, and the presence or absence of market power, that affect an innovator's capacity to access capital.[16]

3.2. *Hierarchy*

Hierarchy arose to help in the administration of military, religious, and governmental activities.[17] While hierarchies are old, deep hierarchies are relatively new. Anthropologists point out that most tribes, clans and agricultural enterprises have rather flat hierarchies. The Roman Catholic Church, for instance, has only four levels. Centralizing and decentralizing are not genuine alternatives for organization; the key issue is to decide the mix. Hierarchies can accomplish complex organizational tasks, but they are often associated with organizational properties inimical to innovation, such as slow (bureaucratic) decision making and weak incentives.

3.2.1. Bureaucratic decision making.

Decision making processes in hierarchical organizations almost always involve bureaucratic features. In particular, a formal expenditure process involving submissions and approvals is characteristic. Decision making is likely to have a committee structure, with top management requiring reports and written justifications for significant decisions. Moreover, approvals may need to be sought from outside the organizational unit in which

[16]For an expanded discussion, see Day et al. (1993), part I.

[17]Hierarchical subdivision is not a characteristic that is peculiar to human organization. It is common to virtually all complex systems of which we have knowledge (Simon, 1973, p.202). The advantages of hierarchy are well understood. In particular, among systems of a given size and complexity, hierarchical systems require much less information transmission among their parts than do other types of systems.

the expenditure is to take place. While this may ensure a matching up of expenditures to opportunities across a wider range of economic activity, it unquestionably slows decision making and tends to reinforce the status quo.

The latter characteristic follows from committee decision making structures, which almost always tend toward balancing and compromise. But innovation is often ill served by such structures, as the new and the radical will almost always appear threatening to some constituents. Put differently, representative structures, bureaucratic or political, often tend to endorse the status quo. Strong leaders can often overcome such tendencies, but such leaders are not always present and their capacities are often thwarted by the organization.[18]

One consequence is what Williamson (1975) has referred to as a "program persistence bias," and its corollary the "anti-innovation bias." Program persistence refers to the funding of programs beyond what can be sustained on merits, and follows from the presence or influence of program advocates in the resource allocation process. This proclivity almost automatically has the countervailing effect of reducing funds available to new programs, which are unlikely to be well represented in the decision making process. As Anthony Downs points out, "the increasing size of the bureau leads to a gradual ossification of operations – since each proposed action must receive multiple approvals, the probability of its being rejected is quite high – its cumbersome machinery cannot produce results fast enough, and its anti-novelty bias may block the necessary innovation" (p.160).

The sharpening of global competition, and diversification (organizationally and geographically) in the sources of new knowledge compels firms to make decisions faster, and to reduce time to market in order to capture value from technological innovation. It seems clear that to accomplish such responsiveness, organizations need new structures and different decision-making protocols to facilitate entrepreneurial and innovative behavior. Burgelman (1984) identifies a menu of such arrangements which include: special business units, new ventures department, new venture divisions, and independent business units. Clearly, all of these designs imply smaller, flatter and more specialized structures within which to conduct activities where speed and responsiveness are critical. In the limit, the spinoff or spinout of a new division signifies that the enterprise's (or at least the individuals associated with it) chances of success are greater outside rather than inside an established hierarchy. In addition to the creation of semi-autonomous units, firms can attempt to "delayer" by stripping out layers of middle management. But flattening organizations need not fundamentally redefine the relationships between people and functions in the organizations. Functions may still work sequentially, with decisions being made from fragmented perspectives.

---

[18]Crozier (1964, p.225) puts it this way: "People on top theoretically have a great deal of power and often much more power than they would have in other, more authoritarian societies. But these powers are not very useful, since people on top can act only in an impersonal way and can in no way interfere with the subordinate strata. They cannot, therefore, provide real leadership on a daily basis. If they want to introduce change, they must go through the long and difficult ordeal of a crisis. Thus, although they are all-powerful because they are at the apex of the whole centralized system, they are made so weak by the pattern of resistance of the different isolated strata that they can use their power only in truly exceptional circumstances."

In essence, the organizational challenge appears to be that activities are not as decomposable as they used to be, and that cross-functional interaction must take place concurrently, rather than sequentially, if firms are to cut time-to-market for new products and processes. Cross-functional and cross-departmental networks must be strengthened without causing information overload. Computer networks can assist cross-functional interaction by project teams, concurrent engineering teams, network teams, task forces and the like. If such activity becomes completely unstructured, it augments rather than displaces bureaucracy. Instead of random ad hoc approaches, what is needed are well-defined cross-functional teams, which can be redefined as needed. With organizational subunits cross-linked in this way, authority occurs as much from knowledge as position in the organizational hierarchy. The challenge is to develop a culture which supports the establishment of cross-functional teams which draw on the requisite knowledge, wherever it may be located.

3.2.2. Low-powered incentives.

As they grow, organizations often become characterized by what Williamson (1985, p.153) calls "low powered incentives." Low-powered incentives can be defined as those where the co-variance of employee compensation with business unit performance is low. One reason is that compensation structures inside large organizations need to be sensitive to relative as well as absolute levels of compensation. If the compensation structure itself has value through the relativities it establishes, then the enterprise will be reluctant to disturb the structure to support innovation. Another reason is that stock options cannot be granted to reflect divisional performance since it is generally the case that the division's shares are nontradeable in public markets. The absence of a public equity market for subunit shares thus deprives the firm of the opportunity to provide an objective capital market-based augmentation to compensation.[19] If the employee is rewarded instead through stock in the total enterprise, the impact of divisional, departmental, and individual performance is likely to be severely diluted.[20]

3.2.3. Principal-agent distortions.

Business firms of great size are rarely owner managed. Inasmuch as managers (agents) trade-off enterprise performance for their own welfare, innovation is likely to be impaired. This is because the interests of managers are sometimes at odds with what innovation requires, because the tenure of top management is usually much shorter than the gestation period for major innovations. Moreover, principals must invest in costly information collection and monitoring activities in order to check up on the performance of agents. These costs can be considerable. Moreover, principals may insist on certain expenditure controls which themselves slow decision making and thwart innovation.

[19]Surrogate valuation indexes can sometimes be created based on the use of "yardstick" companies, but they typically do not convey liquidity.

[20]For further discussion on measurement problems, see Holmström (1993, pp.144–146).

3.2.4. Myopia.

Organizations can become closed to changes in the market and business environment and to new sources of technology. Individuals in organizations, including chief executive officers, can fall into the trap of adopting a citadel mentality. The availability of free cash flows can help sustain that mentality and behavior for considerable periods of time. Closed systems may be able to hone existing routines, but they will lose the capacity to engage in new routines. Organizations can become closed through administrative arrangements (as when the firm's boundaries are delimited by its organization chart), through legalistic (rather than relational) contracting with suppliers and customers, and through social and cultural norms which stress the importance of inside rather than outside considerations.

3.3. Scope

The scope of product market activities may impact the innovative performance of firms in at least three ways. One has just been discussed in the context of finance: the multi-divisional multiproduct firm is in a position to re-allocate cash from businesses that have positive cash flow to new businesses with negative cash flow. A second hypothesis, put forward at various times by Joseph Schumpeter, Richard Nelson and others, is that the product market portfolios of multiproduct firms will increase the payoff to uncertain R&D by increasing the probability that new products and processes resulting from corporate R&D can be commercialized inside the firm. Neither of these will be the main focus here.

Instead, it is suggested that multiproduct firms can more readily develop and commercialize "fusion" technologies which involve the melding of technological capacities relevant to disparate lines of business. This fusion – as with mechanics and electronics (what Kodama, 1986 calls "mechatronics") – by no means occurs automatically, and requires internal structures which are flexible and permeable.[21] Indeed, there appears to be less diversity in firms' products than in their technologies (Pavitt et al. 1989). Nevertheless, the multiproduct firm does afford opportunities for economies of scope based on transferring technologies across product lines and melding them to create new products (Teece, 1980, 1982). Despite the path dependent-nature of technological change, the diversity of application areas for a given technology are often quite large, and it is often feasible and sometimes efficient to apply the firm's capabilities to different market opportunities.

Suppose application areas outside of the core business do in fact open up. The question arises as to whether potential scope economies deriving from the application of proprietary know-how in new markets add more to the innovating firm's value if they are served through licensing and related contractual arrangements to unaffiliated firms who then serve the new product markets in question, or by direct investment, either de novo or by merger/acquisition. This is an important question, the answer to which ought to help shape a positive theory of the scope of the firm's activities.

Whether the firm integrates or not is likely to depend critically on four sets of factors: (1) whether the technology can be transferred to an unaffiliated entity at higher or lower

[21] This is discussed in Section 4.3.

cost than it can be transferred to an affiliated entity; (2) the degree of intellectual property protection afforded to the technology in question by the relevant statutes and laws; (3) whether a contract can be crafted which will regulate the sale of technology with greater or less efficiency and effectiveness or whether department-to-department or division-to-division sales can be regulated by internal administrative procedures; and (4) whether the set of complementary competences possessed by the potential licensee can be assessed by the licensor at a cost lower than alternatives. If they are lower, the available returns from the market will be higher, and the opportunity for a satisfactory royalty or profit-sharing arrangement accordingly greater.

These matters are explored in more detail elsewhere (Teece, 1980, 1984, 1986, Chesbrough and Teece, 1996). Suffice to say that contractual mechanisms are often less satisfactory than the alternative. Proprietary considerations are more often than not assisted by integration, and technology transfer is difficult both to unaffiliated and affiliated partners, with the consequence that integration (or multiproduct diversification) is the more attractive alternative, except where incumbents are already competitively established in downstream activities, and are in a position to render de novo entry by the technology-based firms unattractive because of the excess capacity it would generate. Hence, multiproduct firms can be expected to appear as efficient responses to contractual, proprietary and technology transfer problems in an important set of circumstances. Mixed modes, such as joint ventures and complex forms of profit-sharing collaboration, will also be common according to how the set of transactions in question stacks up against the criteria identified above.

3.4. Vertical integration

The characteristics of technological development identified earlier also have important implications for the vertical structure of the firm, and vice versa. Economic historians have long suggested that there may be links between vertical structures and the rate and direction of innovation. For instance, Frankel (1955) has argued that the slow rate of diffusion of innovations in the British textile and iron and steel industries around the turn of the century was due to the absence of vertically integrated firms. Kindleberger (1964) has gone so far as to suggest that the reason why West Germany and Japan have overtaken Britain may be due to "the organization of [British] industry into separate firms dealing with each other at arm's length." This "may have impeded technological change because of the possibility that part of the benefits of that change would have been external to the separate firms" (pp.146,147). Kindleberger also studied the reasons for the failure of the British railroads to abandon the 10-ton coal wagon in favor of the more efficient 20-ton wagon, and concludes (1964) that the reason for the slow rate of diffusion was institutional and not technical. In short, it stemmed from the absence of vertical integration.[22] General Motors' early dominance in the diesel electric locomotive industry

[22]Technical aspects of interrelatedness do not seem to have held up the movement to more efficient size, either through making such a change uneconomic because of the enormity of the investment required or by adding amounts too great for any one firm to borrow. The sums involved were not large, and railway finance was rarely a limiting factor in the period up to 1914. Private ownership of the coal cars by the collieries, on the other hand, posed a type of interrelatedness that was institutional rather than technical.

has also been attributed to the fact that it was integrated into electrical supply while its competitors were not (Marx, 1976). A systematic exploration of the relationship between technological innovation and enterprise boundaries is needed.

For present purposes, it is useful to distinguish between two types of innovation: autonomous (or "stand-alone") and systemic. An autonomous innovation is one which can be introduced without modifying other components or items of equipment. The component or device in that sense "stands alone." A systemic innovation, on the other hand, requires significant readjustment to other parts of the system. The major distinction relates to the amount of design coordination which development and commercialization are likely to require. An example of a systemic innovation would be electronic funds transfer, instant photography (it required redesign of the camera and the film), front-wheel drive, and the jet airliner (it required new stress-resistant airframes).

With systemic innovation, internal organization (integration) can often assist the workings of the market. Integration facilitates systemic innovations by facilitating information flows, and the coordination of investments plans. It also removes institutional barriers to innovation where the innovation in question requires allocating costs and benefits, or placing specialized investments into several parts of an industry.

Comprehensive evidence with respect to these propositions has yet to be assembled. Vignettes can be found in Chesbrough and Teece (1996). The only statistical test performed to date relates to the petroleum industry (Armour and Teece, 1978). These findings indicated that firm and R&D expenditures for basic and applied research in the U.S. petroleum industry between 1951–1975, were statistically related to the level of vertical integration which the enterprise possessed.[23]

3.5. Organizational culture and values

Market power is an element of industrial structure; scale, scope, integration and hierarchy can be thought of as elements of the formal structure of an organization. Of equal if not greater importance is the informal structure of an organization. Organizational culture is the essence of an organization's informal structure. It is "the pattern of beliefs and expectations shared by the organization's members. These beliefs and expectations produce norms that powerfully shape the behavior of individuals and groups" (Schwartz and Davis, 1981, p. 33).

Organizational culture can be thought of as the "central norms that may characterize an organization" (O'Reilly, 1989, p. 305). A strong culture is a system of informal rules that spells out how people are to behave most of the time. By knowing what is expected of them, employees will waste little time deciding how to act in a given situation (Deal and Kennedy, 1982). There need not be consensus within an organization with respect to these beliefs, as the guiding beliefs or vision held by top management and by individuals

[23]Despite the fact that the ultimate objective of R&D programs is to produce innovations, not simply to dissipate resources on R&D activities, expenditure data can be viewed as a useful proxy for innovative performance in that they reveal the intensity of innovative activity. Furthermore, if the discount rate facing non-integrated firms is similar to that facing integrated firms and if similar risk preferences exist across the management of these firms, the higher productivity per dollar of research expenditure posited in vertically integrated firms implies that, ceteris paribus, such firms will devote more resources to R&D.

lower down in the organization may not be congruent. It is the latter, however, which define an organization's culture (O'Reilly, 1989, p.305).

There seems to be an emerging consensus (Deal and Kennedy, 1982, Peters and Waterman, 1982, O'Reilly, 1989) that the following set of norms assists the development and commercialization of new products and processes. With respect to development, these include: the autonomy to try and fail; the right of employees to challenge the status quo; open communication to customers, to external sources of technology, and within the firm itself. With respect to commercialization or implementation, teamwork, flexibility, trust and hard work are considered to be critically important. The right culture is not just an important asset to assist in technological development; it may be a requirement.

With a few notable exceptions (North, 1990), economists have given almost no attention, and little sympathy, to the topic of organizational culture.[24] Occasionally, economists may speak to the importance of trust and consciousness. Thus Arrow (1974, p.28) notes that "social demands may be expressed through formal rules and authorities, or they may be expressed through internalized demands of conscience. Looked at collectively, these demands may be compromises which are needed to increase the efficacy of all."[25] If Arrow is right in his claim that values can increase efficiency, it is unfortunate that the topic has been left to organizational sociologists and psychologists, and that economic science ignores what appears to be an important set of variables in the understanding of organizational performance.

One way for economists to begin grappling with organizational culture is to see it as control on the cheap; reduction in shirking is just one element.[26] If individuals can be motivated and directed without pecuniary incentives (and disincentives) and the exercise of authority, tremendous resource savings can ensue, and innovation processes can avoid the burdens of bureaucracy. Conversely, if a firm's culture and strategy do not align, it is likely to be unable to implement its strategy, especially strategies which involve innovation. For instance, a declaration by top management of a firm that the firm is now going to be more open to external sources of technological ideas will not ensure that the strategy will be successful if there is a well entrenched "not invented here" culture inside the organization. The failure to develop new norms supportive of a particular strategy "means that changes will persist only where they are closely monitored and directly rewarded" (O'Reilly, 1989, p.310).

---

[24]North's discussion (chapter 5) is almost exclusively limited to societal culture rather than organizational culture. He does however note that informal constraints flowing from the broader societal culture are pervasive. Veblen (1972, p.174) notes that "at least since mankind reached the human plane, the economic unit has been not a solitary hunter, but a community of some kind."

[25]Moreover, there is a tendency to squeeze such concepts into "externalities," where it is not clear they belong. Thus Arrow (1974, p.23) notes that: "Trust and similar values, loyalty or truth-telling, are examples of what the economist would call 'externalities'. They are good, they are commodities; they have real, practical economic value; they increase the efficiency of the system, enable you to produce more goods or more whatever values you hold in high esteem. But they are not commodities for which trade on the open market is technically possible or even meaningful."

[26]Alchien and Demsetz (1972).

3.6. External linkages

Economists, as well as many organization theorists, have traditionally thought of firms as islands of hierarchical control embedded in a market structure and interacting with each other through the price mechanism. Indeed, Coase (1937) has referred to firms as "islands of conscious power." Coase's metaphor needs to be transformed from islands to archipelagos to capture important elements of business organization. This is because firms commonly need to form strategic alliances, vertically (both upstream and downstream), laterally, and sometimes horizontally in order to develop and commercialize new technologies.[27] Compared to arm's-length market contracts, such arrangements have more structure, involve constant interaction among the parts, more open information channels, greater trust, rely on voice rather than exit, and put less emphasis on price. Compared to hierarchies, such alliances or networks among firms call for negotiation rather than authority and put great emphasis on boundary-spanning roles. Although firms connected through alliances have a high degree of autonomy, the relationship may well be anchored by a minority equity position. These arrangements can be used to provide some of the benefits of integration while avoiding some of the costs. This undoubtedly helps explain the proliferation of alliances in recent decades.

The variety of such arrangements to link organizations is almost unlimited, and the resultant forms quite diverse. A constellation of licensing, manufacturing and marketing agreements will typically characterize many interorganizational arrangements. R&D joint ventures, manufacturing joint ventures, co-marketing arrangements and consortia are just a few of the resultant forms. Some of these arrangements constitute extremely complex open systems, and some may be unstable. The managerial functions in these interorganizational networks are quite different from the authority relationship which commonly exists in hierarchies. Managers have to perform boundary-spanning roles, and learn to manage in circumstances that involve mutual dependency.

3.7. Assessment

The above discussion of the variables which impact firm-level innovation suggests that economic and organizational research needs a richer framework if the innovation process is to be better understood. Economic research needs to pay greater attention to organizational structure, both formal and informal, and organizational research needs to understand the importance of market structure, internal structure, and the business environment. Fig. 1 is a diagrammatic presentation of the various classes of variables that have been identified, as well as considerations deemed to be important but assumed away in this analysis. For instance, the firm's human resources/capital and the mechanism by which firms attract, train, and hold first-rate people has not been deeply analyzed. Nor has the role of government in the support of the scientific and technological infrastructure been analyzed. Another major omission has been the strategy by which firms identify

[27]Imai (1988, p.2) notes that "corporate networks in a broad sense are the vital economic institution which has led the Japanese economic development. The long history of cooperation between firms may be a crucial factor to explain the special adaptability of the Japanese economy." Imai uses the term, as it is used in this paper, to indicate interfirm relationships in general, including zaibatsu and business groups.

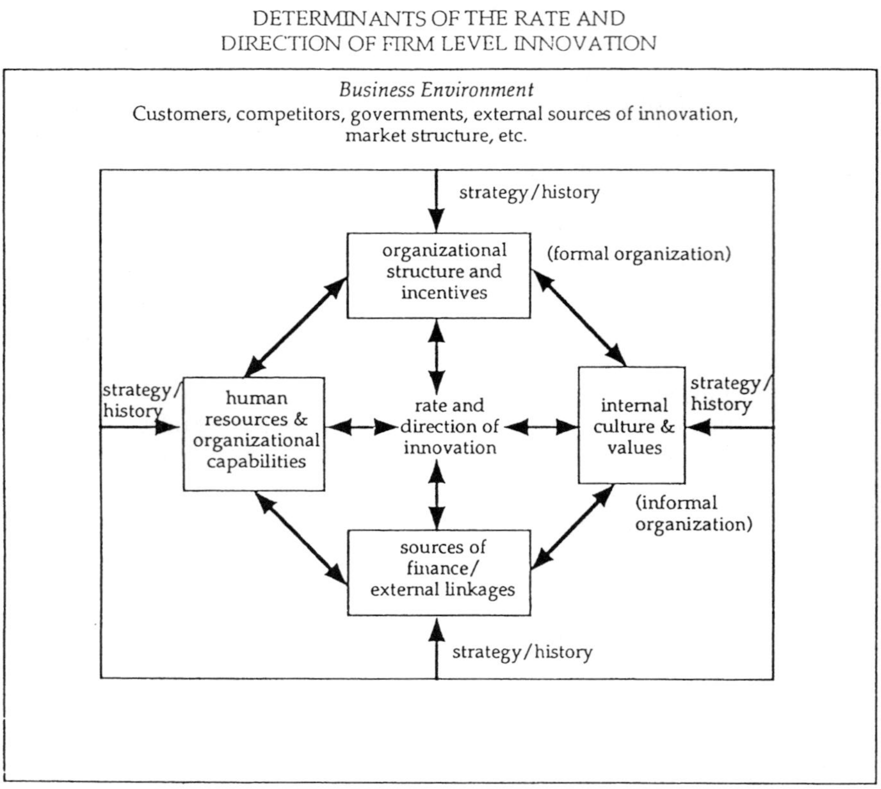

Fig. 1. A determinants of the rate and direction of firm level innovation.

what projects to engage and what assets to build or buy in order to commercialize technology.

In Sections 4 and 5 consideration is given to identifying particular organizational forms that have distinct implications for certain types of innovation. The treatment is illustrative and not comprehensive. It suggests that there are a variety of organizational modes that can support innovation, but that there are important differences amongst organizations in the types of innovation they can support.

4. Distinctive governance modes (Archetypes)

In the previous section, various organizational characteristics were identified. Distinctive governance modes arise when these characteristics are represented to greater or lesser degrees. The specification of the governance mode requires attention to at least four classes of variables: firm boundaries, internal formal structure, internal informal structure, and external linkages. What immediately becomes clear is that for purposes of considering the innovative potential of various organizational forms, one can no longer simply specify the type by reference to one or two aspects of structure. For example, it is no longer meaningful to discuss the innovative potential of conglomerates, vertically

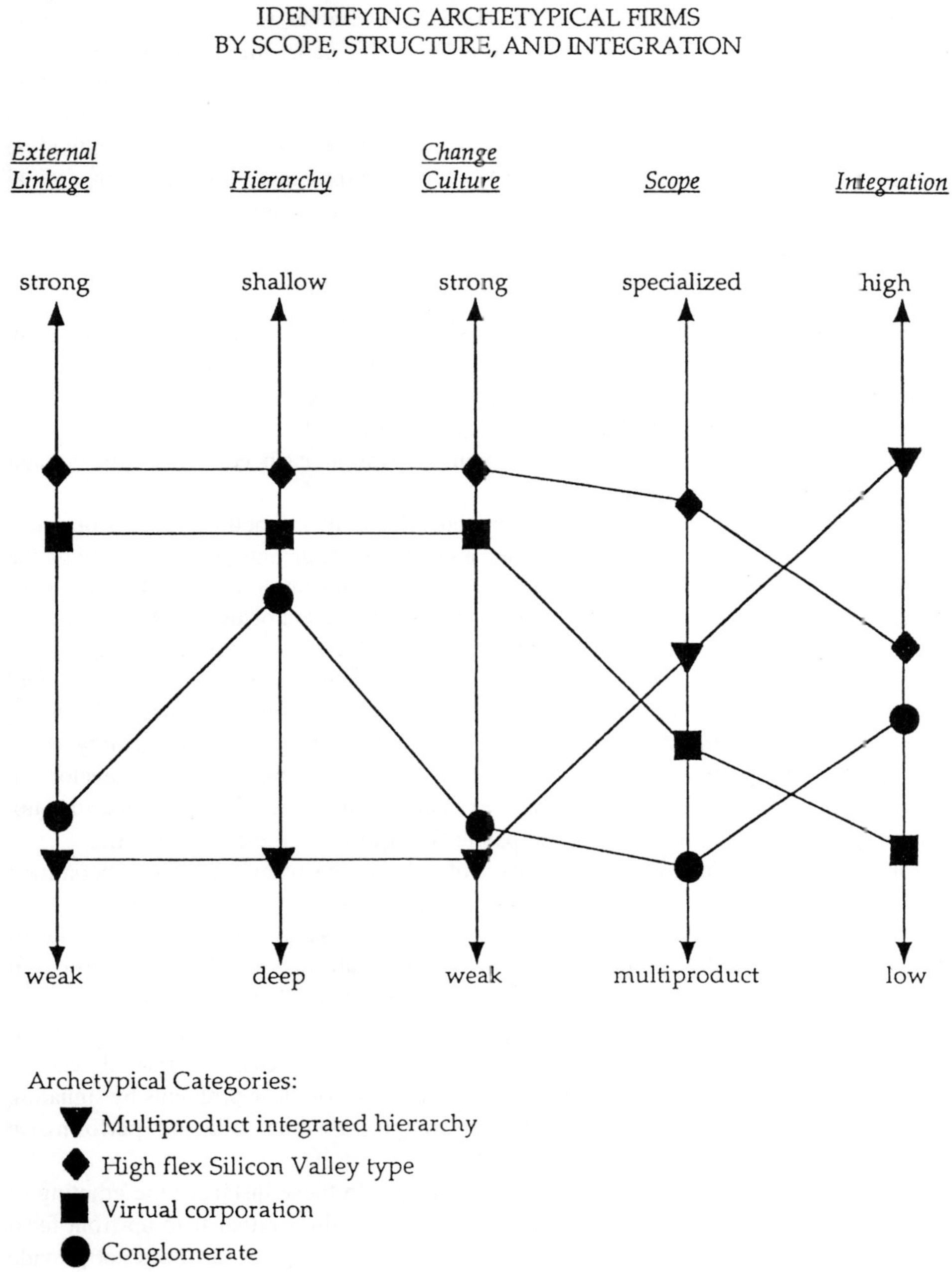

Fig. 2. Identifying archetypical firms by scope, structure, and integration.

integrated firms, etc. without specifying much more. Rather than specify all possible permutations and combinations of these variables in this paper, the focus will be on the following archetypes: (1) stolid, multiproduct, integrated hierarchies; (2) high flex "Silicon Valley"-type firms; (3) hollow corporations of various types; and (4) conglomerates of various types. There will also be a brief discussion of the individual inventor (not really an organizational form). Fig. 2 graphs these structures on ordinal scales measuring various structure variables plus scope and external linkages.

4.1. The individual inventor and the stand-alone laboratory

Many still cling to the notion that the individual inventor, standing outside of an organization, is responsible for the lion's share of innovation in today's economy. This myth springs in part from the first industrial revolution when invention was the province of the individual or pairs. But since the last quarter of the 19th century and the emergence of R&D labs, and more recently venture capital, innovation has become more the domain of organizations, not individuals.[28]

The problems that the inventor-entrepreneurs have in extracting value from new technology are considerable. However, when an inventor (or an enterprise) can rely on the instruments of intellectual property protection to protect invention from imitation, theory suggests that the inventor can appropriate a substantial fraction of the invention's market value. When property rights are weak (the normal case), the inventors' ability to capture value are dramatically circumscribed (Teece, 1986). In the case where the individual inventor has a patent but little else, then the patent holder's options include: (i) licensing the technology to incumbent firms who already have the necessary complementary assets in place; (ii) using the patent as collateral to raise debt funds to help develop an organization to exploit the technology; (iii) exchanging the patent for equity in a start-up, equity-funded firm; (iv) exchanging the patent for equity in an established firm.

None of these options avoid the problem of valuing the patent and the concomitant leakage problems which this process exposes. Valuation is likely to require disclosure and the triggering of what Arrow (1971, p.152) has referred to as the *fundamental paradox* of information: Its value for the purchaser is not known until one has the information, but then one has in effect acquired it without cost. While this problem is somewhat softened when there is good patent protection, most non-industrial providers of funds are going to need technical experts to evaluate the technology, in which case the risk of leakage remains. A wealthy inventor can of course overcome some of these problems by signaling value to financiers and joint venture partners through providing collateral, performance guarantees, or by co-investing.

If imitation is easy, the problems are more difficult. In those instances the granting of low royalty, non-exclusive output-based licenses (i.e. royalties rather than up-front fees) are likely to yield higher rents to the inventor. In this way, the inventor does not provide much incentive for firms to invent around (in the case of a weak patent) or otherwise invest resources in imitation; these costs can be extracted in part by the licensor.

---

[28]That is not to say that individual inventors are not sometimes very important and very successful. My argument is that when they are successful commercially, it is generally in an organizational setting.

Even when the valuation problem is overcome the parties must meet another challenge – transferring the technology to the buyers. As discussed earlier, the tacit nature of knowledge (which helps make imitation difficult) also makes transfer difficult (Teece, 1980, 1982). Hence the circumstances where imitation is difficult are also the circumstances where transfer is often difficult. The only clear circumstance where the inventor can succeed alone is when (1) the technology is well protected by intellectual property law, (2) the technology can be transferred from the inventor to an organization, and (3) the inventor already has great wealth. The circumstances where these factors occur together is likely to be relatively rare.

The stand-alone research laboratory faces many of the same challenges as the individual inventor. The main difference is that the laboratory can bring multiple organizational skills to bear on the R& D process, and the probabilities of fusing multiple technologies is likely to be enhanced from the bringing together of multiple research disciplines. Moreover, if scale economies exist in R&D, the laboratory is better able to capture these than the individual. But the framework would suggest that stand-alone laboratories cannot be viable, unless they happen to work in areas where strong intellectual property protection is assured.[29]

4.2. Multiproduct, integrated, hierarchical firms

It is not uncommon to find such enterprises on the industrial scene. N.V. Phillips and General Motors in the 1980s were good examples. Hierarchical is meant to signal the presence of bureaucratic decisions, and absence of a powerful change culture and high-powered incentives. Such enterprises are also likely to be internally focused. As a consequence, external changes in the market as well as in the science and technology establishment are unlikely to get recognized in a timely fashion. Decision making is slow and ponderous.

However, if such organizations are able to achieve what Downs (1967, p.160) calls "breakout" – where a new organization, possibly a new venture division, is set up for a special task – it may be able to overcome the anti-innovation bias, at least temporarily. Burgelman (1984) has argued that "autonomous strategic behavior" can take place inside large firms, if management sets up the appropriate internal structures. The range of enabling structures is quite large and includes venture teams' "skunk-works," new venture divisions and the like. The suitability of these various structures depends on a variety of technological, market and organizational factors which will not be explored here.

Nevertheless, integrated firms overcome some basic problems associated with relying on an economy of Lilliputian firms. Integrated firms can readily support systemic innovation as discussed earlier. They can also adapt to uncertainty (Williamson, 1975) in a sequential fashion as events unfold. (Managerial hierarchies are often better at adjudicating disputes inside the firm than courts are at adjudicating

[29]Even setting aside protection issues, stand-alone R&D laboratories have problems in developing information channels to their sponsors to understand their sponsors' needs, and in transferring technology back to the sponsor if in fact useful technology is developed. Moreover, because of leakage problems, competitors are likely to be reluctant to use a common R&D laboratory.

disputes between and among firms.) Large multiproduct, multidivisional integrated firms can take on large projects and can help set standards important to the continued evolution of a technology. In the early years of the PC industry, IBM drove floppy-disk drive capacity. The 5.25 in. floppy diskettes initially held 180 KB each when IBM introduced its PC in 1981. By 1983, capacity had doubled to 360 KB, and a year later had increased to 1.44MB. But they stayed stuck at 1.44 MB for over a decade. The explanation does not lie in inherent limits to the technology, but in IBM's declining ability to coordinate choices of follow-along standards. A new standard requires PC manufacturers to agree to accommodate it in their machines, that diskette manufacturers tool up for it, and that software publishers agree to supply programs in the new format. IBM's leadership is no longer sufficient to convert the industry over.[30]

There are also appropriability benefits. If it is a process technology which is at issue, the vertically integrated firm is capable of using the technology in-house and taking profits not by selling the technology directly, but by selling products that embody or use the process. Thus inasmuch as this type of firm does not have to utilize the market for know-how to capture value from the technology, the appropriability problem is softened. Inasmuch as contracting is internal, specialized assets are protected and recontracting hazards are attenuated. The technology transfer process is likely to be internal, so the tacitness problem is eased considerably, as the redeployment of personnel internally raises far fewer default issues than does external redeployment.

Such firms are likely to need alliance structures in order to tap into external sources of new knowledge. If large integrated firms are able to successfully team up with other firms[31] that have the entrepreneurial structures in place to promote creativity, then such firms are likely to be able to access a pipeline of new product and process concepts. The benefits here are a corollary to the benefits associated with strategic alliances.[32] However, the absence of a change culture and an outward orientation mean that such relationships may not be sought.

4.3. High flex "Silicon Valley"-type firms

The distinguishing features of such firms are that they will possess a change culture upon which there is great consensus.[33] They will have shallow hierarchies and significant

[30] I wish to thank Henry Chesbrough for helping develop the facts on this point.

[31] Such as the one described in VI.3.

[32] It is important to recognize, based on historical experience in the United States in the period up to 1980, that the acquisition of a multiproduct, integrated, hierarchical company by high-flex "Silicon Valley"-type company is often extremely difficult to achieve without destroying the creative and entrepreneurial capacity of the small companies. This is because the organizational controls of the large organization tend to destroy the innovative capacities of small firms, as discussed earlier.

[33] Indicative of this spirit is a statement by Andy Grove, CEO of Intel, "You need to try to do the impossible, to anticipate the unexpected. And when the unexpected happens, you should double your efforts to make order from the disorder it creates in your life. The motto I am advocating is, *Let chaos reign, then rein in chaos*. Does that mean that you shouldn't plan? Not at all. You need to plan the way a fire department plans. It cannot anticipate fires, so it has to shape a flexible organization that is capable of responding to unpredictable events."

local autonomy. Such firms will resist the hierarchical accouterments of seniority and rank found in Category 2 above, and they will resist functional specialization which restricts the flow of ideas and destroys the sense of commonality of purpose. Examples of firms that started this way and still reflect much of this style are Intel, Hewlett Packard, Sun Microsystems, Motorola, Raychem, Genentech, and 3M.

Decision making in these firms is usually simple and informal. Communication and coordination among functions is relatively quick and open. Early on in their development, one or two key individuals, typically the founders, make the key decisions. In the early stages, these firms, however, typically do not have a steady stream of internally generated cash with which to fund new opportunities. Hence, connections to the venture capital community or to other firms with cash available are important. These firms are likely to be highly innovative. But they are also likely to be cash-constrained. Those that are not, are likely to do very well.

The highly specialized nature of such firms and the absence of good intellectual property protection create strategic risks. The ability to capture the rents from innovation is by no means assured. But if such firms are able to develop and manage their external relationships without losing their distinct culture and responsive structures, then many of the problems stemming from uncertainty,[34] indivisibilities,[35] inappropriability,[36] asset specificity,[37] and tacitness[38] can be overcome, while organizational failure issues are held at a distance because much is outsourced and alliances are used frequently. By providing considerable autonomy and strong

---

[34]Primary uncertainty can never be reduced, but organizations can adapt to it. Secondary uncertainty, due to ignorance of complementary investment plans, can obviously be much reduced through bilateral agreements which involve mutual commitments and the maintenance of reciprocity through the exchange of hostages (Williamson, 1985).

[35]It is perhaps in the realm of indivisibilities that bilateral exchange comes closest to the perfect solution of a market failure problem. As discussed elsewhere (Teece, 1980, 1982), interfirm agreements are a relatively straightforward way to access complementary assets, particularly if they are already in place, are in excess capacity, and do not involve a high degree of asset specificity. Even when asset specificity is involved, the incentives for opportunistic recontracting can be attenuated by reputation effects, repeat contracting, or exchange of hostages.

[36]Inasmuch as firms can use bilateral contracts to access existing industry capacities so that new capacity does not have to be put in place de novo, product commercialization time can be reduced and lead time lengthened. Thus a major strategic advantage, lead time, can often be enhanced through the use of bilateral contracts. While the innovator may have to share part of the rent stream with the provider of complementary assets, investment risk for the innovator is typically reduced and imitators can be outpaced.

[37]Bilateral contracts enable specialized assets to be protected. While the degree of protection may not be as great as is provided under vertical integration, it is likely to be significantly higher than under unilateral contracts. A "hostage," or its economic equivalent, including specific investments which are mutually dependent, can be used to help support exchange. Thus if a manufacturer installs dedicated equipment to serve the developer, and the developer makes specialized investments which dovetail with the manufacturer, both can be assured that transactions will have a better chance of continuing in the face of adversity or superior opportunities.

[38]Tacitness is less a problem if a bilateral relationship exists, particularly if it is supported by equity. If repeated transactions are contemplated, spillovers and costs associated with seconding technical staff are less severe as adjustments can be made in subsequent transactions, as long as spillovers and costs are perceived similarly by both parties.

incentives, this organizational form is likely to be able to support many different types of innovation.

4.4. Virtual corporations

The term *virtual corporation* has been used in business parlence in the 1980s and 1990s to refer to business enterprises that subcontract anything and everything. A key question is whether the innovative capacities of such companies are impaired by the absence of in-house manufacturing and other capabilities. Virtual corporations are of course smaller than they might otherwise be (by virtue of the absence of vertical integration) and thus generally have shallow hierarchies. They might well have innovative cultures and external linkages to competent manufacturers.

Defined this way, virtuals have the capacity to be very creative and to excel at early-stage innovation activities. If they do indeed establish a strong alliance with a competent manufacturer, they may also have the capacity to be first to market, despite the absence of the requisite internal capabilities.

The hazards associated with *virtual* structures are not unlike the hazards facing the individual inventor. The problem is that unless the firm is operating in a regime of tight appropriability, the innovator may not be able to capture value from the innovation, and the manufacturer, by integrating into research and distribution, is likely to become the firm's competitor (Teece, 1986). Accordingly, the virtual corporation is not seen to be a viable long-run organizational form, except in limited circumstances.

The RCA color television experience demonstrates the downside of the *virtual* approach to innovation.[39] When RCA developed the color television, it made no attempt to keep the innovation to itself.[40] Rather, it licensed its color TV technology aggressively, and outsourced the manufacturing of key components of the television itself. It utilized a network of retailers to market the sets. Its licensees, however (particularly the Japanese licensees), made major investments in the integration of the television components, and then integrated forward and made the entire television set. In the 1970s, RCA had to abandon the manufacturing and development of color television sets, leaving the Japanese as world leaders in consumer electronics.

The RCA experience is not an anomaly. There are real risks in contracting everything out to the market and functioning as the hub or nexus of contracts. Research and development markets, in particular, are fraught with contractual hazards that undercut the ability of firms to coordinate arms-length purchases of R&D through markets (Teece, 1988). Consider the problem of using fixed price contracts to develop new products and processes. With fixed price contracts, one hazard is the inability to adequately specify in advance the desired output of the contract. Another hazard is that, if the R&D supplier shares too much knowledge ex ante, the buying firm can appropriate the knowledge without any payment. A third concern is the specific nature of most R&D activity. The R&D supplier and buyer confront hold-up hazards from each other, and there are strong lock-in

[39]See Yamamura and Vandenberg (1986). I am not suggesting that RCA was or is a virtual organization; but merely that at various times it has embraced key elements of the virtual approach to innovation.

[40]Peters (1990), for example, advises firms to *License your most advanced technology* as well as *Subcontract anything and everything.*

effects once a relationship is begun. These hazards are softened with cost-plus contracts, but these contracts create other problems. The supplying R&D firm has no incentive to control costs, which creates the possibility of a blank check agreement for the buyer.

For these reasons, R&D is usually linked with manufacturing inside the firm.[41] For a similar set of reasons, the marketing function is also joined with these two functions. The desired output of an R&D process depends critically on the perceived user requirements for the product. This is highly impacted information, which cannot be contracted for in advance. And usually there is iteration between the emerging design and user reactions to the design, requiring an ongoing flow of information between the marketing and R&D functions. As a result, firms find it necessary to combine these complementary functions, not through the high-powered incentives of the market, but through the low-powered incentives of the firm.

Another dimension of this is that while markets are very efficient at coordinating adjustments where the technological coordination is low in terms of the interdependence of one technology on another, as the technologies become more interdependent, the hazards of coordinating through the market rise quickly. Internal organization often incurs greater costs than markets when technological interdependence is low. While coordination costs do rise as interdependence grows, those costs rise much more slowly than the market's costs. Accordingly, as interdependence rises, more integrated structures displace virtuals. Clearly, virtual is not always virtuous (Chesbrough and Teece, 1996).

4.5. Conglomerates

In the framework developed here, the conglomerate is not an especially distinctive organizational form. It is likely to be decentralized, and this favors the innovation process. It can also use the internal capital market to fund the development of new technologies. However, the importance of this is likely to be reduced the more (i) access to capital, including venture capital, is available for new stand-alone businesses, and (ii) headquarters management acts much like external capital market agents. Accordingly, on grounds of access to capital and diversity of activities, one would not expect the conglomerate to look too different from a portfolio of stand-alone firms with respect to its innovative capacity.[42]

However, there are two ways in which one might expect the conglomerate to underperform a portfolio of stand-alone firms with respect to innovation. One is that it is difficult for conglomerates to develop distinctive company-wide corporate cultures.

---

[41]Mowery (1982) also finds that the contract research laboratories in the first half of the 20th century confined their work to simple testing and materials analysis, while internal R&D laboratories conducted the more sophisticated and firm-specific research.

[42]There has been very little discussion of the relationship between the conglomerate and technological innovation. The arguments advanced by Williamson (1975) that conglomerate firms possess miniature capital markets would suggest that the conglomerate is an ideal form for identifying new investment opportunities, including process and product innovations, and funding them until they become cash-flow positive. In the absence of market-for-venture capital, this argument would seem to imply that the conglomerate form ought to be associated with a stream of new product and process launches. The best evidence (Kline, 1995) indicates that relative to appropriately defined yardsticks, conglomerates did not underinvest in R&D, and indeed before 1970 they may well have overinvested.

Accordingly, it may be quite difficult to build a strong internal change culture at the corporate level. Certainly, as compared to a stand-alone firm, getting across to employees the notion that the unit must ultimately "stand on its own bottom" will be quite a challenge. As a consequence, free riding may well be accentuated. Likewise, the design of high-powered incentives for top management and employees will be hindered by the absence of an equity instrument geared to divisional performance. In short, the conglomerate does not appear to offer distinctive advantages in environments characterized by rapid technological change.

4.6. Alliance enterprise

We define an alliance enterprise as a *virtual corporation* that has developed strong commitments to other enterprises, usually through equity-based links to affiliated enterprises lying upstream, downstream, horizontal, and lateral from its core business. Such structures include consortia (e.g. Airbus, Sematech) as well as semipermanent teaming arrangements that transcend particular projects. Many new biotech firms in the United States are heavily alliance-dependent to fund their R&D and move drugs to the market.

The viability and desirability of alliances and other external linkage arrangements depend, not just on the efficacy of this form of contract, but also on the resources/ capabilities which can be accessed in this fashion. Alliances were essential in the 1980s and 1990s to the pharmaceutical industry as a mechanism to tap into the drug development capabilities of new biotech firms. Since the biotechnology revolution has occurred outside the organizational gambit of the established pharmaceutical industry, alliances have been embued with virtues they might not otherwise possess. Put differently, the value of a contract can easily be confused with what it enables one to access. The comparative institutional approach used here imputes to the alliance only that which it can uniquely access as compared to other arrangements.

5. Matching innovation and organizational archetypes[43]

The diversity of organizational forms observed is semipermanent and not a transitional feature of modern industrial economies. The diversity of observed forms in and of itself suggests that different organizational arrangements are suited to different types of competitive environments and differing types of innovation.

One cannot possibly expect to be comprehensive in developing a taxonomy of innovations and organizational archetypes. However, illustrations are developed below which involve matching organizational form to the locus of existing capabilities, and to the type of innovation (autonomous or systemic).

As the interdependence between technologies increases, pure market forms are less effective at achieving the requisite coordination. The more systemic the innovation, the greater the interdependence. Exposure to recontracting hazards is likely to be frequent.

[43]This section draws in part on Chesbrough and Teece (1996).

The discussion in Section 4.4 noted the distinction between autonomous and systemic innovation. Autonomous innovations (Teece, 1984) create improved products and processes that fit comfortably into existing systems. These innovations not only fit well within current industry standards, but they reinforce those standards. An example would be the introduction of a faster microprocessor using the same architecture, such as the Intel 80×86/Pentium family. Systemic innovations, however, change technological requirements and offer new opportunities so that the resulting configuration of both the innovation and its related technologies (which comprise a system of technology) are different; for example, audio CD plays require the abandonment of vinyl records and the manufacture of CD discs. Innovations of this type require that the design of the subsystems be coordinated in order for the gains from the innovation to be realized. Since these innovations span current technology boundaries, a complex coordination problem arises.

The other key dimension in organization form is the extent to which the capabilities needed to exploit the innovation exist within the firm already, and if not, whether those capabilities are available outside the firm. It has been argued elsewhere (Teece et al. 1994) that the firm is best regarded as a bundle of distinctive capabilities that enables it to perform functions more efficiently than its historical competitors. The presence or absence of critical complementary assets affects the prospects for appropriating the gains from innovation when the appropriability regime is not tight (Teece, 1986). These two dimensions motivate the simple framework presented in Fig. 3.

5.1. Autonomous innovation

Alliances and virtual structures will work well when the technology can be sourced externally, and the high-flex Silicon Valley-type will work well if it must be developed internally. When the technology can be sourced externally, the required coordination takes place with known technologies, so that no special hazards in contracting arise and adjustments to related technologies to realize the benefits of innovation are minimal. Indeed, when firms use bureaucratic centralized structures inappropriately to manage autonomous innovations, small firms and more decentralized large firms are likely to outperform them.

As noted, there are important nuances with respect to the particulars of how to organize for autonomous innovation. (1) The first circumstance is when the technology exploits capabilities already present within the firm. In these circumstances, internal development by high-flex Silicon Valley-type firms will work well. (An example here would be the introduction of a faster microprocessor using the same architecture, such as the Intel 80×86/Pentium family.) (2) The second is when innovation remains autonomous, but exploitation requires the firm to access capabilities outside its boundaries. Innovating firms must craft relational structures such as alliances to obtain access, thereby sharing the gains from innovation. Such firms must also overcome hold-up problems between the innovator and the owner of relevant outside capabilities. This is where virtual structures are often virtuous. Sun Microsystems pursued this strategy quite successfully with its SPARC microprocessor architecture. Sun defined the basic architecture, and then licensed out the design to other firms. This strategy induced enough entry by SPARC licensees to

	Type of Innovation: autonomous	Type of Innovation: systemic
Capabilities Exist Inhouse	**S**	**M**
Capabilities Exist Outside	**V**	**A**
Capabilities Must be Created	**A,S**	**S**

Key:

S= Silicon Valley type

M= Multiproduct integrated

A= Alliances (virtual with equity)

V= Virtual (outsource everything & anything)

Fig. 3. A proposed matrix of innovation, capabilities and preferred organization forms.

develop a standard, and attracted outside software developers to support the architecture. Sun later split its organization so that its hardware was separated from its software and its microprocessor design.[44]

(3) A third circumstance is when innovation remains autonomous, but requires new capabilities to be created to exploit its potential. Virtual structures are not quite enough. Consider biotechnology. New products can continue to pass through the same regulatory procedures and sell to physicians through the same marketing and distribution channels as yesterday's ethical drugs. But the underlying technology draws from a different science base. Here, internal development or alliances with equity are required to manage contractual problems between the young biotechnology companies developing new products and the older pharmaceutical companies seeking to add new products to their

[44]Note that Sun's strategy of defining a standard microprocessor architecture, and then licensing it, created a well-defined technical interface and resulted in transforming what would have been a systemic innovation to an autonomous innovation. Sun's approach facilitated decentralized innovation around its standard. The downside to this approach was experienced by IBM in the PC market.

lineup. These additional structures provide a credible ex ante basis for dividing the gains from innovation between the two types of firms.[45]

5.2. Systemic innovations

By their nature, systemic innovations require coordinated adjustment throughout the system to realize the gains from innovation. The potential from systemic innovations cannot be fully realized until adjustments are made throughout the system. Unaffiliated enterprises with weak internal integration will not suffice. The problem is that tight coordination is needed. Without the close integration of personnel, necessary coordination may be forsaken. Property rights issues may also arise if multiple enterprises are involved. Information sharing can be reduced or biased, as each seeks to get the most at the expense of the other. The party that commits first can be held-up by the other parties, while the party that waits until the others have committed themselves can extract more rents from the other partners. Even if such opportunism is contained, the rate of advance of complementary technologies may not be properly matched, so that product release dates slip and co-investment schedules of the parties are mismatched.

Lockheed's failure in the wide-bodied civilian airliner market can be attributed in part to the debacle it had with Rolls Royce. Rolls Royce committed to develop the RB-211 engine to power the L-1011, but technical problems that took time to be revealed to Lockheed caused major delays in commercialization, and took both Lockheed and Rolls Royce into bankruptcy. Similarly, GM's success with the diesel electric locomotive, and the failure of the GE/Alco co-development efforts can properly be tied to the inability of GE and Alcoa to properly coordinate and integrate their development efforts (Marx, 1976).

What is needed to successfully develop and commercialize systemic innovations are institutions with low-powered incentives, where information can be freely shared without worry of expropriation, where entities can commit themselves and not be exploited by that commitment, and where disputes can be monitored and resolved in a timely way. This is precisely what multi-product integrated firms achieve.

While systemic innovation favors integrated structures from a coordination perspective, it may nevertheless be the case that the relevant technological capabilities are resident in unaffiliated enterprises.[46] Alliances are then the best arrangement. Virtuals and even small Silicon Valley-type firms will not survive. NeXT and MIPS, for example, were dependent of the autonomous decisions of other firms in order to be able to realize the benefit of their technology. When these other firms delayed their supporting investments, each company was forced to narrow its focus and retrench.[47] This is where larger firms may have an advantage, by being able to secure minority investment positions in smaller firms with necessary capabilities, or by using their scale to create

---

[45]Pisano (1988) detailed the practice of using joint equity agreements to facilitate these relations in the biotechnology industry.

[46]For a discussion of capabilities, see Teece and Pisano (1994), Teece et al. (1997).

[47]NeXT has completely withdrawn from the hardware side of its workstation business, focusing entirely on its NeXTStep operating system. MIPS was acquired by Silicon Graphics, and is no longer a significant player as a workstation microprocessor architecture developer.

sufficient momentum so that complementary innovations are developed.[48] Japanese *keiretsu* commonly leverage strong relationships to access needed capabilities outside the firm, and they fit this model. For example, Toyota's successful introduction of the *kanban* production system (a truly systemic innovation) required tremendous coordination with its network of suppliers. Since Toyota was much larger than its suppliers, and because until recently it was the largest customer of virtually all of its suppliers, it had sufficient leverage to compel its suppliers to make radical changes to their business practices without exposing itself to hold-up.[49]

Another circumstance considered is where the systemic innovation requires entirely new capabilities in order for the innovation's potential to be realized. This is precisely the situation Chandler (1990) describes in *Scale and Scope* for the leading industries of the late 19th and early 20th centuries. The leading industries of that era – chemicals, steel, and railroads – were all transformed by systemic innovation (Teece, 1993). The winners were the companies that made major investments to shape the markets, and simply did not rely upon them. Today one sees leading companies like Intel and Microsoft making extensive investments to enhance their current capabilities and spur the creation of new ones.[50] Network arrangements among unaffiliated enterprises are exactly the wrong organizational strategy when firms are trying to commercialize and appropriate the gains of systemic innovation. An excellent illustration of this is an example often held to support the benefits of decentralization – the IBM PC. While the PC's early years highlighted the benefits of the *virtual* approach, the passage of more time has revealed the downside of that approach. This example is examined elsewhere (Chesbrough and Teece, 1996).

The process of matching organizational form to underlying technological and market conditions is of course a dynamic one. This aspect is somewhat expressed in Fig. 4. An illustration of some of the dynamics is provided by Motorola which continues to innovate in hand-held communication devices, including cellular phones. Future improvements on cellphone designs, and in particular, weight reduction and extended operation, requires lighter and more long-lived batteries. Motorola is in a position to advance these technologies through its own internal R&D programs, which have historically been very productive. As the older more established battery technologies like Nickel Cadmium have

---

[48]The contrast between MIPS and DEC's Alpha chip is one example. By committing to supply Alpha on its own workstations, DEC is gathering greater commitment from third-party developers than did MIPS. Apple and IBM's PowerPC chip are garnering even more support, as Apple and IBM together claim to have shipped over one million systems with the PowerPC chip. Scale and integration are the key differences here.

[49]According to Gerlach (1992), Japanese manufacturers also place managers on the boards of its supplier firms and usually share their main bank with them. This allows a manufacturer such as Toyota to wield control over the supplier's strategic decisions and control the supplier's access to capital without requiring Toyota to have complete ownership of its suppliers. It is this which creates functional control. These structures provide a blend of market incentives (high-powered incentives, external measures of prices, profits and value added) with internal coordination capability.

[50]Intel, for example, has just committed itself to building the largest fab in the world in New Mexico to improve its manufacturing prowess. Microsoft is competing in virtually every segment of the PC software business with internally-developed products, rather than licensing or buying outside products. It recently announced plans to spend over $900 million in its current fiscal year, a 50% increase over last year's spending *(San Francisco Chronicle*, September 10, 1994, p.D2).

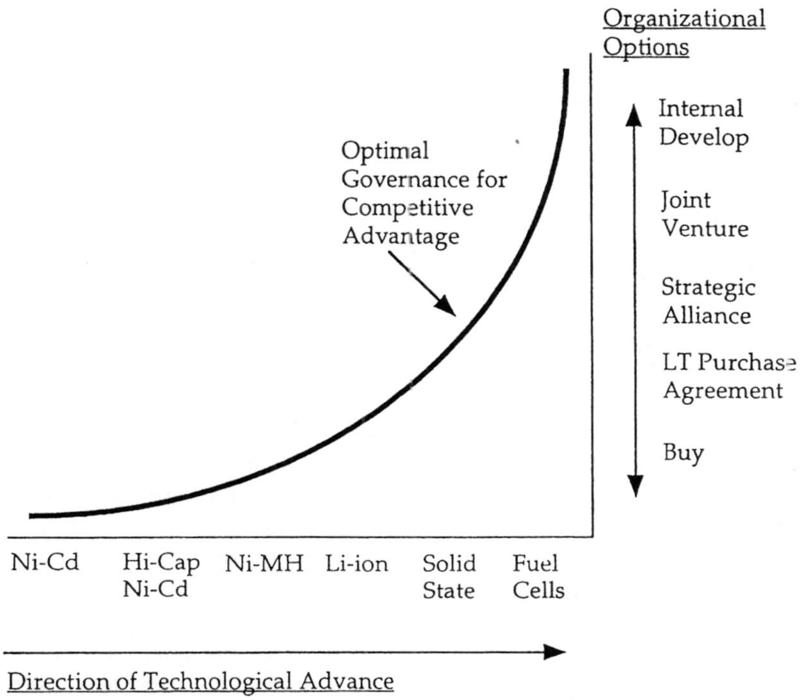

Fig. 4. Available technologies and organizational options for Motorola in battery cell technology.

been widely diffused, Motorola can reasonably rely on outsourcing from numerous existing suppliers to access its requirements in the Ni–Cd domain. However, solid state and fuel cells are still in their infancy as technologies to support personal communication devices. Moreover, Motorola is as well placed as others to advance the development of such technologies. With reliance on unaffiliated parties leading to obvious contractual hazards (Teece, 1988), internal development, or at the minimum joint venture development, is suggested for such technologies. In Fig. 4, this suggests that desirable governance arrangements will migrate toward *internal development* as the technology becomes more *state-of-the-art* and the population of outsider vendors diminishes.[51] It is likely to do so as one confronts more advanced technological options.

6. Conclusions

If this analysis is correct, it has rather strong implications for theory building, for management, and for public policy. With respect to theory building, it suggests the inadequacy of standard economic approaches that have market structure as the key if not the only determinant of the rate and direction of innovation. Clearly, such approaches are

[51]LT purchase agreement in Fig. 4 is an abbreviation for long-term purchase agreement.

poor guides to policy. At the minimum, firm boundaries (the level of integration), the structure of financial markets, and formal and informal organizational structure must be recognized as major determinants. This paper indicates that firm organization (not just product market structure) is an important determinant of innovation, a point made by Williamson (1975) that has largely gone unheeded by industrial organization economists.

The framework developed here is designed to shift the market structure–innovation debate in industrial organization beyond the domain where Schumpeter (1942), Galbraith (1952), Mansfield (1968), Scherer (1980) and others have put it, and into a new domain where internal structure, interfirm agreements, and capital market structures attain new significance. This also has obvious policy significance. The opening up of financial markets and the emergence of a vibrant venture capital industry have "provided new forms of finance for innovative activity again on a scale never seen before, effectively reducing barriers to innovative, competitive entry, across the industrialized world" (Rybczynski, 1993). Put differently, product market structure is not the main and possibly not the key factor in determining the rate and direction of innovation.

The framework also has strong implications for business history. It suggests the possible viability of new hybrid organizational arrangements – such as complex forms of interfirm agreements linking firms with complementary capabilities and capacities – over both the integrated alternatives and the small firm alternatives. These organizational forms may well represent a new and dramatic organizational innovation in business history. Firms are continuing to learn how and when to use them, and scholars are trying to understand them. In retrospect, the emergence and growth of these new forms, dating from about 1970, may turn out to be as significant an organizational innovation as the moving assembly line and the multidivisional firm.

References

Alchien, A. and H. Demsetz, 1972, Production, information cost, and economic organizations, American Economic Review, 62, 777–795.

Armour, H. and D.J. Teece, 1978, Organizational structure and economic performance: A test of the multidivisional hypothesis, The Bell Journal of Economics, 9(2), 106–122.

Arrow, K.J., 1971, Essays in the theory of risk-bearing, (North-Holland, Amsterdam, New York).

Arrow, K.J., 1974, The limits of organization, (Norton, New York).

Arrow, K.J., 1996, Technical information and industrial structure, Industrial and Corporate Change, 5(2).

Burgelman, R.A., 1984, Designs for corporate entrepreneurship, California Management Review, 26(3)(Spring), 154–166.

Chesbrough, H.W. and D.J. Teece, 1996, When is virtual virtuous: Organizing for innovation, Harvard Business Review, January–February.

Coase, R.,, Nature of Firm, Economica, 4, 386–405.

Crozier, M., 1964, The bureaucratic phenomenon, (University of Chicago Press, Chicago).

Day, R.H., G. Eliasson and C. Wihlborg, Eds., 1993, The markets for innovation, ownership and control, (North-Holland, Amsterdam, New York).

Deal, T.E. and A. Kennedy, 1982, Corporate culture, (Addison-Wesley, Reading, MA).

Dosi, G., 1982a, Technological paradigms and technological trajectories, Research Policy.

Dosi, G., 1982b, Technological paradigms and technological trajectories: A suggested interpretation of the determinants and directions of technical change, Res. Policy, 11(3) (June), 147–162.

Dosi, G., C. Freeman, R. Nelson, G. Silverberg and L. Soete, 1988, Technical Change and Economic Theory, (Pinter, London).

Downs, A., 1967, Inside Bureaucracy, (Little Brown, Boston).
Fama, E., 1970, Efficient capital markets: A review of theory and empirical work, American Economic Review, 60, 163–174.
Frankel, M., 1955, Obsolence and technological change in a maturing economy, American Economic Review.
Galbraith, J.K., 1952, American Capitalism, (Houghton-Mifflin, Boston).
Gerlach, M.L., 1992, Alliance capitalism: The social organization of Japanese Business, (University of California Press, Berkeley).
Holmström, B., 1993, Agency costs and innovation, In: R.H. Day et al., The Markets for Innovation, Ownership and Control, (North-Holland, Amsterdam, New York).
Imai, 1988, Japan's corporate networks, unpublished working paper, (Hitotsubashi University, Tokyo).
Jensen, C., 1989, Eclipse of the public corporation, Harvard Business Review, 67(5) (September–October), 61–74.
Kahneman, D. and D. Lovallo, 1990, Timid decisions and bold forecasts: A cognitive perspective on risk taking, In: R. Rumelt, D. Schendel and D. Teece, Ed., Fundamental issues in strategy, (Harvard Business School Press, Boston, MA).
Kay, N., Industrial structure, rivalry, and innovation: Theory and evidence, undated working paper, Department of Economics, (Herriot-Watt University, Edinburgh, Scotland).
Kindleberger, C.P., 1964, Economic growth in France and Britain, 1851–1950, (Harvard University Press, Cambridge, MA).
Kline, P., 1995, Conglomerate organizations and economic performance: Evidence from the 1960s, unpublished doctoral dissertation, (University of California, Berkeley).
Kodama, F., 1986, Japanese innovation in mechatronics technology, Science and Public Policy, 13(1), 44–52.
Koopmans, T.C., 1957, Three essays in the state of economic science, (McGraw-Hill, New York).
Levin, R., A. Klevorick, R. Nelson and S. Winter, 1987, Appropriating the returns from industrial R&D, Brookings Papers on Economic Activity.
Little, A.D., 1985, Management perspectives on innovation, (Harvard University Press, Cambridge, MA).
Mansfield, E., 1968, The economics of technical change, (Norton, New York).
March, J. and Z. Shapira, 1987, Managerial perspectives on risk and risk taking, Management Science, 33(11), 1404–1418.
Marx, T., 1976, Vertical integration in the diesel-electric locomotive building industry: A study in market failures, Nebraska Journal of Economics and Business, 15(4)(Autumn), 37–51.
Mowery, D., 1982, The relationship between contractual and intrafirm forms of industrial research in American manufacturing, 1900–1940, Exploration in Economic History (October), 351–374.
Nelson, R.R. and S.G. Winter, 1977, In search of a useful theory of innovations, Res. Policy, 6(1), 36–76.
Nelson, R.R. and S.G. Winter, 1982, An evolutionary theory of economic change, (Harvard University Press, Cambridge, MA).
North, D., 1990, Institutions, institutional change, and economic performance, (Cambridge University Press, Cambridge).
O'Reilly, C.A., 1989, Corporate culture considerations based on an empirical study of high growth firms in silicon valley, Economia Aziendale, Vol. III, 3.
Pavitt, K., M. Robson and J. Townsend, 1989, Accumulation, diversification and organisation of technological activities in U.K. companies, 1945–1983, In: M. Dodgson, Ed., Technology Strategy and the Firm, (Longman, London).
Peters, T., 1990, Get innovative or get dead, California Management Review (Fall), pp.9–21.
Peters, T. and R. Waterman, 1982, In search of excellence.
Pisano, G.P., 1988, Innovation through markets, hierarchies, and joint ventures: Technology strategy and collaborative arrangements in the biotechnology industry, unpublished Ph.D. dissertation, Haas School of Business, (University of California, Berkeley).
Polanyi, M., 1962, Personal knowledge: Toward a post critical philosophy, New York, Harper and Row.
Richardson, G.B., 1990, Information and Investment, (Oxford University Press, Oxford).
Rybczynski, T., 1993, Innovative activity and venture financing: Access to markets and opportunities in Japan, the U.S. and Europe, In: Day et al., The Markets for Innovation, Ownership and Control, (North-Holland, Amsterdam, New York).

Scherer, F.M., 1980, Industrial market structure and economic performance, 2nd edn., (Rand McNally, Chicago).

Schumpeter, J.A., 1934, The theory of economic development, (Harvard University Press, Cambridge, MA).

Schumpeter, J.A., 1942, Capitalism, socialism and democracy, (McGraw-Hill, New York).

Schwartz, H. and S. Davis, 1981, Matching corporate culture and business strategy, Organizational Dynamics, 30–48.

Simon, H., 1973, Decision making and organizational design, In: D.S. Pugh, Ed., Organizational Theory, (Penguin, London).

Teece, D.J., 1977, Technology transfer by multinational firms: The resource cost of transferring technological know-how, Economic Journal, 87(346)(June), 242–261.

Teece, D.J., 1980, Economics of scope and the scope of the enterprise, Journal of Economic Behavior and Organization, 1(3), 223–247.

Teece, D.J., 1982, Toward an economic theory of the multiproduct firms, Journal of Economic Behavior and Organization, 3(1), 39–63.

Teece, D.J., 1984, Economic analysis and strategic management, California Management Review, 26(3)(Spring), 87–110.

Teece, D.J., 1986, Profiting from technological innovation, Research Policy, 15(6), 285–306.

Teece, D.J., 1988, The nature of the firm and technological change, In: G. Dosi et al., Eds., Technical Change and Economic Theory, (Pinter, London).

Teece, D.J., 1993, The dynamics of industrial capitalism: Perspectives on Alfred Chandler's scale and scope, Journal of Economic Literature, 31 (March).

Teece, D.J. and G. Pisano, 1994, The dynamic capabilities of firms: An introduction, Industrial and Corporate Change, 3:3.

Teece, D.J., G. Pisano and A. Shuen, forthcoming 1997, Dynamic capabilities and strategic Management, Strategic Management Journal..

Teece, D.J.,R. Rumelt,G. Dosi and S. Winter, 1994, Understanding corporate coherence: Theory and evidence, Journal of Economic Behavior and Organization, 23, 1–30.

Veblen, T., 1972, Professor Clark's economics, In: E.K. Hunt and J. Schwartz, A Critique of Economic Theory.

Williamson, O.E., 1975, Markets and hierarchies, (Free Press, New York).

Williamson, O.E., 1985, The economic institutions of capitalism, (Free Press, New York).

Williamson, O.E., 1988, Corporate finance and corporate governance, Journal of Finance, 43 (July), 567–591.

Williamson, O.E., 1996, The mechanisms of governance, (Oxford University Press, New York).

Winter, S.G., 1987, Knowledge and competence as strategic assets, In: D. Teece, Ed., The competitive challenge: Strategies for industrial innovation and renewal, (Ballinger, Cambridge, MA).

Yamamura, K. and J. Vandenberg, 1986, Japan's rapid growth policy on trial: The television case, In: G. Saxonhouse and K. Yamamura, Law and trade issues of the Japanese economy: American and Japanese perspectives, (University of Washington Press, Seattle, WA).

Economica 46, 187–196

Overseas Research and Development by US-Based Firms

By EDWIN MANSFIELD, DAVID TEECE and ANTHONY ROMEO

University of Pennsylvania, Stanford University, University of Connecticut

INTRODUCTION

In recent years, the overseas research and development activities of US-based firms have become the focus of controversy. Some observers view such activities with suspicion, since they regard them as a device to "export" R and D jobs, or as a channel through which American technology may be transmitted to actual or potential foreign competitors.[1] Others, particularly the governments of many developing (and some developed) countries, view them as highly desirable activities that will help to stimulate indigenous R and D in these countries. Indeed, the United Nations Group of Eminent Persons recommended that host countries require multinational corporations to contribute towards innovation of appropriate kinds, and to encourage them to do such R and D in their overseas affiliates.[2] Although the amount of controversy in this area might lead one to believe that the nature of existing overseas R and D activities of US-based firms has been studied quite thoroughly, this is far from the case. The unfortunate truth is that economists have devoted little or no attention to even the most basic questions concerning these activities.[3] In this paper, we try to fill part of this gap.

I. OVERSEAS R AND D EXPENDITURES: 1960–1980

How big are the overseas R and D expenditures of US-based firms, now and in the past, and how big do firms expect them to become by 1980? To help answer this question, we constructed a sample of 55 major manufacturing firms, this sample being divided into two parts. The first subsample, composed of 35 firms, included major US-based firms in the chemical, petroleum, electrical equipment, metals and machinery, drugs, glass, food and paper industries. The second subsample, composed of 20 firms, included major manufacturing firms in the southern New England and Middle Atlantic states. Table 1 shows the percentage of R and D done overseas by these firms, during 1960–1974 (for the first subsample) or 1970–1974 (for the second subsample), as well as the estimated value of this percentage in 1980. In each subsample, about 10 per cent of the total amount spent on R and D by these firms was carried out overseas in 1974. Based on the 35-firm subsample, it appears that this percentage grew substantially during the 1960s and early 1970s.[4] Based on the estimates provided by the firms in the sample, this growth will continue, but at a reduced rate, during the rest of the 1970s; by 1980 they estimated that about 12 per cent of their R and D expenditures will be made overseas.

Because of the importance in the innovation process of close communication and cooperation among R and D, marketing, production and top management, Vernon (1974) and others have argued that a firm's R and D activities will tend to be centralized near its headquarters. Why then do these US-based firms spend

TABLE 1

PERCENTAGE OF COMPANY-FINANCED R AND D EXPENDITURES CARRIED OUT OVERSEAS, 1960–1980: 55 FIRMS

	1960*	1965†	1970‡	1972†	1974	1980§
35-firm subsample:						
Weighted mean	2	6	6	8	10	10
Unweighted mean	2	4	5	7	8	8
Standard deviation	3	7	7	8	10	8
20-firm subsample:						
Weighted mean	—	—	4	—	9	14
Unweighted mean	—	—	5	—	8	11
Standard deviation	—	—	7	—	10	14

* Data were not available for 4 firms in the 35-firm subsample.
† Data were not available for 1 firm in the 35-firm subsample.
‡ Data were not available for 1 firm in the 35-firm subsample and 1 firm in the 20-firm subsample.
§ Data were not available for 9 firms in the 35-firm subsample.
Source: see Section I.

about 10 per cent of their R and D dollars overseas? There are a variety of possible reasons, including the presence of environmental conditions abroad that cannot easily be matched at home, the desirability of doing R and D aimed at the special design needs of overseas markets, the availability and lower cost of skills and talents that are less readily available or more expensive at home, and the greater opportunity to monitor what is going on in relevant scientific and technical fields abroad. In our sample, practically all of the firms doing R and D overseas say that the principal reason is to respond to special design needs of overseas markets. In their view, there are great advantages in doing R and D of this sort in close contact with the relevant overseas markets and manufacturing units of the firm.

II. FACTORS INFLUENCING THE PERCENTAGE OF A FIRM'S R AND D EXPENDITURES CARRIED OUT OVERSEAS

What determines the percentage of its R and D that a firm conducts overseas? Given the fact that overseas laboratories seem to be so closely geared to the special design needs of foreign markets (and the firm's overseas plants), we would expect that the percentage of a firm's R and D expenditures carried out overseas would be directly related to the percentage of the firm's sales that is derived from abroad. Firms with relatively small foreign markets would be expected to spend relatively little on overseas R and D. Further, we would expect that the percentage of a firm's R and D expenditures carried out overseas would be more closely related to the percentage of its sales from foreign subsidiaries than to its percentage of sales from exports. This is because much overseas R and D is in support of foreign manufacturing operations.[5]

Holding constant the percentage of a firm's sales that come from abroad, we would expect that the percentage of a firm's R and D expenditures that is carried out overseas would be directly related to the firm's size. Economies of scale require that R and D laboratories be a certain minimum size if they are to be relatively efficient. If it is going to establish an overseas laboratory, the firm must have a big enough prospective market (in the area served by this laboratory) to support a laboratory of minimum economic scale. If the percentage of a firm's sales that comes from abroad is held constant, the probability that this prospective market will be of the requisite size is an increasing function of the absolute size of the firm.

Further, we would expect that, holding constant both the firm's sales and its percentage of sales coming from abroad, there would be inter-industry differences in the percentage of a firm's R and D expenditures carried out overseas. For example, we would expect this percentage to be relatively high in the pharmaceutical industry because some firms, according to industry sources at least, have moved a substantial amount of R and D abroad to avoid Food and Drug Administration regulations. Also, foreign regulations sometimes require that R and D be done locally. Because of these regulatory considerations, as well as other factors discussed below, we might expect the drug firms in our sample to carry out a relatively high percentage of R and D overseas.

In addition, we would expect that, if the firm's sales, its percentage of sales coming from abroad and its industry are held constant, there will be differences over time in the percentage of a firm's R and D carried out overseas, owing to changes in the profitability of locating R and D overseas rather than in the United States (as well as bandwagon effects).[6] In general, during the period covered here, we would expect these effects of time to be positive, since cost differentials and other factors favoured the expansion of overseas R and D.

To test these hypotheses, we carried out two sets of computations. First, we pooled the 1970 and 1974 data for the 35-firm subsample, and regressed each firm's percentage of R and D expenditures carried out overseas on its percentage of sales from abroad, its sales, an industry dummy variable, and a time dummy variable. The results are

$$(1) \qquad A_{it} = \underset{(0.44)}{-1.13} + \underset{(0.34)}{0.73T_t} + \underset{(1.81)}{0.15Q_{it}} + \underset{(3.09)}{0.004S_{it}} + \underset{(5.41)}{16.10D_i}, \qquad (\bar{R}^2 = 0.50;\ n = 51)$$

where A_{it} is the percentage of the ith firm's R and D expenditures carried out overseas in year t (1970 or 1974), T_t is a dummy variable that equals 1 if t is 1974 and 0 if t is 1970, Q_{it} is the percentage of the ith firm's sales derived from abroad in year t, S_{it} is the ith firm's sales (in millions of dollars) in year t, and D_i is a dummy variable that equals 1 if the ith firm is in the drug industry and 0 otherwise.[7] Each regression coefficient's t-ratio is given in parentheses.

Second, we pooled the 1970 and 1974 data for the 20-firm subsample, and regressed each firm's percentage of R and D expenditures carried out overseas on the same variables as in equation (1), except that we split the percentage of sales from abroad into two parts—the percentage of sales from foreign subsidiaries, and the percentage of sales from exports—and we redefine the industry dummy to include both chemicals and drugs, not drugs alone.[8] The results are:

$$(2) \qquad A_{it} = \underset{(1.16)}{2.79} + \underset{(1.91)}{4.40T_t} + \underset{(4.47)}{0.322F_{it}} - \underset{(2.32)}{0.539E_{it}} - \underset{(1.13)}{0.00162S_{it}} + \underset{(1.76)}{4.52C_i}, \qquad (\bar{R}^2 = 0.54;\ n = 39)$$

where F_{it} is the percentage of the ith firm's sales from foreign subsidiaries in year t, E_{it} is the percentage of the ith firm's sales from exports in year t, and C_i is a dummy variable that equals 1 if the ith firm is in the drug or chemical industries and 0 otherwise.

The econometric results generally are in accord with our hypotheses. As expected, equation (1) shows that there is a direct and statistically significant relationship between a firm's percentage of sales derived from abroad and its percentage of R and D expenditures carried out overseas. And when sales derived from abroad are disaggregated in equation (2), a firm's percentage of sales from foreign subsidiaries has a highly significant positive effect on its percentage of R and D expenditures carried out overseas, while its percentage of sales from exports has a significant negative effect, which suggests that these firms' exports may be more R and D-intensive than their domestic sales.[9] With regard to our hypothesis that A_{it} would be directly related to S_{it}, the results of equation (1) bear this out, but in equation (2), S_{it} does not have a statistically significant effect (and its regression coefficient has the wrong sign). As expected, most of the industry and time dummies are statistically significant.[10]

III. Overseas R and D: Minimum Economic Scale and Relative Cost

As noted above, many governments, particularly of developing countries, favour the establishment in their nations of overseas R and D laboratories by US-based firms. One factor influencing the practicality of establishing a laboratory of a certain type in a particular overseas location is the extent of economies of scale in such laboratories. If the minimum economic scale for a laboratory of this type is quite large, a firm must be prepared to shift considerable R and D resources abroad if the laboratory is to be competitive.[11] Despite the fact that data concerning the minimum economic scale of R and D laboratories of various types would be of value to many kinds of microeconomic studies, practically no information is available on this score. In this section, we present the estimates (obtained from 27 members of the 35-firm subsample) of the annual R and D expenditures for an overseas laboratory of minimum economic scale. Although these estimates should be treated with caution, they are of considerable interest, since they seem to be the first systematic evidence on this topic.[12]

The results, shown in Table 2, indicate that the minimum economic scale tends to be quite substantial in most industries. On the average, for a single product line it was estimated that the expenditures per year for an R and D facility of minimum economic scale would be about $1 million in pharmaceuticals and glass, about $2 million in electrical equipment and petroleum, and about $5 million in chemicals. However, the minimum economic scale seems to vary considerably, depending on the responsibilities of the laboratory. It is less for a laboratory that performs either research or development than for one that performs both, and less for a laboratory that deals with a single product line than for one that deals with several product lines. For a laboratory that is concerned entirely with minor product changes, the average estimated expenditure per year for an R and D facility of minimum economic scale is only about $500,000 per year—and in some industries it is substantially less. In interpreting the results in Table 2, the dispersion among the estimates is almost as interesting as the averages. The estimates in each industry vary enormously, reflecting the fact that the minimum economic scale of an R and D laboratory depends on the specific type of work to be done, as well as the fact that opinions differ on this score even among experts.[13]

TABLE 2

ESTIMATED ANNUAL R AND D EXPENDITURE FOR OVERSEAS LABORATORY OF MINIMUM ECONOMIC SCALE, 1975: 27 FIRMS ($million)

	Single product line				Several product lines			
Industry*	Research	Development	Minor product changes	Research and development	Research	Development	Minor product changes	Research and development
Chemicals ($n = 7$)								
Mean	2·42	3·27	0·72	4·57	2·50	2·46	1·39	3·31
Range	1·0–5·0	0·1–14·0	0·03–2·0	0·46–20·0	1·5–3·5	0·26–6·0	0·24–3·5	0·16–6·7
Petroleum ($n = 6$)								
Mean	1·64	1·46	0·28	2·30	2·60	2·23	1·18	3·40
Range	0·25–3·0	0·25–3·0	0·15–0·50	0·40–5·0	0·40–5·0	0·40–4·5	0·25–3·0	0·6–7·5
Drugs ($n = 2$)								
Mean	3·25	0·50	0·12	1·00	6·00	1·30	0·35	†
Range	1·5–5·0	0·5–0·5	0·10–0·13	0·50–1·5	5·0–7·0	0·60–2·0	0·19–0·50	†
Electronics and electrical equipment ($n = 5$)								
Mean	1·00	1·15	0·40	2·00	2·42	3·95	0·77	6·75
Range	0·8–1·2	0·36–2·0	0·20–0·50	1·25–2·75	1·0–5·0	0·54–8·0	0·51–1·0	1·25–10·0
Glass ($n = 2$)								
Mean	0·70	0·85	0·42	1·38	0·73	1·53	0·71	1·88
Range	0·1–1·3	0·4–1·3	0·08–0·75	0·50–2·25	0·20–1·25	0·80–2·25	0·16–1·25	1·0–2·75
Total ($n = 27$)								
Mean	1·58	1·82	0·47	3·00	2·62	2·83	1·23	4·10
Standard deviation	1·34	2·78	0·45	4·10	1·87	2·76	1·30	3·68

* n is the number of firms that provided estimates.
† An estimate was obtained from only one firm.
Source: see Section I.

According to many observers, one major reason why US-based firms have carried out R and D overseas is that costs have tended to be lower there. However, very little information has been published concerning the extent of this cost differential, and how it has varied over time. To help fill this gap, we obtained data from the 35-firm subsample concerning the ratio of the cost of R and D inputs in Europe, Japan and Canada to those in the United States in 1965, 1970 and 1975.[14] The results, shown in Table 3, indicate that there was a very substantial cost differential in 1965: on the average, the cost of R and D inputs seemed to be about 30 per cent lower in Europe, 20 per cent lower in Canada and 40 per cent lower in Japan than in the United States. And although there was some increase in R and D costs relative to those in the United States during 1965–1970, the cost differential remained quite substantial in 1970.

TABLE 3

MEAN RATIO OF COST OF R AND D INPUTS IN SELECTED OVERSEAS LOCATIONS TO THAT IN THE UNITED STATES, 1965, 1970 AND 1975.* 35-FIRM SUBSAMPLE†

	Location		
Year	Europe‡	Japan	Canada
1965	0·68	0·56	0·82
1970	0·74	0·60	0·86
1975	0·93	0·90	0·96

* Note that there are many costs of communication and coordination in a multinational network of laboratories. See Mansfield (1974).

† Usable data were obtained from 19 firms. Many of the rest had no overseas R and D experience.

‡ There are considerable differences within Europe in the level of R and D costs. According to a number of firms in our sample, costs tend to be relatively low in the United Kingdom and relatively high in West Germany.

However, between 1970 and 1975 the situation changed drastically. Owing in part to the depreciation of the dollar relative to other currencies between 1970 and 1975, the cost differential was largely eliminated for many firms. On the average, the cost of R and D inputs was estimated to be about 10 per cent lower in Japan, and about 5 per cent lower in Europe and Canada, than in the United States in 1975. Of course, this helps to explain the fact (noted in Section I) that the percentage of R and D carried out overseas was expected to increase less rapidly between 1974 and 1980 than in the period prior to 1974. Since the cost differential between overseas and domestic R and D was smaller, it is quite understandable that firms would expect this percentage to grow less rapidly than in earlier years.[15]

IV. OVERSEAS R AND D: NATURE OF WORK AND RELATION TO DOMESTIC R AND D

Some observers, as we have seen, are suspicious of overseas R and D because they fear that it may be a channel through which American technology may "leak out" to foreign competitors. The extent to which such a leakage is likely to occur depends in part on the nature of the work being carried out in the overseas laboratories of US-based firms. For example, if such work is focused largely on the modification and adaptation of products and processes for the local market, there is less need to transfer much of the firm's most sophisticated technology overseas than if the work is focused on major product or process developments intended for a worldwide market. Based on information obtained from 23 firms in our sample, it appears that these firms' overseas R and D activities tend to focus on development rather than research, on product and process improvements rather than on new products and processes, and on relatively short-term, technically safe work.

Specifically, on the average the percentage of overseas R and D going for basic research is about 6 percentage points less than the percentage of domestic R and D going for basic research; the percentage of overseas R and D going for applied research is 10 percentage points less than the percentage of domestic R and D going for applied research; while the percentage of overseas R and D going for development is 16 percentage points greater than the percentage of domestic R and D going for development. Moreover, about three-fourths of these firms' overseas R and D expenditures are aimed at product or process *improvements* and *modifications*, not at entirely new processes or products. This percentage is much higher than for all domestic R and D.

Firms seem to differ considerably in the extent to which they have integrated their overseas R and D with their domestic R and D.[16] Worldwide integration of overseas and domestic R and D exists in almost one-half of the firms (with overseas R and D) in our sample, according to the firms. On the other hand, about one-sixth say that they attempt no such integration, and the rest say that some limited integration is attempted.

Finally, of how much value is overseas R and D to a firm's US operations? Policy-makers are interested in this question because it must be considered in any full evaluation of the effects of overseas R and D (and foreign direct investment) on America's technological position *vis-à-vis* other countries. Unfortunately practically no evidence exists on this score. To shed a modest amount of light on this question, we obtained estimates from 27 firms in our sample concerning the percentage of their 1975 overseas R and D expenditures with no commercial applicability to their US operations. The results indicate that, on the average, about one-third of these firms' overseas R and D expenditures have no such applicability. Also, we asked each firm to estimate the amount that it would have to spend on R and D in the United States to get results of *equivalent value to its US operations* as a dollar spent overseas. The results, which are only rough, indicate that, on the average, a dollar's worth of overseas R and D seems to result in benefits to these firms' US operations that are equivalent to about 50 cents' worth of R and D carried out in the United States.

V. Conclusions

Overseas R and D expenditures by US-based firms topped the billion-dollar mark in the early 1970s. In 1974 they amounted to about one-tenth of total domestic company-funded R and D expenditures, and the firms in our sample reported that they expected them to amount to a larger proportion of their domestic R and D expenditures by 1980.[17] When compared with the total R and D expenditures in various host countries, their size is perhaps even more striking. In the early 1970s about one-half of the industrial R and D performed in Canada and about one-seventh of the industrial R and D performed in the United Kingdom and West Germany was done by US-based firms.[18]

Yet despite the magnitude and importance of these overseas R and D activities, little is known about their purpose, nature or effects. Indeed, the very existence of such activities is ignored in all current econometric studies carried out to estimate the effects of R and D on US productivity growth. Our purpose in this paper has been to present some basic information concerning the size, nature, minimum economic scale and relative cost of overseas R and D. The limitations of these findings should be recognized. In particular, our results pertain to a sample of 55 firms, and some of the data obtained from the firms were necessarily rough. None the less, we believe that these results, although only a first step, shed substantial new light on this topic.

ACKNOWLEDGMENTS

The work on which this paper is based was supported by a grant to Edwin Mansfield from the Division of Policy Research and Analysis of the National Science Foundation. Of course, the views expressed here are not necessarily those of the Foundation. We are grateful to the more than 50 firms that provided us with data concerning their overseas R and D activities. Some of the results were presented in papers given by Mansfield at Johns Hopkins, Vanderbilt and Yale Universities, and at the National Science Foundation, as well as the 1978 annual meeting of the American Economic Association.

NOTES

[1] For discussion of this point of view, see David (1974) and Conference Board (1976).

[2] United Nations (1974). *The Impact of Multinational Corporations on Development and on International Relations*. New York, p. 70.

[3] Caves (1974), Hufbauer (1974), Mansfield (1974) and Stobaugh (1974) have pointed out the need for work concerning this and related aspects of international technology transfer. For some interesting case studies, see Ronstadt (1975). Also, see Mansfield, Romeo and Wagner (1979) for some related findings regarding international technology transfer.

[4] The Conference Board (1976) has estimated the total overseas R and D expenditure of US-based multinational firms in 1971–1973. According to its estimates, overseas R and D constituted about 9–10 per cent of total R and D expenditures carried out by US firms during these years. This agrees quite well with our results for 1972 and 1974. The proportion of firms in our sample with no overseas R and D is somewhat lower than that reported by the Conference Board for firms of comparable size, but this may be due to different industry mix, the later year, or sampling error. The US Department of Commerce (undated) has estimated the total overseas R and D expenditure of US firms in 1966. According to its figures, overseas R and D constituted about 7 per cent of all R and D expenditures carried out by US firms in 1966 (see Conference Board, 1976). This agrees reasonably well with our result for 1965. In 1978 the National Science Foundation published data for 1976 which indicated that overseas R and D constituted about 7 per cent of all R and D expenditures by US manufacturing firms.

[5] Suppose that a firm's desired R and D expenditures in a given year equal

$$R = a_u S_u + a_f S_f + a_e S_e$$

where S_e is its export sales during the relevant year, S_f is its sales through foreign subsidiaries, S_u is its sales from domestic plants to domestic customers, a_u is the proportion of sales to domestic customers that it wants to devote to R and D, a_f is the proportion of sales through foreign subsidiaries that it wants to devote to R and D, and a_e is the proportion of export sales that it wants to devote to R and D. If only the R and D in support of foreign subsidiaries is done overseas, it follows that the proportion of its R and D expenditures carried out overseas equals

$$P = \frac{a_f F}{a_u U + a_f F + a_e E}$$

where F is the proportion of its sales from foreign subsidiaries, U is the proportion of its sales to domestic customers, and E is the proportion of its sales that are exports. Under these circumstances, it can be shown that $\partial p/\partial F$ is always positive, but whether or not $\partial p/\partial E$ is positive depends on whether or not $a_u > a_e$. Of course, this model is a polar case, but it illustrates the point in the text.

[6] In terms of the highly simplified model in n. 5, the a's are a function of time. (Also, as indicated previously in the text, they are a function of the firm's size.)

[7] Other industry dummies were tried in equation (1), but D_l was the only one that was statistically significant. The reason why n is less than 70 is that data could not be obtained concerning the percentage of sales from abroad for all firms in both years.

[8] The reason for this redefinition is that none of the firms in this subsample is really an ethical drug fim. (There are several such firms in the other subsample.) The closest we could come to ethical drugs is the chemical firms, some of which do some work in the drug area (broadly defined).

[9] This result concerning E_{lt} would be expected if $a_e > a_u$ in n. 5, and if the extremely simple model given there were valid. However, although it may be a reasonable approximation to regard some firms' overseas R and D as being entirely in support of foreign subsidiaries, this is far from the case in other firms.

[10] The industry dummy is much larger and more highly significant in equation (1) than in equation (2) because, as pointed out in n. 8, none of the firms in equation (2) is really an ethical drug firm. Because of sampling variation, the estimate of D_l in equation (1) is probably too large. One of the drug firms in our sample carried out an unusually large percentage of its R and D overseas.

[11] By minimum economic scale we mean the smallest scale that realizes all, or practically all, of the relevant economies of scale.

[12] Freeman *et al.* (1965) have presented some relevant data concerning the electrical equipment industry. Eight members of the 35-firm subsample could not provide estimates, sometimes because they had no experience on which to base such estimates.

[13] These figures help to explain why, holding other factors constant, smaller firms in equation (1) tend to carry on a smaller percentage of their R and D overseas than bigger firms. But they should not be interpreted as saying that smaller firms are squeezed out completely. The estimates in each industry vary enormously. In most industries, at least some of the respondents felt that research and development could be carried out effectively with an annual budget of $500,000, and that minor product changes could be carried out with one of about $100,000. Although these levels of expenditure are hardly trivial, they are within the reach of many firms other than the billion-dollar giants.

Needless to say, these results in no way contradict the finding by many economists that small firms and independent inventors continue to play an important role in the inventive process. Their contribution is frequently in the earlier stages of the inventive process, where the costs are relatively low. Further, according to some observers, costs tend to be lower in smaller organizations, and the figures in Table 2 reflect the perceptions of large firms. For some relevant discussion, see Mansfield *et al.* (1977).

[14] The relative cost of R and D inputs is the ratio of the annual cost of hiring an R and D scientist or engineer (together with the complementary amount of other inputs) in various overseas locations to do the sort of work carried out there to the annual cost of hiring a comparable R and D scientist or engineer (together with the complementary amount of other inputs) to do the same sort of work in the United States. Each firm was asked to estimate this ratio for each year. Many of the estimates were based on studies the firms seem to have carried out in recent years on this topic.

[15] If very significant differences exist between the productivity of US and overseas R and D personnel, they may offset the observed differences in the relative costs of inputs. About 80 per cent of the firms in our sample regarded the productivity of their R and D personnel in Canada, Europe and Japan to be no lower than those in the United States. Thus, this factor cannot offset the observed difference in the relative cost of R and D inputs in the great majority of firms in our sample.

[16] By integration, we mean that the firm's worldwide R and D is viewed as a whole, and laboratories are given worldwide missions, if this seems desirable.

[17] In early 1977, the United States Treasury put into effect a new regulation (1.861-8) that, according to some observers, may increase the amount of R and D done overseas by US-based firms. Since the forecasts in Table 1 were made before this new regulation was announced, they do not take this factor into account. Also, they do not take account of recent changes in exchange rates, which may have tended to discourage overseas R and D.

[18] See Conference Board (1976, p. 86).

REFERENCES

CAVES, R. (1974). Effect of international technology transfers on the U.S. economy. *The Effects of International Technology Transfers on U.S. Economy*, Washington, DC: National Science Foundation.

CONFERENCE BOARD (1976). *Overseas Research and Development by U.S. Multinationals, 1966–1975*. New York: Conference Board.

DAVID, E. (1974). Technology export and national goals. *Research Management*, **17**, 12–16.

FREEMAN, C., HARLOW, C. and FULLER, J. (1965). Research and development in electronic capital goods. *National Institute Economic Review*, **34**, 40–91.

HUFBAUER, G. (1974). Technology transfers and the American economy. In *The Effects of International Technology Transfers on U.S. Economy*. Washington, DC: National Science Foundation.

MANSFIELD, E. (1974). Technology and technological change. In *Economic Analysis and the Multinational Enterprise*, (J. Dunning, ed.). London: George Allen and Unwin.

—— *et al.* (1977). *The Production and Application of New Industrial Technology*. New York: W. W. Norton.

——, ROMEO, A. and WAGNER, S. (1979). Foreign trade and U.S. research and development. *Review of Economics and Statistics*, forthcoming.

RONSTADT, R. (1975). R and D abroad: the creation and evolution of foreign R and D activities of U.S.-based multinational enterprises. Unpublished DBA thesis, Harvard University.

STOBAUGH, R. (1974). A summary and assessment of research findings on U.S. international transactions involving technology transfers. *The Effects of International Technology Transfers on U.S. Economy*. Washington, DC: National Science Foundation.

US DEPARTMENT OF COMMERCE (undated). *U. S. Direct Investments Abroad, 1966*, Part II. Washington, DC: US Government Printing Office.

VERNON, R. (1974). The location of economic activity. In *Economic Analysis and the Multinational Enterprise* (J. Dunning, ed.). London: George Allen and Unwin.

MANAGERIAL AND DECISION ECONOMICS, SPECIAL ISSUE, 35–42 (1989)

Inter-organizational Requirements of the Innovation Process

DAVID J. TEECE

School of Business Administration, University of California at Berkeley, CA, USA

INTRODUCTION

At the heart of the innovation process in the United States is the modern corporate R&D facility, found in giant corporations and in small start-up enterprises. However, the institutional structure of innovation in capitalist economies is extremely variegated and involves a complex network of backward, forward, horizontal, lateral relationships and linkages within, among and between firms and other organizations such as universities.

Studies designed to explain innovation and to understand the success of particular firms have been somewhat remiss in not analyzing these linkages, which, we propose, have a significant impact on the productivity and profitability of R&D. In this paper attention will be given to inter-firm and inter-organizational relationships and linkages. An attempt will be made to assess their importance to the development and profitable commercialization of new technology. In addition, an effort will be made to identify appropriate organizational designs for managing them. Before doing so, it will be useful to specify key transactions among the various groups and individuals typically involved in the innovation process.

MODELS OF THE INNOVATION PROCESS

Because of the complexity of the innovation process it is important to build some kind of conceptual model of what is involved in the activity that we commonly refer to as innovation. The dominant model, having its origins in the description and analysis of 'big technology' projects, is described below. Its inadequacies are then addressed in the light of the more 'simultaneous' nature of the process, particularly in certain industries experiencing high rates of technological change.

The 'Serial' Model

In the traditional description of the innovation process it is not uncommon to break the process down into a number of stages which proceed sequentially, albeit with significant overlaps if the project is on a fast track. The innovation process thus proceeds from research to development, to design, to production and finally to marketing, sales and service (see Fig. 1) In simple versions of this model there is no feedback or overlap between and among stages. This view of the process may represent an appropriate summary, in some very limited instances, such as where a very modest product improvement is at issue.

The serial or assembly line model of product development with feedback and overlaps has been a reasonable approximation of what went on in many large US corporations at home and abroad for many years. The approach worked well in many areas, including aircraft, automobiles and telecommunications (especially central office switches). Examples of products developed in this fashion include the 7 × 7 series of Boeing aircraft, the Mercury and Apollo programs, many large DoD programs, the IBM 360 and the Xerox 9000 family of high-speed copiers. This method of management had its advantages whenever large idiosyncratic fixed investments were required for successful commercialization. It is a method of organizaiton in which many large companies are comfortable, as it includes clear reponsibility for each part of a complex delivery system.

However, this assembly line or pipeline representation of a new product delivery system is increasingly a caricature of what actually takes place, or what needs to take place. At the center of the innovation process is design, not science. Research is often stimulated by the problems associated with trying to get the design right. Contrary to the view inherent in the serial model, technology is not merely applied science. Any technological development draws on an array of science, not only that which is embedded in one or two recent findings.

Moreover, important technological breakthroughs can often proceed even when the underlying science is not understood well (for example, the IUD for birth control). Products can often be made to work without much knowledge of why. Airframe design in the aircraft industry, for instance, has a large empirical component. Certain designs are known, from experimentation, to have certain performance features. However, the underlying scientific understanding of

0143–6570/89/SI0035–08$05.00

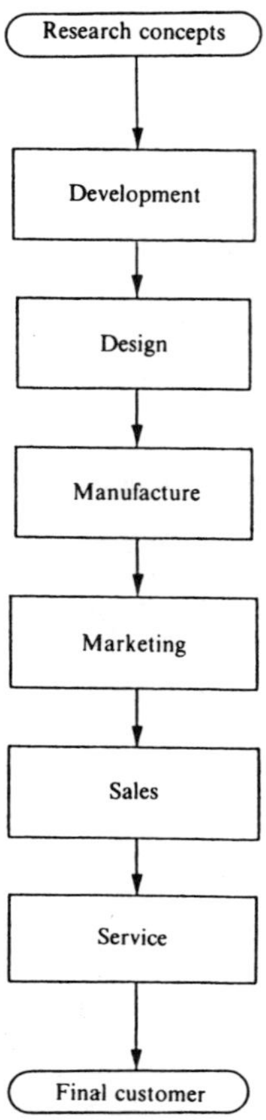

Figure 1. Simplified serial product-delivery process.

airframe design is rudimentary. Accordingly, wind tunnel testing is still an essential part of the development process.

The serial model also underemphasizes the importance of process innovation. Indeed, if the scheme fits innovation at all it is perhaps only for product innovations. Process innovations often do not require marketing, and may not even require new tooling. Nor does the serial model highlight the many small but cumulatively important incremental innovations which are often at the heart of technological change. The serial model has a macro project orientation; but as we know, this is not the way in which most innovation proceeds.

Serial Model with Links and Feedback

In reality, with uncertainty, learning and short product lifecycles, innovation requires rapid feedback, mid-course corrections to designs, and redesign. Feedbacks and trials are essential, whether it is incremental or radical innovation which is at issue. The demands of innovation often lead to scientific developments just as often as innovation draws on science.

Kline and Rosenberg (1986) have proposed what they call a chain-linked model as an alternative (Fig. 2). This model recognizes aspects of the linear one—such as a flow of activity through design to development production and marketing—but also recognizes constant feedback between and among 'stages'. Moreover, 'the linkage from science to innovation is not solely or even preponderantly at the beginning of typical innovations, but rather extends all through the process . . . science can be visualized as lying alongside development processes, to be used when needed' (Kline and Rosenberg, 1986, pp. 290–91).

The identification of needs is critical to the profitable expenditure of R&D dollars. R&D personnel must thus be closely conntected to the market and to marketing personnel. Scientists must have one foot in the laboratory and one in the marketplace. Knowing *what* to develop and design, rather than just how to do it, is absolutely essential for commercial success. Developing this understanding involves a complex interplay between science and engineering, manufacturing and marketing in order to specify product functions and features. It is not just a matter of identifying user needs and assessing engineering feasibility. One must also separate those user needs which are being met by competition and those which are not. This may not become clear until the product is introduced, in which case the ability to redesign quickly and efficiently may be of the utmost importance.

This model recognizes the existence and exercise of tight linkages and feedback mechanisms which must operate quickly and efficiently. These linkages must exist within, among and between firms and other organizations, such as universities. Of course, the positioning of the firm's boundaries (for example, its level of vertical integration) determines in part whether the required interactions are intrafirm or interfirm.

Parallel (Simultaneous) Model

Both versions of the serial model have embedded within them elements of a sequential process. In some circumstances, however, the sequential nature of activity can and should be compressed to such a degree that a parallel model is in fact required. (Fig. 3).

The 'parallel' model, it should be recognized, has been facilitated by a lowering of the cost of changes in digital electronic circuits. While in some domains such as aircraft engines the costs of development have increased, in others (for example, computers) the costs of development per unit of performance have fallen considerably. As a result of this and the great amount of innovation which has occurred and is still continuing in components, major new computer companies such as Apple, Sun, Apollo, Commodore and others have been born and have grown extremely fast. Generally, the hallmark of these new companies is the ability to develop new products quickly. In these companies the linear model has been collapsed and

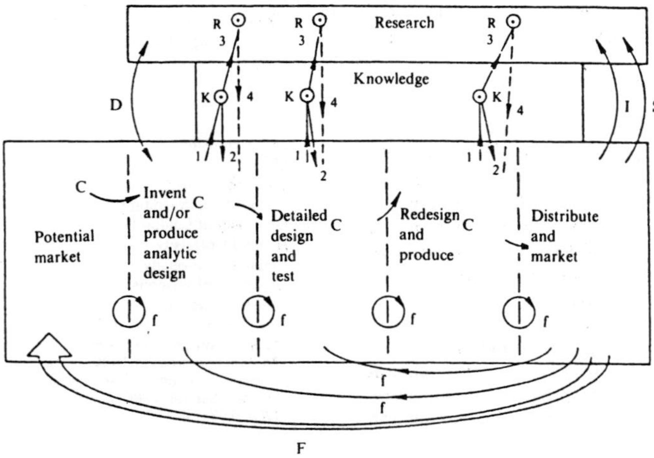

Figure 2. Model showing flow paths of information and co-operation (Kline and Rosenberg, 1986, p. 290.)

Symbols used on arrows:

C: Central chain of innovation
f: Feedback loops
F: Particularly important feedback

K–R: Links through knowledge to research and return paths. If problem solved at node K, link 3 to R not activated. Return from research (link 4) is problematic—therefore dashed line.
D: Direct link to and from research from problems in invention and design.
I: Support of scientific research by instruments, machines, tools and procedures of technology.
S: Support of research in sciences underlying product area to gain information directly and by monitoring outside work. The information obtained may apply anywhere along the chain.

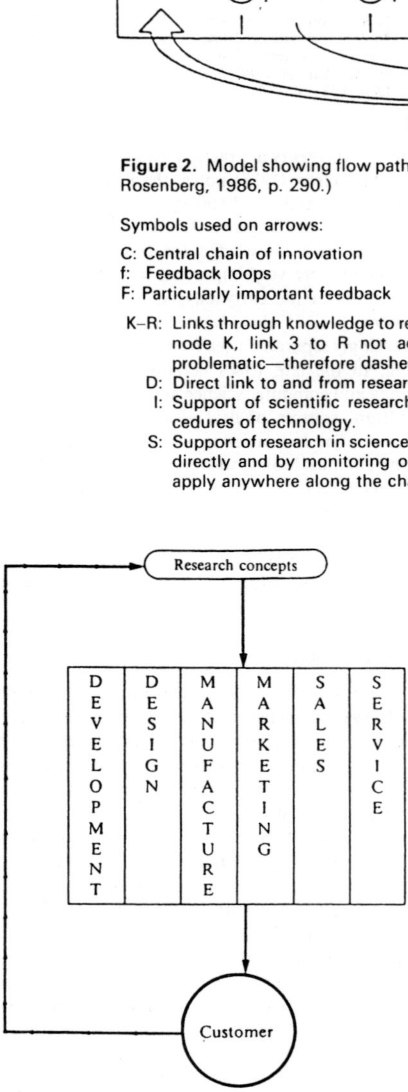

Figure 3. Parallel product-delivery process.

some of the work in product developmen is done in parallel, a process which requires an entirely new approach to management. New information systems and collaboration with outside vendors anc partners is often essential.

Clearly, a parallel approach to development requires intense communication, because various downstream activities may be in progress befcre product features and specifications are finally determined. Hence the amount of information which must be communicated among the various activities is necessarily large, as it is often changing. Workstations and modern communications facilitate the necessary timely technology transfer. Moreover, the simulation of manufacturing facilities can lead to early feedback with respect to the cost ramifications alternative product designs.

LINKAGES TO EXTERNAL INSTITUTIONS

Whether innovation proceeds according to either the serial or the parallel model, it is likely to require access to capabilities which lie beyond the innovating entity. These may be found in universities, other parts of the enterprise or in other unaffiliated enterprises. The roles of some of these key organizational units are now explored.

University–industry Interaction

The US university has made an important contribution to the innovation process since the late nineteenth century, and university science and engineering and science-based industries grew up together in the United States. Universities have provided industry with technical people, and with ideas about product and process innovation (Nelson, 1988).

University research finds its way to industry through a variety of channels, including graduate students, publications, research contracts, technical conferences, industry affiliate programs, personnel exchanges (for example, adjunct professors), consultants, as well as in some cases through directed mechanisms such as research consortia and university industrial liaison offices and licensing offices which market proprietary university technology at universities like Stanford and MIT. Most universities, however, are relatively passive with respect to directed efforts to market university technology, patented or unpatented, to industry. UC Berkeley and UCLA, which have great technological resources, are cases in point.

Two somewhat contradictory threads have been observed in the business perspectives on the commercial applicability and value of university technology. On the one hand, there is a body of evidence that US industry on the whole does not consider universities a viable source of new products. Peters and Fusfeld (1982) examined 464 examples of university–industry research collaboration, but could rarely identify instances where a commercially marketable product or process was an immediate and direct outcome of research interactions or processes. On the other hand, industrial enterprises are often eager to develop good relationships with universities, most probably to get access to the best graduate students, to attract away faculty as partners in new business ventures, and to encourage faculty and students to become familiar with their equipment (as in computers). Even if there are few new products and processes which can be attributed directly to the university, the university connection enables firms to preserve an open window on science and technology and to be alert to changing opportunities and threats.

That university–industry connections rarely result in new patentable products and processes is evident from the minuscule (in relationship to university research budgets) royalties which universities earn (there are some examples where universities have taken equity in exchange for technology and have had a payoff in terms of capital gains or dividends). Table 1 shows annual royalty income for nine US universities and demonstrates, for instance, that the entire university of California (nine campuses) earned only $4 million in gross royalty income in 1986/7.

It would appear that the intensity and necessity for industry–university interaction varies by field and over time. As Nelson (1988) notes, there is evidence that academic research in chemistry and electrical engineering has, over the years, diminished as a source of important new knowledge for industry. Academic researchers were very important to technological developments in the early days of the semiconductor industry, but, as time went by, R&D in industry increasingly separated itself from what academics were doing. In the late 1980s certain areas of academic biology and computer science and metallurgy are very important sources of new ideas and techniques for industry. The industries in which these sciences are important must look to universities for new knowledge and techniques, as well as training. In recent years a number of new models have been developed to facilitate these linkages at some campuses. However, few tight linkages exist, even today. Nevertheless, close links to the university are critical requirements for success in certain industries at certain times.

Table 1. Estimates of Royalty Income at Certain Universities, 1984.

Institution	Annual royalty income ($000)
Johns Hopkins	90
MIT	1500
Stanford University	2500[a]
University of Washington	120
University of California	1700[b]
Harvard University	50
Columbia University	Minimal
University of Wisconsin	6000[c]
Cornell University	1300

[a] $6.5 million gross in 1986/7. (Source: W. Mitchell.)
[b] $4.0 million gross in 1986/7. (Source: W. Mitchell.)
[c] Investment income is a substantial portion. Comparable number for 1986/7 was $10 million. (Source: OTA, 1984, p. 412.)

Vertical Linkages

In a series of important publications Von Hippel (1977, 1988) has presented evidence that, in some industries, industrial products judged by users to offer them a significant performance improvement are usually conceived and prototyped by users, not by the manufacturers. The manufacturers' role in the innovation process in these industries is somehow to become aware of the user innovation and its value and then to manufacture a commercial version of the device for sale to user firms. This pattern of innovation by product users is contrary to the usual assumption that product manufacturers are responsible for the innovation process from finding to filling the need. Figure 4 from Von Hippel (1977, p. 61) contrasts such user-dominated involvement with the more typical manufacturer-dominant innovation. The transfer from user to manufacturer may be initiated by the user, or by the manufacturer, who may have already hired individuals with experience from the user firms.

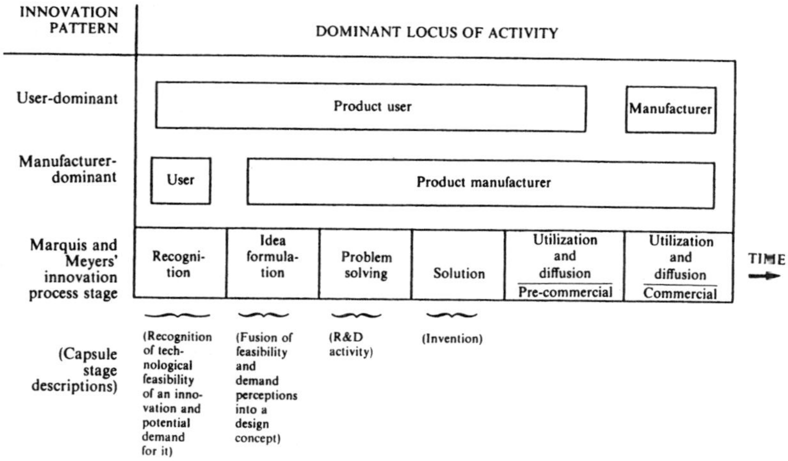

Figure 4. User and manufacturer roles in the innovation process charted against Marquis and Meyers' (1969) innovation process stages (Von Hippel, 1977).

Von Hippel found that user-dominated innovation accounted for more than two-thirds of first-to-market innovations in scientific instruments and in process machinery used in semiconductor and electronic subassembly manufacture. Successful management of the process requires that product engineering (rather than R&D) skills be resident in the manufacturer, and that manufacturers search to identify user solutions rather than user needs.[1] A further implication is that there may be a symbiotic relationship between users and equipment manufacturers which depends upon social and geographical proximity. If user industries migrate offshore, then the manufactures of equipment will disappear from the domestic scene as well. Hence, the decline of the US semiconductor industry can be expected to threaten the US semiconductor equipment industry unless close ties can be developed between foreign users (domestic and offshore) and domestic manufacturers.

Balancing the role that users play in stimulating innovation upstream is the role that suppliers play in stimulating downstream innovation. Much of the innovation in the automobile industry, including fuel injection, alternators and power steering, has its origins in upstream component suppliers.[2] The challenge then becomes how to 'design in' the new components and avoid sole source dependency on the part of the automotive companies. As discussed below, deep and enduring relationships need to be established between component developer–manufacturer and supplier to ensure adoption and diffusion of the technology.

A related set of vertical relationships involving innovation has been commented upon by Rosenberg (1972) in his treatise on technology and US economic growth. The machine tool industry in the nineteenth century played a unique role both in the initial solution of technical problems in user industries, such as textiles, and as the disseminator of these techniques to other industries, such as railroad locomotive manufacture. Rosenberg's description seems to suggest that the users played some role in the development of new equipment. He notes that before 1820 in the United States one could not identify a distinct set of firms who were specialists in the design and manufacture of machinery. Machines were either produced by users or by firms engaged in the production of metal or wooden products (pp. 98–9). Machinery-producing firms were thus first observed as adjuncts to textile factories. However, once established, these firms played an important role as the transmission center in the diffusion of new technology (Rosenberg, 1972, p. 102).

Horizontal Linkages

Successful innovation often requires horizontal as well as vertical co-operation. Horizontal linkages can help reduce spillover externalities and unnecessary duplication of research efforts and can also assist in the definition of technical standards for systemic innovation.

Horizontal linkages can help firms to overcome the appropriability (spillover) problems because the set of firms receiving the benefits is likely to include a greater portion of firms which have incurred R&D costs. The effect of greater appropriability is, of course, to encourage greater investment in new technology It is well understood that competitive markets tend to under-invest in new technology because those firms which support R&D have limited capacity to extract 'fees' from the imitators (free riders).

In addition, collaborative reduces, if not eliminates, what William Norris, CEO of Control Data Corporation, refers to as 'shameful and needless duplication of effort' (Davis, 1985, p. 42). Independent research activities often proceed down identical or near-identical technological paths. This is often wasteful and can be minimized if research plans are coordinated.[3]

This is not to imply that research activity in the United States has not traditionally displayed important co-operative elements. Von Hippel (1988, Ch. 6) has observed informal knowhow trading between engineers from different companies[4] who have common professional interests, and he describes the process as follows:

> In general such trading networks appear to be formed and refined as engineers get to know each other at professional conferences and elsewhere. In the course of such contracts, an engineer builds his personal informal list of possibly useful expert contacts by making private judgments as to the areas of expertise and abilities of those he meets. Later, when engineer A encounters a difficult product or process development problem, A activates his network by calling Engineer B—an appropriately knowledgeable contact who works at a competing (or non-competing) firm—for advice.

The manifold benefits from broad-scale collaboration activity have been recognized abroad, and more recently in the United States as well.[5] One assessment of the United States is that 'up until now, however, we have taken it for granted as an article of faith that no co-operation should be permitted, that it is best that we keep companies apart from one another' (Ouchi, 1984, p. 103). Meanwhile, other countries have adopted different models. For instance, Japanese co-operative activity is ubiquitous and is not only in the form of R&D joint ventures but also R&D collaboration. By the end of 1971 the entire Japanese computer industry (six firms) was paired in order to compete with IBM and its System 370 (Ouchi, 1984, p. 105). While the research was done in existing co-operative laboratories there was intense interaction and information sharing. Another celebrated example of Japanese collaboration was the VLSI (Very Large Scale Integrated Circuit) Research Association, which was an R&D joint venture formed in 1975 with the capital contributed by NEC, Toshiba, Hitachi, Mitsubishi and Fujitsu. At the successful conclusion of the project in 1979 the laboratory was dissolved and the scientists went back to their sponsoring companies.

Since the National Co-operative Research Act of 1984, which limits but does not eliminate antitrust risks, interest has been shown in R&D joint ventures in the United States. Well-known ventures include the Microelectronics and Computer Technology Corp. (MCC) in Austin, Texas; the Semiconductor Research Corporation (SRC) in Research Triangle Park, North Carolina; and Bell Communication Research (Bellcore). Collaboration under the Act is restricted at some undefined point as the product approaches commercialization. MCC's objective is to engage in advanced long-term research and development in computer architecture, semiconductor packaging and interconnect, software technology, VLSI and CAD. Its members include AMD, Boeing, Control Data, Harris, Motorola, Sperry and others. (IBM is conspicuously absent, as are Japanese and European-based firms).

The only form of co-operative research which receives special antitrust treatment (but not exemption) in the United States is the R&D joint venture. Other forms of collaborative R&D activity, such as the pooling of R&D projects and the sharing of development data, are subject to ordinary antitrust scrutiny. Co-operation using existing facilities is thus perceived to be exposed to serious antitrust risk, particularly if participating firms have significant market shares.

However, the successful commercialization of technology often requires collaboration among horizontal competitors that have different capabilities. For instance, the relevant manufacturing capacities need not be resident within the firm responsible for the other activities in the innovation process. In order to capture value from the innovation it may therefore be necessary for a number of firms to collaborate, with different firms being responsible for different activities (Teece, 1986). In some cases these firms may be horizontal competitors and antitrust may block desirable collaboration.

A further area in which horizontal linkages may be critical is where network technologies exist, as is currently the case with teletext and videotext. A careful reconsideration of US antitrust policy is clearly warranted. Subscribers will tend to wait in the hope that others will achieve the cost of achieving compatibility with them. Collaboration among horizontal competitors can then assist in the derivation and implementation of a network standard which would speed adoption of the network technologies.

ORGANIZATIONAL IMPLICATIONS: CO-OPERATION VERSUS INTEGRATION

The need for several kinds of linkages among different organizational units and groups raises basic questions about the organizational structures which should govern these linkages if they are to be developed and maintained in the most (dynamically) efficient way. All kinds of market failures can exist if these interactions are left to unassisted markets fettered by antitrust policies uninformed by the requirements of dynamic competition. (Teece, 1981, 1988). Vertical integration between R&D and production can help overcome a whole set of contracting problems which can be predicted to emerge in pure market transactions. Indeed, in a capitalist economy characterized by weak intellectual property protection and perfect competition, neoclassical markets and specialized (non-integrated) firms, product innovation could not be supported.[6] Indeed, at a highly abstract level a socialist economy should be expected to overcome many of the critical information and exchange problems implicit in the above analysis of interaction and linkages.

However, in real-world situations markets are characterized by a variety of supporting relationships and non-market exchanges. Trust relationships can develop, supported by norms of reciprocity. Generally, the smaller the amount of committed capital which is at stake, the looser the governance structure required. Equity ties can be expected in environments characterized by fewer potential partners and transaction-specific investments in physical and human capital.

Hence technological change itself, as well as changing competition, suggests that different organizational models may be appropriate for different innovation projects. Unfortunately, space constraints do not afford the opportunity for anything but a superficial treatment of the issue. As mentioned earlier, the lowering of the costs of digital electronics by complex silicon-integrated circuits has opened up new possibilities. Many new products, such as the Sun workstations, can be developed using off-the-shelf components and technologies. 'Ferraris' can be made out of spare parts, according to Sun's CEO, Scott McNealey. This reduces the sunk costs in certain development activities and suggests not only a parallel type development process but also one that need not involve high levels of organizational integration. Hence a high range of less hierarchical structures can be used to commercialize new technologies in certain industries at certain times. Collaboration with component suppliers, can thus substitute for integration in many important circumstances. When feasible, it can also lead to a reduction in development costs, by drawing on established competences.

In some instances equity linkages may not be feasible or desirable, as with linkages to the university. However, enduring relationships can and must be built using mechanisms which support stable and close relationships. De facto integration must be attained.

CONCLUSIONS

Innovation requires complex interactions and de facto integration among a multiplicity of organizational units. These interactions are vertical, horizontal and lateral and require special infrastructure to proceed smoothly and efficiently.

Capitalist economies have, over time, developed much of the organizational and incentive apparatus that is required, so that it is not surprising that innovation proceeds most efficiently in capitalist economies. However, advanced capitalist economies are not all identical, and a variety of models within and across economies exist by which to organize innovative activity.

The linkages discussed in this paper facilitate the process of coupling user needs to technological opportunities, and of commercializing new products and processes in a timely fashion. These linkages can either be intra-organizational (within firms) or inter-organizational (between and among firms). If attainable with non-hierarchical decentralized modes of organization, the former is generally preferred, as it increases the probability that innovators rather than follower-imitators will capture value from new technology. Nevertheless, collaboration among unaffiliated enterprises will often suffice, and is imperative in instances where firms conducting R&D must catch up, as where hierarchical internal structures would otherwise suffocate the innovation process.

Acknowledgements

The author is especially grateful to William Spencer of Xerox Corp. for valuable insights and comments. The intellectual debt to Richard Nelson, Nathan Rosenberg and Oliver Williamson will be readily apparent to the reader. Will Mitchell provided helpful assistance and comments.

NOTES

1. Note that user innovation requires two kinds of technology transfer: first from user to manufacturer, and then from manufacturer to the developer–user and other users.
2. Bendix and Bosch developed fuel injection and Motorola the alternator.
3. Needless to say, uncertainty often requires that multiple (but not identical) technological paths be pursued simultaneously. See, for example, Nelson (1984, Ch. 2).
4. The companies he observed were in the US steel minimill industry.
5. According to William Norris, US corporations were not willing to give collaborative research a try until 'these companies had the hell scared out of them by the Japanese' (Davis, 1985, p. 42).
6. With vertical integration, product innovations would be converted to process innovations and would not have to be revealed to others in order to be profitably commercialised.

REFERENCES

D. Davis (1985). R&D consortia. *High Technology* October.

S. J. Kline and N. Rosenberg (1986). An overview of innovation. In N. Rosenberg and R. Landau (eds), *The Positive Sum Strategy*, Washington, DC: National Academy Press.

D. Marquis and S. Meyers (1969) *Successful Industrial Innovations*, National Science Foundation, May.

R. R. Nelson (1984). *High Technology Policies*, Washington, DC: American Enterprise Institute.

R. R. Nelson (1988). The roles of firms in technical advance: a perspective from evolutionary theory. Unpublished manuscript, Columbia University, January.

OTA (1984). *Commercial Biotechnology: An International Analysis*, Washington, DC: Office of Technology Assessment, January.

L. Peters and H. Fusfeld (1982). Current U.S. university/industry research connections. *University/Industry Research Relationship*, National Science Foundation.

N. Rosenberg (1972). *Technology and American Economic Growth*, New York: Harper & Row.

D. Teece (1981). The market for know-how and the efficient international transfer of technology. *Annals of the American Academy of Political and Social Science* November, 81–96.

D. Teece (1986). Profiting from technological innovation: implications for integration, collaboration, licensing, and public policy. *Research Policy* **15**(6) December, 285–305.

D. Teece (1988). Technological change and the nature of the firm. In G. Dosi, C. Freeman, R. Nelson, G. Silverberg and L. Soete (eds), *Technical Change and Economic Theory*, (London: Pinter publishers).

E. Von Hippel (1977). The dominant role of the user in semiconductor and electronic subassembly process innovation *IEEE Transactions on Engineering Management* **EM-24**, 2, May.

E. Von Hippel (1988). *The Sources of Innovation*, New York: Oxford University Press.

STRATEGIC ALLIANCES AND INDUSTRIAL RESEARCH

David C. Mowery and David J. Teece

FOR THE PAST FIFTEEN YEARS, U.S. industrial research has been in the throes of a restructuring that has changed the position of industrially funded in-house research within the corporate innovation process. A number of central corporate research laboratories have undergone significant cutbacks or, in a few instances, have been eliminated entirely. Since 1980, as Nelson and Rosenberg's chapter in this book and numerous other studies have noted, U.S. firms have expanded their funding for and relationships with university-based research. In addition, numerous domestic and international intercorporate alliances that span R&D, manufacturing, and marketing have sprung up. This chapter considers the motives for and some implications of these trends, which reflect efforts by many U.S. firms to "externalize" a larger share of the industrially financed R&D that formerly was performed within their boundaries.

Some analyses (for example, the National Science Foundation's 1992 study) have speculated that the recent expansion in external research relationships has reduced growth in spending on in-house research and is responsible, at least in part, for declines in the rate of growth in industry-funded R&D expenditures. Although the growth in alliances and research consortia certainly has affected the role of in-house research, we believe that those who think that alliances and consortia can fulfill all of the functions of in-house R&D are mistaken.

The growth in strategic alliances in R&D is part of a broad restructuring of the U.S. national R&D system that involves change in the funding and functions of industry, universities, and government agencies. United States firms were among the pioneers in the development of in-house industrial research laboratories in the late nineteenth and

Reprinted with permission from *Engines of Innovation: U.S. Industrial Research at the End of an Era* (Harvard Business School Press, 1996), edited by Richard S. Rosenbloom and William J. Spencer, pp. 111–129.

early twentieth centuries. For decades, industrial research laboratories stood at the "heart of the system" of public and private institutions that financed and managed the creation, commercialization, and adoption of new technologies within the U.S. economy (Nelson 1991). Industry accounted for roughly 50 percent of the national R&D investment during much of the postwar period, but it was responsible for performing more than 70 percent of the nation's R&D (National Science Foundation 1992). Thus, today's growth in strategic alliances should be seen in its broader context of declining rates of growth in U.S. industry-financed R&D.[1]

Industrial research laboratories were first established within many large corporations as part of an effort to strengthen central, strategic functions within the firm; that is, to prevent long-range planning and investment decisions from being dominated by day-to-day operating concerns (Chandler 1962; 1977; 1990; see also Teece 1977; 1988). To be successful, however, R&D, like other operations of central corporate management, has to be integrated effectively with both day-to-day and long-range decision making. Many of the problems that have contributed to recent managerial skepticism of corporate research laboratories stem from the failure to integrate R&D strategies with corporate strategies in today's environment in which the demands for rapid response are far more compelling than they were in the beginning of the century.

By itself, outsourcing R&D does not address this failure. Instead, corporate strategists need to manage external and internal R&D activities as complementary activities within a coherent research program that links R&D and corporate strategy. In-house R&D can monitor, absorb, and exploit the results of research performed in research consortia and at other external sites, including universities. Corporate managers also must improve their management of technology transfer and absorption from joint development projects with erstwhile competitors or suppliers. Better management of these relationships can raise the returns to R&D investments.

In contrast to the arguments that portray knowledge transfer and exploitation as virtually costless (Teece 1988; Mowery 1983; Mowery and Rosenberg 1989; Cohen and Levinthal 1990), these arguments are built on a portrayal that emphasizes the costs and importance of managing the transfer and exploitation of scientific and technological information. Finally, as Teece, Pisano, and Shuen (1992) and Prahalad and Hamel (1990) have noted, there is a need for a dynamic view of the firm and the competencies or capabilities for the enduring competitiveness of firms. Although innovation is prominent among the sources of dynamic competitive advantage, the integration of R&D and firm strategy requires an understanding of the role of technology and a discriminating commitment to its support by senior corporate management.

EXTERNAL SOURCING OF R&D

In several respects, today's efforts by R&D managers to expand their links to external sources of new technology have revived an important function of corporate R&D laboratories during much of the period preceding 1940. Early research facilities of such firms as DuPont, Eastman Kodak, General Electric, and AT&T were expected to monitor technological advances occurring elsewhere within their industries and to advise senior management on the acquisition of technologies from other firms and independent inventors (Hounshell and Smith 1988; Reich 1985; Mueller 1962; and Jenkins 1975). In many cases, in-house R&D laboratories modified and commercialized patents or technologies acquired from external sources. In addition, as Nelson and Rosenberg's chapter has pointed out, a number of pre–World War II corporate research laboratories worked with researchers at U.S. universities

After World War II, however, the outward orientation of many large corporate research laboratories changed. Several factors influenced this shift in R&D strategy, which has been discussed most thoroughly by Hounshell and Smith (1988) for the DuPont Corporation. The wartime demonstration of the power of organized engineering and innovation and the postwar surge in federal R&D contracts led many large firms to develop or expand central research facilities that had weak links with operating divisions. Having been encouraged or requested to do so by federal funders of classified R&D projects, some corporations created autonomous central research "campuses." University researchers also benefited from the expansion in federal research funding and in doing so shifted their attention and fund-raising efforts away from industry (Swann 1988). The tough antitrust policy that emerged in the late 1930s under Thurman Arnold and remained in place through much of the 1970s also made a number of large firms reluctant to seek external sources for new technologies, which had been a key element of their R&D strategies.

Thus, the "golden years" of corporate research described in the introduction to this book were associated with the inward orientation of industrial R&D. While the growth of central corporate research may be associated with improvements in the basic research capabilities of many corporations, all too often, as Hounshell and Smith (1988), Graham (1986), and anecdotal histories of such facilities as the Xerox Corporation's Palo Alto Research Center (Uttal 1983) suggest, this research was not linked effectively to a corporate strategy for its exploitation. In our view, any reorganization or "externalization" of corporate research that does not include restructuring the relationship between corporate strategy and the firm's internal and external R&D investments will not improve innovative performance.

During the 1970s and 1980s, a series of events contributed to the decision of many firms to seek alternatives to exclusive reliance on in-house sources of expertise in the innovation and technology commercialization processes. First, the U.S. antitrust climate changed significantly during the 1980s, as illustrated by the National Cooperative Research Act of 1984 and the 1993 relaxation of federal antitrust restrictions on joint production ventures. Second, the costs of R&D, especially those associated with the development of new products, grew dramatically (in commercial aircraft, for example, new product development costs grew at an average annual rate of more than 10 percent throughout the postwar period), and became increasingly difficult for individual firms to shoulder in an economic environment characterized by high capital costs and intense competition from other domestic and foreign firms. Third, the recovery from the global political and economic upheaval that occurred between 1914 and 1945 meant that the capability to develop and commercialize new technologies had diffused throughout the world by the 1970s (Nelson 1991). Many of the U.S. firms that had dominated R&D and innovation in their industries during the 1950s and 1960s now faced more technologically sophisticated competitors, which increased the financial penalties associated with slow commercialization. Widespread distribution of the technological and nontechnological (marketing and manufacturing) assets needed to bring a new product to market meant that low-cost access to these complementary assets could be achieved most effectively through collaboration with other firms. Finally, scientific and technological advances increased the demands on firms to develop expertise in a wide array of technologies. Firms in food processing and pharmaceuticals, for example, confronted the challenges of biotechnology; telecommunications and computer technologies virtually merged; and advanced materials increased their importance in a broad range of manufacturing industries. Even the largest U.S. corporations, many of which also faced demands for improved financial performance, could not shoulder the costs of in-house development of expertise in an expanding array of strategic technologies. Thus, other firms, consortia, or universities offered possibilities for sharing the costs of developing the required new capabilities.

These forces and others that are specific to each type of collaborative activity have influenced the development of three broad forms of R&D collaboration during the past 15 years. These forms can be characterized as international strategic alliances, precommercial research consortia, and university-industry research collaborations. Collaborative ventures between U.S. and foreign firms now focus on activities, such as joint product development, that did not figure prominently in many of the international joint ventures of the 1950s and 1960s. Domestic consortia of U.S. firms, such as the Microelectronics and Computer Technology Corporation (MCC) have been organized during the past decade to carry out "precommercial" research. University-industry research collaborations now involve larger flows of funds and more U.S. firms.

Each of these forms of collaboration differs somewhat in activities, strategy, and goals. Therefore, the effects of each on in-house corporate research usually differ. International strategic alliances focus mainly on development, production, and marketing rather than precommercial research.[2] Thus far, most domestic collaborations among U.S. firms have concerned research that is not closely linked to specific commercial products. Despite the aspirations of several of them at the time of their founding, these collaborations have rarely focused on basic research. University-industry research collaborations appear to incorporate scientific and engineering research that extends downstream from basic research but typically is not specific to a single commercial product. Thus, two of the three forms of research collaboration do not concern development, the D of R&D that accounts for more than two-thirds of all U.S. R&D spending. In other words, a considerable portion of the research collaboration occurring in U.S. firms involves a relatively small share of their R&D investment.

International Strategic Alliances

International joint ventures have long been common in extractive industries such as mining and petroleum production (Stuckey 1983) and have accounted for a significant share of the foreign investment by U.S. manufacturing firms since World War II.[3] Recently, however, the number of collaborations (between U.S. and foreign enterprises) has grown.

Furthermore, such collaborations now appear in a wide range of industries.[4] The activities that are central to many of these recent ventures, including research, product development, and production for world markets, were absent from most of the ventures of the pre-1975 era, which focused primarily on production and marketing for the domestic market of the non-U.S. firm.

While these ventures are primarily responses to the rising costs and risks of unassisted product development, the growth of technological strengths within foreign firms, the prominent role of nontariff trade barriers in world markets, and government support for the development of advanced technologies, there are other important reasons for their creation. Many recent domestic and international alliances have been formed in the effort to create "bandwagons" behind a particular technical standard. For example, Matsushita's victory over Sony in the Beta-VHS competition in videocassette recorders, for example, owed much to the firm's success in gaining the allegiance of other Japanese and foreign firms to its VCR architecture (Cusumano et al., 1992; Grindley 1990).[5] On the other hand, U.S. firms have created international strategic alliances to improve their access to foreign markets, especially high-technology markets in which governments are engaged in managing trade flows.[6] The search for foreign capital and technological resources also has motivated U.S. firms to enter international joint ventures in industries ranging from semiconductors to steel.[7]

Firms engaged in international strategic alliances need to maintain a strong intrafirm competence in technologies related to the joint venture, for several reasons. Although the central activities of many international joint ventures usually are focused on the development and/or manufacture of specific products or technologies, they provide many opportunities for all parties to learn from their collaborators. According to some scholars (Prahalad and Hamel 1990), some firms, such as NEC of Japan, have developed technology-based core competences relatively inexpensively through their use of joint ventures as learning opportunities. Thus, focusing solely on the completion of the development project may limit opportunities for learning from joint venture partners. Moreover, because many joint development projects produce intellectual property by-products, it is important for firms to negotiate carefully the provisions governing the valuation, exploitation, and sharing of any revenues associated with these by-products.[8]

Exploiting learning opportunities requires an intrafirm capacity to absorb and apply the fruits of the venture. Joint venture participants should create mechanisms for absorbing technology transferred from their partners. The creation of these mechanisms often requires complementary in-house investments. One such investment is the rotation of research and engineering personnel from the firm through a collaborative project. However, it is not enough for an employee simply to capture knowledge (both codified and tacit) or skills from other firms. The employee must be given opportunities to communicate that knowledge to others within the parent firm. This can be done through parallel development and engineering activities within the parent firm.

Because international joint ventures act as vehicles for technology transfer and skills acquisition, the value of the knowledge or capabilities contributed by any single partner depreciates, *ceteris paribus*. As technology is transferred through a collaborative venture, learning by the other participants reduces the value of the technological capabilities that were originally unique to one or another participant. Depreciation may be even more rapid in ventures in which one firm contributes its marketing knowledge and network or other "country-specific" expertise. Although an alliance may be an essential means to gain access to new knowledge, as the other participants improve their knowledge of the markets in which this partner has specialized, they are likely to choose to continue without it.[9] This fact has played a role in the breakup of a number of collaborative ventures between Japanese and U.S. producers of auto parts. As the Japanese partners in these ventures have gained knowledge about local markets and production conditions (particularly when selling to Japanese transplant operations in the United States), they have withdrawn from the joint venture to continue independently (Phillips 1989). Although technology-based assets are likely to depreciate more slowly, especially if technology transfer is closely managed, Hamel, Doz, and Prahalad (1989) suggest that process technologies are less easily exploited by other participants than product technologies, which are more transparent to venture partners.[10]

Depreciation in the value of assets within a joint venture is no less inevitable than depreciation of physical capital assets within a manufacturing plant. In both cases, participants must take steps to reduce erosion in the value of their contribution and, at the same time, seek ways to offset the consequences of such depreciation. Intrafirm investments in technology development are essential to the creation or maintenance of the quality of the technological assets contributed to the joint venture. Therefore, participant firms must sustain in-house technology development activities in product lines and technologies that are related to the joint venture and managers need to pursue ways to offset the depreciation that will occur, for example, by exploiting learning opportunities. If the collaborative venture aids in the establishment of a product design as a market standard or a venture with an established firm provides an endorsement of a technology, some of the detrimental effects of this depreciation also can be reduced. When a firm provides a static asset like market access, the collaborative venture may function most effectively as a means for exiting the industry or as a temporary channel for learning process and product technologies.

For many U.S. firms, joint ventures involve closer work with suppliers. In these user-supplier ventures, more responsibility for the development of components to meet specific performance parameters is delegated to the supplier and the risks and costs of development are shared. In the semiconductor industry, these ventures often team U.S. and Japanese firms and generate considerable product-specific and technology-specific know-how and intellectual property. Shuen's research (1993) suggests that the failure of some U.S. semiconductor firms to invest sufficient resources in monitoring and absorbing jointly developed intellectual property has reduced their returns from these relationships. It would appear that U.S. managers need to broaden the channels through which they obtain technologies from external sources. Know-how and technological capabilities do not come exclusively from formal "horizontal" joint ventures. They also flow from marketing, supplier, and numerous other relationships. Moreover, as firms come to rely more heavily on these relationships, they need better channels for transfer and absorption.

In addition to improving market access, reducing risk, and lowering the costs and time required for new product development, international joint ventures in product development can raise the efficiency of a firm's internal R&D. These ventures allow firms to exercise greater selectivity in their in-house technology investments. However, selectivity must be based on a careful analysis of the firm's strengths and weaknesses, and the long-run impact of reducing investment in specific technologies on corporate strategy must be clearly laid out. In other words, entry into an international joint venture should be based on an integrated analysis of technological and firm strategy. Such an assessment should include some evaluation of the competitive sensitivity of specific parts—the firm's "crown jewels"—of its technology portfolio. The uncertainties associated with technology-based competition mean that any such analysis is likely to rely more heavily on the construction and evaluation of scenarios than on the illusory precision of quantitative estimates. Moreover, the firm pursuing this selective approach still will require in-house expertise to evaluate the strengths and weaknesses in the technologies of its prospective collaborators.

Precommercial Research Consortia

Research consortia, funded entirely or in part by industry funds and focusing on precommercial research activities, are a recent innovation in U.S. firms. In most cases, these consortia have involved U.S. firms only (some consortia that are funded in part from public sources, for example, SEMATECH, have formal policies excluding foreign firms). More than 450 such ventures have been registered with the U.S. Justice Department through 1994 under the terms of the 1984 National Cooperative Research Act (Evan and Olk 1990; Werner 1992; Link 1995).

Several of the most widely publicized consortia founded during the first half of the 1980s, such as MCC, were established in response to Japanese cooperative research programs (Peck 1986), particularly the VLSI program of the 1970s and the Fifth Generation Computing project that was undertaken during the 1980s as a successor to the VLSI program. Computer industry executives in the United States concluded that Japanese cooperative programs supported the type of long-range research that no single firm would undertake; projects like MCC were created to fill this void in the U.S. computer industry.

The short history of MCC and the experiences of consortia in other industries (for example, the Electric Power Research Institute, which serves the U.S. electric utility industry) suggest that research consortia rarely sustain a long-range focus but rather shift their focus to research on generic technology issues of more immediate interest to member firms. Both SEMATECH and the National Center for Manufacturing Sciences are now pursuing technology-focused research that seeks to improve vertical relationships between users and suppliers of capital equipment (Grindley et al. 1994). Interestingly, the Japanese industry consortia that sparked U.S. concern have rarely undertaken basic research. Instead, they have focused on technology development and dissemination among their members. Japan's Fifth Generation Computing project, which pursued a longer-term research agenda, has been relatively unsuccessful.

Like international strategic alliances, most research consortia focus on technology development. In contrast to alliances, however, agendas of consortia usually are not highly product-specific. Indeed, the example of SEMATECH suggests that agreement among consortium members on an agenda that focuses on specific (and often proprietary) product or process technologies may be difficult if not impossible to achieve (Grindley et al. 1994). For this reason, the near-term competitive risks from participating in a consortium may be less significant than the risks associated with an international joint venture. The financial costs of consortium participation also are likely to be lower than the costs of a joint venture. Therefore, decisions on consortium participation, management, and so forth may raise fewer long-term corporate strategy issues and do not require the same degree of senior management participation as decisions on international joint ventures.

Although there are a number of important differences in their goals and structure, the requirements for maximizing technology-based benefits from consortia closely resemble the requirements associated with strategic alliances. Complementary investments in R&D within member firms, the creation of channels of communication and technology transfer with the consortium, and the development of an in-house receptor are necessary to increase the returns from participation. For example, MCC's reliance on its own research staff, in contrast to SEMATECH's use of assignees from member firms, made it difficult for member firms to absorb the results of MCC research. The complex structure of MCC, which established barriers to some firms' access to certain research areas, further impeded technology transfer.

University-Industry Research Collaboration

Much of the recent expansion in domestic research collaboration involves a renewal of the link among state governments, publicly supported universities, and industry that languished during the post-1945 period (Mowery and Rosenberg 1993). The huge size, decentralized structure, and research-intensive character of the American universities are unique and increase the potential payoff from collaboration between universities and industry. Nevertheless, clearly defined "deliverables" often are of secondary importance in successful university-industry collaborations (Mowery and Rosenberg 1989; Rosenberg and Nelson 1994).

Thus far, university-industry collaborations appear better suited to the support of long-term, precommercial research than interfirm consortia. This tentative conclusion is based on the tendency of consortia to shift their agenda toward near-term research. The personnel flows between firm and research laboratory that often figure prominently in university-industry collaborations also aid communication between university and corporate research establishments.

Because U.S. universities include education as well as research in their activities, they are important sources of scientific and engineering personnel for industry. Firms can use collaborative ventures as filters for hiring research personnel, since the ventures allow them to observe the performance of potential researchers before making employment commitments. Moreover, the hiring of graduates of these programs facilitates the transfer of knowledge and technology even more effectively than does the rotation of industry personnel through university research facilities. Given the interdisciplinary character of current industrial technological and research challenges, the training of research personnel is an especially important benefit to industry that may emerge from industry-university collaborations. Firms in the semiconductor, biotechnology, or robotics industries now need individuals with interdisciplinary research training. Industrial funding, like federal government research support during the 1960s and 1970s, can aid in the establishment of university interdisciplinary research and education programs, which are notoriously difficult to develop without external funds.[11]

Thus, through interdisciplinary research and education programs, collaboration with universities can provide firms with "windows" for monitoring novel research areas and filters for hiring research and technical personnel. While the results of university-industry research collaborations may rarely be applied directly to commercial innovation, by improving access to university research, they can improve the efficiency of in-house research activities. As Nelson and Levin (1986) and David, Mowery, and Steinmueller (1992) have argued, many of the economic benefits of university research and other basic research are realized through the ability of the research findings to improve the efficiency of applied research; that is, basic research results lead to a better informed, and therefore more efficient, "search" process in technology development. It is the general knowledge produced by this research, rather than any specific discoveries, that provides many of the economic benefits.[12]

What does this characterization of the economic payoffs from these collaborations imply for managers who wish to increase the competitive benefits arising from such relationships? As in the case of strategic alliances and research consortia, the creation and maintenance of good channels of communication and transfer are critical and require both the hiring of graduates and the rotation of firm personnel through university research facilities. Managers also must maintain "receptors" within the firm to absorb and apply university research findings to technology development. One of the few empirical studies of the role of external basic research in innovative performance found that pharmaceuticals firms with strong intrafirm "academic," or basic, research capabilities more successfully exploited such research than firms lacking these capabilities (Gambardella 1992). In other words, university-based research must complement in-house research activities. Without some capability to understand and exploit the results produced in collaborative research relationships, the returns to these external investments are likely to be low. University research collaborations may allow for greater selectivity in the in-house basic research agenda, but they cannot be effective without complementary in-house research activities.

CONCLUSION

This chapter has provided a taxonomy for understanding the external research relationships that have recently sprung up within U.S. industrial research and has suggested some ways in which the management of these relationships influences in-house research activities and innovative performance. Although some recent experiments in strategic alliances represent a revival of the earlier, outward-oriented R&D strategy followed by many of the pioneers in U.S. industrial research, the outcomes of these new undertakings remain uncertain. The restructuring of U.S. corporate research is likely to continue for some time because of severe competitive and financial pressures on U.S. firms.

In order to utilize research collaboration as an effective competitive solution, managers must define the problem they are addressing. Undertaking external R&D relationships primarily or solely as a means of reducing the corporate R&D budgets, for example, may do little to improve the long-term returns to corporate R&D investments. The disappointing returns to many R&D investments reflect a frequent failure to maintain links between R&D priorities and corporate strategy rather than excessive levels of R&D investment. This problem is not addressed by external R&D, and without a solution to it, external R&D

may well fail. Research collaboration provides opportunities for R&D cost reduction or improved market access, but entry into such ventures solely for these purposes is ill-advised. Collaborative research ventures should be undertaken and managed for their potential to strengthen the capabilities that underpin firms' competitive strength.

While external research ventures can support learning from other firms or research institutions and a more focused in-house R&D agenda and budget, to do so successfully corporate strategy and R&D priorities must be well integrated. The failure to integrate may then prove to be counterproductive causing R&D collaboration to erode, rather than strengthen, corporate competitive advantage. To improve the payoff from external research ventures, a firm must invest in activities that facilitate the inward transfer of knowledge and technology, but this is insufficient by itself. Complementary investments in intrafirm R&D are also necessary to provide opportunities for the exploitation and "absorption" of the fruits of external research. As noted earlier, managing these external research relationships as complements to an in-house research portfolio can facilitate a more efficient allocation of intrafirm R&D investments among technologies or strategic opportunities.

Successful management of external research relationships also requires a good fit between the type of external research venture and corporate or business unit goals. Using a university-industry collaboration to accelerate the development of a product or using a precommercial research consortium to strengthen basic research capabilities is likely to prove disappointing. Furthermore, the failure to recognize that these activities may yield multiple types of benefits can reduce their payoffs. Joint product development ventures often yield significant learning by-products and may develop intellectual property in related areas or technologies. Capturing these benefits and managing their exploitation requires careful consideration of different approaches to organizing and negotiating the terms of a venture.

As noted earlier, the focus of many of these collaborations on the lower-cost phases of industrial innovation means that their effects on in-house R&D spending may have been overstated. Moreover, as noted repeatedly, the successful exploitation of external R&D requires complementary in-house R&D investments. In light of these conclusions, recent flat trends in industrially funded R&D spending may not reflect improvements in efficiency or productivity as a result of collaborations. Indeed, policies that seek to improve industrial competitiveness by encouraging externally based research networks (for example, tax incentives or direct subsidies) may not offset the effects of declining intrafirm R&D if the impact of these collaborations on innovative performance depends on intrafirm R&D investment. These policies also cannot address performance problems that reflect deficiencies in the integration of corporate strategy and R&D management and priorities.

Notes

1. Real annual growth rates in industry-financed R&D spending have declined since 1984. Rates reached zero in 1986–87 and in 1990–91 (National Science Foundation 1992). The share of GDP accounted for by industry funded R&D in the United States has lagged behind the share of GDP in both Germany and Japan by a widening margin during the past decade.
2. A number of regional programs in Western Europe, such as ESPRIT and EUREKA, focus on precommercial research. For purposes of this discussion, however, these consortia within an economically unified region are treated as similar to domestic research consortia.
3. Indeed, although their relative importance has declined, the absolute size and number of joint ventures in the extractive industries remain substantial and may have increased. Karen Hladik's analysis (1985) of data from the Harvard Multinational Enterprise Project concluded that 39 percent of the number of foreign subsidiaries established by U.S. manufacturing firms from 1951 to 1975 were joint ventures. Benjamin Gomes-Casseres (1988) analyzed these data and found a significant decline in the share of joint ventures within U.S. multinationals' international subsidiaries from 1961 to 1968, followed by a resumption of growth in the share of joint ventures from 1969 to 1975.
4. Kathryn Harrigan (1984) concluded that domestic joint ventures involving U.S. firms had grown during the previous decade. In the 1960s, joint ventures were concentrated in the chemicals, primary metals, paper, and stone, clay, and glass industries, but they now extend beyond these sectors. Karen Hladik (1985) found significant growth from 1975 to 1982 in the number of international joint ventures involving U.S. firms. This trend has almost certainly continued through the present.
5. These alliances are not always successful, however. The Advanced Computer Environment (ACE) failed to establish the RISC microprocessor architecture developed by MIPS Computer Systems as a standard. See Jonathan Khazam and David C. Mowery (1994).
6. U.S. firms are not the only ones to use joint ventures in this manner. Foreign firms have found that joint ventures can improve their access to U.S. markets.
7. Historically, in industries such as telecommunications, technical standards have been established through multilateral or plurilateral negotiations

among governments. These negotiations are heavily affected by governments' perceptions of the effects of a particular standard on the competitive fortunes of "national champions," many of which are government-owned or controlled. By establishing a network of international alliances, U.S. telecommunications firms have sought to gain advantage in negotiations over government-sponsored technical standards. See David C. Mowery (1989).

8. David J. Teece (1992) argues that the importance of such by-products and the development of provisions for their exploitation may favor the use of a shared-equity ownership structure for joint ventures.

9. Michael E. Porter and Mark B. Fuller (1986) have observed that collaborative ventures centered on marketing "may be particularly unstable, however, because they frequently are formed because of the access motive on one or both sides. For example, one partner needs market access while the other needs access to product. As the foreign partner's market knowledge increases, there is less and less need for a local partner" (p. 334).

10. "The type of skill a company contributes is an important factor in how easily its partner can internalize the skills. The potential for transfer is greatest when a partner's contribution is easily transported (in engineering drawings, on computer tapes, or in the heads of a few technical experts); easily interpreted (it can be reduced to commonly understood equations or symbols); and easily absorbed (the skill or competence is independent of any particular cultural context). . . . Western companies face an inherent disadvantage because their skills are generally more vulnerable to transfer. The magnet that attracts so many companies to alliances with Asian competitors is their manufacturing excellence—a competence that is less transferable than most" (Gary Hamel, Yves Doz, and C. K. Prahalad 1989, 136). The converse is also true. A central technological asset contributed by Boeing to its collaborative ventures with Japanese firms is its expertise in production technology and in the management of fluctuations in production volume for commercial airframes (Mowery 1987).

11. The departmental structure of U.S. universities appears to make this task easier than it is in many Western European universities: "Among the factors cited to explain West Germany's slow entry into commercial biotechnology is an educational system that prevents the kind of interdisciplinary cooperation that is viewed by most experts as essential to the development of this field. In particular, the traditional separation of technical faculties from their arts and sciences counterparts means that process technicians, usually located in the technical schools, rarely come into contact with colleagues holding university appointments in biochemistry or microbiology" (U.S. House of Representatives, Office of Technology Assessment 1985, 424).

12. D. M. Hercules and J. W. Enyart (1983, 7) report that the following four areas of collaboration had very high potential payoffs and currently lacked sufficient activity: (1) lectureships by academic scientists at industrial sites; (2) student interns at industrial sites; (3) continuing education programs at

industrial sites; and (4) corporate support for employees to obtain advanced degrees. Note that none of these areas involve significant transfers to industry of intellectual property or other deliverables.

References

Chandler, Alfred D., Jr. 1962. *Strategy and Structure.* Cambridge, Mass.: MIT Press.

Clark, Kim B. 1989. "What Technology Can Do for Strategy." *Harvard Business Review* (November–December): 94–8.

Cohen, Wesley M., and David A. Levinthal. 1990. "Absorptive Capacity: A New Perspective on Learning and Innovation." *Administrative Sciences Quarterly* 35: 128–52.

Cohen, Wesley, Richard Florida, and Richard Goe. 1994. *University-Industry Research Centers in the United States.* Pittsburgh: Carnegie Mellon University.

Cusumano, Michael, Yorgis Mylonadis, and Richard Rosenbloom. 1992. "Strategic Maneuvering and Mass-Market Dynamics: The Triumph of Beta over VHS." *Business History Review* 66: 51–94.

David, Paul A., David C. Mowery, and W. Edward Steinmueller, "Analyzing the Economic Payoffs to Basic Research." *Economics of Innovation and New Technology* 2: 73–90.

Evan, William M., and Paul Olk. 1990. "R&D Consortia: A New Organizational Form." *Sloan Management Review* 31 (Spring): 37–46.

Feller, I. 1990. "Universities as Engines of R&D-Based Economic Growth: They Think They Can." *Research Policy* 19: 335–48.

Gambardella, A. 1992. "Competitive Advantages from In-House Scientific Research: The U.S. Pharmaceutical Industry in the 1980s." *Research Policy* 21: 391–407.

Ghemawat, P., M. E. Porter, and R. A. Rawlinson. 1986. "Patterns of International Coalition Activity." In *Competition in Global Industries* edited by Michael E. Porter. Boston, Mass.: Harvard Business School Press.

Gomes-Casseres, Benjamin. 1988. "Joint Venture Cycles: The Evolution of Ownership Strategies of U.S. MNEs, 1945–75." In *Cooperative Strategies in International Business,* edited by F. J. Contractor and P. Lorange. Lexington, Mass.: Lexington Books.

Graham, Margaret B. W. 1986. *RCA and the Videodisc: The Business of Research.* New York: Cambridge University Press.

Graham, Margaret B. W., and Bettye H. Pruitt. 1990. *R&D for Industry: A Century of Technical Innovation at Alcoa.* New York: Cambridge University Press.

Grindley, P. 1990. "Winning Standards Contests: Using Product Standards in Business Strategy." *Business Strategy Review* 1 (Spring): 71–84.

Grindley, P., David C. Mowery, and B. Silverman. 1994. "Sematech and Collaborative Research: Lessons in the Design of High-Technology Consortia." *Journal of Policy Analysis and Management* 13: 723–58.

Hamel, Gary, Yves Doz, and C. K. Prahalad. 1989. "Collaborate with Your Competitors—and Win." *Harvard Business Review* (January–February): 133–39.

Harrigan, Kathryn R. 1984. "Joint Ventures and Competitive Strategy." Working paper, Graduate School of Business, Columbia University, New York.

Hercules, D. M., and J. W. Enyart. 1983. "Report on the Questionnaire on Current Exchange Programs Between Industries and Universities." Council on Chemical Research, University-Industry Interaction Committee.

Hladik, Karen. 1985. *International Joint Ventures.* Lexington, Mass.: D.C. Heath.

Hounshell, David A., and John Kenly Smith. 1988. *Science and Corporate Strategy: Du Pont R&D, 1902–1980.* New York: Cambridge University Press.

Jenkins, Reese V. 1975. *Images and Enterprise: Technology and the American Photographic Industry, 1839–1925.* Baltimore: The Johns Hopkins University Press.

Khazam, J., and D. C. Mowery. 1994. "The Commercialization of RISC: Strategies for the Creation of Dominant Designs." *Research Policy* 23: 89–102.

Link, A. N. 1995. "Research Joint Ventures: Patterns from *Federal Register* Filings." Economics working paper, Center for Applied Research, Bryan School of Business and Economics, University of North Carolina-Greensboro.

Mowery, David C. 1983. "Economic Theory and Government Technology Policy." *Policy Sciences* 16, no. 2: 27–43.

———. 1987. *Alliance Politics and Economics: Multinational Joint Ventures in Commercial Aircraft.* Cambridge, Mass.: Ballinger.

Mowery, David C., and N. Rosenberg. 1989. *Technology and the Pursuit of Economic Growth.* New York: Cambridge University Press.

———. 1993. "The U.S. National Innovation System." In *National Innovation Systems: A Comparative Analysis,* edited by Richard R. Nelson. New York: Oxford University Press.

Mueller, W. F. 1962. "The Origins of the Basic Inventions Underlying Du Pont's Major Product and Process Inventions, 1920 to 1950." In *The Rate and Direction of Inventive Activity,* edited by Richard R. Nelson. Princeton, N.J.: Princeton University Press.

National Science Foundation. National Science Board. 1992. Committee on Industrial Support for R&D. *The Competitive Strength of U.S. Industrial*

Science and Technology: Strategic Issues. Washington, D.C.: U.S. Government Printing Office.

Nelson, Richard R. 1990. "U.S. Technological Leadership: Where Did It Come From and Where Did It Go?" *Research Policy* 19: 117–132.

———. 1991. "Capitalism as an Engine of Progress." *Research Policy* 20: 193–214.

Nelson, Richard R., and Richard C. Levin. 1986. "The Influence of Science University Research and Technical Societies on Industrial R&D and Technical Advance," Research Program on Technological Change Policy discussion paper 3, Yale University, New Haven, Conn.

New York Times. 1992. "University of California Proposes Laboratory-to-Marketplace Link," 11 December, A14.

Pavitt, Keith. 1991. "What Makes Basic Research Economically Useful?" *Research Policy* 20: 109–19.

Peck, Merton J. 1986. "Joint R&D: The Case of the Microelectronics and Computer Technology Corporation." *Research Policy* 15: 219–32.

Phillips, S. 1989. "When U.S. Joint Ventures with Japan Go Sour." *Business Week* (24 July): 30–31.

Porter, Michael E., and Mark B. Fuller. 1986. "Coalitions and Global Strategy." In *Competition in Global Industries*, edited by M. E. Porter, Boston: Harvard Business School Press.

Prahalad, C. K., and Gary Hamel. 1990. "The Core Competence of the Corporation." "Harvard Business Review" (May–June): 79–91.

Reich, Leonard S. 1985. *The Making of Industrial Research: Science and Business at GE and Bell, 1876–1926*. New York: Cambridge University Press.

Rosenberg, Nathan. 1990. "Why Do Firms Do Basic Research (With Their Own Money)?" *Research Policy* 19: 165–74.

Rosenberg, Nathan, and Richard R. Nelson. 1994. "American Universities and Technical Advance in Industry." *Research Policy* 23: 323–48.

Servos, John W. 1994. "Changing Partners: The Mellon Institute, Private Industry, and the Federal Patron." *Technology and Culture* 35 (April): 221–57.

Shuen, Amy S. 1993. "Co-Developed Know-how Assets in Technology Partnerships." Haas School of Business, University of California, Berkeley.

Stuckey, J. S. 1983. *Vertical Integration and Joint Ventures in the Aluminum Industry*. Cambridge, Mass.: Harvard University Press.

Swann, John P. 1988. *Academic Scientists and the Pharmaceutical Industry: Cooperative Research in Twentieth-Century America*. Baltimore: The Johns Hopkins University Press.

Teece, David J. 1977. *The Multinational Corporation and the Costs of International Technology Transfer*. Cambridge, Mass.: Ballinger.

———. 1988. "Technological Change and the Nature of the Firm." In *Technical Change and Economic Theory,* edited by G. Dosi, C. Freeman, R. Nelson, G. Silverberg, and L. Soete. London: Frances Pinter.

———. 1992. "Competition, Cooperation, and Innovation: Organizational Arrangements for Regimes of Rapid Technological Progress," *Journal of Economic Behavior and Organization* 18, no. 1: 1–25.

Teece, David J., Gary Pisano, and Amy Shuen. 1992. "Dynamic Capabilities and Strategic Management." Working paper, Haas School of Business, University of California, Berkeley.

U.S. House of Representatives. 1985. Office of Technology Assessment. *Commercial Biotechnology: An International Analysis* (Washington, D.C.: U.S. Government Printing Office, 1985).

Uttal, Bro. 1983. "The Lab that Ran Away from Xerox." *Fortune* 108 (5 September): 97–102.

Wall Street Journal. 1991. "U.S.'s DNA Patent Moves Upset Industry," 22 October, B4.

Werner, Jerry. 1992. "Technology Transfer in Consortia." *Research-Technology Management* 35, no. 3: 38–43.

Organizing for Innovation: **When Is Virtual Virtuous?**

by Henry W. Chesbrough and David J. Teece

Advances in information technology have made it easier for companies to exchange data and coordinate activities. That has given rise to a radical new vision of corporate organization–one in which individual companies outsource many of their activities to an array of partners. Such virtual enterprises may be more efficient, but what are the broader strategic implications of rampant subcontracting?

Henry Chesbrough and David Teece sound a note of caution. When it comes to innovation, they argue, virtuality often does more harm than good. Loose partnerships of companies inevitably produce more conflicts of interest than do centrally managed corporations, and those conflicts can hamper the kind of complex, systematic innovation that creates valuable business breakthroughs. Innovation is a destabilizing force and will therefore be resisted by companies wary of upsetting a comfortable status quo.

Chesbrough and Teece acknowledge that some degree of outsourcing can further corporate creativity and that virtuality makes sense under certain conditions. But every company, they contend, needs to tailor its organization to its own operations and its unique sources of innovation. Blindly following fads is a recipe for disaster.

Outsourcing may make you more flexible. But it could also undermine your ability to innovate.

CHAMPIONS of virtual corporations are urging managers to subcontract anything and everything. All over the world, companies are jumping on the bandwagon–decentralizing, downsizing, and forging alliances to pursue innovation. Why is the idea of the virtual organization so tantalizing? Because we have come to believe that bureaucracy is bad and flexibility is good. And so it follows that a company that invests in as little as possible will be more responsive to a changing marketplace and more likely to attain global competitive advantage.

There is no question that many large and cumbersome organizations have been outperformed by smaller "networked" competitors. Consider, for instance, the eclipse of IBM in PCs and of DEC in workstations. But while there are many successful virtual companies, there are even more failures that don't make the headlines. After many years of studying the relationship between organization and innovation, we believe that the virtues of being virtual have been oversold. The new conventional wisdom ignores the distinctive role that large integrated companies can play in the innovation process. Those rushing to form alliances instead of nurturing and guarding their own capabilities may be risking their future.

What's Special About Virtual?

What gives the virtual company its advantage? In essence, incentives and responsiveness. Virtual companies coordinate much of their business through the marketplace, where free agents come together to buy and sell one another's goods and services; thus virtual companies can harness the power of market forces to develop, manufacture, market, distribute, and support their offerings in ways that fully integrated companies can't duplicate. As William Joy, vice president of research and development at Sun Microsystems, puts it, "Not all the smart people [in the workstation industry] work for Sun." Because an outside developer of workstation software can obtain greater rewards by selling software to Sun customers than by developing the same software as a Sun employee, he or she will move faster, work harder, and take more risks. Using high-powered, market-based incentives such as stock options and attractive bonuses, a virtual company can quickly access the technical resources it needs, if those resources are available. In situations where technology is changing rapidly, large companies that attempt to do everything inside will flounder when competing against small companies with highly trained and motivated employees.

What gives the virtual company its advantage? In essence, incentives and responsiveness.

But the incentives that make a virtual company powerful also leave it vulnerable. As incentives become greater and risk taking increases, coordination among parties through the marketplace becomes more and more difficult, precisely because so much personal reward is at stake. Each party to joint development activity necessarily acts in its own self-interest. Over time, innovation can generate surprises that work to the advantage of some parties and to the disadvantage of others. The result: Once friendly partners may be unwilling or unable to align strategically, and coordinated development activity falters. In contrast, integrated, centralized companies do not generally reward people for taking risks, but they do have established processes for settling conflicts and coordinating all the activities that are necessary for innovation.

This trade-off between incentives and control lies at the heart of the decision that managers must make about how to organize for innovation. (See the exhibit "Finding the Right Degree of Centralization.") If virtual organizations and integrated companies are at opposite ends of the spectrum, alliances occupy a kind of organizational middle ground. An alliance can achieve some of the coordination of an integrated company but, like players in a virtual network, the members of an alliance will be driven to enhance their own positions, and over time their interests may diverge. The challenge for managers is to choose the organizational form that best matches the type of innovation they are pursuing.

Types of Innovation

When should companies organize for innovation by using decentralized (or virtual) approaches, and when should they rely on internal organization? The answer depends on the innovation in question.

Some innovations are autonomous – that is, they can be pursued independently from other innovations. A new turbocharger to increase horsepower in an automobile engine, for example, can be developed without a complete redesign of the engine or the rest of the car. In contrast, some innovations are fundamentally systemic – that is, their benefits can be realized only in conjunction with related, complementary innovations. To profit from instant photography, Polaroid needed to develop both new film technology and new camera technology. Similarly, lean manufacturing is a systemic innovation because it requires interrelated changes in product design, supplier management, information technology, and so on.

The distinction between autonomous and systemic innovation is fundamental to the choice of organizational design. When innovation is autonomous, the decentralized virtual organization can manage the development and commercialization tasks quite well. When innovation is systemic, members of a virtual organization are dependent on the other members, over whom they have no control. In either case, the wrong organizational choice can be costly.

Consider what happened to General Motors when the automobile industry shifted from drum brakes to disc brakes, an autonomous innovation. GM was slow to adopt disc brakes because it had integrated vertically in the production of the old technology. GM's more decentralized competitors relied instead on market relationships with their suppliers – and the high-powered incentives inherent in those relationships. As a re-

Henry W. Chesbrough is an assistant professor of business administration and the Class of 1961 Fellow at Harvard Business School in Boston. This is the first of three HBR articles on innovation he has coauthored. He can be reached at henry@chesbrough.com. David J. Teece is the Mitsubishi Bank Professor of International Business and Finance and the director of the Institute of Management, Innovation, and Organization at the Haas School of Business at the University of California, Berkeley.

sult, they were able to beat GM to market with the new disc brakes, which car buyers wanted. When companies inappropriately use centralized approaches to manage autonomous innovations, as GM did in this case, small companies and more decentralized large companies will usually outperform them.

To understand why the two types of innovation call for different organizational strategies, consider the information flow essential to innovation. Information about new technologies and products often develops over time as managers absorb new research findings, the results of early product experiments, and initial customer feedback. To commercialize an innovation profitably, a tremendous amount of knowledge from industry players, from customers, and sometimes from scientists must be gathered and understood. This task is easier if the information is codified.

Codified information – for example, specifications that are captured in industry standards and design rules – can often be transferred almost as effectively from one company to another as it can within a single company. Because such information is easily duplicated, it has little natural protection. Sometimes, bits and pieces can be protected by intellectual property rights, but those pieces, especially trade secrets and patents, are small islands in a broad ocean of knowledge.

Other information does not travel as easily between companies. Tacit knowledge is knowledge that is implicitly grasped or used but has not been fully articulated, such as the know-how of a master craftsman or the ingrained perspectives of a specific company or work unit. Because such knowledge is deeply embedded in individuals or companies, it tends to diffuse slowly and only with effort and the transfer of people. Established companies can protect the tacit knowledge they hold, sharing only codified information. They can be quite strategic about what they disclose and when they disclose it.

The information needed to integrate an autonomous innovation with existing technologies is usually well understood and may even be codified in industry standards. Systemic innovations, on the other hand, pose a unique set of management challenges regarding information exchange. By their very nature, systemic innovations require information sharing and coordinated adjustment throughout an entire product system. Here is where a market-based, virtual approach to innovation poses serious strategic hazards. Unaffiliated companies linked through arm's-length contracts often cannot achieve sufficient coordination. Each company wants the other to do more, while each is also looking for ways to realize the most gain from the innovation. Information sharing can be reduced or biased, as each seeks to get the most at the other's expense. In most cases, the open exchange of information that fuels systemic innovation will be easier and safer within a company than across company boundaries. The inevitable conflicts and choices that arise as a systemic innovation develops can best be resolved by an integrated company's internal management processes.

The Case of Industry Standards

Coordinating a systemic innovation is particularly difficult when industry standards do not exist and must be pioneered. In such instances, virtual organizations are likely to run into strategic problems. Consider how technical standards emerge. Market participants weigh many competing technologies and eventually rally around one of them. There are winners and losers among the contestants, and potential losers can try to undermine the front-runner or to fragment the standard by promoting a rival. Until a clear winner emerges, customers may sit on the sidelines rather than risk making the wrong choice.

Finding the Right Degree of Centralization

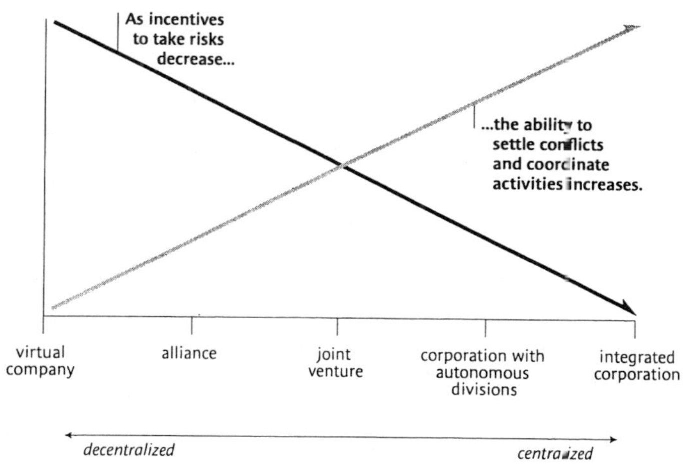

By virtue of its size and scope, an integrated company may be able to advance a new standard simply by choosing to adopt a particular technology. If a large company commits itself to one of a host of competing technologies, consumers as well as companies promoting rival technologies will probably be persuaded to follow suit. Virtual companies, however, which may be struggling to resolve conflicts within their networks, won't be able to break a deadlock in a complicated standards battle. Players in a network won't be able to coordinate themselves to act like a large company.

Once a standard has been established, virtual organizations can manage further innovation quite well. But when an industry begins to advance technology to a new level, the cycle can begin anew. Again, technically feasible choices present new strategic trade-offs. Suppliers, competitors, and customers may fail to agree on a common path. Unless a big player emerges to break the logjam among rival technologies, the existing standard will prevail long past its usefulness.

Today, computer floppy disks are frozen in an old standard because no single company has been able to establish a new one. IBM pioneered the 3.5-inch hard-case diskette in 1987 when it introduced its PS/2 personal computers. Within two years, the memory capacity of 3.5-inch diskettes roughly doubled from 720 kilobytes to 1.44 megabytes, where it has remained ever since.

Why? The technical capability to expand diskette capacity is available, but no company has the reputation and strength to set a new standard. Through the 1980s, IBM was large enough to coordinate standards among the key participants in the industry: personal computer manufacturers, diskette makers, and software publishers. If IBM told the industry it would use a particular capacity on its next generation of machines, others did the same. But in the 1990s, IBM's leadership of the PC market came to an end, perhaps permanently. Today, IBM is not strong enough to move the industry by itself, and it won't move ahead of the other industry players and risk being stranded if they don't follow.

The very reliance of virtual companies on partners, suppliers, and other outside companies exposes them to strategic hazards.

A simple rule of thumb applies: When innovation depends on a series of interdependent innovations – that is, when innovation is systemic – independent companies will not usually be able to coordinate themselves to knit those innovations together. Scale, integration, and market leadership may be required to establish and then to advance standards in an industry.

The IBM PC: Virtual Success or Failure?

IBM's development of the personal computer is a fascinating example of both the advantages and disadvantages of using virtual approaches to pursue innovation. When IBM launched its first PC in 1981, the company elected to outsource all the major components from the marketplace. By tapping the capabilities of other companies, IBM was able to get its first product to market in only 15 months. The microprocessor (the 8088) was purchased from Intel, and the operating system (which became PC-DOS) was licensed from a then-fledgling software company, Microsoft. In effect, the IBM PC had an open architecture: It was based on standards and components that were widely available. The high-powered incentives of the marketplace could coordinate the roles of component manufacturers and software vendors. IBM successfully promoted its open architecture to hundreds of third-party developers of software applications and hardware accessory products, knowing that those products would add to the appeal of the PC.

IBM also relied on the market to distribute the product. Although IBM launched its own IBM Product Centers as retail storefronts and had its own direct sales force for large corporate customers, the majority of the company's systems were distributed through independent retailers, initially ComputerLand and Sears. Eventually, there were more than 2,000 retail outlets.

By using outside parties for hardware, software, and distribution, IBM greatly reduced its investment in bringing the PC to market. More important, those relationships allowed IBM to launch an attack against Apple Computer, which had pioneered the market and was growing quickly. The IBM PC was an early success, and it spawned what became the dominant architecture of the entire microcomputer industry. By 1984, three years after the introduction of the PC, IBM replaced Apple as the number one supplier of microcomputers, with 26% of the business. By 1985, IBM's share had grown to 41%. Many observers attributed the PC's success to IBM's creative use of outside relationships. More than a few business analysts hailed the IBM PC development as a model for doing business in the future.

Indeed, IBM's approach in its PC business is exactly the kind of decentralized strategy that commentators are urging large, slow-moving companies to adopt. The early years of the IBM PC illustrate many of the benefits of using markets and outside companies to coordinate innovation: fast development of technology and tremendous technological improvements from a wide variety of sources.

With the passage of time, though, the downside of IBM's decentralized approach has become apparent. The company failed to anticipate that its virtual and open approach would prevent it from directing the PC architecture it had created. The open architecture and the autonomy of its vendors invited design mutinies and the entry of IBM-compatible PC manufacturers. At first, competitors struggled to achieve compatibility with IBM's architecture, but after several years, compatibility was widespread in the industry. And once that happened, manufacturers could purchase the same CPU from Intel and the same operating system from Microsoft, run the same application software (from Lotus, Microsoft, WordPerfect, and others), and sell through the same distribution channels (such as ComputerLand, BusinessLand, and MicroAge). IBM had little left on which to establish a competitive advantage.

To maintain technological leadership, IBM decided to advance the PC architecture. To do that, IBM needed to coordinate the many interrelated pieces of the architecture – a systemic technology coordination task. However, the third-party hardware and software suppliers that had helped establish the original architecture did not follow IBM's lead. When IBM introduced its OS/2 operating system, the company could not stop Microsoft from introducing Windows, an application that works with the old DOS operating system, thereby greatly reducing the advantages of switching to OS/2. And third-party hardware and software companies made investments that extended the usefulness of the original PC architecture. Similarly, Intel helped Compaq steal a march on IBM in 1986, when Compaq introduced the first PC based on Intel's 80386 microprocessor, an enhancement over the earlier generations of microprocessors used in IBM and IBM-compatible machines. Even though IBM owned 12% of Intel at the time, it couldn't prevent Intel from working with Compaq to beat IBM to market. That was the beginning of the end of IBM's ability to direct the evolution of PC architecture.

By the third quarter of 1995, IBM's share of the PC market had fallen to just 7.3%, trailing Compaq's 10.5% share. Today, its PC business is rumored to be modestly profitable at best. Most of the profits from the PC architecture have migrated upstream to the suppliers of the microprocessor (Intel) and the operating system (Microsoft) and to outside makers of application software. The combined market value of those suppliers and third parties today greatly exceeds IBM's.

IBM's experience in the PC market illustrates the strategic importance of organization in the pursuit of innovation. Virtual approaches encounter serious problems when companies seek to exploit systemic innovation. Key development activities that depend on one another must be conducted in-house to capture the rewards from long-term R&D investments. Without directed coordination, the complementary innovations required to leverage a new technology may not be forthcoming.

The Virtuous Virtuals

How have the most successful virtual companies accomplished the difficult task of coordination? The virtual companies that have demonstrated staying power are all at the center of a network that they use to leverage their own capabilities. Few virtual companies that have survived and prospered have outsourced everything. Rather, the virtuous virtuals have carefully nurtured and guarded the internal capabilities that provide the essential underpinnings of competitive advantage. And they invest considerable resources to maintain and extend their core competencies internally. Indeed, without these companies'

Matching Organization to Innovation

To organize a business for innovation, managers must first determine whether the innovation in question is autonomous (it can be pursued independently) or systemic (it requires complementary innovations). They must also assess whether the capabilities needed to produce the innovation can be easily obtained or must be created.

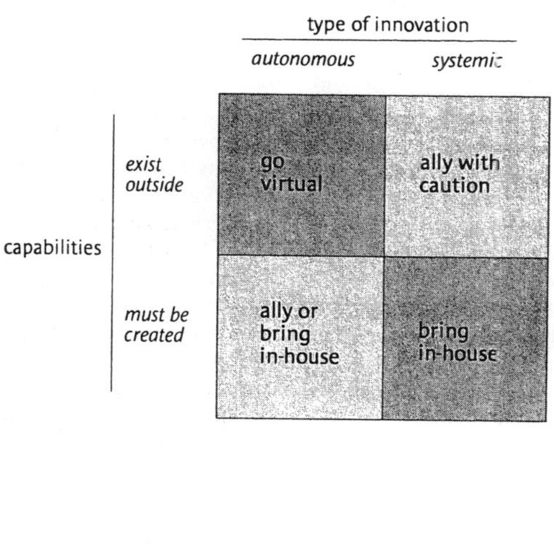

unique competencies and capabilities, their strategic position in the network would be short-lived.

Consider the well-known battle between MIPS Technologies and Sun Microsystems for control of workstation processors. MIPS was trying to promote its Advanced Computing Environment (ACE) against Sun's Scalable Processor Architecture (SPARC). Sun had strong internal capabilities, whereas MIPS tried to compete as a more virtual player, leveraging the competencies of partners such as Compaq, DEC, and Silicon Graphics. MIPS had a good technical design, but that was literally all it had, and this hollowness left the company at the mercy of its partners. As soon as DEC and Compaq reduced their commitment to the ACE initiative, the network fell apart and pulled MIPS down with it. The very reliance of virtual companies on partners, suppliers, and other outside companies exposes them to strategic hazards. Put another way, there are plenty of small, dynamic companies that have not been able to outperform larger competitors. In particular, a hollow company like MIPS is ill equipped to coordinate a network of companies. Although Sun also worked with alliance partners, it had strong internal capabilities in systems design, manufacturing, marketing, sales, service, and support. As a result, Sun can direct and advance the SPARC architecture, a dominant technology in the industry.

Many companies with superior capabilities have prospered as the dominant player in a network. Japanese *keiretsu* are structured that way. Consider Toyota, whose successful introduction of the lean production system – a truly systemic innovation – required tremendous coordination with its network of suppliers. Because Toyota was much larger than its suppliers, and because, until recently, it was the largest customer of practically all of them, it could compel those suppliers to make radical changes in their business practices. In a more egalitarian network, suppliers can demand a large share of the economic benefits of innovations, using what economists call holdup strategies. Strong central players like Toyota are rarely vulnerable to such tactics and thus are in a better position to drive and coordinate systemic innovation.

The most successful virtual companies sit at the center of networks that are far from egalitarian. Nike may rely on Asian partners for manufacturing, but its capabilities in design and marketing allow it to call all the shots. In the computer industry, Intel has effective control of the 80X86 microprocessor standard, Microsoft dominates PC operating systems, and Sun is driving the SPARC architecture. Those companies control and coordinate the advance of technologies in their areas, and in this regard they function more like integrated companies than like market-based virtuals.

Choosing the Right Organizational Design

Today, few companies can afford to develop internally all the technologies that might provide an advantage in the future. In every company we studied, we found a mix of approaches: Some technologies were purchased from other companies; others were acquired through licenses, partnerships, and alliances; and still other critical technologies were developed internally. Getting the right balance is crucial, as IBM's disastrous experience in PCs illustrates. But what constitutes the right balance?

Consider how a successful innovator such as Motorola evaluates the trade-offs. Motorola, a leader in wireless communications technology, has declared its long-term goal to be the delivery of "untethered communication" – namely, communication anytime, anywhere, without the need for wires, power cords, or other constraints. In order to achieve that goal, Motorola must make important decisions about where and how to advance the required technologies. Those decisions turn on a handful of questions: Is the technology systemic or likely to become systemic in the future? What capabilities exist in-house and in the current supplier base? When will the necessary technologies become available?

For Motorola, battery technology is critical because it determines the functionality that can be built into a handheld communications device and the length of time that the device can be used before recharging. Batteries have been a pacing technology in this area for many years.

As Motorola scans the horizon for improved battery technology, it encounters a familiar trade-off between the degree of technological advancement and the number of reliable volume suppliers. Conventional battery technologies such as nickel cadmium (Ni-Cd) have become commodities, and there are many suppliers. But few if any suppliers can offer the more advanced technologies Motorola needs. And the most exotic technologies, such as fuel cells and solid-state energy sources, are not yet commercially viable from any supplier. How should Motorola organize to obtain each of the technologies it might need? Under what circumstances should the company buy the technology from a supplier, and when should it form alliances or joint ventures? When should Motorola commit to internal development of the technology? (See the exhibit "Matching Organization to Innovation.")

For Ni-Cd technology, the clear choice for Motorola is to buy the technology or to use the market to coordinate access to this technology, because Motorola can rely on competition among many qualified suppliers to deliver what it wants, when needed, for a competitive price. Motorola faces a more complex decision when it comes to fuel cells and solid-state battery technologies. Should Motorola wait until those technologies are more widely available, or should the company opt for a joint venture or internal development?

Before Motorola decides whether or not to wait for cutting-edge battery technologies to be developed, it must consider three issues. The first is that Motorola could lose the ability to influence the direction of the technologies; the early commercial forms may be designed for applications that do not benefit Motorola, such as electric automobiles. The second problem is that Motorola might lose the ability to pace the technologies – in other words, to bring them to market at a competitively desirable time. The third issue is that if such technologies are – or become – systemic, and Motorola has no control over them, the company may not be able to advance related technologies and design features to achieve its ultimate goal of untethered communication.

Those issues suggest that Motorola cannot simply wait for the technologies to be provided by the market. Rather, Motorola needs to build strong ties to suppliers with the best capabilities, thus increasing its ability to direct the path of future systemic innovation. Where Motorola itself has strong capabilities, the company should pursue the technologies on its own.

To retain its leadership over the long term, Motorola must continue to develop the critical parts of its value chain internally and acquire less-critical technologies from the market or from alliances. Although networks – with their high-powered incentives – may be effective over the short term for an unchanging technology, they will not adapt well over the long term as technology develops and companies must depend on certain internal capabilities to keep up. The popularity of networked companies and decentralization arises, in part, from observations over a time horizon that is far too short. Remember the enthusiasm that greeted IBM's early success in PCs.

Scale and Scope

Business history presents us with a lesson of striking relevance to the organizational decisions managers face today. In the classic *Scale and Scope,* Alfred Chandler details how the modern corporation evolved in the United States, Germany, and Great Britain at the end of the nineteenth century. Managers who invested the capital to build large-scale enterprises blazed the trail for the leading industries of the second industrial revolution. Markets in railroads, steel, chemicals, and petroleum were developed and shaped by major companies, not the other way around. The most successful of those companies were the first in their industries to make the massive investments in manufacturing, management, and distribution that were needed to realize the gains from innovation.

Companies that failed to make such coordinated, internal commitments during this period were soon thrust aside. The experience of British companies provides a cautionary tale for the champions of the virtual company. Many enjoyed early technological leads in their industries, but the reluctance of those family-run companies to relinquish control to outside investors prevented them from investing to build the capabilities they needed to commercialize their technologies. When German or U.S. competitors made the requisite investments, British companies lost their leadership position. In chemicals, for example, the British lead in the 1870s was completely lost by 1890. History even provided British chemical companies with a second chance when Germany's defeat in World War I temporarily cost German chemical manufacturers their plants and distribution networks. But by 1930, German chemical companies regained the lead because the British again failed to invest adequately. The lesson is that companies that develop their own capabilities can outperform those that rely too heavily on coordination through markets and alliances to build their businesses.

The leading industries of the late nineteenth and early twentieth centuries – chemicals, steel, and railroads – all experienced rapid systemic innovation. The winners were the companies that made major internal investments to shape the markets rather than those that relied on others to lead the way. While business conditions have certainly changed, many of the principles that worked a century ago still pertain.

Today, leading companies like Intel and Microsoft make extensive investments to enhance their current capabilities and spur the creation of new ones. Because so many important innovations are systemic, decentralization without strategic leverage and coordination is exactly the wrong organizational strategy. In most cases, only a large company will have the scale and scope to coordinate complementary innovations. For both the chemicals industry 100 years ago and the microcomputer industry today, long-term success requires considerable and sustained investment within a company. The lessons of the second industrial revolution apply to the third: Adept, well-managed companies that commit the right internal resources to innovation will shape the markets and build the new industries of the twenty-first century.

The popularity of networked companies and decentralization arises, in part, from observations over a time horizon that is far too short.

Reprint R0208J
To place an order, call 1-800-988-0886.

Mitigating Procurement Hazards in the Context of Innovation

JOHN M. DE FIGUEIREDO AND DAVID J. TEECE
(Haas School of Business, University of California, Berkeley, CA 94720-1930 USA)

This paper extends the transaction cost economics framework to examine the contractual hazards that arise in the course of technological innovation. We identify three main strategic hazards related to future technological opportunities that may develop in business transactions: loss of technological pacing possibilities on the technological frontier, loss of technological control at or behind the frontier, and design omissions. In examining these hazards we focus on the increasingly common phenomenon of vertically integrated firms supplying downstream competitors. We then analyze how constellations of safeguards, particularly relational safeguards, can augment transaction-specific safeguards in many instances to ensure high-powered incentives are maintained. We also consider under what conditions downstream divestiture is a desirable economizing option. Supportive illustrations are drawn from the desktop laser printer and telecommunications industries.

> I have spoken to many audiences about the benefits of AT&T being a vertically integrated business that had both services and equipment. … There have been many advantages to our current structure. But the dramatic changes in our markets driven by our customers, new technologies like wireless, and public policy decisions, have opened up so many new opportunities that we need to simplify and more sharply focus our businesses to respond swiftly and effectively to those opportunities. It is not a secret that our Network Systems business has been affected by the conflicts that our Communication Services Group has been having with the RBOCs both in the public policy arena and increasingly in the marketplace as we entered the intra-LATA market. These conflicts foreshadow similar issues with some PTTs around the world. In recent months, it has become clear that the advantage of our size … is starting to be offset by the amount of time, energy, and expense it takes to manage conflicting business strategies . . So, in this spirit, we prepare to launch three strong businesses. …
>
> (Bob Allen, Chairman, AT&T.
> Message to Employees on the Restructuring of AT&T,
> 20 September 1995).

Industrial and Corporate Change Volume 5 Number 2 1996

1. *Introduction*

Mitigating contractual hazards is one of the core functions of the business enterprise (Williamson, 1975, 1985, 1996a; Teece, 1980a, 1986). Vertical integration and other complex organizational arrangements often arise to safeguard transactions against the hazards of *ex post* opportunism. Nevertheless, hazards that arise in the course of business vary in gravity and magnitude. These can be severe when technological innovation is involved. Yet there have been few attempts to elucidate and enumerate innovation-driven hazards in a systematic way.

In this paper, we focus on the contractual hazards and organizational mechanisms that arise in the context of pacing, controlling and directing, current and future technological development and the products that emanate from innovation. We discuss three specific hazards that may occur in high technology transactions: poor sequencing (pacing) of developments in complementary technologies, loss of control over knowledge and intellectual property, and technological 'foreclosure'.

We limit our analysis to a class of cooperative and competitive transactions that have been largely neglected in the strategic management literature to date — vertically integrated firms supplying their downstream, non-integrated competitors. Antitrust analysis has often been suspicious of such types of transactions, viewing them as mechanisms to effectuate predatory acts, such as vertical price squeezes, by the integrated firm. However, if the integrated supplier achieves upstream efficiencies through some economies of scale or core capabilities that are difficult for independent suppliers to replicate, then it may be efficient and desirable for a non-integrated downstream competitor to source from such vertically integrated competitors. The problem that arises for the buyer is that there may be incentives for the integrated supplier, if it is dominant upstream, to disadvantage the buyer at contract renegotiation or to otherwise handicap the buyer. In many cases, the incentives of the integrated firm to act strategically may be higher than those of independent suppliers. Even if the supplier does not behave strategically, buyers may be concerned that the supplier might do so at some future date and act accordingly.

Assessing the magnitude of technological hazards for the buyer requires examination of two factors: first, whether the upstream firm possesses market power and/or the ability to appropriate quasi-rents, and second, whether there is complexity and unobservability in the transaction, in that the performance dimensions of the intermediate good being supplied is only over time revealed, thus creating monitoring problems for the customer. If these two conditions hold in a transaction involving leading edge technology, then technological hazards suggest particular organizational arrangements which provide strong safeguards.

In some contexts, transaction-particular safeguards may not suffice. Rather, individual, transaction-specific safeguards need to be augmented by a constellation of safeguards developed within the context of a broader business relationship that spans numerous transactions. Malfeasance in one transaction may then have repercussions on other transactions in the current relationship, or severe reputation effects with other exchange partners. Thus, in a number of circumstances, relational safeguards that span multiple transactions may be employed to protect against technological hazards, thereby facilitating exchange. However, if complexity, uncertainty and unobservability are sufficiently high, situations may arise when even a constellation of safeguards will not be adequate protection against contracting hazards. In these situations, vertical integration—with the associated possible loss of upstream economies—is sometimes necessary to protect against the hazards of incomplete contracting and the problematic incentives which are sometimes occasioned when a downstream firm sources from a competitor. Alternatively, integrated firms can take the 'ultimate' form of credible commitment not to behave strategically in the downstream market. The upstream integrated firm may decide to simply divest (or not integrate downstream) in order to signal that it will not behave strategically.

It should be noted at the outset that this paper does not address public policy responses. Rather, it examines how managerial solutions accomplished through contractual safeguards and organizational design can act to protect transactions against technological hazards that may arise. Its purpose is to elucidate the hazards that ensue in the context of an integrated firm supplying a non-integrated, downstream competitor, and to examine safeguards which are employed to support contractual relationships in such circumstances. The next section reviews the phenomenon of cross-competitive supply and examines the development and shortcomings of the current literature. Section 3 unravels some of the types of technological hazards that occur in the context of high technology transactions. Section 4 reviews the safeguards that firms employ to protect against these hazards in cross-competitive supply relationships. Section 5 closes with some concluding remarks.

2. *Background*

The Phenomenon: Buying From Integrated Competitors

Buying from one's competitor is not uncommon, and may be increasing in frequency. In this paper, we examine sourcing by non-integrated firms from

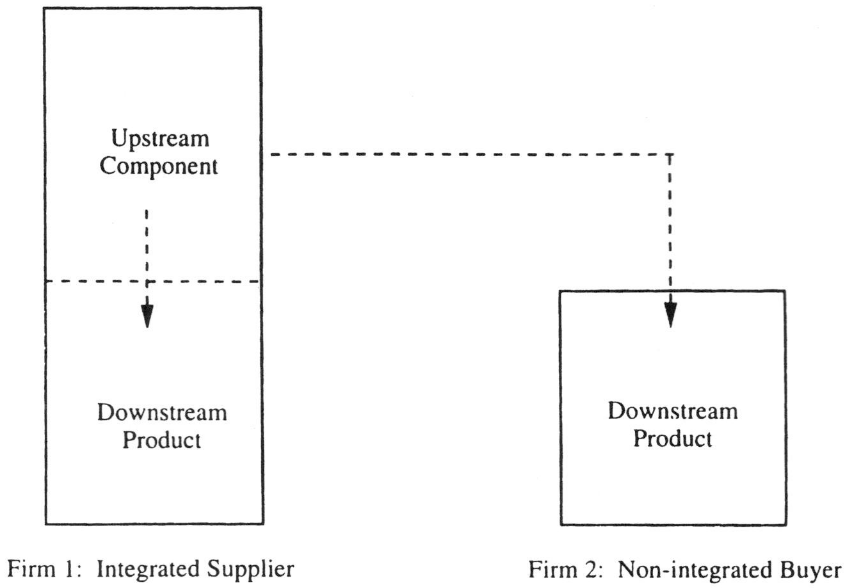

FIGURE 1. Supply.

integrated competitors in high technology markets. Figure 1 illustrates this relationship. While Firms 1 and 2 both compete in the downstream market, only Firm 1 is vertically integrated into the upstream component market. However, Firm 2 sources components, sometimes exclusively, from Firm 1. Often there may be independent component suppliers who provide feasible alternative sources of the component at the outset, yet are not awarded the supply contract by firm 2.

There are a number of reasons a firm would want to source from a competitor. The most advanced or most reliable technology may be possessed by the competitor. The cheapest component may be provided by a competitor. Mere transportation cost considerations may call for competitor supply. These and other reasons often make competitor supply a sensible strategy, provided contractual safeguards can be erected.

Examples of this abound. Consider the laser printer industry. The engines of laser printers determine, among other things, print speed and print quality; the laser printer industry contains many firms that make engines, some firms which are present only downstream, some firms which are present only upstream. It is common to see vertically integrated firms supplying engines on a contractual basis to non-integrated companies against whom they compete. Figure 2, illustrates this point. The left-hand column identifies some of the engine manufacturers. The right-hand column identifies some of the firms that are present in the desktop laser printer market. The arrows indicate

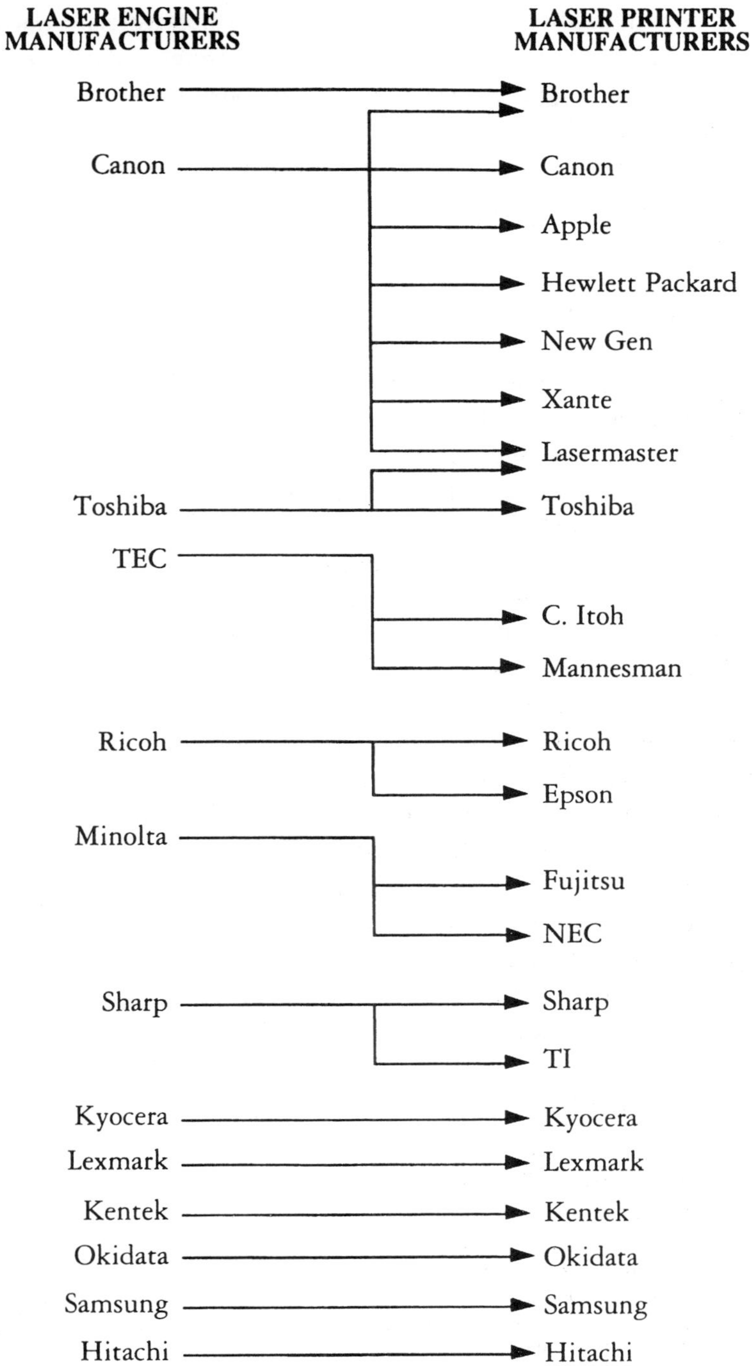

FIGURE 2. Supply chain in the desktop laser printer industry, circa 1990

which engine manufacturers supply which printer manufacturers. Clearly, several types of relationships exist. Perhaps the most striking example of a vertically integrated firm supplying a competitor (and one to which we will return later), is the Canon-Hewlett-Packard (HP) relationship. Canon makes both engines and printers and supplies nearly all of (non-integrated) HP's very significant volume engine of requirements. This contractual relationship has persisted between the two firms for over a decade, despite the apparent strategic hazards.

Cross-competitor supply relationships can also been seen in electronics, semiconductors, and telecommunications. One of the reasons enumerated for the recent split of AT&T into three separate, publicly traded companies was the problem of strategic conflicts between AT&T Technologies (the manufacturing arm) and the regional Bell operating companies (RBOCs), who were its traditional customers. The recent divestiture will separate the communications systems (AT&T) from the equipment manufacturing (Network Systems). As AT&T and the RBOCs evolved into competitors, the RBOCs were reluctant to buy telecommunications equipment, especially switches, from AT&T. Overseas, companies such as Deutsch Telekom in Germany and France Telecom were also wary of buying equipment from AT&T since they expected AT&T to emerge as a competitor in their national market. All this meant that competitors were passing their equipment dollars to other vendors, even when AT&T equipment might have been superior.[1]

These examples, and those that follow, serve to elucidate a common phenomenon that has been little studied—vertically integrated firms supplying competitors in the downstream market in high technology industries. We examine the conditions under which this is a stable outcome and inquire as to when contractual safeguards will be inadequate to sustain this type of relationship. Although both parties may face contractual hazards, our focus is on understanding the strategic hazards that the non-integrated firm faces.

Theoretical Antecedents

Chicago School analysts conducted the first systematic theoretical investigation into cross-competitive supply (for example, Posner, 1976, pp. 196–

[1] 'There's always tension when buying a majority of equipment from somebody who is also your biggest competitor', said Robert Barada, vice-president for corporate strategy and development, for Pacific Telesis Group. Spinning off the equipment arm 'will alleviate that [problem] in a big way'. Moreover, Brian Adamik, an analyst at Yankee Group in Boston, commented, 'After the break-up, Network Systems could now even sell to MCI, which wouldn't buy equipment from AT&T in the past because of the its long-distance arm.' The AT&T example has been drawn from: John Keller, 'Defying Merger Trend, AT&T Plans to Split into Three Companies,' *Wall Street Journal*, 21 September 1995, pp. A1, A16; Leslie Cauley, 'AT&T's Rivals Shrug, but Not Analysts,' *Wall Street Journal*, 21 September 1995, p. A6; Jeff Pelline, 'Giant AT&T to Break into 3 Companies,' *San Francisco Chronicle*, 21 September 1995, pp. A1, A15.

207; Bork ,1978; Posner and Easterbrook, 1981).[2] They pointed out that because monopoly rents could only be taken once in the supply chain, a firm with monopoly (or market) power had two options to retain monopoly rents. First, it could vertically integrate and charge monopoly prices in the downstream market. Equivalently, it could set the price for the intermediate good at the monopoly level and extract all the rents without forward integration. Thus, a vertically integrated firm with upstream monopoly power has little incentive to exclude downstream rivals, since forward integration is not necessary to extract upstream rents. Accordingly, vertical mergers pose no problems. If upstream Firm A and downstream Firm B merge, the merger ought not change concentration or prices. Thus monopoly power will not and cannot be extended. Vertical leveraging is simply impossible.

This view of cross-competitor supply, as it relates to business strategy, has been extended on two fronts. First, game theorists have argued that this Chicago School analysis is not a subgame perfect equilibrium (Hart and Tirole, 1990; Ordover *et al.*, 1990). Other firms in the industry may respond to a competitor's integration decision with integration and price strategies of their own. Under a variety of strategies, the subgame perfect equilibrium can result in inefficiency and a refusal to deal (Hart and Tirole, 1990, p. 212). One insight that is gleaned from this literature is that vertically disintegrated structures or parallel vertical integration may be equilibrium outcomes.

A second extension was the development of transaction cost economic analysis (Williamson, 1971, 1975, 1983, 1985). In *Markets and Hierarchies* (1975), Williamson introduced the notion of contractual hazards and contractual safeguards, primarily flowing from the possibility of *ex post* opportunism in an incomplete contracting regime. His analysis revised and enriched the Chicago approach, and added an efficiency dimension. Williamson (1975) and later Klein *et al.* (1979), showed how asset specificity could cause contractual hazards. Common ownership of such assets could ease those hazards and facilitate efficient investment. Williamson (1983) subsequently highlighted the importance of reciprocity and 'exchange of hostages' as ways to attenuate hazards which might otherwise destroy the basis for exchange. The transaction cost economics framework,

[2] This phenomenon is quite different from the voluminous writing on strategic alliances (for example, Parkhe 1993; Contractor and Lorange 1988; Gulati 1995). With strategic alliances, partners typically have expertise in different functional or technological areas, and the strategic alliance facilitates teaming to create new products. The success of the alliance hinges on strategic alignment and the ability of the two firms to combine their resources and capabilities to build a single product. If either firm fails, both firms lose. Our analysis differs from the strategic alliance literature in that we are primarily concerned with those transactions where the firms are competitors in the downstream market and only one firm possesses the upstream facilities which it supplies to the downstream competitor. One firm, in our analysis, possesses the resources to build and possibly to market the product before the two firms come together.

because of its focus on hazards and safeguards, is a useful point of departure for our analysis. We show that transaction cost economics can in turn be extended to embrace organizational questions when changing technology is at issue.

In much of the work of transaction cost economics, hazards are developed in a rather particular manner, often illustrated by price renegotiation, and quality degradation. These usually augment costs. We seek to expand the framework by examining contracting hazards related to the (usually partial) denial by one party of future technological and associated market opportunities.

Transaction cost economics posits that firms can protect against hazards through contractual safeguards such as penalty for premature termination, dispute resolution mechanisms, and bilateral exchange of hostages (Williamson, 1985, pp. 33–34, 1996b). While these types of safeguard do assist transactions, we go beyond the analysis of strictly discrete transactional safeguards that frequently preoccupies transaction cost economics. In cases when vertical integration is costly, a constellation of safeguards transcending the traditional, discrete transaction safeguards may be effective.[3] This constellation may include transaction specific safeguards in the traditional transaction cost sense, but also may encompass broader relational and multi-transactional structures that safeguard groups of transactions. It is the relational safeguards that may insure adequate protection of transactions, and void the need for vertical integration.

Relevance for Strategic Management

One conceptual argument we put forward is that transaction cost economics can be usefully informed by expanding the notion of hazards and safeguards. Specifically, we view contractual relationships as generating bundles of hazards, and governance structures as providing bundles of safeguards. The task of management is to identify the hazards that are attendant to a business relationship and then to create and insure that the economizing bundle of safeguards necessary for effective execution of transactions and exploitation of opportunities are in place.

We examine the contractual hazards that arise when developing and commercializing new technology, and show that discrete transactional safeguards may be usefully supplemented by multi-transactional relational safeguards. This is especially so when non-integrated firms buy from integrated competitors. Though many of our concepts are applicable to a variety of transactional arrangements, we limit our analysis to cross-competitor supply.

[3] Nickerson (1996) explores how hazards are interrelated in the trucking industry.

3. *Technological Hazards in Competitor Supply*

Several well recognized contractual hazards occur in supply situations. First, when a downstream firm must make some specific investment in order to support efficient production, it exposes its non-redeployable assets to *ex post* recontracting hazards (Williamson, 1985). If investment in specific assets is not required for efficient production, then the non-integrated downstream firm can costlessly switch suppliers and avoid all hazards. If the buyer must invest in non-redeployable assets, it could avoid the hazards associated with *ex post* recontracting through vertical integration.

Second, the characteristics of the intermediate component that is being delivered to the downstream firm might not be readily observable. This may occur because of the extreme complexity of the product or may be due to other non-observable characteristics. In short, there may be an acute monitoring problem for the downstream firm that only resolves itself over the long-run. While these two classes of problems are important, we focus instead on circumstances where new technology is at stake.

In business transactions when new technology is at stake, a less understood set of hazards may arise. This class of contracting hazards stems not so much from the extraction of quasi-rents, but from the guarding of future strategic opportunities. Integrated suppliers can, for example, exclude firms from immediate access to new knowledge, and future possibilities for technological progress. These types of situations may arise when an integrated firm has the ability to use its upstream technological prowess to exclude a downstream rival from a transaction that will open up future technological and commercial opportunities[4]. The seller's ability to pace, direct, control, and guard the development of new products and technologies are all hazards to the buyer that arise in this context. It is these technological hazards that we examine in more depth.

Out-Sourcing and Competitive Advantage: Key Components and Pacing Technology at the Frontier[5]

Technological hazards exist if a downstream firm has failed to accumulate the capabilities needed to make critical components, or has opted, for other (normally sound) business reasons, such as scale, cost, or risk, to forego production of these components. The firm may then have no choice but to pur-

[4] This notion can also be viewed as a dynamic extension of the raising rival's cost literature (Salop and Scheffman, 1983).

[5] When we speak about the frontier, we refer to a component or service being procured which enlists technology that is not ubiquitously employed in the industry. Frontier technologies are those leading edge innovations being incorporated into subsystems and components.

chase critical components from its competitors. Some components, whether commodity or customized to the producer's own downstream requirements, are simply not all that important to competitive advantage. The only strategic advantage to the integrated supplier-competitor flowing from opportunistic behavior with respect to these components is the one time benefit associated with capturing quasi-rents associated with irreversible investments already made. Loss of quasi-rents associated with employing idiosyncratic physical assets may have considerable short-run implications, but if the firm's balance sheet is strong, it need not impair long-run competitive advantage.

Contrast this to new components based on leading edge technology that convey strategic advantage. If firms wish to pace or direct the evolution of new products, then these types of components should, *ceteris paribus*, be developed internally. If such products are outsourced, the supplier gets to pace and direct the development of the technology. Transactions cost and recontracting hazards are not the core considerations; rather, it is the failure to accumulate critical competences important to the firm's overall new product development strategy which matters.

Consider Motorola as it attempts to develop battery technology for their mobile communications products.[6] Motorola can obtain nickel-cadmium (Ni-Cd) batteries from a host of suppliers. However, because the bottleneck technology for Motorola is the battery, there is benefit to accelerating/controlling the development of more long-lasting and lighter weight batteries (e.g. fuel cells). Motorola can safely source standard Ni-Cad batteries externally, but more advanced batteries may need to be developed and sourced internally (see Figure 3). The reasons do not flow from exposure to recontracting hazards as such. Rather, control of the development process is critical to coordinate and accomplish the roll-out of new products.

Microsoft provides another illustration of how an integrated firm can pace technological development downstream. Microsoft not only retains control over the operating system (Windows) for the majority of personal computers, but also commercializes some key applications downstream. Other independent application designers rely on Windows for their applications to run. Thus, Windows acts as a constraint on some of the technological features of the downstream applications (e.g. speed and protocol of data exchange). Microsoft's ability to pace the upstream technology and its ability to exploit its operating system technology in its applications software, has helped it to become one of the dominant players in applications.

A second source of competitive advantage related to technological pacing (that may require integration) is the difficulty of accomplishing what

[6] See Chesbrough and Teece (1996) for a more detailed description of Motorola case.

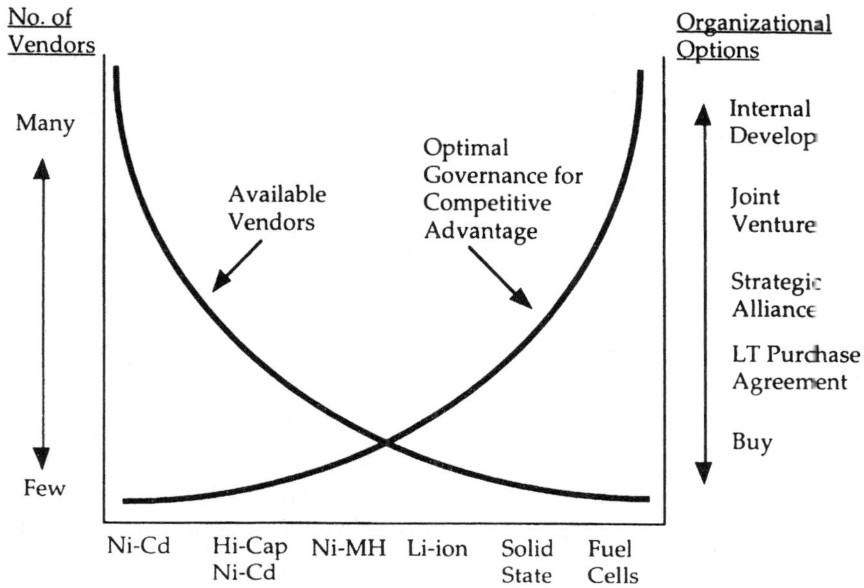

FIGURE 3. Available sources and organizational options for Motorola in battery cell technology.

Richardson (1960) and Williamson (1975) have called 'convergence of expectations'. Investment (in R&D) must be coordinated between upstream and downstream entities, and this is difficult to effectuate using contractual mechanisms. Teece (1988) has characterized innovations as either autonomous or systemic. Autonomous innovations are those that do not require coordinated activities between parties to the innovation. Each innovation can occur within its own walls and can then be 'plugged into' the bigger project. These types of innovations are pervasive when standards are present, such as the open architecture of the IBM personal computer. Systemic innovations are those developments that require coordinated action by the parties to the innovation (e.g. the development of new cameras and film which instant photography required). In order to achieve the coordinated outcomes systemic innovation requires, it is often necessary to vertically integrate (Teece, 1980b). Chesbrough and Teece (1996) have argued that coordinating the development of complementary technologies is difficult when pursued contractually. Delays are frequent and need not be strategic; they may simply flow from uncertainty and divergent goals amongst the parties. MIPS encountered this with their failed attempted to promote their Advanced Computing Environment (ACE) to compete with Sun's Scalable Processor Architecture (SPARC). MIPS set up alliances with Compaq, DEC, Silicon Graphics and other firms to pursue a RISC-based computing standard. However, as DEC and Compaq announced that they were going to reduce their commitment to ACE, the alliance soon fell apart because MIPS could not pick up the slack in some of the upstream activities. It failed to understand and develop competencies in key aspects of the technology, and was not able to create a common expectation for the alliance (Gomes-Casseres, 1994).

Technological Control at or Behind the Frontier

A second hazard that arises where new technology (at or even behind the frontier) is utilized, is the leakage of technology to competitors. Arrow (1962) first brought to light the disclosure problem in the market for know-how, and others have since elaborated on this and related technology transfer problems (Goldberg, 1977; Teece, 1981, 1985, 1986). Appropriability hazards are of concern when property rights are difficult to establish and enforce and when the knowledge is not retained entirely at the organizational level (Oxley, 1995). The leakage we have in mind can occur vertically (upstream and downstream) as well as horizontally (Silverman, 1996).

Proprietary knowledge that leaks from buyer (supplier) to supplier (buyer) in the course of fulfilling a purchase contract is especially problematic when the supplier (buyer) is integrated downstream (upstream). The argument is of course symmetric. Although an independent supplier who obtains knowledge from the buyer may choose to integrate into the downstream product, the likelihood that this will occur is small. However, a firm which is already vertically integrated downstream and supplies a downstream competitor may be able to take the know-how that has leaked to its upstream division and incorporate it into the downstream products and processes relatively quickly.[7] Downstream investments in manufacturing have already been made and small modifications may be all that is necessary to enhance competitiveness in the downstream operations. Staff engineers who understand the downstream process and product can take the know-how and improve upon the current product or process.

In the digital switching business, AT&T has had difficulty selling switches to some of the RBOCs because of fear that it will misuse commercial secrets it learns from the RBOCs. This happens because many switches are configured to the needs of each customer. In order to customize the switch, the RBOC must reveal its telephone traffic patterns, capacity utilization, infrastructure layout, and other extremely sensitive information to the switch supplier. While contractual mechanisms can be designed to protect

[7] When we use the term leakage, we do not mean to imply that intellectual property rights have necessarily been violated. We have in mind the quite legal imitation and emulation that takes place in the normal course of business.

this information, information flows through informal channels and internal movement of personnel, combined with the cost and difficulty of enforcement of such contractual clauses, create hazards for the buyer.[8] The ability of a competitor to use and improve upon know-how which has been inadvertently leaked may be higher for integrated suppliers than for non-integrated suppliers.

Guarding Technological Capabilities

Another disadvantage buyers may experience when market power exists upstream and intellectual property regimes are strong (Teece, 1986), is that the integrated supplier may be in a position to guard its know-how advantage and if it chooses, simply refuse to sell to its competitors. The integrated firm may have incentives to exclude its rivals from its most advanced components if this will hinder the ability of the downstream rival to participate in related or future business. By protecting its competitive advantage in this way, the integrated firm can advance its competitive position downstream. Thus, integrated suppliers may have a higher incentive to withhold (or degrade) supply than would independent firms.[9] The downstream division of the integrated firm may be able to advance its position in related and emerging markets if the upstream division does not assist downstream competitors.

For example, in digital telecommunications equipment supply, the switches hold the key to competitive advantage for many future service innovations (such as value-added network services). The purchase of switches and other equipment from AT&T has the potential for creating many hazards for the RBOCs, who may be foreclosed from the opportunity to participate in the future technological innovations if their switch supplier chooses not to add features that would enhance the RBOC's ability to compete against AT&T in telecommunication service markets. The incentives of an integrated AT&T, if it has some market power in switch manufacturing, may be to act in a manner that might disadvantage the RBOCs in domains where they currently compete or are likely in the future to compete with AT&T. The problem evaporates, however, if there are abundant actual or potential competitive supply alternatives. Vertical integration into manufacturing is just one among many possible solutions.

[8] We thank James Dalton of Bell South for making this point.

[9] We stress features rather than an entire product. If an independent supplier could monopolize the downstream market, then it might integrate downstream or it might charge a price for the intermediate good so that it obtains a monopoly rent. This result is less likely if only certain features of the product are not available, as in the digital switching example. For an example of this at the national level, see General Accounting Office, 'International Trade: US Business Access to Certain State-of-the-Art Technology,' *GAO Report* (GAO/NSIAD-91-278), September 1991, p. 33.

4. *Safeguarding Against Technological Hazards*

The need to safeguard incomplete contracts against *ex post* opportunistic behavior is not a new or novel concept. The need to monitor effort under technological non-separability (Alchian and Demsetz, 1972), assert control when property rights are not well-defined (Coase, 1960; Demsetz, 1967), and protect streams of quasi-rents in a competitive contracting process (Klein *et al.*, 1978) has long been recognized. Most of these analyses, however, have been concerned with vertical integration as a way to safeguard specific investments. Indeed, the paradigmatic safeguard, and the one which has received the majority of attention in the literature, is vertical integration (Williamson, 1975).

Although vertical integration is normally an effective safeguard, it is not always desirable. When poorly implemented, vertical integration can be fraught with low-powered incentives and bureaucratic costs which will impair performance and make it unattractive (Williamson, 1975, 1985, 1991; Teece, 1976, 1996). A superior alternative in many instances is to rely on contracts and develop a constellation of safeguards, enveloping many aspects of the firm's business with such protections, so that a contractual exchange that might otherwise be considered risky, can flourish.

Constellations of Safeguards

In the original concept of hostages as safeguards (Williamson, 1983), each firm places a hostage that has an *ex ante* (screening) or *ex post* (bonding) effect to support exchange. The parties' incentive to perform when bonds are held is high because exit would entail forfeiture of the hostage.[10] Thus hostages can, in many circumstances, be more cost effective than integration in supporting exchange.

However, incomplete contracting and bounded rationality make hostage exchange difficult in many circumstances where there is uncertainty as to outcomes, such as with innovation. The amount of the bond to be placed in custody is difficult to determine *ex ante* because unforeseen contingencies may render the hostage inefficiently large or, of more concern, too low in value. This is especially likely to be a problem when technological innovation and new product development is involved. In these circumstances, firms may exit the transaction—appropriating the technology, and leaving the insufficient hostage on the table for the other party. Alternatively, the

[10] Williamson (1983, p. 527) has noted: '...a king who is known to cherish two daughters equally [one beautiful and one ugly] and is asked, for screening purposes, to post a hostage is better advised to offer the ugly one'.

hostage may itself become the subject of the opportunistic bargaining as the ante is renegotiated.[11] While hostages may support market transactions, they do so over a limited range of situations, or in conjunction with other safeguards.

Our interest in this paper is in demonstrating the efficacy of constellations of safeguards in cross-competitive supply situations, including ones where technological hazards exist. Although contracts give guidance as to how transactions can proceed, constellations of arrangements can create a larger context in which contracts can operate. In this sense, multilateral and multi-transactional relationships are akin to relational contracts (Macneil, 1978) where contracts occur against the background of the relationship, where changes to a contract are adopted 'only in the overall context of the whole relation' (p. 890) and where often 'preservation of the relation' (p. 895) is the concern.

The Canon-HP relationship, with special reference to the laser printer industry, is illustrative. As mentioned earlier, Canon, a vertically integrated firm that makes both laser engines and laser printers, supplies all of HP's engine requirements. At first glance, it would seem there would be many contractual hazards facing HP. However, the Canon-HP transaction is embedded in a larger relationship that displays a constellation of safeguards.

Relational safeguards. The Canon-HP relationship began long before the advent of the desktop laser engine.[12] Throughout much of the 1970s and early 1980s, Canon not only supplied laser engines for HP's midrange laser printer systems, but also engaged in some joint development with HP of floor-model printer technology. The move from midrange to desktop laser printers was only one of many steps in the printer market that the two firms took together. This historical relationship provides important context. Canon and HP had (and still have) a web of ties in numerous transactions and technologies that serve as the basis for exchange. We understand that these discrete transactions are linked through a position created in each firm referred to as the 'relationship manager'.[13] Although this manager does not have formal authority to compel divisions to cooperate, this person does have tremendous informal authority, reports to a division head, and has responsi-

[11] Sequential hostages often face the same problems. Some type of contractual and hybrid relationships require a number of hand-offs of a project between parties in order to achieve the goal. Both firms are better off if the project goal is reached, but incentives may exist to expropriate the other partner at each stage of development. Although the hostage is the project itself, interim knowledge gained may prove to be more valuable than the successful completion of the project. Moreover, in the last stage of the project, the last firm has the incentive to expropriate the partner because the game has come to an end. The game then unravels under most conditions.

[12] See Mowery and Beckman (1995) for a more detailed discussion of HP's strategy.

[13] We thank Lee Rhodes of Hewlett-Packard for bringing this to our attention.

bility for the smooth running of the broader strategic relationship. Moreover, as overseer of the strategic relationship between the two firms, this manager also becomes a mediator should disputes develop. Through this central mechanism, transactions are 'bundled' together, albeit imperfectly, for what is a relational governance scheme. Opportunistic behavior and conflict in one transaction may change how the other contracting party views the potential for opportunism in other transactions. Bundling transactions therefore raises the cost of cheating. There is the threat that there will be loss of the entire relationship if there is substantial malfeasance in the specific transaction. Usually, loss of the entire relationship is more costly than the gains from expropriation available from a single transaction. As firms enter into larger types of relationships that involve multiple transactions, the individual transaction becomes subsumed in a web of current and future transactions. Indeed, Williamson has noted, '... interdependencies among series of related contracts may be missed or undervalued as a consequence [of examining each trading nexus separately.] Greater attention to the multilateral ramifications of contract is sometimes needed' (1985, p. 393).

Transaction-specific safeguards. In addition to these relational safeguards, HP possesses some discrete transactional safeguards that support exchange. One flows from the fact that HP purchases annually hundreds of millions of dollars in printer engines from Canon, representing nearly 50% of Canon's engine output. Hence, it is highly unlikely that Canon would act opportunistically with HP in laser engine supply because of the volume of business that HP represents. While this capacity is not technically transaction-specific, there is little doubt that at least some assets that support it would be stranded if HP suddenly stopped buying engines from Canon. Thus, if the non-integrated buyer represents a significant portion of the purchases of the integrated supplier, and there does not exist excess demand in the market, then it is unlikely that the supplier will opt to engage in expropriation at contract renegotiation.[14]

A second transaction-specific safeguard that protects HP from opportunism is the reputation effects that Canon would suffer if it acted opportunistically to expropriate HP or provide HP with a consistently sub-standard product. If HP effectively publicized and detailed malfeasance by Canon in laser engines, Canon might lose a large portion of its remaining customers. At best, remaining customers might update their probabilities of opportunistic behavior (Crocker and Reynolds, 1993) in their own transac-

[14] Williamson (1985, p. 93) addresses this in the context of 'dedicated assets' which he defines as a 'discrete investment in generalized (as contrasted with special purpose) production capacity that would not be made but for the prospect of selling a significant amount of product to a specific customer'.

tions with Canon. Given the short lifetime of a laser printer model (1–3 years), product model turnover acts to favor laser printer manufacturers. The printer manufacturer is therefore only committed to the current laser engine for the current printer (though there is the need for some continuity between models). If malfeasance occurs, the downstream buyer can obtain a new supplier for its subsequent models. The result in the extreme case, then, is that Canon could lose its entire customer base in 2–3 years. This is extreme, but illustrates how short product turns (with reasonable discount rates) enhance the need for suppliers to maintain a good reputation. Williamson discusses reputation effects and argues, 'Suffice it to observe here that reputation effect are no contracting panacea' (1985, pp. 395–396). Indeed, we argue that reputation acts in concert with other safeguards to ensure exchange.

Mixed-mode safeguards. HP's proprietary control of most aspects of its ink-jet technology provides HP's laser printer division with considerable protection against possible opportunism from Canon. The possibility that an altogether different technology can be developed, at reasonable cost, which effectively substitutes for the technology of the transaction will serve to keep the contract sustainable. Most core capabilities for the rapid roll-out of the alternative technology must exist within the boundaries of the non-integrated firm if this is to be a credible threat The degree of protection such a safeguard provides is directly related to the degree and cost of substituting the alternative technology. The ink-jet technology that HP has developed was not created as an intentional safeguard to laser technology, but rather as an alternative technology with an independent marketing strategy. However, ink-jet technology has evolved. If Canon were to refuse to supply engines or to escalate prices, HP might credibly shift its focus from laser jet technology to the improvement of the current ink-jet technology. No doubt, HP would prefer to stay with laser printers. However, it might be able to push ink-jet technology so as to seriously challenge the laser technology. Ink-jet printers already cannibalize laser printers on the lower end of the market. The impetus given to ink-jet technology could have a substantial detrimental effect on the laser printer market and specifically on Canon, the largest supplier of laser engines in the world.

Canon has many customers for its laser engines, yet HP has (asymmetrically) exposed itself to contractual hazards with Canon if HP is looked at in isolation. However, by employing constellations of relational, transaction-specific, and mixed-mode safeguards, HP appears to have protected itself against possible opportunism by Canon and, market exchange continues between the two firms. Thus, a more global view of safeguards appears to be warranted.

Vertical Integration, Divestiture and Credible Commitments

Maintenance of high powered incentives and avoidance of bureaucratic costs are benefits retained if safeguards short of vertical integration will suffice (as described above). However, sometimes contractual and relational safeguards do not suffice. In such instances, both integration (by the purchaser) and divestiture (by the seller) serve to correct some of the hazards we have identified. They are not perfect alternatives, but they do paradoxically provide some degree of remedy to the (strategic) hazards we have identified. We consider circumstances where vertical integration may be the better solution.

Controlling the rate and direction of innovation. Where there is a need to be on the frontier of the technological possibility curve, technological pacing becomes most important. The firm that chooses (for whatever reasons) to eschew vertical integration, yet endeavors to direct the rate and direction of development using contracts, may find it difficult. As noted earlier, reliance on one's competitor for supply is problematic because of the misaligned incentives of the integrated firm. In some cases, even independent supply could be problematic. In these instances, vertical integration (internal development) may be the best way to confidently pace interrelated technological developments. This is particularly true when solutions inside an existing technological paradigm (Dosi, 1982) will suffice. This is because investment in R&D will yield a degree of predictability as to outcomes. When this is not the case (i.e. required solutions require radical advances unlikely within the paradigm), vertical integration (internal development) may be ineffective.[15]

To illustrate this point, we accumulated data on a sample of PC compatible, desktop laser printers introduced in 1992 and 1993[16]. Two characteristics often cited as determinants of the proximity of a laser printer to the technological frontier are print speed, measured by pages per minute (PPM), and print resolution, measured by the number of dots per inch (DPI) it prints. High performance is associated with greater speed and greater dot density. Figure 4, places all of these printers on a scatter plot graph. Each point is assigned an ▲, ❑ or ○, which designates who makes the engine for

[15] For a related discussion of competency-enhancing and competency-destroying innovation in centralized versus decentralized firms, see Bercovitz *et al.*, 1996.

[16] This data set is being compliled from trade magazines such as *PC World* and *PC Magazine*

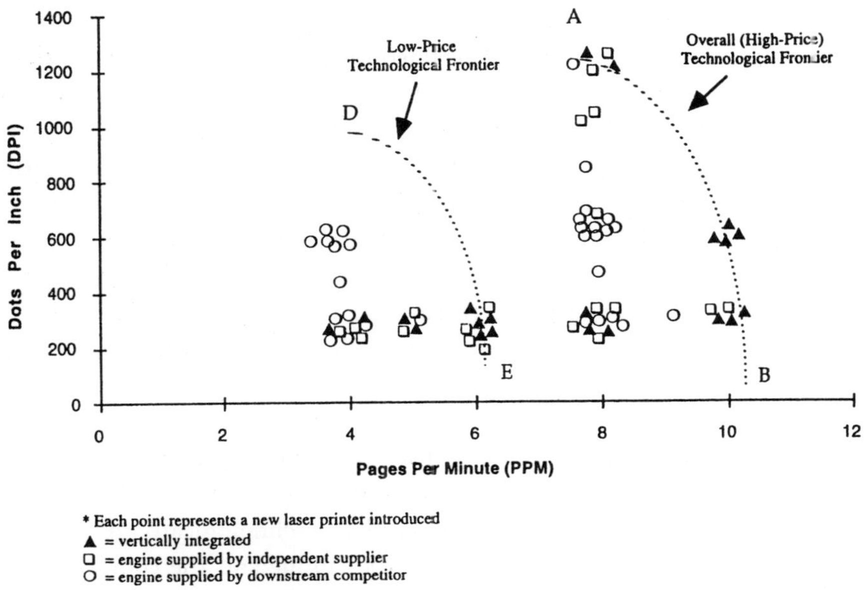

FIGURE 4. Desktop laser engine–laser printer transaction: governance decisions, 1992–1993.

the printer: the same firm that makes the printer (▲ = hierarchy), an independent upstream supplier that does not make printers (❑ = independent), or a vertically integrated competitor that makes engines and printers (O = competitor). On the outer technological frontier of desktop laser printers (represented by the AB curve), there is a high proportion of firms vertically integrated into engine and printer production.[17] This suggests that technological pacing (as well as technological control) may play a role in this industry. The second largest number of printers on the frontier source engines from independent suppliers. Only one of the 14 'frontier' printers has an engine manufactured by a competitor. This is consistent with our analysis that competitor relationships are fraught with hazards that independent supply, and certainly integrated supply, avoid.

It is also interesting to note that there is a second group of firms near 6 PPM which share the same governance characteristics as the outside frontier firms. This seems to occur because the desktop laser market has split into two segments—a high end, full feature, high-priced segment; and a low end, low feature, low-priced segment.[18] In the low-price segment, firms

[17] The 14 printers on the frontier represent the integration strategies of eight firms. We drew the frontiers by analyzing each printer by its list price and features. We then derived the technological frontiers by constructing a curve for each segment that approximated the leading edge technology available at each price.

[18] Beginning in 1990, the desktop laser printer market began to diverge into two segments. Preliminary analysis of the data indicates that the lower-priced printers have fewer features, less processing paper, and, in some cases, longer duty cycles.

seem to be trying to create a printer that is technologically advanced with as many features as possible, but constrained by an upper limit on list price of approximately $2,000. Hierarchical firms (and independent suppliers) show up en masse on this frontier of the low end printers (represented by the secondary 'frontier' CD), followed by a high preponderance of competitor supply. Thus, on both frontiers, firms that seem to be pacing the development of the technologies tend to be vertically integrated in the first instance, and tend not to source engines from a competitor. We must stress, however, that while suggestive, these results are highly preliminary. The data set is not complete, and controls for other independent variables have not been included.

Sustaining technological capabilities. Maintaining technological control, even when behind the technology frontier, sometimes requires vertical integration. The two key components that technologically differentiate products in the laser printer market are the engine and the controller card. From 1984 to 1988, HP outsourced the controller cards for its laser printers to Canon. Although HP designed these cards, Canon actually manufactured the cards on its own board line in Japan. However, in 1988, HP chose to integrate this activity into their printer division in Boise. Two concerns seem to have led to this decision. First, HP felt integration would allow them to respond to changes in technological and market conditions in a real-time manner.[19] Second, we understand that HP feared that if the controller cards and engines for its laser printers were both sourced from Canon, it might lose its technology to Canon and, in effect, create a formidable competitor. The need to retain control over its own technology and ensure sustained access to key components seem to be key motivating factors in the change of HP strategy. Moreover was also the time when DRAM chips were in short supply. This may also have had an influence on HP's decision to internalize controller card production.

Design omissions. Finally, vertical integration may be desired to guard against strategic product design manipulation. It is not surprising that in the digital telecommunications switching equipment industry with continued deregulation bringing AT&T and the RBOCs into competition, that non-integrated suppliers have emerged (e.g. DSC and Motorola). It is also not surprising to see RBOCs abandon traditional purchase relations with AT&T in favor of purchases from Northern Telecom, Ericson, NEC, and others for whom there is no strategic conflict downstream. Indeed, with the

[19] 'Hewlett Packard begins making component of top-selling laserjet printer,' *Business Wire*, 7 December 1988.

impending entry of the inter-exchange carriers into the local exchange, and the local exchange companies into the inter-exchange business, AT&T may have the incentive to guard its technology jealously, if by doing so it can delay the RBOCs offering new services which would require the use of AT&T product technology. Whether meritorious or not, concerns along these lines cost AT&T significant equipment sales, and appears to be a major factor behind its tripartite divestiture.

5. *Conclusion*

Thinking of contracts and governance structures as generating hazards and providing safeguards enriches our understanding of the theory of the firm. In this paper, we argued that identifying technological considerations that arise in the context of innovation gives a different perspective on the organizational problems. We have discussed a variety of safeguards. Although these may be transaction specific, many times relational or mixed-mode safeguards will be employed that span transactions and serve as a cost-effective mechanism of governance. When the parties are linked by a broad constellations of contracts, vertical integration can be avoided and high-powered incentives retained. There will be times, however, when even a rich constellation of safeguards will be insufficient. In these circumstances, we can expect to see both parallel vertical integration, or in the alternative, divestiture.

Our focus on the future opportunities and appropriability considerations that arise in technology transactions suggests a need to move to a more dynamic and evolutionary approach to organizations. Indeed, changes in governance structures over time may reflect not only the shifts in traditional economic incentives to integrate, but also exposure to strategic hazards and the need to accumulate capabilities necessary for current and future competitive advantage. An industry's structure could change over time as knowledge accumulation and diffusion change the hazards associated with sourcing from a competitor. Thus, vertical structures may be reshaped as these conflicting forces wax and wane.

Acknowledgements

We wish to acknowledge the helpful comments and assistance of Sara Beckman, Janet Bercovitz, Glenn Carroll, David Mowery, Jackson Nickerson and Oliver Williamson. A much earlier draft of this paper was presented at the 'Firms, Markets and Organizations' Conference held at Berkeley, 6–8 October 1995. We also wish to thank the Sloan Foundation and the Ameritech Foundation for their generous financial support.

References

Alchian, A. A. and Demsetz, H. (1972), 'Production Costs, Information Costs, and Economic Organization,' *American Economic Review*, 62, 777–795.

Arrow, K. J. (1962), 'Economic Welfare and the Allocation of Resources of Invention,' in National Bureau of Economic Research (ed.), *The Rate and Direction of Inventive Activity: Economic and Social Factors*. Princeton University Press: Princeton, NJ, pp. 609–625.

Bercovitz, J. E. L., de Figueiredo, J. M. and Teece, D. J. (1996), 'Firm Capabilities and Managerial Decision-Making: A Theory of Innovation Biases,' in R. Garud, P. Nayaar and Z. Shapira (eds), *Technological Learning, Foresights, and Oversights*. Cambridge University Press: New York, forthcoming.

Bork, R. H. (1978), *The Antitrust Paradox*. Basic Books: New York.

Chesbrough, H. W. and Teece, D. J. (1996), 'When is Virtual Virtuous: Organizing for Innovation,' *Harvard Business Review*, 74, 65–73.

Coase, R. H. (1960), 'The Problem of Social Cost,' *Journal of Law and Economics*, 3, 1–44.

Contractor, F. J. and Lorange, P. (eds) (1988), *Cooperative Strategies in International Business*. Lexington Books: Lexington, MA.

Crocker, K. J. and Reynolds, K. J. (1993), 'The Efficiency of Incomplete Contracts: An Empirical Analysis of Air Force Engine Procurement,' *Rand Journal of Economics*, 24, 253–264.

Demsetz, H. (1967), 'Toward a Theory of Property Rights,' *American Economic Review*, 57, 347–359.

Dosi, G. (1982), 'Technological Paradigms and Technological Trajectories,' *Research Policy*, 11, 147–162.

Goldberg, V. (1977), 'Competitive Bidding and the Production of Precontract Information,' *Bell Journal of Economics*, 8, 250–261.

Gomes-Casseres, B. (1994), 'Group Versus Group: How Alliance Networks Compete,' *Harvard Business Review*, (July), 62–74.

Gulati, R. (1995), 'Does Familiarity Breed Trust—The Implications of Repeated Ties for Contractual Choice in Alliances,' *Academy of Management Journal*, 38, 85–112.

Hart, O, and Tirole, J. (1990), 'Vertical Integration and Market Foreclosure,' *Brookings Papers: Microeconomics*, 205–276.

Klein, B., Crawford, R.A. and Alchian, A.A. (1978), 'Vertical Integration, Appropriable Rents, and the Competitive Contracting Process,' *Journal of Law and Economics*, 21 (October), 297–326.

Macneil, I. R. (1978), 'Contracts: Adjustments of Long-Term Economic Relations Under Classical, Neoclassical, and Relational Contract Law,' *Northwestern University Law Review*, 72, 854–906.

Mowery, D. C and Beckman, S. L. (1995), 'Corporate Change and Competitiveness: The Hewlett-Packard Company,' CCC Working Paper No. 95–13, University of California, Berkeley,

Nickerson, J. (1996), 'Strategy and Structure: The Role of Transactional Interdependencies on Organizational Form,' PhD dissertation, University of California, Berkeley.

Ordover, J. A., Saloner, G. and Salop, S. (1990), 'Equilibrium Vertical Foreclosure,' *American Economic Review*, 80, 127–142.

Oxley, J. E. (1995), 'International Hybrids: A Transaction Cost Treatment and Empirical Study,' PhD dissertation, University of California, Berkeley.

Parkhe, A. (1993), 'Strategic Alliance Structuring — A Game Theoretic and Transaction Cost Examination of Interfirm Cooperation,' *Academy of Management Journal*, 36, (August), 794–829.

Posner, R. A. (1976), *Antitrust Law*. University of Chicago Press: Chicago, IL.

Posner, R. A. and Easterbrook, F. H. (1981), *Antitrust: Cases, Economic Notes, and Other Materials*, 2nd edn. West Publishing: St. Paul.

Richardson, G.B.H. (1990), *Information and Investment*, Oxford University Press: Oxford.

Salop, S. and Scheffman, D. (1983), *Raising Rival's Costs*, American Economic Review, 73 (May), 267–271.

Silverman, B. S. (1996), 'Technical Assets and the Logic of Corporate Diversification,' Ph.D. Dissertation, University of California, Berkeley.

Teece, D. J. (1976). *Vertical Integration and Divestiture in the US Oil Industry*. The Stanford University Institute for Energy Studies: Stanford, CA.

Teece, D. J. (1980a), 'Economies of Scope and the Scope of the Enterprise,' *Journal of Economic Behavior and Organization*, 1, (September), 223–245.

Teece, D. J. (1980b), 'Vertical Integration and Procurement,' Unpublished Working Paper, University of California, Berkeley.

Teece, D. J. (1981), 'The Market for Know-How and the Efficient Transfer of Technology,' *The Annals of the Academy of Political and Social Science*, 81–96.

Teece, D. J. (1985), 'Multinational Enterprise, Internal Governance and Industrial Organization,' *American Economic Review*, 75, 233–238.

Teece, D. J. (1986), 'Profiting from Technological Innovation: Implications for Integration, Collaboration, Licensing, and Public Policy,' *Research Policy*, 15, 285–305.

Teece, D. J. (1988), 'Technological Change and the Nature of the Firm,' in G. Dosi (ed), *Technical Change and Economic Theory*. Pinter Publishers: London, pp. 256–281.

Teece, D. J. (1996), 'Firm Organization, Industrial Structure, and Technological Innovation,' *Journal of Economics, Behavior, and Organization*, forthcoming.

Williamson, O. E. (1971), 'The Vertical Integration of Production: Market Failure Considerations,' *American Economic Review*, 61 (May), 112–123.

Williamson, O. E. (1975). *Markets and Hierarchies: Analysis and Antitrust Implications*. The Free Press: New York.

Williamson, O. E. (1983), 'Credible Commitments: Using Hostages to Support Exchange,' *American Economic Review*, 73 (September), 519–540.

Williamson, O. E. (1985), *The Economic Institutions of Capitalism*. The Free Press: New York.

Williamson, O. E. (1991), 'Comparative Economic Organization: The Analysis of Discrete Structural Alternatives,' *Administrative Science Quarterly*, 36, 269–296.

Williamson, O. E. (1996a), *The Mechanisms of Governance*, Oxford University Press: New York.

Williamson, O. E. (1996b), 'Revisiting Legal Realism: The Law, Economics and Organization Perspective,' *Industrial and Corporate Change*, 5, 383-420.

Part IV

THE MULTINATIONAL ENTERPRISE

TECHNOLOGY TRANSFER AND R&D ACTIVITIES OF MULTINATIONAL FIRMS:

SOME THEORY AND EVIDENCE

David J. Teece

I. INTRODUCTION

The significance and visibility of the multinational enterprise (MNE) in the world economy continues to grow. However, confusion still surrounds the way foreign investment is evaluated, as evidenced by the schizophrenic nature of policies toward this mode of economic organization. Many host countries encourage foreign investment in one period, while thwarting it in the next. Part of this ambivalence is no doubt due to swings in nationalist fervor, but I submit that part of the confusion stems from an inability to appreciate the distinctive properties of the multinational firm and its technology transfer activities. Thus technology importers gather to draft codes of conduct designed to regulate the importation of

Research in International Business and Finance, Vol. 2, pp. 39–74.

ISBN: 0-89232-140-7

technology because of its high cost and disruptive effects while some labor and industry groups in the United States simultaneously meet to discuss technology export controls because of concern that valuable knowhow is being "given away." Clearly, the multinational enterprise (MNE) is yet to be fully understood, despite an enormous outpouring of scholarly research. Unfortunately, much of the received literature has been concerned with the market power aspects of multinational enterprise, to the neglect of efficiency considerations. However, recent advances in the theory of contracting provide the analytical structure needed to examine the efficiency properties of large firms. An application of these tools to the study of multinational enterprise serves to highlight transactions cost considerations.

One purpose of this chapter is to explore the efficiency and market power characteristics of multinational enterprise. Another is to place the international R&D and technology transfer activities of the MNE in perspective. Particular emphasis is placed on the role of the MNE on technological choice in the host country. A concluding section identifies avenues for further research.

II. A THEORY OF MULTINATIONAL ENTERPRISE

Introduction

Recent developments in the theory of the firm make it possible to delineate a theory of multinational enterprise and to introduce some rigor and consistency to the analysis of the efficiency and competitive implications of direct foreign investment. The starting point is the conceptualization of the multinational firm as an organization that internalizes various international transactions that could conceivably take place in a market. While this observation is almost self-evident and has been made by others (Hymer, 1970; McManus, 1972; Buckley and Casson, 1976), the implications for economic efficiency, economic development, and competition have not been fully developed. Furthermore, to the extent that the economics of internalization have been addressed, efficiency and monopoly power considerations have often been confounded, leaving the public policy implications of direct foreign investment more ambiguous than is necessary.

In order to delineate a theory of multinational enterprise, it is desirable to identify some general properties of markets and some general properties of internal organization. This will be done by outlining the Markets and Hierarchies (M&H) approach recently developed by Oliver Williamson (Williamson, 1975). A theory of multinational enterprise emerges once the nature of the transactions typically internalized by multinational firms are delineated and related market failure considerations are explored.

The Economics of Internalization

The M&H approach developed by Williamson[1] attempts to assess the properties of various organizational modes by comparing their relative efficiency characteristics. The market is considered to be one way of organizing economic activity, and the desirability of using market processes is assessed by comparing the efficiency properties of markets with alternative organizational modes. The concept of market failure embedded in the M&H framework is therefore not one based principally on pareto optimality considerations; it rests instead on relative efficiency considerations. In order to assess the relative efficiency properties of various organizational modes, Williamson, following Commons (1934), selects the transaction as the basic unit of analysis and suggests that the relative efficiency of various organizational structures can be gauged by examining the transactions costs properties of each. Transactions costs are not explicitly defined by Williamson but seem to embrace all the costs associated with organizing in order to accomplish tasks. In a market context, for instance, transactions costs include the costs of discovering with whom one wishes to strike a deal, to inform market agents that one wishes to deal and on what terms, to conduct negotiations leading up to the bargain, to draw up the contract, to undertake the inspection needed to make sure that the terms of the contract are being observed, and so on (Coase, 1960, p. 15). In a nonmarket context, transactions costs would similarly include the costs of identifying exchange opportunities and the costs of employing administrative processes to direct and monitor human and nonhuman resources. Whether one is assessing market or nonmarket (internal) modes of organization, in essence what is involved is an analysis of the facility with which contracts can be written, executed, and enforced. This is because:

> explicitly or implicitly, the institutional device by which transactions are organized is the contract . . . to be sure, not every problem can usefully be posed as a contracting problem. More can be described in these terms than is generally realized. Not only can an actual or implicit contract be described for every market interface, but many nonmarket relations can be expressed in contracting terms as well. For every problem that arises as or can be transformed into a contracting relation without emasculating its main features, transactions cost analysis is appropriately brought to bear. Whatever the contractual context, moreover, such an analysis ultimately reduces to an examination of the manner in which human agents cope with complex events in the face of uncertainty (Williamson and Teece, 1979).

This is the analytic framework that is now brought to bear upon the study of multinational enterprise.

The emergence of multinational enterprise represents a response to a number of incentives. These can be divided into three groups: circumventing or minimizing taxes and controls, monopoly, and efficiency. Whereas all three categories yield private gain, only the last is likely to be associated with a net gain to economic welfare. The first category includes adapting to or circumventing tariffs and differential profit taxes and other distortions. For instance, vertical integration coupled with transfer price manipulation to minimize tariffs can yield private gain without necessarily enhancing economic efficiency. Monopoly power considerations may also be important, and they have been heavily emphasized in the literature. The treatment is in the spirit of the inhospitality tradition, which finds monopoly power embedded in many poorly understood business phenomenon (Coase, 1972). Considering that the multinational is a complex organizational form, it is unsurprising that some observers have interpreted it in inhospitable terms.'' More interesting, possibly more important, and certainly less well understood are the efficiency consequences of the organization of economic activity by multinational firms.'' (Williamson and Teece, 1979, p. 14.) This will be the primary focus of this paper, although efforts will also be made to delineate the circumstances under which multinational firms can be vehicles for anticompetitive behavior.

In order to proceed with the analysis of the multinational firm using the transactions cost framework developed by Williamson, it is first necessary to identify the markets that are being internalized by this particular institutional mode. For purposes of analysis, it is possible to identify several different categories of international markets that multinational firms typically span. They are the following: intermediate product markets, markets for proprietary and nonproprietary know-how, and international capital markets. A multinational firm—which can be defined as an enterprise that owns and controls activities in different countries—may internalize aspects of one or all of the above markets. The first category will involve international vertical integration. The second category will involve either international horizontal or lateral integration. The third category could involve vertical, horizontal, lateral, or conglomerate integration. The efficiency properties associated with internalizing these markets will now be examined.

Vertical Integration Across National Boundaries

The emergence of multinational firms is often traced to the sourcing of raw materials such as oil, copper, and aluminum. If intermediate product markets for these commodities are well developed on an international scale and work in the frictionless textbook fashion, the efficiency gains from MNEs internalizing these markets might be problematic. However, as was explained earlier, intermediate product markets often do not operate according to the textbook ideal, in which case vertical integration is likely to have compelling efficiency properties. These are summarized below. The treatment closely follows Williamson (1975, Chapters 5, 6, and 7).

One important attribute to vertical integration is that it is an efficient method for achieving supply reliability. Thus, consider a production facility in one country that utilized raw materials and/or intermediate products produced or to be produced in another. Assume, to be quite general, that changes are required from time to time in the quantities or type or grade or design of the inputs and that the requirements are specific to each production facility. These changes are necessitated, it is assumed, by changes in the economic environment. The following question is now posed: What is the most efficient way to arrange for supply under these circumstances? At least three different solutions can be envisaged. The firm could simply rely on spot purchases to satisfy its requirements. Alternatively, the firm could negotiate various kinds of short- and long-term contracts with upstream suppliers, or it could integrate backwards into the production of the input and arrange all of its supply requirements internally.

Consider the possibility of relying on the spot market. Because of the assumption that the requirements of the production facility are subject to a degree of uncertainty and that the requirements are specific to a given production facility, spot purchases need not provide an assured supply in a timely and reliable fashion. Because the firm's requirements are, by assumption, specific to itself, the relevant intermediate product markets are likely to be quite thin. The result is that the spot market may from time to time "vanish" in the sense that the requisite inputs may not be available at the right time and at the right place. Long- or short-term contracts, however, do offer the possibility of surmounting this difficulty. Consider, therefore, the possibility of writing a long-term contract or a series of short-term contracts to secure the necessary inputs from upstream suppliers.

Clearly it is impossible, or prohibitively costly, to attempt to write a comprehensive contract in which contingent supply relations are exhaustively stipulated. Furthermore, incomplete long-term contracts pose obvious trading risks, as there is an incentive for each party to bargain opportunistically when contractual ambiguities develop. Haggling is likely to ensue, and when the stakes are high there is an incentive to invest considerable resources in the haggling process. Eventually, the haggling process may involve the courts. Besides haggling costs arising from genuine ambiguities in the contract, it very often turns out that it is quite impossible or prohibitively expensive to enforce a valid contract. This is especially true if the terms of trade change significantly against one of the parties. Thus, both complete and incomplete once-and-for-all contracts experience difficulties. Consider whether short-term contracts could be employed instead. These, presumably, would permit terms to be redrawn at the contract renewal interval; new information could be appropriately taken into account as events unwind and only a short-term forecast of the immediate future would be required.

While short-term contracts have advantages in these respects, they pose additional problems if 1) efficient supply requires investment in special-purpose, long-life equipment or 2) the winner of the initial contract requires a cost advantage by reason of first-mover position. Therefore, resort to short-term contracts as a means of fulfilling supply requirements is unlikely to prove satisfactory. The difficulty with 1) is that optimal investment considerations favor the award of a long-term contract so as to permit the supplier to amortize his investment confidently. The difficulty with 2) is that unless the total supply requirements for the first and all subsequent contracts are stipulated, a "buying-in" strategy would be risky for the supplier. Accordingly, even if bidders for the supply contract capitalize anticipated future first-mover gains (in other words, the bid price and anticipated loss on the first contract is set so as to exactly offset anticipated monopoly returns, appropriately discounted, on subsequent contracts) a risk premium will most likely be attached to the initial bid price. Short-term contracts thus experience what may be serious limitations in circumstances where nontrivial first-mover advantages obtain.

Vertical integration can attenuate or avoid the problems associated with market contracting. Divergent interests between parties can be reconciled by common ownership of the upstream and downstream facilities. Vertical integration harmonizes interest and offers internal control procedures for reconciling differences. This permits an adaptive, sequential, and therefore more efficient decision process to be utilized. Furthermore, vertical integration facilitates investment planning. It permits interdependent units to adapt to unforeseen contingencies in a coordinated fashion. If firms in each of the various primary, intermediate, and final stages of an industry take their own observations on how events are changing (including how other parties are expected to adapt) and act accordingly, there is a risk that the resulting set of decisions will be made in a jointly incompatible manner. Vertical integration reduces this hazard, as internal planning mechanisms can be used to achieve coordination.

The transactions cost arguments favorable to vertical integration, be it domestic or international, must eventually confront the following dilemma. If vertically integrated firms have such attractive properties for economizing on transactions costs, why don't they grow until they pre-empt the intermediate product markets quite generally? The reasons, according to Williamson, are that the distinctive powers of internal organization are impaired and transactional diseconomies are incurred as firm size and the degree of vertical integration are progressively extended, with organization from held constant. Although appropriate internal reorganization can sometimes shift the point at which the marginal costs of administering the incremental transaction begin to exceed those of completing transactions through the market, diminishing returns can eventually be expected. Nevertheless, the existence of limits to vertical integration do not upset the qualitative aspects of the thesis propounded.

Williamson has identified three types of diseconomies that are commonly associated with internal organization (1975, p. 119). They can be labelled an internal procurement bias, an internal expansion bias, and a program persistence bias. The internal procurement bias has its roots in the existence of an internal source of supply that tends to distort procurement decisions. Divisional or subgroup goals may be given too much weight in relation to objective profitability calculations. The internal expansion bias is driven by knowledge that the reward system for top management is often positively correlated with firm size, inducing management to engage in decisions to expand the size of the firm, even if it is not in the stockholders' best interest to do so. Finally, the existence of a base for cross-subsidization enables the managers of large firms to maintain uneconomic dividions where the market would lead to their much quicker elimination. In this respect, Drucker contends that "while no institution likes to abandon anything, budget based institutions are more prone to persist with unproductive or obsolete projects than are revenue-based institutions since the necessary support for the latter will be removed by the market" (Drucker, 1973). Supporting all three biases are the distortions that can occur in strategic information passing horizontally and vertically within the firm. Although the intention of internalization is to harmonize the incentives of the various individuals and groups in the firm, individuals within the firm can sometimes seek to promote personal goals by distributing false or misleading information through the firm's communication channels.

It is thus apparent that internal and market organizations are both subject to opportunism. Shifting a transaction from the market to the firm is significant not because a small-numbers exchange relation is eliminated, but rather because the incentives of the parties are transformed. In the case of internal organization, opportunism takes the form of subgoal pursuit and information distortion; and "although internal auditing and experience rating serve to check distortions, the general management of the firm is nevertheless severely limited in information impactedness respects" (Williamson, 1975, p. 125). It is simply too expensive or perhaps infeasible for the general management to be appraised of everything that is going on. These problems are likely to become more severe as the firm becomes progressively more complex. Hence, the distortion-free exchange is a fiction. When a transaction is shifted from the market to a vertically integrated firm, then a trade-off is implicitly made in which it is understood that some irreducible degree of opportunism will remain.

Williamson's perspective on vertical integration outlined above perhaps concedes too little in terms of the inefficiency of market processes, particularly when the intermediate product in question is to be supplied from a less-developed country in which the infrastructure necessary to support market processes is absent or poorly developed. Thus, if entrepreneurship is not ubiquitous or if the know-how needed to produce the item in question is absent, then price signals need not induce the necessary reallocation of resources. That is, raising (lowering) the price of an intermediate product need not call forth additional (less) output from indigenous suppliers, in which case the efficiency properties of international vertical integration should appear to be all the more compelling. Administrative processes could then be called upon to transfer know-how and to establish productive capacity abroad at least until development processes are sufficiently advanced that intermediate product markets can function with an acceptable degree of efficiency. Accordingly, vertical integration may also be associated with the internalization of markets for know-how. The internalization of know-how markets turns out to be an important aspect of multinational enterprise and is addressed in the following section.[2]

Markets for Know-how and the International Transfer of Technology by Multinational Firms

Perhaps the most important efficiency property of the multinational firm is that it is an organizational mode capable of transferring know-how in a relatively efficient fashion. The importance of this property looms large once it is acknowledged, first of all, that lack of know-how is one of the principle features of underdevelopment. The corollary is that modernization requires the utilization of the world stock of available knowledge. Second, even among countries of similar levels of economic development, the differential distribution of know-how and expertise means that mutually advantageous trading in know-how is profitable but will take place only if institutional modes exist to provide the appropriate linkage mechanisms and governance structures to surround the transactions. Third, markets are seriously faulted as institutional devices for facilitating trading in many important kinds of technological know-how. As points one and two are common knowledge, it is the third point that appears to need the most elaboration.

Thus, consider the modern business enterprise characteristic of developed Western economies. A principal feature of the enterprise is that it is an organizational entity possessing technological, organizational, and managerial know-how. Know-how of these kinds has some of the characteristics of a public good in that it can sometimes be used in a subsidiary without its value to the parent being substantially impaired. Furthermore, the marginal cost of employing knowhow in a subsidiary is likely to be much less than its average cost of production and transfer. Accordingly, although know-how is not a pure public good[3] the transfer of proprietary information abroad is likely to be profitable if organizational modes can be discovered to conduct the transfer at low cost. In this regard, the relative efficiency properties of markets and organizations—such as the multinational enterprise—need to be assessed. If reliance on market processes is surrounded by special difficulties, and hence costs, then internal organization, and in particular foreign ownership of production facilities, may be preferred.

An examination of the properties of markets for know-how readily leads to the identification of several difficulties. They can be summarized in terms of 1) recognition, 2) disclousre, and 3) team organization. Consider a firm that has accumulated know-how that can potentially find application in foreign markets. If there are firms abroad that can apply this know-how with profit, then according to received microtheory, trading will ensue until Pareto Optimality conditions are satisfied. Or, as Calabresi has put it, "if one assumes rationality, no transactions costs, and no legal impediments to bargaining, all misallocations of resources would be fully cured in the market by bargains" (Calabresi, 1968). However, one cannot in general expect this happy result in the market for proprietary know-how. Not only are there high costs associated with obtaining the requisite information, but there are also organizational and strategic impediments with using the market to effectuate transfer.

Consider, to begin with, the information requirements associated with using markets. In order to carry out a market transaction it is necessary to discover with whom one wishes to deal, to inform people that one wishes to deal and on what terms, to conduct negotiations leading up to the bargain, to draw up the contract, to undertake the inspection needed to make sure that the terms of the contract are being observed, and so on (Coase, 1960, p. 15). Furthermore, the opportunity for trading must be identified. As Kirzner has explained:

> for an exchange transaction to be completed it is not sufficient merely that the conditions for exchange which prospectively will be mutually beneficial be present; it is necessary also that each participant be aware of his opportunity to gain through exchange. . . . It is usually assumed . . . that where scope for [mutally beneficial] exchange is present, exchange will in fact occur. . . . In fact of course exchange may fail to occur because knowledge is imperfect, in spite of conditions for mutually profitable exchange (Kirzner, pps. 215–16).

The transactional difficulties identified by Kirzner are especially compelling when the commodity in question is proprietary information, be it of a technological or managerial kind, and where the trading partner is located in a LDC. The protection of the ownership of technological know-how often requires suppressing information on exchange possibilities. By its very nature, industrial R&D requires disguising and concealing the activities and outcomes of the R&D establishment. As Marquis and Allen point out, industrial laboratories, with their strong mission orientation, must

> cut themselves off from interaction beyond the organizational perimeter. This is to a large degree intentional. The competitive environment in which they operate necessiates control over the outflow of messages. The industrial technologist or scientist is therefore essentially cut off from free interaction with his colleagues outside of the organization (Marquis and Allen, 1966, p. 1055).

Except insofar as production or marketing specialists within the firm perceive the transfer opportunity, transfer may fail by reason of nonrecognition.

Even where the possessor of the technology recognizes the opportunity, markets may break down because of the problems of disclosing value to buyers in a way that both is convincing and does not destroy the basis for exchange. A very severe informational problem exists, on which account the less-informed party (in this instance the buyer) must be wary of opportunistic representations by the seller. If, moreover, there is sufficient disclosure, including veracity checks thereon, to assure the buyer that the information possesses great value, the "fundamental paradox" of information arises: "its value for the purchaser is not known until he has the information, but then he has in effect acquired it without cost" (Arrow, 1971, p. 152).

Suppose that recognition is no problem and that buyers concede value and are prepared to pay for information in the seller's possession. Occasionally this may suffice. Thus a formula for a chemical compound or the blueprint for a special device may be all that is needed to effect the transfer. However, more is frequently needed. Know-how has a strong learning-by-doing character, and it may be essential that human capital in an effective team configuration accompany the transfer. Sometimes this can be effected through a one-time contract to provide a "consulting team" to assist start-up. Although such contracts will be highly incomplete, and the failure to reach a comprehensive agreement may give rise to dissatisfaction during execution, this may be an unavoidable, which is to say irremediable, result. Plainly, foreign investment is an extreme response to the needs of a one-time exchange. In the absence of a superior organizational alternative, one-time, incomplete contracting for a consulting team is likely to prevail.

Where a succession of exchanges is contemplated, however, and as two-way communication is commonly needed to promote the recognition and disclosure of opportunities for information transfer as well as the actual transfer itself, reliance on repeated contracting is less clearly warranted. The parties in these circumstances are effectively joined in a bilateral monopoly trading relation; as discussed by Williamson, such contracting is shot through with hazards (Williamson, 1975). A more cooperative arrangement for joining the parties here enjoys a greater comparative institutional advantage. Specifically, intrafirm transfer to a foreign subsidiary, which avoids the need for repeated negotiations and attenuates the hazards of opportunism, has advantages over autonomous trading. Better disclosure, easier agreement, better governance, and more effective team organization and reconfiguration all result. Here lies an incentive for foreign direct investment. Furthermore, the need for other firms, the potential licensees, to evaluate the firm's knowledge base is attenuated because management, by fully employing resources internally, economizes on information requirements.

The above arguments, while couched in the context of technological know-how, are in fact quite general and extend to many different kinds of proprietary information. Technological know-how, manageral (including organization) know-how, and goodwill (including brand loyalty) all represent types of assets for which markets may falter as effective exchange mechanisms and for which a foreign investment strategy is therefore suggested.

International Capital Markets

Until the late 1960s the prevailing view of direct foreign investment by multinational firms was that it represented one process by which rates of return to capital were equalized across countries. It was subsequently realized that this interpretation was wanting, in that the equalization of rates of return did not imply that ownership of the means of production was required. A portfolio position would appear to achieve the same objectives at lower cost. Nor would pure diversification advantages seem to be involved. As modern financial theory has demonstrated, a security's risk and return can be decomposed into two elements: 1) what is specific to each company and called "unsystematic" because it can be diversified away and 2) what is "systematic" because it is common to all securities and hence is nondiversificable. As the unsystematic risk can be eliminated through simple portfolio diversification, the investor does not need widely diversified multinational companies to eliminate this risk. Hence, it would be surprising if capital transfers were a significant feature of foreign direct investment. Indeed, the available evidence indicates that only a small percentage of foreign affiliates capital expenditures are financed by the parent companies. A study of 115 foreign subsidiaries for the U.K. revealed that local sources (specifically, retained earnings, depreciation allowances, liquid assets, and local bank loans) accounted for 80% of the subsidiaries' local investment (Brook and Remmers, 1970). A similar suggestion appears in the investment expenditures of majority-owned foreign affiliates of U.S. companies as compared to net capital outflows from the U.S. to all foreign incorporated and unincorporated affiliates. The ratio is typically about 3:1. In short, it is doubtful that the international transfer of capital is an especially significant attribute of foreign direct investment.

Still, there may be a number of efficiencies associated with the internalization of international capital markets. If capital markets abroad are not well developed, then an efficent internal capital market within the multinational firm may well be able to assist in the efficient worldwide allocation of resources. Funds will be allocated to high-yield uses, irrespective of national boundaries. The multinational firm can thus perform as an effective substitute for capital markets where these are poorly developed.

Potential Anticompetitive Consequences of Multinational Enterprise

An alternative explanation of multinational enterprise emphasizes the role of oligopoly and monopoly in explaining the overseas growth of MNEs. According to Hymer (1970), direct foreign investment has a dual nature:

> It is an instrument which allows business firms to transfer capital, technology, and organizational skill from one country to another. It is also an instrument for restraining competition between firms of different nations. . . . The important point is to note that the general presumption of international trade economists in favor of free trade and free factor movements, on the grounds of allocative efficiency, does not apply to direct foreign investment because of the anticompetitive effect inherently associated with it. . . . A restriction on direct investment or a policy to break up a multinational corporation may be in some cases the only way of establishing a higher degree of competition in that industry. . . . Given the oligopolistic front maintained by the firms from developed countries, the underdeveloped countries need to

devote an important share of their scarce resources to building up national enterprises which they can control and use in bargaining with foreign oligopolists.

Given the currency of these views, it is important to ascertain the degree to which MNEs can be vehicles for anticompetitive behavior. It will also prove necessary to distinguish the concept of strategic advantage from the concept of monopoly power, as it appears that Hymer and others have imputed monopoly concerns to any kind of competitive advantage the MNE may possess.

It is accordingly of interest to contrast the "market failure" view of the MNE developed above with the imperfect competition view originating with the work of Hymer (1960) and extending through Kindleberger (1969) and Caves (1971, 1974). According to this latter view, the existence of MNEs is predicated upon some monopolistic advantage. MNEs have unique assets, perhaps innovations or differentiated products, that enable them to offset the advantages that indigeneous firms have in operating in their own familiar environment. This conceptualization of the firm is quite consistent with the conceptualization accepted here. Firms do have know-how and other major assets that they transfer abroad. But to say that firms are not identical and that they have unique assets is not to say that rivalry is suppressed and that public policy intervention is thereby occasioned.

Thus, it appears that the emphasis on the "special assets" the multinational firm possesses and the "monopoly advantage" thus conferred has been interpreted by some [Hymer, 1970; Newfarmer, 1979] to imply that MNEs, by their very nature, are vehicles for anticompetitive behavior. There seems to be no other way of explaining Hymer's remark that there is an anticompetitive effect inherently associated with direct foreign investment and that "a restriction or direct investment or a policy to break up a multinational corporation may in some cases be the only way of establishing a higher degree of competition in that industry." However, the relationship between the "special advantage" that a firm might possess and the "anticompetitive effect" supposedly inherent in the multinational firm is a subtle one that must be interpreted with great care. In this regard it should be apparent that in the real world firms, like individuals, are unique. There are very few if any markets where the assumptions of perfect competition are not violated. The dynamics of competition in an enterprise economy involve first the creation and then the destruction of a firm's competitive advantage; and while the entering of multinational firms into foreign markets may be explained in terms of their possession of unique assets, this does not imply that consumer welfare is injured by this process; just the contrary is generally to be expected.

In this regard, it is of interest to examine the concept of competition and how it might be interpreted for public policy purposes. In economic theory, competition is typically defined to be a state of the market in which the individual buyer or seller does not influence the price of his purchases or sales.[4] Competitive markets will normally arise when there are large numbers of buyers and sellers, product homogeniety, perfect knowledge, and divisibility of output. There may not, however, be any real world market that meets the assumptions of the competitive model; but that does not imply that public policy intervention will secure superior results for the consumer. Bork has suggested an alternative definition of competition, which appears to have merit as a guide for public policy. Specifically, he suggest (p. 61) that:

> "Competition" may be read as a shorthand expression, a term of art, designating any state of affairs in which consumer welfare cannot be increased by moving to an alternative state of affairs by judicial decree. Conversely, "monopoly" and "restraint of trade" would be terms of art for situations in which consumer welfare could be so improved, and to monopolize or engage in "unfair competition" would be to use practices inimical to consumer welfare (Bork, 1978).

If the multinational firm possesses distinctive competitive advantages in the form of unique assets, then the exploitation of this advantage will typically enhance consumer welfare. The principal considerations that arise with respect to multinational as compared to indigeneous enterprises relate merely to the distribution of the rents associated with the employment of the firms' unique assets; but as, by assumption, these assets are transferred from abroad, it would not appear that host country welfare would be reduced by foreign direct investment unless deleterous externalities accompany the technology transfer process. But while appropriate public policy towards multinationals can be expected to share many features common to indigeneous firms, the international context does raise a few special considerations that need to be identified. The discussion here is restricted to issues raised by firm structure, particularly vertical, horizontal, and conglomerate integration.

Vertically Integrated Multinationals

Firms based in one country can engage in backward or forward integration into foreign markets. The anticompetitive aspects of vertical integration have received considerable elucidation in the past decade and the circumstances under which vertical integration can have anticompetitive consequences are now fairly well understood and seem to be narrowly circumscribed. Hence while leverage theories of vertical integration—by which vertically integrated firms in one market are supposedly able to extend their monopoly power into upstream and downstream markets—were once commonplace, the application of economic reasoning has largely discredited these notions. Thus, imagine an industry with two stages, component manufacture and assembly. If component manufacture is monpolized and assembly is competitive, the question arises as to whether the monopolist can increase profits by buying out the assemblers. If the manufacturer buys out the assemblers and increases the markup on assembly operations, then the markup on components must be decreased by the same amount if profits are to be maximized. Because the demand of components is derived from the demand for the assembled product, the component manufacturer cannot maximize his profits by charging a price above the monopoly price determined with reference to all relevant costs, including the costs of assembly. In short, vertical integration does not permit a monopolist to extend its market power, at least insofar as it does not facilitate price discrimination or create barriers to entry of a contrived kind, a consideration that is now examined.

Consider the possibility that vertical integration into assembly will delay entry (or otherwise make it more costly) at the component stage by making it necessary for a new entrant to enter at both stages. Suppose a potential entrant has developed competitive components. Because, by assumption, the component monopolist is integrated into assembly, the new entrant must come in at both the component and the assembly stage or independent new entrants must occur simultaneously at the assembly stage. One view of the matter is that vertical integration by the monopolist will not impede entry. Bork, for example, contends that:

> In general, if greater than competitive profits are to be made in an industry, entry should occur whether the entrant has to come in at both levels or not. I know of no theory of imperfections in the capital market which would lead suppliers of capital to avoid areas of higher return to seek areas of lower return (1969, p. 148).

Clearly, integrated entry occasions an increase in financial requirements, and the issue, as Williamson points out, is whether an increase of the financial requirements is attended by an adverse alternation of the terms under which capital becomes available (Williamson, 1975, p. 110). If it is the case that the potential entrant is not competitive in assembly, because it has no direct or related experience, then it is evident that an efficient capital market will adjust the terms of finance against the new entrant or, to quote Williamson, "to contend that the terms of finance are the same . . . implies that the capital market has equal confidence in the new entrants' qualifications to perform (assembly) activities as it does in firms that are already experienced in the business. Except in circumstances where experienced firms are plainly inept, this is tantamount to saying that experience counts for nought" (Williamson, 1975, p. 111)[5]

Whatever the merits of this position, it is clear that difficulties arise if capital markets are poorly developed or if entrepreneurial skills are absent. For instance, Posner suggests that if the existing component producer owns all of the existing assembly facilities, then a new entrant ought not have to engage in entry at both stages, because if entry at one stage is anticipated, new firms will enter at the other stage in order to provide a market for the new product (Posner, 1976, p. 198). There seems to be two basic requirements in order for this to occur. One is that market processes must be able to coordinate the investment expectations of the potential upstream and downstream producers. Posner recognizes that this ought not be a serious concern. The other requirement, which is related to the first, is that entrepreneurs will in fact come forth to take advantage of the opportunity. Though these assumptions might appear reasonable in the context of developed Western economies, they are clearly inappropriate for many less-developed countries. Accordingly, entry at both stages may well be necessary, and the additional capital may just not be available to the potential entrant because of market failure or institutional considerations. Perfect capital markets are the exception rather than the rule once an international perspective is adopted.

Accordingly, public policy towards the vertically integrated multinational enterprise ought not be neutral. Vertical integration can be used as a barrier to entry where there is horizontal market power at one stage and where capital markets experience disabilities. Whether the appropriate policy involves eliminating or encouraging the exercise of monopoly will depend on the objectives of the nation states, the ownership of the multinational, and the markets in which it is selling. Nevertheless, it is important to reiterate that, in the absence of horizontal market power, the vertically intergrated firm is without anticompetitive implications and may in fact have strong procompetitive attributes.

Another type of antisocial consequence may result from vertical integration: circumventing taxes and tariffs. Vertical integration affords opportunities for transfer price manipulation and profit-shifting. This may be an especially important concern in the context of international trade, as differential tariffs and taxes characterize much of international commerce. Vertical integration completed simply to take advantage of such considerations may yield pecuniary economies without accompanying efficiency gains.

The Horizontal Expansion of Multinational Firms

A significant portion of direct foreign investment is of the horizontal kind, that is, investment abroad is directed at producing goods and services that are currently produced domestically. Whether the potential for anticompetitive behavior exists with respect to investment of this kind will depend importantly on the scope of the relevant markets. If the relevant markets are national or regional in scope so that foreign production has no impact on the domestic market, then foreign investment will have a substantially neutral impact on competition at home. The impact on competition abroad cannot be assessed without examining the particular circumstance of the foreign market. If the investment represents new entry, then competition is generally enhanced unless entry is on such a massive scale that a dominant firm or monopoly position in an otherwise competitive market is the result. As Caves points out,

> it is seldom recognized that the multinational company is a favored entrant to industries with high barriers to entry. . . . Thus the multinational company is a likely potential entrant into national industries that might otherwise be cloistered by . . . entry barriers. . . . Besides its ability to enter a market, there is also some possibility that the multinational may be an entrant particularly disruptive of an oligopolistic consensus, especially in the early period of its presence. Its alien status may make it initially less sensitive to signals about an oligopolistic consensus emanating from established native firms (Caves, 1979, p. 60–61).

However, the case for a general policy of openness to market entry by MNEs is not completely clear-cut. "If the multinational company is good at scaling existing industrial barriers to the entry of new firms, it is also good at building up such barriers. The resources required to contrive such barriers . . . are often found in the portfolios of multinational companies" (Caves, 1979, p. 61).

Evaluation

It seems clear that the MNE has few distinctive characteristics that ought to be the focus of special public policy initiatives. If offensive forms of market behavior do exist, the most direct approach is to regulate or prohibit the behavior directly rather than blocking foreign investment. As most of the monopolistic abuses typically attributed to MNEs can also characterize indigeneous firms with no extraterritorial investments,[6] nondiscriminatory regulatory approaches would appear to be desirable. Conduct designed to reduce the competitiveness of host country markets is socially undesirable no matter the identity of the firm engaging in it.

In short, it is not clear that the MNE should be considered a distinctive organizational form with respect to antitrust issues. To the extent that social abuse is attributable to the MNE, the same abuses are typically open to indigeneous firms as well. In this regard, consider the Brazilian electrical industry (Newfarmer, 1979). Newfarmer identifies what he considers to be seven forms of MNE conduct: interlocking directorates, mutual forebearances, control of supply channels, cross-subsidization of production, formal and informal collusion, formal political ties, and MNE acquisition behavior. He concludes that MNEs "exhibit strong propensities towards organizing and preserving various forms of market power in host economies. These tactics are often based on the advantage of global financial strength or perceived international and local interdependence" (Newfarmer, 1979, p. 135). Although Newfarmer is able to provide persuasive evidence of collusion in the Brazilian electrical industry, the identified abuses might equally well have been characteristic of Brazilian firms. His recommendation that "as one tool of national planning, serious measures to curb restrictive practices, to prevent non-socially beneficial takeovers, and to generally counterbalance the disadvantaged position of domestic entrepreneurs" (1979, p. 136) would not seem to be a well-founded proposal if it is directed at MNEs alone.

III. THE R&D AND TECHNOLOGY TRANSFER ACTIVITIES OF MULTINATIONAL FIRMS

It is clear from the preceding discussion that an important attribute of the business enterprise is that it is an organizational mode capable of transferring know-how among its various parts in a relatively efficient and effective fashion. Given the opportunities that apparently exist for international trade in know-how, it should not be surprising to find MNEs frequently engaging in the internal transfer of technology. Whether MNEs have an incentive to engage in R&D abroad is a different, but nevertheless related matter, which has not been examined from a theoretical perspective. (The paper by Thomas Pugel in this volume is an exception.) In this section attention is focused on both the theoretical and the empirical dimensions of the issue.

Choice of Technology Transfer Channel

Section II identified, from a theoretical perspective, some of the difficulties occasioned by the use of markets for trading in proprietary know-how. However, it should not be supposed that markets are always faulty. It is to be expected that, when problems of recognition, disclosure, and team transfers are not severe, market processes can be utilized, in which case the licensing of know-how among nonaffiliated enterprises will be observed. Recognition, disclosure, and team transfer problems will be modest, it would seem, when: 1) the know-how at issue is not recent in origin so that knowledge of its existence has diffused widely, 2) the know-how at issue has been commercialized several times so that its important parameters and performance in different situations is well understood, thereby reducing the need for start-up assistance, and 3) the receiving enterprise has a high level of technological sophistication. Some evidence supportive of these propositions has recently been presented. Mansfield, employing a sample of 23 multinationals, discovered that foreign subsidiaries were the principal channel of transfer during the first five years after commercialization. For the second five-year period after commercialization, licensing turned out to be more important (Mansfield, 1978, p. 211–12). Larger firms also tended to rely more on internal transfer than did smaller firms, although this might not reflect relative efficiency considerations but rather the sunk costs that larger firms have already made in foreign subsidiaries. Mansfield also concluded from his field studies that joint ventures, licensing, or exports were typically not good substitutes for internal transfer (Mansfield, 1978, p. 272). These findings are entirely consistent with the theoretical propositions advanced earlier.

Technology Transfer to Overseas Subsidiaries by MNEs: Lead Time and Resource Cost Considerations

Typical of the debate surrounding the economic impact of direct foreign investment is the disparate contentions commonly advanced with respect to the returns that multinational firms are able to obtain from the international transfer of technology. Receiving countries often complain that exorbitant profits are garnished by transferors while it is not uncommon to hear concerns expressed in some countries, such as the United States, that the gains from international transfer of technology are being captured primarily by the host countries. Though these issues are unlikely ever to be resolved, research results are now available that shed some light on the issues, and additional research is in progress to augment the available evidence on these matters.[7]

With respect to the age of the technology transferred abroad, the standard view, possibly derived from the product cycle model of international trade and investment, was that transferors typically did not transfer technology abroad until the product had entered the mature phase of its life cycle, at least domestically. However, Baranson (1976) has recently presented some case studies that suggest that U.S.-based firms, driven by competitive necessity, are transfering their newest technology abroad more frequently than in the past. To investigate this issue, Mansfield and Romeo (1979) recently obtained information concerning the age of the technology transferred abroad in a sample of 65 transfers taken from 31 U.S.-based multinationals. As shown in Table 1, they have found that the mean age of the technologies transferred to overseas subsidiaries in developed countries was about 6 years, which was significantly less than the mean age of technologies transferred to overseas subsidiaries in developing countries (about 10 years). Table 1 also suggests that the mean age of the technologies transferred through licenses, joint ventures, and channels other than subsidiaries are commonly higher than the mean age of the technologies transferred to subsidiaries. This is consistent with the results previously presented that indicated that firms tend to transfer their newest technology overseas through wholly owned subsidiaries rather than via licenses or joint ventures, but the latter channels become more important as the technology becomes older. Manfield and Romeo also tested whether there had been a tendency for the average age of the technology transferred overseas to decline and concluded this to be the case for subsidiaries located in LDCs but not for subsidiaries located in developing countries or for technology transferred through channels other than subsidiaries.

Another concern of countries that generate new technology is that the transfer of technology to overseas subsidiaries will hasten the time when foreign producers have access to this technology. Some evidence has recently become available

Table 1. Mean and Standard Deviation of Number of Years between Technology's Transfer Overseas and Its Initial Introduction in the United States, 65 Technologies

Channel of Technology Transfer	*Mean (years)*	*Standard Deviation (years)*	*Number of Cases*
Overseas subsidiary in developed country	5.8	5.5	27
Overseas subsidiary in developing country	9.8	8.4	12
Licensing or joint venture	13.1	13.4	26

Source: E. Mansfield and A. Romeo, "Technology Transfer to Overseas Subsidiaries by U.S.-Based Firms." Working paper, Department of Economics, University of Pennsylvania, 1979.

on the speed with which technology "leaks out" and the extent to which international transfer actually hastens its "leaking out." The evidence, based on a sample of 26 technologies transferred abroad, indicates that the mean lag between the transfer and the time when foreign firms had access to the technology was about four years (Mansfield and Romeo, 1979). In over half the cases, the technology transfer was estimated to have had no effect at all on how quickly foreign competitors had access to the technology. On the other hand, in about one fourth of the cases, it was estimated to have hastened their access to the technology by at least three years. Technology transfer hastened the spread of process technologies to a greater degree than it did the spread of product technologies. According to the study, the most frequent channel by which the technology "leaked out" was reverse engineering. That is, foreign competitors took apart and analyzed the new or modified product to gain insights into the relevant technology. Clearly, this evidence gives only a very sketchy impression of the level and nature of the returns from international technology transfer. However, there is little evidence that the technological lead of the U.S. in various industries is about to disappear as a result of the technology transfer activities of U.S. firms. Indeed, there is some evidence, admittedly of a conjectural nature, that the international transfer of technology stimulated R&D activities by multinational firms (Manfield, Romeo, and Wagner, 1979).

The cost of transfer is also another important dimension to the technology transfer process. There are both transmittal and receiving costs as technology is more than just a stock of blueprints immediately accessible to all. Very often, considerable resources have to be deployed by both the transferor and the transferee in order to ensure the success of a project. To the extent that these resources are nontrivial in their amount, the notion cannot be sustained that know-how is a public good available at zero marginal cost. An empirical investigation of these issues based upon a sample of 26 international transfers indicated that the resource cost of international transfer is nontrivial (Teece, 1977a). Transfer costs ranged from 2.25% to 59.00% of total project costs, with a mean of 19.16%. They declined with each subsequent application of the technology and were typically lower the greater the amount of related manufacturing experience possessed by the transferee. Experience with transfer and experience with the technology appear to be key considerations with respect to the ease with which technology can be transferred abroad. However, much about the technology transfer process is poorly understood, and much more research is needed before the significance of these findings can be fully evaluated.

Overseas R&D by MNEs

Why should MNEs engage in offshore R&D? If the principal feature of an MNE is to internalize the transfer of know-how, how is it possible to explain the nontrivial percentage of R&D expenditures carried out abroad by U.S. MNEs?

Table 2. Percentage of Company-Financed R and D Expenditures Carried Out Overseas, 1960–1980: 55 Firms

	*1960**	*1965†*	*1970‡*	*1972†*	*1974*	*1980§*
35-firm subsample:						
Weighted mean	2	6	6	8	10	10
Unweighted mean	2	4	5	7	8	8
Standard deviation	3	7	7	8	10	8
20-firm subsample:						
Weighted mean	—	—	4	—	9	14
Unweighted mean	—	—	5	—	8	11
Standard deviation	—	—	7	—	10	14

Notes:
* Data were not available for 4 firms in the 35-firm subsample.
† Data were not available for 1 firm in the 35-firm subsample.
‡ Data were not available for 1 firm in the 35-firm subsample and 1 firm in the 20-firm subsample.
§ Data were not available for 9 firms in the 35-firm subsample.
Source: Mansfield, Teece, and Romeo (1979).

(see Table 2). The answer is not difficult to find, as it is apparent that adaptation is frequently required if an innovation is to earn its maximum return. Accordingly, one would expect to observe that the percentage of a firm's R&D expenditures carried out overseas would be directly related to the percentage of the firm's sales that is derived from abroad. One would also expect that the percentage of a firm's R&D expenditures carried out overseas would be more closely related to the percentage of its sales from foreign subsidiaries than to its percentage of sales from exports, as the existence of separate foreign manufacturing facilities will open the opportunity as well as occasion the need (in the case of process technology) for adaptation. Empirical support for these propositions has recently been discovered using data from 55 major U.S.-based manufacturing concerns (Mansfield, Teece, and Romeo, 1979).

IV. MNEs AND ECONOMIC DEVELOPMENT: ISSUES RELATING TO THE CHOICE OF TECHNIQUE

The choice of technique in less-developed countries is one of the central issues in development economics and perhaps also in the process of economic development itself. The multinational firm has been catapulted onto center stage because it is an important actor in the technology choice process. The results are often considered to be unfortunate, as the MNEs are said to transfer technology inappropriate to the needs of the developing countries. As a result of this perceived misconduct, the United Nations Group of Eminent Persons has recommended that host countries require multinational corporations to contribute towards innovation of appropriate kinds and to encourage them to do such R&D in their overseas affiliates.

Given the importance of this issue, and the central role played by the MNE, it appears necessary to analyze the technology choice process, placing the activities of the MNE into perspective. It turns out that the *process* of technology choice is very poorly understood and that much further research is needed before an adequate understanding of the technology choice process and the role of MNEs can be obtained. Until further research has improved understanding of these issues, a humble posture for researchers and commentators alike would seem to be warranted. Some elements of a needed research strategy are identified next.

Paradigm for Technology Choice

Existing models of technology choice suffer from a misplaced emphasis. While the *results* of technological choice decisions have received considerable attention, the *process* of choice has been almost completely neglected. "Yet as economic analysis acquires a broader concern with the dynamics of choice under uncertainty, it will become more and more essential to consider choice processes" (Simon, 1978, p. 2). Fortunately,

> in the past twenty years there have been important advances in our understanding of procedural rationality. . . . The importation of these theories of the processes of choice into economics could provide immense help in deepening our understanding of the dynamics of rationality and of the influences upon choice of the institutional structure within which it takes place (Simon, 1978, pp. 2, 3).

Simon's suggestion needs to be taken seriously, for without understanding the process of technological choice and the institutional structure within which choice takes place there can be little hope for designing effective policies to bring technological choices into line with national objectives. Accordingly, the broad outlines of a technology choice paradigm are developed next.

Technology consists of a collection of techniques. Each nation, no matter its level of development, draws to some degree from a world stock of knowledge (Teece, 1976a, 1977a). However, each nation receives (or has access to) only a subset of this world stock. The information acquisition and processing capabilities of organizations and individuals necessarily restrict a nation's ability to draw comprehensively from the available stock. The decisions and activities of "gatekeepers" (multinational firms, the indigeneous science community, and the like) determine that portion of the stock of knowledge that is available for domestic use. Entrepreneurs or individuals performing entrepreneurial functions then pick from the menu of offered techniques, using particular decision procedures. The choices made depend not only on the procedures used but on the goals of the decision-makers and the constraints imposed by the economic environment.

The essence of this paradigm is that the choice of technique can be thought of as being influenced by three basic sets of factors. First, the availability of techniques determines the menu from which the user can choose. The techniques available depend on the mechanisms that couple the international scientific and engineering "community" to the user and on the capabilities of the indigeneous research, development, and engineering community. Second, the economic policies of individual nation states will determine the costs and benefits associated with utilizing particular techniques and accordingly determine the contextual framework for individual decision-making. Third, the characteristics and goals of individual decision-makers determine the choices actually made within the constraints imposed by the first and second sets of factors. The choice of technique can accordingly be thought of as the result of applying two sets of screens to the potentially available set of techniques. The first screen is that imposed by domestic economic policy, and the second is that proposed by the goals and capabilities of the decision-makers involved in choosing among available techniques. The paradigm is presented schematically in Figure 1. Although this paradigm is very simple, it serves to highlight and differentiate the various factors that determine technology choice outcomes.

Linkage Mechanisms: MNEs and the Indigenous Science Community in LDCs

As the internal generation of technology in any one less-developed country is miniscule by world standards, a critical factor determining technological choice is the manner in which decision-makers are connected to the world stock of knowledge. Because the world stock of technical and organizational knowledge is so vast and so diverse, the particular fashion in which LDCs are connected with advanced countries is likely to have a significant effect on the menu of techniques to which the LDC has ready access. At least four principal linkage mechanisms can be identified: 1) multinational firms, 2) equipment vendors, 3) foreign aid agencies, and 4) the indigenous science community.

The multinational firm is clearly one of the most significant organizational linkages between LDCs and external technological and managerial know-how. The MNC often not merely supplies technology (as when a licensing agreement is made) but may constitute a more or less permanent organizational linkage facilitating the continuous flow of technological information among various parts of the enterprise located in different nations. Because different MNCs are linked to different sources of technology, it can be expected that the menu of choices potentially available to MNC licensees and subsidiaries will vary according to the national origin of the firm and its technology transfer strategies and policies.

In addition to the licensing and direct investment activities of international firms, it is important to recognize that the equipment vending activities of multinational and domestic firms may also provide an important connection to the world stock of techniques. Besides the technology embodied in machines, there is equally important user information that may be provided along with the hardware. The advertising and other information-dissemination activities of

Figure 1. A Technology Choice Paradigm

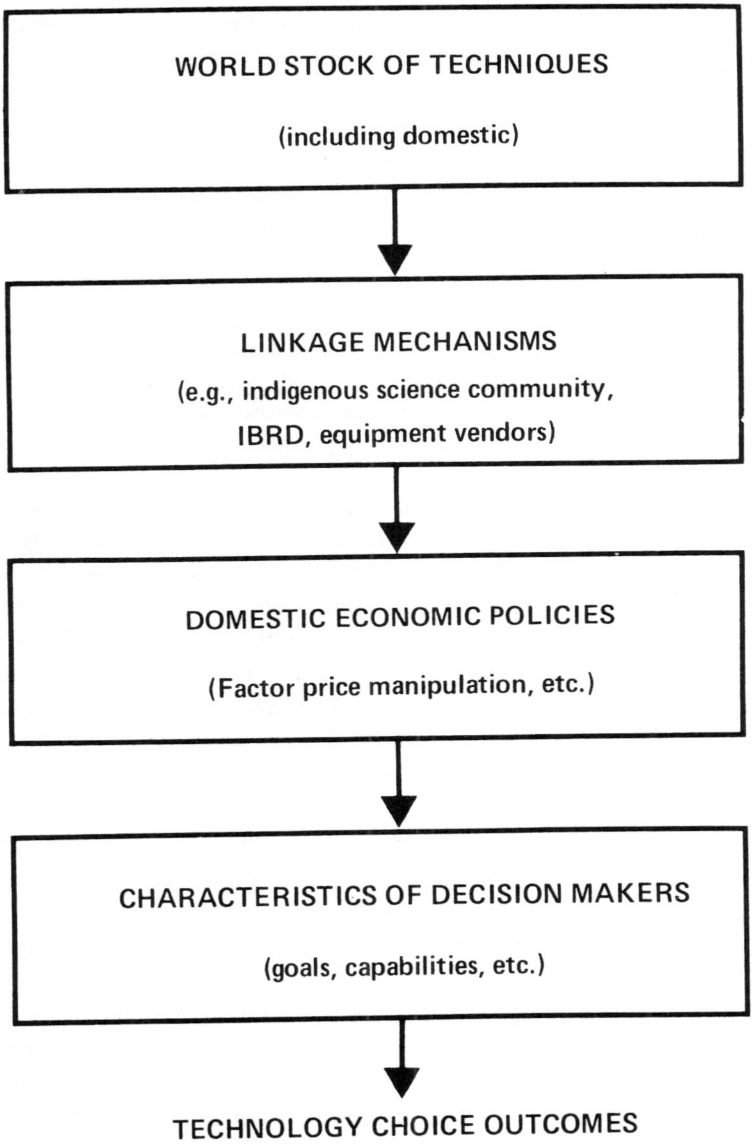

equipment vendors is often an important source of information on available techniques.

Aid agencies, government procurement agencies, and the IBRD also provide linkages to the world stock of techniques. They filter and screen the available techniques and in some cases stipulate the technology to be employed and the country from which it must be purchased.

The last linkage mechanism identified above is the indigenous scientific community, which includes the universities and R&D organizations, public and private. These organizations play a potentially critical role in technology choice for they sit between the international scientific community and technology decision-makers in LDCs. Ideally, they are linked to both and screen and adapt technology for local needs. However, the performance of indigenous industrial and agricultural R&D facilities in LDCs has not been spectacular. The source of the difficulty does not appear to be that linkages to the international community are weak; rather, the difficulty seems to be the inadequate linkage to indigenous users (Sunkel, 1971; Vietorisz, 1973). As Sunkel notes:

> For a technoscientific literature to influence the process of development there must be close institutional links with the productive structures of society as well as with the political and administrative structures that control decisions in the field of development policy. The lack of these links, rather than the actual amount of research, is the main characteristic of technical and scientific underdevelopment.

Perhaps the problem can be traced to the organization of indigenous R&D. The following appear to characterize the organization of R&D in many less-developed countries: 1) practically all R&D is funded by the government, and 2) most R&D is conducted in government and university laboratories. In short, the potential users are typically not involved in the R&D process. Channels of communication between the R&D establishment and users are typically nonexistent. It is not therefore surprising that a field researcher who visited 50 research centers in 13 developing countries concluded that in general their activities were not relevant to domestic problems (Blackledge, 1972). Domestic firms apparently did not have confidence in the results the centers could produce, and the research centers were not aware of the needs of potential users. This also appears to be true of research in the universities, which is oriented towards appealing to the international scientific community. Local problems and specifically the problems associated with social and economic development receive insufficient attention (Harrera, 1973).

The few basic research centers of quality are generally closely connected with the scientific systems of advanced countries. They function as isolated enclaves that do little to encourage technological research to suit local needs. However, one of the very few in-depth studies of scientific communities in developing countries (McCarthy, 1972) discovered that scientists located at regional universities in the Philippines were more likely to engage in research with a Philippine focus. These scientists were relatively isolated from the international scientific community and instead had contacts with industry, large agriculturalists, and professional and community groups. While there is no attempt to evaluate the relative success of the different universities, the study concluded that the presence of a "mutual delivery system," linking scientists to other segments of society, had positive attributes.

Organizational issues also seem to fault the Indian scientific establishment. A report of the government of India concluded that India's chain of national laboratories has failed to produce a significant impact on the country's technological progress. Very little of the research output is of use to industry. The report also pointed out that there is no arrangement for enabling scientists to visit industrial establishments and acquire a detailed knowledge of industry problems. It is an interesting contrast that India has had more success in promoting technological development in areas such as atomic energy and defense, where the number of users is small and concentrated in the public sector. It would appear, therefore, that the coupling of the producers and users of technology is an important yet neglected dimension of technological development in LDCs.

The Contextual Framework: Economic Policies of LDCs

National governments, knowingly and unknowingly, influence technological choices by the policy decisions that take or do not take, by the economic infrastructure they mold, and by the organizational framework adopted (e.g., planned economy, market economy, public versus private ownership, etc.). As Sagasti notes:

> A basic issue is whether policy instruments used to implement industrial development policies have any impact on the behavior of productive units. Unfortunately, at times the array of policy instruments has been designed with little knowledge, or with a very naive understanding, of the nature of industrial productive activities, the rationality of entrepreneurs, or the forces orienting the expansion of industry. The result is that policies and policy instruments are formally superimposed on an industrial structure that does not respond to the prescriptive, motivating, or coercive measures they contain: although policy instruments are designed and implemented according to an assumed or perceived reality by government, often industry operates according to a different logic and responds to different stimuli (Sagasti, 1978).

What is needed is a more complete specification and identification of the elements of domestic policy that influence technological choices. Economists have focused almost exclusively on factor prices and factor price distortions, as this is what neoclassifical theory identifies as the critical determinant of choice of technique in a world of well-informed maximizers. Obsession with factor price considerations has deflected attention away from other more critical factors, including a) foreign exchange convertibility, b) import controls, c) bilateral trading arrangements, d) restrictions on royalty payments and remittances, e) patent policies, and f) product quality standards and regulation. These considerations are potentially as important if not more important than factor price considerations.

Characteristics of Decision-makers

With a few notable exceptions, the analysis of choice of technique in LDCs has typically been conducted under the assumption that entrepreneurs and farmers are rational, well-informed profit (or utility) maximizers. But this is not the case. As Steward points out:

> Those who introduce techniques into underdeveloped countries thus make a choice among the techniques available; the choice actually made depends on the nature of the decisionmakers and their objectives, the economic circumstances in the economy concerned, and the characteristics associated with different techniques, bearing in mind that their choice is confined to the techniques they know about, and that knowledge may often be incomplete or inaccurate (Stewart, 1977, p. 22).

In a similar spirit, Ranis notes also that:

> a basic limitation of the traditional approach which became increasingly apparent during the course of our research is the identification of the entrepreneur as a rational profit maximizer concerned entirely with the economic environment and the economic calculus within that environment . . . it is the understanding of decision-making which goes beyond the short-term profit maximizing calculus which is likely to illuminate what actually transpires (Ranis, 1978, pp. 11, 13).

If an improved understanding of technology choices is to be obtained, alternative theories of choice must be examined, for clearly the neoclassical approach is wanting. One of the primary ingredients of a new choice theory must be the notion of "bounded rationality." Since the early writings of Herbert Simon (Simon, 1957), bounded rationality—by which is meant the limited ability of human decision-makers to process, store, and retrieve information (i.e., behavior that is intentionally rational but only limitedly so)—has come to be recognized as an accurate portrayal of much choice behavior and as a sensible interpretation of the costs and character of information gathering and processing procedures used by human beings (Radner, 1975a, 1975b; Radner and Rothschild, 1975; Connolly, 1977). Because of bounded rationality, decision-makers very often develop decision procedures that are sensible, given the constraints, even though they might not be sensible if the constraints were removed. Simon has coined the term "satisficing" as a shorthand label for such procedures. There have been other searches for rationality in decision-making, with the result that a literature has developed containing ideas of limited rationality, contextual rationality, game rationality, and process rationality (March 1978, pp. 591–92). This literature is relevant to understanding technological choice in less developed countries. For instance, researchers from the University of Strathclyde found that the technology chosen for sugar production in Ghana (the vacuum pan process) was both more capital intensive and less efficient than an available alternative (the open pan process). The researchers concluded that

> in its strongest and most challenging form, the puzzle which has to be resolved—as a prelude to policy action—is why, given this technological flexibility and demonstrated scope for improvement even in the face of factor price distortions appraisal of developing country projects does not produce more rational results than those now obtained (Pickett, Forsyth, and McBain, 1974, p. 51).

Resolution of this puzzle is clearly of some importance. ''On the assumption that the findings for sugar and footwear in Ethiopia and Ghana can stand substantial generalization to other industries and countries, it is important to ask why decision-makers have hitherto been imprudent'' (Picket, et al., p. 51). The Strathyclyde researchers explain these imprudent decisions in ''the malign influence of the engineer and in the conceit of the economist.'' They go on to explain:

> The investment decision, as it is widely made, is an engineering one subject to a broad economic constraint. A decision is taken, for example, to establish a plant of some productive capacity in a developing country. Engineers trained according to developed country curricula are asked to design the plant. They produce blueprints for a limited number of alternatives, each of which is a variant on current ''best-practice'' technique. The alternatives are chosen, and another capital intensive technological plant is established (Pickett et al., p. 5).

This explanation is, however, unsatisfactory in that it does not seem to be supported by the rest of the analysis. The Strathclyde group claimed that decision-makers were aware of the alternative technologies (Pickett, p. 53), but they suggest that muddled thinking and analysis may well have been involved. The engineering/economist debacle implies a higher degree of rationality than the ''muddling through'' paradigm, which is also advanced.

Nevertheless, the hypothesis that engineers offer ''economists'' (perhaps ''decision-makers'' is a more appropriate term in this context) a restricted set of techniques from which to choose is consistent with the view of technological choice outlined earlier: organizations and individuals essentially prescreen techniques. This is not, of course, a directed endeavor. Rather, the activities of multinational firms and the domestic scientific and engineering community generate selective exposure to techniques on the part of the indigeneous entrepreneurs. For instance, multinational firms are more attuned to technologies available within the corporation and in the home country than they are to technologies external to the enterprise. Information about the technological possibilities appears to be critical to the technological choice process.

The central role of information in determining technological choices has also been advanced as an explanation for dualistic development (Ghandour and Muller, 1977). This view is in contrast with the traditional labor surplus/agricultural surplus absorption approach advanced in the 1950s and 1960s (Lewis, 1954; Fei and Ranis, 1964; Jorgensen, 1969). According to this paradigm the central process in economic development is the extraction and disposition of surplus labor and/or agricultural surplus from the backward sector. The paradigm assumes that the basic structural characteristics of the two sectors can be attributed to differences in the relative availabilities of factors of production. The nonutilization of reproducible capital in the backward sector and its utilization in the advanced sector are due to differences in the capital intensities of the productive process and differences in relative prices. Accordingly, the central process in economic development is the absorption of the backward sector by the advanced sector.

An alternative interpretation of the dual economy is that "entrepreneurs" in the two sectors have access to different information sets regarding available techniques. They also have different knowledge endowments such that entrepreneurs in the backward sector do not receive information on alternative techniques and if they did they would perhaps not know how to utilize this information. Hence, "because of the differences in the sectoral information sets, the production functions are so different that even with identical price ratios, production patterns in the two sectors will be distinctly different" (Ghandour and Muller, 1977, p. 633). So far, only limited empirical evidence has been assembled to support his proposition (Shapiro and Muller, 1977), and the available evidence does not seem to make it clear why farmers do not use all the information to which they have access (Leibenstein, 1977), or whether different farmers simply have access to different information sets.

The discussion above related to different methods of producing the same product. If products are differentiated by quality, then it may well be that high quality products require the utilization of modern capital-intensive methods of production (Stewart, 1977). If this proposition is correct, and there is reason to believe that it is, then choice of technique is severely circumscribed if not eliminated once the choice of product is made. Conversely, choice is extended if it is possible to make choice of product a decision variable. Thus, if products are defined in a functional sense (e.g., water conduit), the choice of technique may be quite broad (e.g., clay pipes, bamboo pipes, steel pipes, plastic pipes). Stewart argues that once a modern enclave sector has been established, it creates demands for other products, and traditional products and techniques are put at a disadvantage. Consumer demand, through advertising and the creation of elites, is shifted toward "later" products and techniques. Accordingly, older and more labor-intensive techniques find it difficult to compete with newer techniques because the traditional products that are the outcomes of such processes are no longer in demand by the modern sector elite, although they may still satisfy consumers in the traditional sector. Once again, information considerations are the key to the arguments. Imitating modern advanced-country consumption habits drives out the demand for traditional goods and severely constrains the choice of technique. However, although there may not be a proliferation of efficient techniques to produce a well-defined commodity, there could well be some degree of flexibility. Different developed countries might well have perfected efficient techniques to meet the special size and factor endowment positions of their domestic markets. There may be one best technique for each of a number of major developed countries, so that LDCs actually have a variety of alternative techniques from which to choose. Clearly, this is an empirical matter that can be resolved only by further research.

Elements of an Agenda for Empirical Research

A considerable amount of empirical research has already been done on the choice of technique, but it has mainly been of the case study variety. As White points out, "Large gaps still remain in our knowledge of appropriate factor use. The microeconomic studies of efficient alternative factor proportions have been completed for only a handful of industries" (White, 1978, p. 56). It appears that the most neglected areas of empirical inquiry are the linkage issues. To delineate the issues further and to explore the empirical validity of the above framework, it is necessary to focus on the coupling and transmittal mechanisms by which technical information is made available to less-developed countries, and then to focus on the choice of technique from among the "effectively available" techniques.

Several elements of a research strategy now become apparent. As multinational firms based in various nation states are likely to be coupled to different scientific, technological, and economic communities, firms of different national identity engaged in technology transfer will transfer different kinds of technology—technology that to a large degree reflects their own domestic market circumstances. Hence, if different transferors transferring technology to similar receivers are examined, systematic differences should be apparent in the choice of technique. If there is only one best way, then no differences should be apparent when the end product is held constant. When the end product is not held constant, but the functional end use is held constant, then according to Stewart, indeterminancy evaporates as traditional goods would then become competitive. However, it might also be the case that different advanced countries have different products for satisfying the same end use (e.g., flexible wrapping material could be met by plastic, cellophane, or paper products). It would also be of interest to check for systematic differences across transferors according to choice of product introduced to satisfy a particular end use.

An analysis is also needed of the role of indigeneous R&D in the developing countries. In view of the critical role, at least potentially, of these organizations both as "gatekeeper," and for adapting technology to suit local needs, possible sources of organizational failure need to be examined. Elements of an operational framework for the examination of these three issues is laid out next.

Differences Across MNEs: U.S.-Based vs. Japanese Multinationals

The preceding considerations suggest that one might observe systematic differences across transferors with respect to choice of technique. Indeed, there are assertions in the literature to that effect. For instance, a Japanese observer claims that one can identify a "Japanese type" and an "American type" technology transfer (Kojima, 1978, Chapter 7). According to Kojima, Japanese direct investment is characterized by the orderly transfer of technology, while

American-direct investment is characterized by "technology transfer in reverse order." Furthermore, the characteristics of Japanese direct investment makes it

> easier to transplant new industry having large spillover effects in the host country and makes the new industries competitive in the international market.... In short, the most suitable manufacturing industry for developing countries is traditional industry which is labor intensive, well-standardized and price-competitive. This is the kind of industry which Japanese-direct investment has transplanted (Kojima, 1978, pp. 134–5).

Kojima goes on to claim that "Japanese technology transferred to developing countries is not so much specific production techniques but rather knowhow or general industrial experience involving not the latest but mature techniques...." Because of the nature of technology transferred (i.e., mostly production techniques of standardized products and general industrial experience), no fundamental technical change seems to be required of a given technology so transferred (Kojima, 1978). The argument seems to be that the *products* which the Japanese transfer are more appropriate in that they are more mature. They are labor-intensive products driven out of Japan by rising wages. In contrast, he claims that

> American-direct investment has been concentrated on highly technology-intensive products not only in developed economies but also in developing areas.... There can be hardly any possibility that an American entrepreneur will launch into direct investment to developing nations in the field of standardized products such as textiles and steel. Why is this the case with America? Perhaps American firms are not interested in direct foreign investments in these standardized products because they do not generate any monopolistic or oligopolistic advantage or profits (Kojima, 1978, p. 139).

The result is the claim that

> American transfer of technology and direct investment is different in all key respects from its Japanese counterpart. Big American firms transfer technology and products which do not exist in developing countries or for which there is a very big technological gap between them. This might be called "reverse-order" transfer of technology. This type of technology transfer and direct investment, heavily protected by the patent system, is motivated by monopolistic or oligopolistic objectives (Kojima, 1978, p. 143).

In contrast, Japanese companies engage in "orderly technology transfer" (Kojima, 1978, p. 144) by transferring technology only where the gap between transferor and transferee is smallest.

The question arises as to whether the same efficiency and appropriability issues which appear to drive American direct foreign investment also drive Japanese direct foreign investment. If Kojima's empirical generalizations are correct, does it reflect the fact that Japan's comparative advantage with respect to technology is different, or does it perhaps reflect that the Japanese multinational corporations "listen" more carefully to host country concerns?

A research program is needed to compare the characteristics of Japanese and American investments in specific industries in selected countries. It would be too burdensome to examine the patterns of Japanese direct foreign investment at the aggregate level. A number of micro studies would seem more manageable. Within selected industries and countries it would be of great interest to compare Japanese and American direct foreign investment with respect to 1) choice of products, 2) choice of technique, 3) scale, and 4) input requirements, and other relevant variables. At the time of going to press, Professor R. Hal Mason, of the Graduate School of Business Administration at UCLA, had such a project underway. Preliminary results from Brazil do not appear to support the Kojima thesis.

An additional research issue that could be usefully addressed is an examination of 1) technology transfer to less-developed countries by the *subsidiaries* of U.S. multinationals (e.g., U.S. firms in Australia), 2) technology transfer to less-developed countries by indigenous firms with small domestic markets (e.g., Australia, New Zealand), and 3) technology transfer from indigenous firms in developing countries to indigenous firms in other developing countries (e.g., Korea and Brazil as transferors).

Consider, for example, the case of foreign-based multinational firms in Australia. During a study of direct foreign investment in Australian manufacturing (Parry and Watson, 1978), it became apparent that the Australian subsidiaries of multinational firms were involved in significant outflows of domestic technology. Specifically, 32 firms licensed to foreign firms for royalty or knowhow fees and 39 firms had technology outflows as "swap arrangements." There is evidence, then, that at least in the Australian case significant domestic technology is generated within the foreign-owned manufacturing sector, and this technology enters international trade. Recipient nations include the Philippines, Indonesia, Malaysia, Singapore, and Hong Kong. The interesting research question, recognized by Parry, is the extent to which there is technological adaptation in a medium-sized, advanced market such as Australia for the smaller developing markets of South and East Asia. It would also be of interest to compare the outflow from indigenous Australian firms and subsidiaries of foreign multinationals to see the extent to which MNE subsidiary status affects the nature of the technology adaptation and the direction of the transfer. In both cases it would be possible and of interest to compare the extent of further adaptation undertaken or still needed in the Asian markets. (It might also be possible to examine whether the expectation of the "re-export" of technology affects the nature and extent of the importation of technology into countries such as Australia). Similar studies of the adaptation process can be made with respect to the outflow of technology from countries such as New Zealand (agricultural technology), Brazil, Mexico, and India.

With respect to transferors based in LDCs such as Brazil, Mexico, and India, it could well be that the strategy motivating technology transfer is quite different. Wells hypothesizes that lack of attention to cost minimization by advanced-country-based MNEs leaves a market niche for MNEs from developing countries (Wells, 1977). However, the implications for the choice of technique are not immediately obvious.

The Indigenous R&D Establishment

Richard Nelson has succinctly laid out a research strategy as follows:

> One of the research tasks to which I would assign high priority would be a detailed examination of industrial R&D in the LDC's, both public and (where it exists) private ... many applied R&D facilities have published lists of their accomplishments. These of course need to be scrutinized, but, more important, their impact needs to be evaluated. The evaluation needs to consider the specific economic benefits such as productivity enhancement and cost reduction, export yield, etc. But, more broadly, it seems important to examine the extent to which a national R&D policy and availability of local engineers and applied scientists can reduce dependence on foreign corporations for modern technology, the relative effects of these two means on employment, income of nationals, exports, etc. Here the Japanese case seems particularly worth examining in detail as well as Mexico's and India's experience with public applied industrial R&D (Nelson, 1974, p. 75).

As was pointed out earlier, examining the role of the indigenous R&D organizations is important with respect to discovering the conditions under which these organizations influence the choice of technique. The kinds of coupling mechanisms that have proved to be the most appropriate in LDCs need to be identified. No doubt the results will vary from sector to sector, so attention must be given to designing a methodology that would control for the relevant external parameters.

V. CONCLUSION

The MNE is an organizational construct that has been extremely controversial for several decades. In this paper some real economies (especially economies from vertical integration and from technology transfer) associated with this organizational mode have been identified.

Of course not all direct foreign investment need be driven by attempts to capture real economies. One suspects, though it is difficult to prove, that pecuniary economies often trigger firms based in one country to establish subsidiaries in another. Whether multinational firms are open to forms of anticompetitive abuse not associated with purely domestic enterprises was also examined. It turns out that there are few anticompetitive abuses uniquely associated with multinational firms. The formulation of national economic policy should take these various considerations into account. Clearly, MNEs based on real economies will enhance global efficiency, although the distributional implications are ambiguous. Although most of the efficiency gains can be expected to accrue to the stockholders (if the firm's objective is to maximize stockholder wealth), the host country will also gain (even assuming that all of the stockholders are nonresidents) because of externalities and the increased tax revenues which the efficiency gains should occasion.

The theoretical analysis also served to focus attention on the potential sources of real efficiency which may be associated with multinational enterprise. Besides vertical integration economies, the transfer of proprietary knowhow is a task that MNEs can be expected to perform more efficiently than alternative modes, especially when the difference in the level of development between transferor and transferee is acute. However, it is often alleged that the technology which is transferred abroad by this channel is inappropriate for the host country, especially when the host country is less developed. Though this may be the case, the process of technology choice is so poorly understood that it is difficult to identify the appropriate policies to redress the problem. Further research is needed, and the elements of an empirical research project were outlined that might illuminate some of the issues. What is missing in this area is imaginative research with solid theoretical and empirical foundations. If this deficiency can be redressed, then a foundation will have been established for fashioning public policy capable of advancing rather than confounding the process of economic development.

ACKNOWLEDGMENT

The financial support of the National Science Foundation and the Center for the Study of Organizational Innovation, University of Pennsylvania, is gratefully acknowledged. This support enabled me to write this chapter while visiting the University of Pennsylvania in 1979. I am also deeply indebted to Oliver Williamson, whose writings on the economics of internal organization have had a pervasive influence on the way in which I view the MNE. Some of the ideas in this paper were worked out together while we were preparing a joint paper on European Integration (Williamson and Teece, 1979).

NOTES

1. This section is based on Williamson (1975) and Williamson and Teece (1979).
2. The following discussion is based on Williamson (1975, Chapters 5, 6 & 7) and partly on Williamson and Teece (1979) and Teece (1980).
3. This is because the value of information often declines with its dissemination, and it cannot be transferred at zero marginal cost.
4. More specifically, each and every buyer faces an infinitely elastic supply curve and each and every seller faces an infinitely elastic demand curve.
5. There is an issue here of whether the potential entrant will have the terms of finance adjusted against him by an amount greater than the monopolist would have experienced when the monopolist integrated into assembly. It would seem that since, by assumption, the monopolist had no initial advantage with respect to assembly, it would have confronted the same capital costs. If entry barriers are defined (following Stigler) as a condition which imposes higher long-run costs of production on a new entrant than are borne by firms already in the market, then it is not clear that the Williamson scenario ought to be considered to generate a barrier to entry.
6. However, a number of forms of abuse are open to vertically integrated MNEs that are not open to vertically integrated firms that are purely domestic in the scope of their activities. Perhaps the most

tional trade. Though wealth transfer considerations are involved, there need not be any competitive implications.

7. A research project is underway at Stanford University which is attempting to measure the returns that U.S. multinationals obtain from the international transfer of their technology.

REFERENCES

Arrow, K. J. (1971), *Essays in the Theory of Risk Bearing*. Chicago: Markham.

Baranson, J. (1976), "International Transfers of Industrial Technology by U.S. Firms and Their Implications for the U.S. Economy." Report to the U.S. Department of Labor.

Blackledge, J. (1972), *The Role of the Research Institute in Industrial Growth*. Unpublished paper, Denver Research Institute.

Bork, R. (1979), "Vertical Integration and Competitive Processes." In J. Fred Weston and S. Peltzmon (eds.), *Public Policy Towards Mergers*. Pacific Palisades, California: Goodyear Publishing Company.

______ (1978), *The Antritrust Paradox*. New York: Basic Books.

Brooks, M. and Remmers, H. (1970), *The Strategy of Multinational Enterprise*. New York: American Elsevier.

Buckley, P. J. and Casson, M. (1976), *The Future of the Multinational Enterprise*. London: Holmes & Meier Publishers, Inc.

Calabresi, G. (1968), "Transactions Costs, Resource Allocation, and Liability Rules: A Comment." *Journal of Law and Economics* (April).

Caves, R. E. (1971), "International Corporations: The Industrial Economies of Foreign Investment." *Economica* No. 149 (February).

______ (1974), "Industrial Organization." In John Dunning (ed.), *Economic Analysis and the Multinational Enterprise*. New York: Praeger.

______ (1979), "International Cartels and Monopolies in International Trade." In R. Dornbusch and J. Frankel (eds.), *International Economic Policy*. Baltimore: The John Hopkins University Press.

Coase, R., (1969), "The Problem of Social Cost." *Journal of Law and Economics* (October).

______ (1972), "Industrial Organization: A Proposal for Research." In V. R. Fuchs (eds.), *Policy Issues and Research Opportunities in Industrial Organization*. (New York: National Bureau of Economic Research.

Commons, J. R. (1934), *Institutional Economics*. Madison: University of Wisconsin Press.

Connolly, T. (1977), "Information Processing and Decision-Making in Organizations." In B. M. Staw and G. R. Salancek (eds.), *New Directions in Organizational Behavior*. Chicago: St. Clair.

Drucker, P. (1973), "On Managing the Public Service Institutions." *The Public Interest* (Fall).

Fei, J. C., and G. Ranis (1964), *Development of the Labor Surplus Economy*. Homewood, Ill. Irwin.

Ghandour, M., and J. Muller (1977), "A New Approach to Technological Dualism." *Economic Development and Cultural Change* (July).

Herrera, A. (1973), "Social Determinants of Science Policy in Latin America: Explicit Science Policy and Implicit Science Policy." *Journal of Development Studies* (October).

Hymer, S. (1960), "The International Operations of National Firms: A Study of Direct Investment. Doctoral dissertation, M.I.T. Cambridge, Mass.

______ (1970), "The Efficiency (Contradictions) of Multinational Corporations." *American Economic Review* 60 (May): 441–48.

Jorgensen, D. (1969), "The Role of Agriculture in Economic Development: Classical vs. Nonclassical Models of Growth." In C. Wharton (ed.), *Subsistence Agriculture and Economic Development*. Chicago: Aldine.

Kindleberger, C. P. (1969), *American Business Abroad*. New Haven, Conn.: Yale University Press.

Kirzner, I. (1962), *Competition and Entrepreneurship*. Chicago: University of Chicago Press.

Kojima, Kiyoshi (1978), *Direct Foreign Investment: A Japanese Model of Multinational Business Operations*. New York: Praeger.

Leibenstein, H. (1977), "X-Efficiency, Technical Efficience, and Incomplete Information Use: A Comment." *Economic Development and Cultural Change* (January).

Lewis, W. A. (1954), "Economic Development with Unlimited Supply of Labor." *Manchester School of Economic and Scoial Studies* (May).

Mansfield, E. (1977), *The Production and Application of New Industrial Technology*. New York: W. W. Norton.

______ (1978), "Statement to the Senate Commerce Committee Concerning International Technology Transfer and Overseas Research and Development." Hearings Before the Subcommittee on International Finance of the Committee on Banking, Housing, and Urban Affairs of the Committee on Commerce, Science, and Transportation, United States Senate, Ninety-fifth Congress, Second Session, Part 7: Oversight on U.S. High Technology Exports, May.

Mansfield, E., D. Teece, and Romeo, A. (1979), "Overseas Research and Development by U.S.-Based Firms." *Economica* 46:187–196.

Mansfield, E., A. Romeo (1979), "Technology Transfer to Overseas Subsidiaries by U.S -Based Firms." Research Paper, University of Pennsylvania.

Mansfield, E., A. Romeo, and S. Wagner (1979), "Foreign Trade and U.S. Research and Development." *Review of Economics and Statistics* 61(1):49–57.

March, James (1978), "Bounded Rationality, Ambiguity, and the Engineering of Choice " *Bell Journal of Economics* (Autumn).

Marquis, D., and T. Allen (1966), "Communication Patterns in Applied Technology." *American Psychologist* Vol. 21.

McCarthy, F. (1972), "Third Cultural Networks of Philippine Physical, Life, and Social Scientists." Ph.D. Thesis, Michigan University.

McManus, J. C. (1972), "The Theory of the International Firm." In Gilles Paquet (ed.), *The Multinational Firm and the Nation State*. Don Mills, Ontario: Collier Macmillan.

Nelson, R. (1974), "Less Developed Countries—Technology Transfer and Adaptation: The Role of the Indigeneous Science Community." *Economic Development and Cultural Change* (October).

Newfarmer, R. S. (1979), "Oligopolistic Tactics to Control Markets and the Growth of MNC's in Brazil's Electrical Community." *Journal of Development Studies* 15(3).

Parry, T., and J. Watson (1978), "Technology Flows and Foreign Investment in the Australian Manufacturing Sector." New York University Graduate School of Business, Research Paper No. 78-44, April.

Pickett, J., D. Forsyth, and N. McBain (1974), "The Choice of Technology, Economic Efficiency and Employment in Developing Countries." *World Development* (March).

Posner, R. (1976), *Antitrust Law*. Chicago: University of Chicago Press.

Radner, R. (1975a), "A Behavior Model of Cost Reduction." *Bell Journal of Economics* (Spring).

______ (1975b), "Satisficing." *Journal of Mathematical Economics* Vol. 2.

Radner, R., and M. Rothschild (1975), "On Allocation of Effort." *Journal of Economic Theory* Vol. 10.

Simon, H. A. (1961), *Administrative Behavior*, 2nd ed. New York: Macmillan.

Sagasti, F. (1978), *Science and Technology for Development*. Ottawa: International Development Research Center.

Shapiro, K., and J. Muller (1977), "Sources of Technical Efficiency: The Role of Modernization and Information." *Economic Development and Cultural Change* (January).

Stewart, Francis (1977), *Technology and Underdevelopment*. Boulder, Colo.: Westview Press.

Sunkel, O. (1971), "Underdevelopment: The Transfer of Science and Technology and the Latin American University." *Human Relations* No. 1.

Teece, David (1976a), *Vertical Integration and Vertical Diverstiture in the U.S. Oil Industry*. Stanford: Institute for Energy Studies.

______ (1976b), *The Multinational Corporation and the Resource Cost of International Technology Transfer*. Cambridge, Mass.: Ballinger.

______ (1977a), "Technology Transfer by Multinational Firms: The Resource Cost of International Technology Transfer." *Economic Journal* (June).

______ (1977b), *R&D in Energy*. Stanford: Institute for Energy Studies.

______ (1980), "Economies of Scope and the Scope of the Enterprise." *Journal of Economic Behavior and Organization* (1).

Vietorisz, T. (1973), "Diversification, Linkage, and Integration: Focus on the Technology Policies of Developing Countries." Seminar on the Role of Small-Scale Industries in the Transfer of Technology OECD, June.

Wells, L. (1975), "The Internationalization of Firms from the Developing Countries." In T. Agmon and C. P. Kindleberger (eds.), *Multinationals from Small Countries*. Cambridge: MIT Press.

White, L. (1978), "The Evidence on Appropriate Factor Propertions for Manufacturing in Less Developed Countries: A Survey." *Economic Development and Cultural Change* (October).

Williamson, O. E. (1975), *Markets and Hierarchies: Analysis and Antitrust Implications*. New York: Free Press.

Williamson, O. E., and D. J. Teece (1979), "European Economic and Political Integration: The Markets and Hierarchies Approach." Research Paper No. 491, Graduate School of Business, Stanford University, June.

Reprinted from *MIT Sloan Management Review*, Spring 1981, pp. 3–17, by permission of publisher.

The Multinational Enterprise: Market Failure and Market Power Considerations

David J. Teece **Stanford University**

The multinational firm continues to attract the attention of scholars and commentators of many persuasions. It has been variously interpreted as an instrument of colonial exploitation and as an efficient vehicle for the economic development of the Third World. However, emphasis on polemics typically obfuscates an understanding of the phenomenon at hand. Indeed, it is only in the last few years that a coherent and persuasive theory of multinational enterprise has begun to emerge. The purpose of this article is to pull together the various elements of a theory of multinational enterprise (MNE). Its starting point is the conceptualization of the multinational firm as an organization that internalizes various international transactions which could conceivably take place in a market. While this observation has been made by others, the implications for economic efficiency, competition, and public policy have not been fully discussed.[1] Furthermore, to the extent that the economics of internalization has been addressed, efficiency and monopoly power considerations have often been confounded, leaving the policy implications of direct foreign investment more ambiguous than is necessary.

In order to delineate a theory of multinational enterprise, it is desirable to identify some general properties of markets and of internal organization. This will be done by following the markets and hierarchies (MH) approach recently developed by Oliver Williamson.[2] A theory of multinational enterprise emerges once the nature of the transactions typically internalized by multinational firms is delineated and once related market failure considerations are explored.

Relative Efficiency Properties of Markets and Hierarchies

The MH approach developed by Williamson attempts to assess the properties of various organizational modes by comparing their relative efficiency characteristics.[3] The market is considered to be one way of organizing economic activity. The desirability of using market processes is assessed by comparing the efficiency properties of markets with those of alternative organizational modes. The concept of market failure embedded in the MH framework, therefore, is not based principally on Pareto optimality considerations. It rests instead on relative efficiency considerations. In order to assess the relative efficiency properties of various organizational modes, Williamson, following Commons, selects the transaction as the basic unit of analysis.[4] He suggests that the relative efficiency of various organizational structures can be gauged by examining the transactions cost properties of each. Transactions costs embrace all the costs associated with organizing the economic system. In a market context, for instance, transactions costs include the costs of:

- Discovering who one wishes to deal with;
- Informing market agents that one wishes to deal and on what terms;
- Conducting negotiations leading up to the bargain;
- Drawing up the contract;
- Undertaking the inspection needed to make sure that the terms of the contract are being observed.[5]

In a nonmarket context, transactions costs

similarly include the costs of identifying exchange opportunities and of employing administrative processes to organize economic activity. Whether one is assessing market or nonmarket modes of organization, the MH approach involves an analysis of the facility with which contracts can be written, executed, and enforced. This is because:

> Explicitly or implicitly, the institutional device by which transactions are organized is the contract . . . to be sure, not every problem can usefully be posed as a contracting problem. More can be described in these terms than is generally realized. Not only can an actual or implicit contract be described for every market interface, but many nonmarket relations can be expressed in contracting terms as well. For every problem that arises as or can be transformed into a contracting relation without emasculating its main features, transactions cost analysis is appropriately brought to bear. Whatever the contractual context, moreover, such an analysis ultimately reduces to an examination of the manner in which human agents cope with complex events in the face of uncertainty.[6]

This analytic framework will now be brought to bear upon the study of multinational enterprise.

Multinational Firms and Market Failure

The emergence of multinational enterprise represents a response to a number of incentives.[7] These incentives can be divided into three groups: circumventing or minimizing taxes and controls, monopoly, and efficiency. Whereas all three categories yield private gain, only the latter is likely to be associated with a net gain to economic welfare. The first category includes such distortions as adapting to or circumventing tariffs and taxes on differential profits. For instance, vertical integration, coupled with transfer price manipulation to minimize tariffs, can yield private gain without necessarily enhancing the efficient allocation of the world's resources. Monopoly power considerations may also be important, and they have been heavily emphasized in the literature. As Coase has observed, monopoly power is often imputed to many poorly understood business phenomena.[8] Considering that the multinational is a complex organizational form, it is not surprising that some scholars have missed important features of multinational enterprise; they are obsessed instead with market power considerations. But, the efficiency consequences of the organization of economic activity by multinational firms are more interesting, possibly more important, and certainly less well understood than these market power considerations. These consequences will be the primary focus of this article, although efforts will also be made to delineate the circumstances under which multinational firms can be vehicles for anticompetitive behavior.

In order to proceed with the analysis of the multinational firm, it is first necessary to identify the markets that are internalized by this particular institutional mode. For purposes of analysis, it is possible to identify several different categories of international markets, which multinational firms typically span. These markets, are intermediate product markets, markets for proprietary and nonproprietary know-how, and international capital markets. A multinational firm — which can be defined as an enterprise which owns assets and controls activities in different countries — may internalize aspects of one or all of the above categories of markets. The first category will involve vertical foreign investment. The second category will involve horizontal foreign investment. The third category could involve either vertical, horizontal, lateral, or conglomerate investment.[9] The efficiency properties associated with internalizing these markets will now be examined.

Intermediate Product Markets and Vertical Direct Foreign Investment

The emergence of multinational firms is often traced to the sourcing of raw materials, such as oil, copper, and alumina.[10] If intermediate product markets for these commodities were well developed on an international scale and worked in a frictionless fashion, there would be very few circumstances where internalizing these markets would yield efficiencies. However, intermediate product markets often do not operate according to the textbook ideal. In such cases vertical integration is likely to have compelling efficiency properties. To understand the incentive for the emergence of vertical foreign investment, it is necessary to understand the incentives for vertical integration more generally, since the reasons for vertical foreign investment are basically the same as for any form of vertical integration. Thus, if the United States were to consist of fifty independent nations, a large number of vertically integrated domestic firms would probably become classified as vertically integrated multinationals.

Perhaps the critical incentive driving backward integration abroad has been the search for reliable low-cost supplies of raw materials. Throughout much of the twentieth century, vertical integration has afforded greater security of supply than has reliance on market contracts with independent nonaffiliated enterprises. Furthermore, when the know-how and capital have not been available to develop production capacity abroad, the need for vertical integration has been even more obvious. However, assuming that an independent foreign enterprise exists abroad and that it is capable of supplying the raw materials or specialized inputs needed in downstream refining, processing, or assembling facilities, it might still be efficient to vertically integrate upstream. One reason to do so would be to avoid the hazards that are sometimes associated with relying on long- or short-term supply contracts. These hazards are sufficiently important to warrant further exploration.

In attempting to explain the complicated contractual details of actual market exchange, one can begin by noting that it is impossible or prohibitively costly to write, execute, and enforce complete, fully contingent contracts. This proposition is obvious to even the casual observer of economic phenomenon. Accordingly, the relationships between transacting parties often cannot be fully described by a court-enforceable formal document that the parties have signed.[11] Sometimes the common law of contracts supplies a body of rules and principles that are read into each contract. In many cases, though, explicit terms (which include these general unwritten terms) remain somewhat vague and incomplete. They are incomplete because uncertainty implies the existence of a large number of possible contingencies. It may be very costly to know and specify in advance responses to all of these possibilities. Therefore, contractual breach may often be difficult to prove to the satisfaction of a third party enforcer — if one happens to exist.

Given the presence of incomplete contracts, opportunistic transactors often have the ability and incentive to renege on the transaction. They do so by "holding up" the other party, in the sense of taking advantage of unspecified and unenforceable elements of the contractual relationship. Such behavior is, by definition, unanticipated. It is not sustainable in the long run.

While various forms of complex contracts can sometimes be engineered to avoid the holdup problem, there are numerous cases where this is not possible. In general, the holdup problem (and hence the incentives for vertical integration on its account) is more likely the more difficult it is to write, execute, and enforce contracts for the service in question. It is also more likely the more asymmetric the relationship, the higher the appropriable quasi rents due to firm-specific investments, and the greater the costs of "contract transfer" (the costs of switching suppliers). These considerations are developed in more detail elsewhere.[12]

By way of example, consider a smelter located in the home country. Assume that there are many different types and grades of ore and that to smelt the ore at the lowest cost the smelter needs to be designed to process a particular grade of ore. If this ore is found only in one or two locations abroad, then it may be very hazardous to rely on long-term supply contracts with another enterprise. Once the investment in idiosyncratic smelting facilities has been made, the enterprise engaged in smelting will be extremely vulnerable to changes made by the ore supplier in the conditions of sale. The supplier of the ore may well behave opportunistically, raising the price above the previously contracted level. Judicial redress may be weak, so to the extent that alternative suppliers are not available, the owner of the smelter will be obliged to honor the new terms. In extreme cases of dependence, the supplier will be able to extract a pecuniary advantage from the owner, up to the value of the smelting facility. Backward vertical integration can eliminate this risk, for if supplier and purchaser are one and the same, the incentive for postcontractual recontracting is attenuated, and trading relations can proceed in a smooth and efficient fashion. The above analysis serves to delineate an important efficiency incentive for vertical integration. But, vertical integration is not itself a costless activity, since there are control costs associated with the vertical (and horizontal) expansion of the enterprise. The international expansion of the firm, assuming organizational structure is held constant, may involve additional hierarchical layers of management. In this case, management will have less contact with personnel at lower levels. Furthermore, vertical integration may itself create various distortions. According to Williamson, three types of distortion are common: an internal procurement bias, an internal expansion bias, and a program persistence bias.[13]

The internal procurement bias has its roots in the existence of an internal source of supply that tends to distort procurement decisions. Divisional or subgroup goals tend to be given too much weight in relation to objective profitability calculations. Norms of reciprocity can develop, and the opportunities for this kind of behavior may be more extensive within the firm than in the market.

The internal expansion bias is driven by the knowledge that the reward system for top management is often positively correlated with firm size, inducing management to engage in decisions to expand the size of the firm, even if it is not in the stockholders' best interest to do so. Finally, the existence of a base for cross subsidization enables the managers of large firms to maintain uneconomic divisions where the market would lead to quicker elimination.

Supporting all three biases are the distortions of strategic information, which passes horizontally and vertically within the firm. Although the intention of internalization is to harmonize the incentives of the various individuals and groups in the firm, individuals within the firm can sometimes seek to promote personal goals by distributing false or misleading information through the firm's communication channels.

Fortunately, the adoption of appropriate organizational and control structures can relieve, if not eliminate, these problems. As a result, it is reasonable to hypothesize that a large, well-managed, integrated multinational enterprise is not subject to diminishing returns as it expands. Robin Marris has summarized the matter as follows:

> Large organizations have many weaknesses. . . . But in the narrowest sense — and this I believe is the basic organizational innovation of our century — they relearned something that was lost with the Roman Empire, namely how to control central loss. Large organizations know they are susceptible to loss of control, and they set up a variety of checks against it. Many of these are crude and have costly side effects. Others are more subtly effective. As a result, in my view, the best hypothesis is that there are generally constant returns to organizational expansion. In other words, I advocate a maxim, "When in doubt, assume constant returns." Some people will be content to agree with this maxim. Others instinctively assume diminishing returns, because they are absolutely convinced that bigness brings badness. Others still feel that where large-scale organization appears to flourish, the explanation must be found in conventional increasing returns. What is not in my view even today sufficiently realized is the extent of the explanatory power of the constant-returns hypothesis. It is in my view sufficient to explain much of the world we actually see.[14]

When the intermediate product in question is to be supplied from a less developed country (in which the infrastructure necessary to support market processes is nonexistent or poorly developed), then there are additional efficiency incentives for vertical integration. Thus, if entrepreneurship is absent or if the know-how needed to produce the item in question is missing, then price signals may not induce the necessary reallocation of resources. That is, raising the price of an intermediate product need not call forth additional output from indigenous suppliers. In this case, the relative efficiency properties of international vertical integration are especially compelling. With vertical integration, administrative processes can be called upon to transfer know-how and to establish productive capacity abroad. These processes can be used at least until development is advanced to the point where intermediate product markets can function with an acceptable degree of efficiency.

However, the transfer of know-how is a characteristic of the multinational firm that is not just associated with vertical investment decisions. Horizontal investment abroad may also involve know-how transfer. Accordingly, this property of the multinational firm must be analyzed in more detail.

Horizontal Direct Foreign Investment and the Market for Know-how

Horizontal direct foreign investment occurs when a firm with production facilities in one country establishes similar facilities in another. This phenomenon is more difficult to explain than vertical foreign investment. However, as with vertical foreign investment, market failure considerations lie at the heart of horizontal foreign investment. If markets operate in the frictionless fashion portrayed in economics textbooks, then all of the advantages from horizontal investment could be captured using contracts. But, "failures" in the market for know-how provide an important incentive for horizontal direct foreign investment. This argument can be made apparent by examining some other possible explanations for foreign investment.

If there are economies of scale in the production of commodities, then one would expect to see production geographically focused, with foreign markets supplied by exports. Clearly, transportation costs and tariffs often dictate that production facilities be located near the markets served. However, despite the importance of tariffs and transportation costs, they do not, in a fundamental sense, explain foreign investment. If one assumes that indigenous firms have (by virture of familiarity with language, customs, regulations, and local markets) a cost advantage over foreign firms in the home market, then one would expect to see very little direct foreign investment, since indigenous firms would be able to meet indigenous market needs. Clearly foreign firms sometimes have unique assets that confer advantages over indigenous firms. Furthermore, the assets possessed may yield their highest returns when utilized within the firm rather than being sold via arm's-length contracts.

The notion that the multinational firm possesses unique assets has been developed by others, including Caves.[15] However, the argument is incomplete unless market failure issues are addressed as well. Indeed, failure to consider this dimension of the multinational firm has led to a serious misinterpretation of the distinctive properties of direct foreign investment. In particular, it has deflected attention from efficiency considerations to monopoly power considerations. Accordingly, an investigation of the efficiency properties of the multinational firm is warranted.

A distinctive attribute of the firm is that it is an organization, which possesses knowledge and skills. Perhaps the most important efficiency property of the multinational firm is that it is an organizational mode capable of transferring this knowledge and skill abroad in a relatively efficient fashion.[16] The importance of this property becomes apparent once it is acknowledged that the world's stock of knowledge is unevenly distributed.[17] Economic growth and modernization require the utilization of the world's stock of available knowledge. The differential distribution of know-how and expertise means that mutually advantageous trading opportunities exist but will be realized only if institutional modes are established to provide the appropriate linkage mechanisms and governance structures to surround and protect transactions. Unfortunately, markets are seriously faulted as institutional devices for facilitating trading in many important kinds of technological and managerial know-how. This factor helps explain why the multinational enterprise is of great significance.

Thus, consider the modern business enterprise that is characteristic of developed Western economies. The know-how possessed by the enterprise has some of the characteristics of a public good, since it can often be used in another enterprise without its value being substantially impaired. Furthermore, the marginal cost of employing know-how abroad is likely to be much less than its average cost of production and transfer. Accordingly (although know-how is not a pure public good), the international transfer of proprietary know-how is likely to be profitable, if organizational modes can be discovered to conduct and protect the transfer at low cost.[18] In this regard, the relative efficiency properties of markets and of the multinational firm need to be examined.

An examination of the properties of markets for know-how readily leads to the identification of several transactional difficulties. These difficulties can be summarized in terms of recognition, disclosure, and team organization. Consider a firm that has accumulated know-how, which can potentially find application in foreign markets. If there are firms abroad that can apply this know-how with profit, then according to traditional microeconomic theory, trading will ensue until the gains from trade are exhausted. Or, as Calabresi has put it, "if one assumes rationality, no transactions costs, and no legal impediments to bargaining, all misallocations of resources would be fully cured in the market by bargains."[19] However, one generally cannot expect this happy result in the market for proprietary know-how. Not only are there high costs associated with obtaining the requisite information, but there are also organizational and strategic impediments associated with using the market to effectuate transfer.

Consider the information requirements associated with using markets. In order to carry out a market transaction, it is necessary to discover who one wishes to deal with and to inform people that one wishes to deal and on what terms. It is also necessary to conduct negotiations leading up to the bargain, to draw up the contract, to undertake the inspection needed to make sure that the terms of the contract are being observed, and so on.[20] Furthermore, the opportunity for trading must be identified. As Kirzner has explained:

> For an exchange transaction to be completed it is not sufficient merely that the conditions for exchange which prospectively will be mutually beneficial be present; it is necessary also that each participant *be aware* of his opportunity to gain through the exchange. . . . It is usually assumed . . . that where such scope is present, exchange will in fact occur. . . . In fact, of course, exchange may fail to occur because knowledge is imperfect, in spite of the presence of the conditions for mutually profitable exchange.[21]

The transactional difficulties identified by Kirzner are especially compelling when the commodity in question is proprietary information—be it technological or managerial — and where the potential trading partner is located in a less developed country. Protecting the ownership of technological know-how often requires the suppression of information on exchange possibilities. By its very nature, industrial R&D requires that the activities and outcomes of the R&D establishment be disguised or concealed. Marquis and Allen point out that industrial laboratories, with their strong mission orientation, must:

> cut themselves off from interaction beyond the organizational perimeter. This is to a large degree intentional. The competitive environment in which they operate necessitates control over the outflow of messages. The industrial technologist or scientist is therefore essentially cut off from free interaction with his colleagues outside of the organization.[22]

Except insofar as production or marketing specialists within the firm perceive the transfer opportunity, transfer may fail by reason of nonrecognition. Nonrecognition might also occur if there are few indigenous enterprises that have the requisite capabilities for absorbing the foreign know-how. In short, opportunities for arm's-length trading might have to await enterprise development in the recipient market.

Even where the possessor of the technology recognizes the opportunity and has the capability to absorb know-how, markets may break down. This difficulty arises because of the problems of disclosing value to buyers in a way that is convincing and that does not destroy the basis for exchange. Due to a very severe informational problem, the less informed party (in this instance, the buyer) must be wary of opportunistic representations by the seller. Moreover, if there is sufficient disclosure (including veracity checks thereon) to assure the buyer that the information possesses great value, the "fundamental paradox" of information arises: "its value for the purchaser is not known until he has the information, but then he has in effect acquired it without cost."[23]

Suppose that recognition is no problem and that buyers concede value and are prepared to pay for information in the seller's possession. Thus, a formula for a chemical compound or the blueprints for a special device may be all that is needed to effect the transfer. However, more is frequently needed. Know-how cannot always be codified, since it has an important tacit dimension: individuals often know more than they are able to articulate. When knowledge has a high tacit component, it cannot be codified. It is, therefore, extremely difficult to transfer it without intimate personal contact, demonstration, and involvement. It is well known, for instance, that the diffusion of crafts from one country to another often depended on the migration of groups of craftsmen, such as when the Huguenots were driven from France by the repeal of the Edict of Nantes under Louis XIV. Indeed, in the absence of intimate human contact, technology transfer is sometimes impossible. As Polanyi has observed, "It is pathetic to watch the endless efforts — equipped with microscopy and chemistry, with mathematics and electronics — to reproduce a single violin of the kind the half literate Stradivarius turned out as a matter of routine more than 200 years ago."[24]

In short, the transfer of knowledge may be impossible in the absence of the transfer of people. Furthermore, it will often not suffice just to transfer individuals. While a single individual may sometimes hold the key to much organizational knowledge, team support is often needed, since the organization's total capabilities must be brought to bear on the transfer problem.[25] In some instances, the transfer can be effected through a one-time contract, which would provide a consulting team to assist in the start-up. Such contracts may be highly incomplete. The failure to reach a comprehensive agreement may give rise to dissatisfaction during execution. This dissatisfaction may be an unavoidable — which is to say an irremediable — result. Plainly, foreign investment is an extreme response to the needs of a one-time exchange. In the absence of a superior organizational alternative, one-time incomplete contracting for a consulting team is likely to prevail.

Reliance on repeated contracting is less clearly warranted, however, where a succession of exchanges is contemplated. It is also less clearly warranted when two-way communication is needed to promote the recognition and disclosure of opportunities for information transfer as well as the actual transfer itself. The parties in these circumstances are effectively joined in a bilateral monopoly trading relation. As Williamson explains, such contracting is full of hazards.[26] A more cooperative arrangement for joining the parties would enjoy a greater comparative institutional advantage. Specifically, intrafirm transfer to a foreign subsidiary (which avoids the need for repeated negotiations and attenuates the hazards of opportunism) has advantages over autonomous trading. Better disclosure, easier agreement, better governance, and more effective team organization and reconfiguration all result. Here lies an incentive for foreign direct investment.

The above arguments, while couched in the context of technological know-how, are in fact general and extend to many different kinds of proprietary information. For instance, managerial (including organizational) know-how and goodwill (including brand loyalty) represent types of assets for which markets may falter as effective exchange mechanisms. A foreign investment strategy is, therefore, suggested.

International Capital Markets and Direct Foreign Investment

Until the late 1960s, the prevailing view of direct foreign investment by multinational firms was that it represented one process by which rates of return to capital were equalized across countries. It was subsequently realized that this interpretation was inadequate, since the equalization of rates of return did not imply ownership of the means of production. A portfolio position would appear to achieve the same objectives at lower cost. Nor are pure diversification advantages involved. Following modern financial theory, a security's risk and return can be decomposed into two elements: (1) risk that is specific to each company (called "unsystematic," because it can be diversified away) and (2) risk that is common to all securities (called "systematic," because it is nondiversifiable). Since unsystematic risk can generally be eliminated through simple portfolio diversification, the investor does not need widely diversified multinational companies to eliminate this risk. Thus, it would be surprising if capital transfers were a significant feature of foreign direct investment. Indeed, the available evidence indicates that only a small percentage of the capital expenditures of foreign affiliates is financed by the parent companies. A study of 115 foreign subsidiaries for the U.K. revealed that local sources (specifically retained earnings, depreciation allowances, liquid assets, and local bank loans) accounted for 80 percent of the subsidiaries' local investment.[27] A similar pattern appears in the investment expenditures of majority-owned foreign affiliates of U.S. companies. The ratio of these expenditures to the net capital outflows from the U.S. to all foreign incorporated and unincorporated affiliates is typically about 3:1. In short, it is doubtful that the international transfer of capital is an especially significant attribute of foreign direct investment. It is mainly incidental to the other considerations identified here.

Still, there may be a number of efficiencies associated with the internalization of international capital markets. If capital markets abroad are not well developed, then an efficient internal capital market within the multinational firm may well be able to assist in the efficient worldwide allocation of resources. Funds will be allocated to high yield uses, irrespective of national boundaries. Thus, the multinational firm can perform as an effective substitute for capital markets where these markets are poorly developed.

Potential Anticompetitive Consequences of Multinational Enterprise

An alternative explanation of the overseas growth of multinational enterprises emphasizes the role of oligopoly and monopoly. According to Hymer, direct foreign investment has a dual nature:

> It is an instrument which allows business firms to transfer capital, technology, and organizational skill from one country to another. It is also an instrument for restraining competition between firms of different nations .... The important point is to note that the general presumption of international trade economists in favor of free trade and free factor movements, on the grounds of allocative efficiency, does not apply to direct foreign investment because of the anticompetitive effect inherently associated with it .... A restriction on direct investment or a policy to break up a multinational corporation may be in some cases the only way of establishing a higher degree of competition in that industry .... Given the oligopolistic front maintained by the firms from developed

> countries, the underdeveloped countries need to devote an important share of their scarce resources to building up national enterprises which they can control and use in bargaining with foreign oligopolists.[28]

Given the currency of these views, it is important to ascertain the degree to which MNEs can be vehicles for anticompetitive behavior. It will also prove necessary to distinguish the concept of competitive advantage from the concept of monopoly power, since it appears that Hymer and others have imputed monopoly concerns to any kind of competitive advantage the MNE may possess. In this regard, it is of interest to contrast the "market failure" view of the MNE developed above with the imperfect competition view, which originated with the work of Hymer and extends through Kindleberger and Caves.[29] According to this latter view, the existence of multinational enterprise is predicated upon some monopolistic advantage. MNEs have unique assets (e.g., innovations or differentiated products) which enable them to offset the advantages that indigenous firms have in operating in their own familiar environment. This view of the firm is consistent with the conceptualization of horizontal direct foreign investment developed earlier. Firms have know-how and other major assets, which they transfer abroad in order to capture quasi rents. But, saying that firms have unique assets does not mean that rivalry is necessarily suppressed and that public policy intervention is thereby called for. Rather, explicit consideration of costs and benefits from foreign investment is necessary, since there may be important incidental benefits to the host country.

However, it appears that the emphasis on the "special assets" that the multinational firm possesses (and on the "monopoly advantage" which those assets confer) has been interpreted by some to imply that MNEs are vehicles for anticompetitive behavior.[30] There seems to be no other way of explaining Hymer's remark that there is an anticompetitive effect inherently associated with direct foreign investment. Nor does there seem to be any other way of explaining his statement that "a restriction on direct investment or a policy to break up a multinational corporation may in some cases be the only way of establishing a higher degree of competition in that industry." It would appear that the relationship between the "special advantage" that a firm might possess and the "anticompetitive effect" supposedly inherent in the multinational firm is a subtle one, which must be interpreted with great care. The dynamics of Schumpetarian competition in an enterprise economy involve first the creation and then the destruction of a firm's competitive advantage. While the entrance of multinational firms into foreign markets may be explained in terms of the unique assets of these firms, consumer welfare is not necessarily injured by this process. Indeed, improvement in consumer welfare is often to be expected.

In this regard, it is of interest to examine the concept of competition and how it might be interpreted for public policy purposes. In economic theory, competition is typically defined to be a state of the market in which the individual buyer or seller does not influence the price of his purchases or sales. Competitive markets will normally arise when there are large numbers of buyers and sellers, product homogeneity, perfect knowledge, and divisibility of output. However, there may not be any real world market that meets the assumptions of the competitive model. Nevertheless, it is not necessarily true that public policy intervention will secure superior results for the consumer. Bork has suggested an alternative definition of competition, which appears to have merit as a guide for public policy. Specifically, he suggests that:

> "Competition" may be read as a shorthand expression, a term of art, designating any state of affairs in which consumer welfare cannot be increased by moving to an alternative state of affairs by judicial decree. Conversely, "monopoly" and "restraint of trade" would be terms of art for situations in which consumer welfare could be so improved, and to monopolize or engage in "unfair competition" would be to use practices inimical to consumer welfare.[31]

If the multinational firm possesses a distinctive competitive advantage in the form of unique assets, then the exploitation of this advantage will typically enhance consumer welfare. The principal considerations, which arise with respect to multinational rather than indigenous enterprises, relate merely to the distribution of the quasi rents associated with the employment of the firms' unique assets. Since, by assumption, these assets are transferred from abroad, it would not appear that host country welfare would be reduced by foreign direct investment unless deleterious externalities accompanied the technology transfer process. Although appropriate public policy towards multinationals can be expected to share many features of policy commonly applied to indigenous firms, the international context raises a few special considerations, which need to be identified. The following discussion is restricted to issues raised by firm structure—particularly vertical, horizontal, and conglomerate integration.

Vertically Integrated Multinationals

Firms based in one country can engage in backward or forward integration into foreign markets. The anticompetitive aspects of vertical integration have received considerable elucidation in the past decade. The circumstances under which vertical integration can have anticompetitive consequences are now fairly well understood and seem to be narrowly circumscribed. Leverage theories of vertical integration, once commonplace, explain how vertically integrated firms in one market are supposedly able to extend their monopoly power into upstream and downstream markets. However, the application of economic analysis has largely discredited these notions. Consider an industry with two stages: component manufacture and assembly. If component manufacture is monopolized and assembly is competitive, the question then is whether the monopolist can increase profits by buying out the assemblers. If the manufacturer buys out the assemblers and increases the markup on assembly operations, then the markup on components must be decreased by the same amount if profits are to be maximized. Since the demand for components is derived from the demand for the assembled product, the component manufacturer cannot maximize his profits by charging a price above the monopoly price (which has been determined with reference to all relevant costs, including the costs of assembly). In short, vertical integration does not permit a monopolist to extend his market power. It fails to do so at least insofar as it does not create contrived barriers to entry.[32] This consideration will now be examined.

Consider the possibility that vertical integration into assembly will delay entry (or otherwise make it more costly) at the component stage by making it necessary for a new entrant to enter at both stages. Suppose a potential entrant has developed competitive components. Because the component monopolist is integrated into assembly, the new entrant must come in at both the component and the assembly stage, or else independent new entrants must occur simultaneously at the assembly stage. One view is that vertical integration by the monopolist will not impede entry. Bork, for example, contends that:

> In general, if greater than competitive profits are to be made in an industry, entry should occur whether the entrant has to come in at both levels or not. I know of no theory of imperfections in the capital market which would lead suppliers of capital to avoid areas of higher return to seek areas of lower return.[33]

Clearly, integrated entry occasions an increase in financial requirements. The issue, as Williamson points out, is whether an increase in the financial requirements is accompanied by an adverse alternation of the terms under which capital becomes available.[34] If the potential entrant is not competitive in assembly (e.g., because it has no direct or related experience), then an efficient capital market will adjust the terms of finance against the new entrant. As Williamson points out: "To contend that the terms of finance are the same . . . implies that the capital market has equal confidence in the new entrants' qualifications to perform [assembly] activities as it does in firms that are already experienced in the business. Except in circumstances where experienced firms are plainly inept, this is tantamount to saying that experience counts for nought."[35]

The issues raised are somewhat technical and have been clarified elsewhere.[36] Nevertheless, it is clear that entry barriers arise if capital markets are poorly developed or if entrepreneurial skills are absent. Posner has argued that if the existing component producer owns all of the existing assembly facilities, then a new entrant would not have to engage in entry at both stages. If entry at one stage is anticipated, new firms will enter at the other stage, in order to provide a market for the new product.[37] However, there seem to be two basic requirements for this situation to occur. One is that market processes must be able to coordinate the investment expectations of the potential upstream and downstream producers. Posner claims that this requirement should not be a serious concern. The other requirement, which is related to the first, is that entrepreneurs come forth to take advantage of the opportunity. While these conditions might appear appropriate in the context of developed Western economies, they are clearly inappropriate for many less developed countries. Entry at both stages may well be necessary. The additional capital just may not be available to the potential entrant due to market failure or institutional considerations. Perfect capital markets are the exception rather than the rule, once an international perspective is adopted.

Accordingly, public policy towards the vertically integrated multinational enterprise need not be neutral. Vertical integration can be used as a barrier to entry where there is horizontal market power at one stage and where capital markets experience disabilities. The appropriate policy will depend upon the objectives of the nation-states, the ownership of the multinational, and the markets in which it is selling. Nevertheless, it is important to reiterate that in the absence of horizontal market power, the vertically integrated firm is generally without anticompetitive implications. It may, in fact, have strong efficiency and procompetitive attributes.

There is another type of antisocial consequence which may result from vertical integration: the circumvention of taxes and tariffs. Vertical integration affords opportunities for transfer price manipulation and profit sharing. This consequence may be an especially important concern in the international context, since differential tariffs and taxes characterize international commerce. Vertical integration that is completed simply to take advantage of such considerations may yield pecuniary economies to the firm without yielding equivalent efficiency gains for the host country. Competitive implications also arise if tax shifting or tax avoidance gives multinational firms a cost advantage over equally efficient indigenous firms. The auditing of transfer prices may well be necessary.

The Horizontal Expansion of Multinational Firms

A significant portion of direct foreign investment is of the horizontal kind that is, investment abroad is directed at producing goods and services that are currently produced domestically. Whether the potential for anticompetitive behavior exists with respect to this kind of investment will depend to a large degree on the scope of the relevant markets. If the relevant markets are national or regional in scope (so that foreign production has no impact on the home market), then foreign investment will have a substantially neutral impact on competition at home. The impact on competition abroad cannot be assessed without examining the particular circumstance of the foreign market. Competition is generally enhanced: (1) if the investment represents new entry and (2) if entry is not on such a massive scale that it results in the firm's achievement of dominance or of monopoly position in an otherwise competitive market. As Caves points out:

> It is seldom recognized that the multinational company is a favored entrant to industries with high barriers to entry . . . . Thus, the multinational company is a likely potential entrant into national industries that might otherwise be cloistered by . . . entry barriers . . . . Besides its ability to enter a market, there is also some possibility that the multinational may be an entrant particularly disruptive of an oligopolistic consensus, especially in the early period of its presence. Its alien status may make it initially less sensitive to signals about an oligopolistic consensus emanating from established native firms.[38]

However, the case for a general policy of openness to market entry by MNEs is not completely clear-cut. "If the multinational company is good at scaling existing industrial barriers to the entry of new firms, it is also good at building up such barriers. The resources required to contrive such barriers . . . are often found in the portfolio of multinational companies."[39]

Evaluation

It seems clear that the MNE has few distinctive characteristics that ought to be the focus of special public policy initiatives. If offensive forms of market behavior do exist, the most direct approach is to regulate or prohibit the behavior directly rather than blocking foreign investment. Since most of the monopolistic abuses typically attributed to MNEs can also characterize indigenous firms with no extraterritorial investments,[40] nondiscriminatory regulatory approaches would appear to be desirable. Conduct designed to reduce the competitiveness of host country markets is socially undesirable, regardless of the identity of the firms involved.

In short, it is not clear that the MNE should be considered a distinctive organizational form with respect to the formulation of competition policy. To the extent that social abuse is attributable to the MNE, then the same abuses are typically open to indigenous firms as well. In this regard, consider the Brazilian electrical industry. Newfarmer identifies what he considers to be seven forms of MNE conduct: interlocking directorates, mutual forbearances, control of supply channels, cross subsidization of production, formal and informal collusion, formal political ties, and MNE acquisition behavior. He concludes that MNEs "exhibit strong propensities toward organizing and preserving various forms of market power in host economies. These tactics are often based on the advantage of global financial strength or perceived international and local interdependence."[41] While Newfarmer is able to provide persuasive evidence of collusion in the Brazilian electrical industry, the identified abuses might equally well have been characteristic of Brazilian firms. His recommendation that "as one tool of national planning, serious measures to curb restrictive practices, to prevent nonsocially beneficial takeovers, and to generally counterbalance the disadvantaged position of domestic entrepreneurs" would not seem to be a well-founded proposal if it is directed at MNEs alone.[42]

Conclusion

The multinational firm appears to have both efficiency and market power properties. However, the market power properties are more narrowly circumscribed than is commonly supposed. The analysis developed here points to the distinctive capabilities of the multinational firm as an organizational mode that is capable of establishing efficient vertical supply relations and of transferring technology in an efficient fashion. While multinational firms can be used to extend market power, their distinctive capabilities in this regard do not appear to be very significant. The general presumption in favor of free trade and free factor movements is not as readily overturned as some observers have suggested. Accordingly, the prescription that Third World countries devote an important share of their scarce resources to building up national enterprises rests on extremely weak theoretical (and empirical) underpinnings.[43]

While the establishment and enforcement

of industrial policies favoring competition will generally have merit, it appears that there are few circumstances where efficiency considerations would support the desirability of policies discriminating between indigenous and international firms. In any event, the internal resource allocation properties of multinational firms need careful examination before momentous policy prescriptions can be derived with confidence.

References

1
See:
P.J. Buckley and M. Casson, *The Future of the Multinational Enterprise* (London: Holmes & Meier Publishers, 1976);
S. Hymer, "The Efficiency (Contradictions) of Multinational Corporations," *American Economic Review* 60 (May 1970): 441-448;
J.C. McManus, "The Theory of the International Firm," in *The Multinational Firm and the Nation State*, ed. G. Paquet (Don Mills, Ontario: Collier Macmillan, 1972).

2
See O.E. Williamson, *Markets and Hierarchies: Analysis and Antitrust Implications* (New York: Free Press, 1975).

3
This section is based on Williamson (1975).

4
See J.R. Commons, *Institutional Economics* (Madison, WI: University of Wisconsin Press, 1934).

5
See R. Coase, "The Problem of Social Cost," *Journal of Law and Economics*, October 1960, p.15.

6
See O.E. Williamson and D. Teece, "European Economic and Political Integration: The Markets and Hierarchies Approach " (University of Pennsylvania, Center for the Study of Organizational Innovation, 1979).

7
This section is based partly on Williamson and Teece (1979) and on D. Teece, "Economies of Scope and the Scope of the Enterprise," *Journal of Economic Behavior and Organization* 1 (1980): 223-247.

8
See Coase (1960).

9
Lateral integration occurs when a firm sells products which are functionally related in production and/or distribution.

10
See R. Vernon, *Sovereignty at Bay: The Multinational Spread of U.S. Enterprises* (New York: Basic Books, 1971), ch. 2.

11
See:
B. Klein, "Transaction Cost Determinants of 'Unfair' Contractual Arrangements," *American Economic Review* (May 1980);
S. Macauley, "Non-Contractual Relations in a Business: A Preliminary Study," *American Sociological Review* 28 (1963): 55-69;
D. Teece, *Vertical Integration and Vertical Divestiture in the U.S. Petroleum Industry* (Stanford, CA: Institute for Energy Studies, 1976); Williamson (1975).

12
See:
K. Monteverde and D. Teece, "Appropriable Rents and Quasi Integration" (Stanford University Graduate School of Business, Working Paper, 1980a);
K. Monteverde and D. Teece, "Supplier Switching Costs and Vertical Integration in the U.S. Automobile Industry" (Stanford University Graduate School of Business, Working Paper, 1980b).
If it is just firm-specific physical capital that is involved, the holdup problem can be avoided by quasi integration: the supplier can require the purchaser to supply the requisite physical capital (such as tools and dies) needed for efficient supply. Hence, supplier switching costs turn out to be critical. See ibid.

13
See Williamson (1975), p. 119.

14
See R. Marris, "The Future of Corporate Society" (Working Paper, Department of Economics, University of Maryland, October 1980).

15
See R.E. Caves, "International Corporations: The Industrial Economies of Foreign Investment," *Economica* 38 (1971).

16
See:
E. Mansfield, D. Teece, and A. Romeo, "Overseas

Research and Development by U.S.-Based Firms," *Economica* 46 (1979): 187-196.
D. Teece, *The Multinational Corporation and the Resource Cost of International Technology Transfer* (Cambridge, MA: Ballinger, 1976).

17
See S. Kuznets, *Modern Economic Growth: Rate, Structure and Spread* (New Haven, CT: Yale University Press, 1966).

18
This is because the value of information often declines with its dissemination, and it cannot be transferred at zero marginal cost.
See D. Teece, "Technology Transfer by Multinational Firms: The Resource of International Technology Transfer," *Economic Journal*, June 1977.

19
See G. Calabresi, "Transactions Costs, Resource Allocation, and Liability Rules: A Comment," *Journal of Law and Economics*, April 1968.

20
See Coase (1960), p.15.

21
See I. Kirzner, *Competition and Entrepreneurship* (Chicago, IL: University of Chicago Press, 1962), pp. 215-216.

22
See D. Marquis and T. Allen, "Communication Matters in Applied Technology," *American Psychologist* 21 (1966): 1055.

23
See K.J. Arrow, *Essays in the Theory of Risk Bearing* (Chicago, IL: Markham Publishers, 1971), p. 152.

24
See M. Polanyi, *Personal Knowledge: Towards a Post Critical Philosophy* (Chicago, IL: University of Chicago Press, 1958).

25
See N. Rosenberg, "Economic Development and the Transfer of Technology: Some Historical Perspectives," *Technology and Culture* 11 (1970): 550-575.

26
See Williamson (1975).

27
See M. Brooks and H. Remmers, *The Strategy of Multinational Enterprise* (New York: American Elsevier, 1970).

28
See Hymer (1970).

29
See:
R.E. Caves, "Industrial Organization," in *Economic Analysis and the Multinational Enterprise*, ed. J. Dunning (New York: Praeger, 1974);
S. Hymer, "The International Operations of National Firms: A Study of Direct Investment" (Cambridge, MA: Massachusetts Institute of Technology, doctoral dissertation, 1960);
C.P. Kindleberger, *American Business Abroad* (New Haven, CT: Yale University Press, 1969);
Caves (1971).

30
See:
R.S. Newfarmer, "Oligopolistic Tactics to Control Markets and the Growth of MNEs in Brazil's Electrical Community," *Journal of Development Studies* 15 (April 1979);
Hymer (1970).

31
See R. Bork, *The Antitrust Paradox* (New York: Basic Books, 1978), p.61.

32
However, when monopsony and monopoly stages are integrated vertically, integration facilitates arriving at the input choice that extracts maximum profits from whatever monopoly power exists at either stage. Decisions regarding how much of an input to use can be guided by the actual marginal cost of the input, rather than by bargaining strategems or by the monopsonist's concern for restraining the volume of its purchases in order to avoid driving up the supply price.

33
See R. Bork, "Vertical Integration and Competitive Processes," in *Public Policy towards Mergers*, ed. J. Fred Weston and S. Peltzman (Pacific Palisades, CA: Goodyear Publishing, 1969), pp. 139-146.

34
See Williamson (1975), p. 110.

35
See Williamson (1975), p. 111.
There is an issue here of whether the potential entrant will have the terms of finance adjusted against him by an amount greater than the monopolist would have experienced when the monopolist integrated into assembly. It would seem that since, by assumption, the monopolist had no initial advantage with respect to assembly, it would have confronted the same capital costs. If entry barriers are defined (following Stigler) as

a condition which imposes higher long-run costs of production on a new entrant than are borne by firms already in the market, then it is not clear that the Williamson scenario ought to be considered to generate a barrier to entry.

36
See G. Saloner and D. Teece, "Vertical Integration and Capital Market Entry Barriers: A Clarifying Note" (Stanford University, Working Paper, April 1980).

37
See R. Posner, *Antitrust Law* (Chicago, IL: University of Chicago Press, 1976), p. 198.

38
See R.E. Caves, "International Cartels and Monopolies in International Trade," in *International Economic Policy*, R. Dornbusch and J. Frankel (Baltimore, MD: Johns Hopkins University Press, 1979), p. 61.

39
See Caves (1979), p. 61.

40
However, a number of forms of abuse are open to vertically integrated MNEs which are not open to vertically integrated firms which are purely domestic in the scope of their activities. Perhaps the most obvious is transfer price manipulation designed to circumvent corporate taxes or tariffs on international trade. While wealth transfer considerations are involved, there need not be any competitive implications unless the tax shifting or tax avoidance gives multinational firms a competitive advantage over equally efficient indigenous firms.

41
See Newfarmer (1979), p. 135.

42
Ibid., p. 136.

43
See Hymer (1970).

The American Economic Review, 75 (May 1985), 233–238

Multinational Enterprise, Internal Governance, and Industrial Organization

By DAVID J. TEECE*

A multinational enterprise (*MNE*) is a firm that controls and manages production establishments located in at least two countries. In a Coasian sense, *MNE*s substitute for the market. They can be divided into two nonmutually exclusive types. One turns out essentially the same line of goods or services from each facility in several different locations, and will henceforth be referred to as horizontally integrated *MNE*s. The other produces outputs in some facilities which serve as inputs into other facilities located across national boundaries. These are vertically integrated *MNE*s.

Two conceptually distinct issues exist which need to be addressed in order to explain the *MNE*.[1] The first is the locational question of why production occurs where it does. The second is why certain production activities occur under the control of foreign enterprises while others do not. The locational issues are explained rather well by the standard theories of comparative costs. The control (internalization) issues cannot be explained by standard theory, yet these are central to a theory of the *MNE* as compared to a theory of international production. The former is concerned with explaining the nationality of the firms engaging in international production, while the latter simply explains the international distribution of the world's productive activities without concern for ownership and control patterns.

Until about two decades ago, economists viewed the *MNE* as simply an arbitrageur of capital, transferring equity capital from countries where returns were low to those where it was higher, earning the arbitrageurs rents and contributing to efficient resource allocation. The capital arbitrage theory of *MNE* predicted that *MNE*s would be headquartered in countries where the domestic marginal productivity of capital was relatively low, from which they will transfer capital to subsidiaries where the marginal productivity was higher.

As Stephen Hymer (1976) first observed, however, there are several features of direct foreign investment (*DFI*) and *MNE* which are inconsistent with this theory. The *MNE*s overwhelmingly finance their host-country operations in host-country capital markets. Furthermore, there are substantial cross flows of direct foreign investment (see Jean-Francois Hennart, 1982), as well as substantial concentration of *DFI* and *MNE*s in particular industries. These observations would be consistent with an arbitrage theory only if domestic capital markets were highly balkanized.

Despite the contrary evidence, Robert Aliber (1970, 1983) has reiterated a version of the capital arbitrage theory based on the identification of separable currency areas. The argument goes approximately as follows: there are substantial differences among countries in nominal and real interest rates. Because nominal interest rate differentials are poor forecasts of future changes in exchange rates, a wedge is introduced between returns on similar securities denominated in different currencies.

> If portfolio managers had "perfectly" priced exchange risk, then the corporate managers would be reluctant to have their firms incur foreign exchange

*Professor of Business Administration, University of California, Berkeley, CA 94720. I thank Robert Aliber, Greg Hawkins, and Jim Wilcox for helpful comments.

[1] John Dunning (1981) develops a useful taxonomy which explains *MNE*s in terms of ownership (of rent-yielding assets), locational, and internalization advantages. For expositional reasons, I collapse what Dunning refers to as ownership factors into locational factors because it is the coupling of the two which explains where production ought to be located.

> risks... the parents of the subsidiaries take on the foreign exchange exposure because the corporate managers believe that the interest rate differentials are significantly larger than from the anticipated rate of change of the exchange rate... The expansion of firms across national borders is consistent with the view that corporate managers internalize the costs and risks of foreign exchange exposures at lower costs than portfolio investors.
>
> [Aliber, 1983, p. 252]

The argument is flimsy and incomplete at best. No explanation of why corporate managers can internalize exchange rate risk at lower costs than portfolio managers is provided or even suggested. Moreover, as mentioned earlier, only a small part of direct foreign investment involves intercompany loans from parents to subsidiaries. Aliber claims otherwise but cites no evidence.[2]

In searching for a plausible theory of *DFI*, Hymer advanced two major tenets. One was that *DFI* was motivated by attempts to remove competition among enterprises in different countries, and the other was the *DFI* was motivated by domestic firms' attempts to increase the returns from the utilization of firms' special advantages (Hymer, 1976, p. 33). With respect to the second category, Hymer drew on Joe Bain (1956) to suggest that the source of the advantage could be in superior production techniques, imperfections in input markets which allow lower buying prices for established firms, and similar first-mover advantages. Possessing such special advantages, a national firm could be profitable outside the home country despite its relative ignorance of local conditions abroad. It would prefer to do so rather than license in order to avoid possible bilateral monopoly situations, to enforce use restriction which could not be imposed contractually, to avoid the haggling between licensor and licensee associated with the evaluation of the value of the technology, and to better protect its advantage from the perils of misappropriation.

These were powerful insights, which laid the foundation for a completely new paradigm of the international firm. In one short thesis, Hymer transported the theory of direct foreign investment out of international trade and finance and into industrial organization. But the field of industrial organization circa 1960 did not have quite the richness it has today, and the library of concepts from which Hymer could borrow was especially sparse with respect to the economics of complex organizations.

Furthermore, Hymer was constrained by the absence of a clear welfare criteria for evaluating the *MNE*. In the absence of alternatives, Hymer seized upon perfect competition and Pareto optimality and inevitably arrived at troublesome conclusions. This is because he saw the raison d'être of the *MNE* as stemming from the "impurities of the market [that] would not arise in competitive industries" (1976, p. 86), which led him to the conclusion that "a restriction on direct investment or a policy to break up multinational corporations may be in some cases the only way of establishing a higher degree of competition in that industry... the underdeveloped countries need to devote an important share of their scarce resources to building up national enterprises... ." (Hymer, 1970, p. 444). However, if the *MNE* is based on a special advantage, as Hymer claims, then the conclusion of this line of logic is that the special cost advantage should be abandoned in order to worship at the altar of perfect competition.

There is no doubt that Hymer's work represents a major contribution to the positive economics of the international firm. However, his thesis is misleading in its emphasis upon market power rather than efficiency, as is explained below. Relatedly, it does not provide a workable framework for assessing host-country control issues. The framework

[2] The Aliber theory may have some validity with respect to shifts in the total levels of foreign investment at the national level. Even here, "safe haven" and economic-growth considerations also appear to be important. The recent drop in direct foreign investment in the United States—$7.0 billion in 1983, $10.8 billion in 1982, 23.2 billion in 1981, 12.2 billion in 1980, and $15.3 billion in 1979—certainly doesn't square well with the theory.

advanced is also static, and unable to grapple with intermediate organizational forms, such as cooperative and teaming agreements.

I. Monopoly vs. Efficiency Interpretations of the Multinational Enterprise

As I have indicate elsewhere (1981, 1983), the essence of the multinational firm varies according to whether its investment abroad is primarily vertical or horizontal. The rationale for vertical direct investment stems from the efficient functioning of internal production and distribution systems when bilateral dependence with attendant strategic maneuvering and costly haggling might otherwise emerge (Oliver Williamson, 1975; myself, 1977, 1983; K. Monteverde and myself, 1982). By substituting an internal governance structure (i.e., vertical integration) capable of circumventing such problems, the multinational firm makes a substantial contribution to economic efficiency. Vertical integration permits specialized cost-saving equipment to be installed in both upstream and downstream locations with less risk that it will be idled by international disputes between enterprises of different nationality facing different incentives. Vertical direct foreign investment ought to be seen primarily as a response to market failure, and its explanation does not require appeal to classical market power considerations. Indeed, except under restrictive assumptions, vertical integration cannot be employed as a mechanism to extract monopoly or monopsony rents. Hymer avoids separate treatment of vertical direct foreign investment, but hints that sequential monopoly may be at its heart, blind to the spurious nature of leverage theories of vertical integration. In any event, Hymer appears unaware of the organizational efficiencies which can be associated with vertical integration, and misinterpreted the welfare implication of the *MNE* on this account.

The explanation of horizontal direct foreign investment requires the coupling of two requirements. One is that the firm possess a rent-yielding asset of some kind (for example, know-how) which warrants utilization in offshore production facilities, and the second is that market transactions (know-how, licensing arrangements) are inferior to direct foreign investment as instruments for appropriating rents from the sale of the services of the asset in foreign markets. This is because of both the revenue-enhancing and cost-saving properties of the *MNE*. Direct foreign investment will be selected where contractual difficulties such as the ability to price know-how, or to write, execute, and enforce use restrictions governing technology transfer arrangements are anticipated. The intra-organizational mode for transferring technology is not only revenue enhancing but also cost saving, as empirical studies have demonstrated (my 1977 article). Hymer tacitly recognized these cost-saving aspects of horizontal direct foreign investment (1976, p. 50) and explicitly recognized its revenue-enhancing properties, which he chose to emphasize, bluntly stating that "monopoly problems pervade any discussion of international operations and direct investment" (1976, pp. 85–86). In a technical sense, Hymer was correct, in that in the presence of perfect competition and frictionless markets, the incentives for direct foreign investment disappear. However, applied welfare economists are unlikely to view any departure from perfect competition with alarm. What is needed, and what was not supplied by Hymer, was an assessment of the *comparative* welfare properties of multinational firms and the market alternatives. Hymer's comparisons of the *MNE* with the frictionless market fiction was inapposite and fueled host-country antagonism towards the *MNE*.

Hymer's thesis has stimulated a flurry of books and articles on the multinational firm, including P. J. Buckley and Mark Casson (1976) and Alan Rugman (1982). As Charles Kindleberger (1984, pp. 180–81) points out, there isn't all that much in this literature that cannot be found in Hymer. What is accomplished, however, is a reemphasis away from market power concerns and towards the economies associated with internalization, a literature which is derived from Williamson (1975). Departures from perfect competition in product markets and transactional advantages from internalization are needed to explain horizontal *DFI*, but Hymer chooses only to emphasize and develop the former. The new literature therefore provides a much-needed balance.[3] The principal deficiency is that the internalization arguments are insufficiently micro analytic, failing to establish a structure which can discriminate between situations where direct foreign investment is more efficient than alternative organizational modes. Clearly, direct foreign investment isn't the most efficient mode of organization in all circumstances where a firm possesses an advantage which is the foundation of a business opportunity abroad. The contingent framework which is needed to create a theory of the *MNE* is sketched in my papers (1981, 1983), where it is shown that the incentives to use nonmarket modes for horizontal transactions swing upon the degree to which the technology in question cannot be evaluated, packaged, and protected using contractual instruments. In the case of vertical transactions, the presence of transaction-specific assets generating *ex post* "lock in" conditions is the key incentive driving foreign direct investment. Needless to say, these conditions don't exist all the time, and arm's length contracts will often suffice.

II. Host-Country Controls

The transaction cost theory embedded in the above explanation of the *MNE* also contains some interesting policy conclusions. Besides maintaining that the *MNE* ought mainly to be regarded in economizing terms, the micro-analytic lens afforded by transaction-cost economics enables one to assess, from an efficiency perspective, the desirability of host countries exercising special controls over the activities of *MNE*s.

The need for special governance over the *MNE* depends in large measure on whether the host country provides transaction-specific assests to support the *MNE*. This might be the case, for instance, if the host country dedicates roads and electric utilities to support a particular *MNE*, or if workers develop special skills to meet the needs of an *MNE* affiliate. When specialized assets are developed for the foreign firm, they cannot be redeployed easily should the *MNE* decide to pull out, or to otherwise adjust its plans opportunistically. The host country has incentives to develop investment safeguards in these circumstances. These can take many forms, including some type of penalty for early withdrawal or delayed expansion, or the creation of specialized governance structures for resolving disputes between the *MNE* and the host country. Alternatively, the *MNE* could be asked to post a bond, engaging in what Williamson (1983) refers to as the exchange of hostages. Despite safeguards of the type indicated, additional benefits might accrue from information disclosure involving relevant strategic plans, or informational representation on the Board of Directors, as this might enable the host country and workers with firm-specific skills to anticipate future developments and plan accordingly. Failure to design such governance structures may result in host countries failing to provide the necessary infrastructure. It bears repeating, however, that special governance or regulatory machinery is not needed unless the host country is deploying specialized assets to support the foreign firm.

The argument is also symmetrical and more empirically relevant when viewed from the *MNE*'s perspective. If *MNE*s are to invest specialized assets dedicated to the host country, immobile plant and equipment being a case in point, investment safeguards are needed. An expropriation price agreed upon *ex ante* which the *MNE* can trigger at its discretion, somewhat analogous to certain types of severance pay in employment contracts, is one such possibility. Arbitration is another. Failure to design machinery of this kind will dissuade direct foreign investment when specialized nonredeployable assets are needed for efficient production. Alternatively, the *MNE* will extract a "fee" to offset this risk. The fee might take the form of

[3] It is noteworthy, however, that subsequent work in the spirit of Hymer, such as Richard Caves (1971), chose to emphasize the monopoly issues and neglected the internalization issues which are mentioned in Hymer but not well developed. Caves (1982) has subsequently redressed this misemphasis and signed on to the transaction cost thesis.

higher prices for goods sold in the host country, and/or lower prices for commodities purchased.

III. Dynamic Considerations

The literature on the *MNE*, whether emphasizing market power or efficiency, suffers from a common deficiency: underemphasis on dynamics. The historical evidence shows American and British *MNE*s transitioning to offshore production after first establishing a sales branch abroad which in turn commonly preceded the establishment of contractual relationship with a foreign sales agent (Steven Nicholas, 1983). Agency contracts were unsatisfactory, containing vague and difficult to enforce performance clauses. Attempts to shore these up by specifying, quantitatively, budgets for traveling, advertising, engineering support, and the like, as well as certain inventory levels on the agents' premises proved unsatisfactory. The transition from agency to branch sales office was facilitated by the manufacturer's gradual accumulation of information about the foreign market, acquired through monitoring its foreign agents, and by expansion in sales volumes to levels which would support a facility of minimum economic size. The establishment of a sales branch also demonstrated to customers a more solid commitment on the part of the manufacturer to support the market in question. Often triggered by the failure or termination of an agent, the establishment of a foreign sales subsidiary subsequently became the platform upon which a manufacturing investment was made (Alfred Chandler, 1977, p. 369). However, the transition to manufacturing depended upon the relationship of production costs abroad to production costs at home plus tariffs and transportation, as well as governance considerations. Nevertheless, as Yair Aharoni (1966) points out, the direct investment process is governed by more than just economic incentives. An initiating force, galvanizing the organization into action, is often needed. The presence of a sales office also assists information collection, thereby significantly lowering perceptions of uncertainty, and raising the probability that a firm will engage in direct foreign investment if the underlying cost conditions permit it. Clearly, transaction-cost economics must be married to organizational decision theory if the dynamics of channel selection are to be better understood.

IV. Conclusion

The Hymer thesis cast the *MNE* and direct investment in a new light. The contribution was considerable, though the analysis was incomplete, and the welfare conclusions misleading. Considerable progress has been made in the last two decades in embellishing the theory and correcting the errors. More work needs to be done in understanding the distinctive characteristics of the *MNE*, its internal governance structure, and its incentive limits. Progress in this area is unlikely to be rapid until our understanding of the internal resource allocation and governance processes within firms begins to match our understanding of how these processes work in markets.

REFERENCES

Aharoni, Yair, *The Foreign Investment Decision Process*, Boston: Division of Research, Harvard Business School, 1966.

Aliber, Robert Z., "Money, Multinationals and Sovereigns," in Charles P. Kindleberger and David B. Audretsch, eds., *The Multinational Corporation in the 1980s*, Cambridge: MIT Press, 1983, ch. 11.

______, "A Theory of Direct Foreign Investment," in C. P. Kindleberger, ed., *The International Corporation*, Cambridge: MIT Press, 1970, ch. 1.

Bain, Joe S., *Barriers to New Competition*, Cambridge: Harvard University Press, 1956.

Buckley, P. J. and Casson, M., *The Future of the Multinational Enterprise*, London: Holmes and Meier, 1976.

Caves, R. E., "International Corporations: The Industrial Economics of Foreign Investment," *Economica*, February 1971, *38*, 1–27.

______, *Multinational Enterprise and Economic Analysis*, Cambridge: Cambridge

University Press, 1982.

Chandler, Alfred, *The Visible Hand*, Cambridge: Harvard University Press, 1977.

Dunning, John, *International Production and the Multinational Enterprise*, London: Allen and Unwin, 1981.

Hennart, Jean-Francois, *A Theory of Multinational Enterprise*, Ann Arbor: University of Michigan Press, 1982.

Hymer, Stephen, *The International Operations of National Firms: A Study of Direct Foreign Investment*, Cambridge: MIT Press, 1976.

______, "The Efficiency (Contradictions) of Multinational Corporations," *American Economic Review Proceedings*, May 1970, *60*, 441–48.

Kindleberger, C. P., *Multinational Excursions*, Cambridge: MIT Press, 1984.

Monteverde, K. and Teece, D., "Supplier Switching Costs and Vertical Integration in the U.S. Automobile Industry," *Bell Journal of Economics*, Spring 1982, *13*, 206–13.

Nicholas, Steven, "Agency Contracts, Institutional Modes, and the Transition to Foreign Direct Investment by British Manufacturing Multinationals Before 1939," *Journal of Economic History*, September 1983, *43*, 675–86.

Rugman, Alan M., *New Theories of Multinational Enterprises*, New York: St. Martins Press, 1982.

Teece, David, "Technological Transfer by Multinational Firms: The Resource Cost of International Technological Transfer," *Economic Journal*, June 1977, *87*, 242–61.

______, "Multinational Enterprise: Market Failure and Market Power Considerations," *Sloan Management Review*, September 1981, *22*, 3–17.

______, "Technological and Organizational Factors in the Theory of Multinational Enterprise," in Mark Casson, ed., *Growth of International Business*, London: Allen and Unwin, 1983, ch. 3.

Williamson, Oliver E., *Market and Hierarchies*, New York: Free Press, 1975.

______, "Credible Commitment: Using Hostages to Support Exchange," *American Economic Review*, September 1983, *73*, 519–40.

Journal of Economic Behavior and Organization 7 (1986) 21–45. North-Holland

TRANSACTIONS COST ECONOMICS AND THE MULTINATIONAL ENTERPRISE

An Assessment

David J. TEECE*

University of California, Berkeley, CA 94720, USA

Received February 1985, final version received June 1985

This paper applies transactions cost principles to the multinational enterprise in order to ascertain its distinctive properties as a mode of economic organization. The analysis helps identify just when and where contractual alternatives to the multinational firm are likely to be viable. This turns out to depend upon the nature of the technology, the regime of appropriability within which the firm operates, and the characteristics of the markets in question. Transactions cost analysis is also extended to the multinational enterprise–host country relationship. Implications for management and public policy are derived.

1. Introduction and background

After briefly reviewing the literature on the theory of the multinational enterprise,[1] this paper uses transactions cost principles to categorize transactions into those which can be supported by unassisted markets and those which need to be embedded within the multinational enterprise if efficient governance is to be provided.[2] The transactions cost framework is also

*I acknowledge a deep intellectual debt, beyond that evidenced by the references, to Mark Casson, Richard Caves, John Dunning, Stephen Hymer, Charles Kindleberger, Alan Rugman, Oliver Williamson and many others who have advanced the theoretical and empirical understanding of the multinational enterprise. I also appreciate the financial support of NSF Grant SRS84-10556 given to the University of California, Berkeley and administered by the Center for Research in Management.

[1]A multinational enterprise is a firm that controls and manages production establishments located in at least two countries. It is thus a multiplant firm with not all of its production capacity located in just one country. Most giant enterprises and many medium and small companies are multinational. Clearly, the minimum overseas 'plant' needed to make a firm multinational is a matter of judgment. For instance, a foreign sales office is not considered here to represent a foreign plant.

[2]In the last decade, there has been flurry of articles and books on the theory of the multinational enterprise, including Hymer (1970, 1976), Buckley and Casson (1976), Dunning (1981), Teece (1981a, b, 1983), Hennart (1982), Rugman (1981, 1982) and Caves (1982). These works have in common their emphasis on the multinational enterprise as an internalizer of economic activity, in the spirit of Coase (1937). According to Kindlberger (1984, ch. 13), there is nothing new in this literature as it is mostly contained in Stephen Hymer's dissertation, completed in 1960, published posthumously in 1976, but widely recognized during the 1970s. Indeed, Kindleberger (1933, p. 181) claims that virtually all the ideas that have emerged as the 'new theories' can be found in Hymer. This assessment is reviewed in Teece (1985).

applied to the multinational enterprise–host country relationship in order to identify the degree to which management contracts are an effective substitute for equity positions in host-country economies. Implications for management and public policy are developed.

Until the work of Hymer surfaced in the 1960s, economists did not consider the international firm or foreign direct investment as a distinctive phenomenon. The multinational enterprise was simply an arbitrageur of capital, moving equity from countries where returns were low, such as the U.S. and the U.K., to those where it was higher, earning the profits of arbitrage while simultaneously contributing to the more efficient world-wide allocation of capital. Hymer, however, recognized that the predictions of the capital arbitrage theory were quite at odds with the resource transfer activities of multinational enterprises. Such enterprises both invest and borrow abroad, and there are substantial cross movements internationally within industries [Erdilek (1985)]. Their activities seem to be focused on certain industries (e.g., autos, electronics, chemicals, petroleum) and less so in others (e.g., textiles, steel, agriculture), and moreover the capital transfer is not from developed to less developed economies, but primarily from developed economies to developed economies.[3]

In order to advance a more plausible theory of direct foreign investment (DFI), Hymer (1970) appealed to oligopoly theory and advanced two major reasons why firms would move beyond their original national borders. One was to attenuate competition by buying it up or somehow displacing it. The other was to employ abroad the firm's special advantages, advantages such as skills, entrepreneurship or access to capital, advantages which 'may be more potent abroad than at home' (p. 43). The firm could, of course, export the commodity in which the advantage is embodied, or it could license it to a foreign enterprise (p. 47). But it will prefer to invest abroad rather than license to avoid possible bilateral monopoly situations (p. 49), to avoid technological misappropriation (p. 50), and to prevent the costly haggling between licensor and licensee. It is the latter cost and control consideration which decides the hierarchical mode of organization.[4]

These were powerful insights, which laid the foundation for a completely new paradigm of the international firm, one rooted in industrial organization rather than international trade and finance. The approach was quickly adopted and refined by others, most notably Kindleberger (1969) and Caves (1971). The framework, however, lacked a clear welfare criteria. Hymer thought the multinational enterprise 'would not arise in competitive indus-

[3]According to the United Nations (1978, p. 237), only about one quarter of the stock of foreign direct investment of developed countries is located in developing countries.

[4]John Dunning has provided supporting evidence and clarified the Hymer argument on these matters in what he refers to as an 'eclectic theory' of the multinational enterprise. See Dunning (1981, ch. 2) and Dunning (1984).

tries' (p. 86), which led him to the conclude that 'a restriction on direct investment or a policy to break up multinational corporations may be in some cases the only way of establishing a higher degree of competition in that industry ... the underdeveloped countries need to devote an important share of their scarce resources to building up national enterprises...' [Hymer (1970)]. But if the multinational enterprise is based on a unique product or special cost advantage, then the advocated policy implies that special cost or product advantages should be abandoned.

Since Hymer, an important 'internalization school' [Buckley and Casson (1976), Rugman (1981), Hennart (1982), and others] has emerged which emphasizes the advantages and costs associated with internalizing economic activity. While several different types of imperfections in intermediate markets are identified (Buckley and Casson, p. 36–38), 'the strongest case (where these kinds of imperfections arise) concerns the markets for various types of knowledge (p. 39). Other markets where internalization is likely to be advantageous are perishable agricultural products, intermediate products in capital-intensive manufacturing processes, and raw materials whose deposits are geographically concentrated' (p. 40).

The main conclusion is that 'markets will tend to be relatively more efficient than firms in handling transactions between a large number of buyers and sellers. Markets will be at a comparative disadvantage when transactions are subject to a high degree of uncertainty and when they consist of long-term exchanges of complex and heterogeneous products between a comparatively small number of traders' (pp. 167–168).

In order to go beyond these general conclusions and to provide a foundation for unambiguous normative criteria useful for management and public policy, the *nature of the transactions which are being internalized* must be given much closer scrutiny. The unit of analysis should be the transaction [Williamson (1975, 1979, 1981)], rather than the firm, and explicit attention must be given to ex ante vs ex post conditions of competition in the markets which the multinational enterprise internalizes. In short, the internalization paradigm developed in the literature to date needs to have transactions cost economics embedded within it if a deep understanding of the multinational enterprise is to evolve.

2. A transactions cost perspective of the multinational enterprise

At one level the internalization 'school' and the transactions cost approach are one in the same. Both see the firm as a response to market 'failure'. Profit-seeking firms internalize operations when by so doing the costs of organizing and transacting business will thereby be lowered.

The paradigms diverge, not in spirit but in emphasis, when it comes to specifics. The transactions cost approach is microanalytic, focusing on the transaction as the basic unit of analysis. It contends that a leading purpose of economic organization is to economize on the costs, broadly defined, of business transactions over time. It regards firms, markets, and hybrid mixed modes as alternative governance structures. The selection of one or the other depends on their relative efficiency properties. This will in turn depend on the nature of the transactions and the type of governance structure that is needed. The most important attribute for assessing whether a transaction requires a special governance structure, beyond that provided by unassisted markets, is the degree to which the parties to a transaction must invest in assets dedicated to the proposed exchange of goods and services [Williamson (1975, 1979), Klein, Crawford and Alchian (1978)]. Dedicated assets are transaction specific because their value in a given transaction is higher than in their next best use. The ability of unassisted markets to safeguard buyers and sellers depends on the degree to which efficient production does not require the deployment or development of dedicated assets, either physical or human, by the buyer or the seller. If dedicated assets are needed, then the parties to an exchange will become locked in once the assets are deployed. The 'lock in' phenomenon almost always results from investment or development of transaction specific assets. There is the strong possibility that in the absence of some kind of safeguards, one or both parties will try to take advantage of the fact that the other can no longer opt out of the arrangement, except at great cost [Williamson (1981), Monteverde and Teece (1982)]. Situations of this kind are remarkably common in business transactions, and they generally require some form of specialized governance structure. Although buyer and seller have a long term interest in adaptations that make both parties better off, each also has an interest in selfishly appropriating as much of the gain as possible. 'Efficient adaptations may thus be made only with costly haggling or may even go unmentioned because one party fears that the other's opportunism will dissipate the gains. The parties, therefore, have an incentive to develop governance structures that help prevent opportunism and infuse confidence' [Williamson (1984, p. 1203)].

The internal governance structure provided within the modern corporation is one such possibility. A transaction can be taken out of the market – that is, it can be internalized – and the specialized governance structure of the firm can be used to shield and protect the transaction and ensure the full utilization of the specialized assets in question. Of course, bringing the transaction out of the market and into the firm markedly changes the incentives of the parties since, from the perspective of the firm's global profits, there is nothing to gain and probably something to lose by having one party (e.g., a foreign subsidiary) gain/lose at the expense of another party (e.g., domestic division or subsidiary). The point, however, is that *transactions cost economics provides a framework for discriminating between those transac-*

tions which need to be internalized and those which do not. Without such a framework, internalization theories of the multinational enterprise must be considered incomplete, and perhaps even tautological.

3. International production and multinational enterprise

In order to apply transactions cost economics to the multinational enterprise, it is necessary to recognize two classes or groups of multinational firms: one group turns out essentially the same line of goods from each plant in several different locations. Such firms can be referred to as *horizontally integrated multinational enterprises.* A second type produces outputs in some of its plants which serve as inputs into others located in different nation states. Such firms will be labeled as *vertically integrated multinational enterprises.* Obviously, vertical and horizontal integration are both possible within a single firm.

Setting aside for the moment the question of whether production should be controlled by a multinational or a domestic firm, it ought to be apparent that if production techniques and knowledge are uniformly distributed internationally, the location of production will simply depend on differences in factor costs, tariffs, taxes, transportation costs, and the size of markets. The theory of comparative advantage reveals how differences amongst countries in comparative costs will cause the international specialization of production and concomitant trade. Economies of scale heighten the degree of specialization and may cause one country to be the world's sole supplier of a particular commodity.

Thus, consider the economic factors determining whether textbooks should be printed in the U.K. to serve both markets, in the U.S. to serve both markets, or whether production destined for each market should be produced in facilities located in each market.

Fig. 1 summarizes how the situation might appear to a U.S. publisher looking to supply the U.K. market via exports. Because a tariff must be paid on books imported into the U.K. and because transportation costs are also incurred, *a–b* is the relevant portion of the cost curve for supplying the U.K. market from the U.S. Should costs for printing $q_b - q_a$ units in the U.K. be higher than is represented by the cost curve segment *a–b*, the least cost approach will involve performing the printing in the U.S. rather than in the U.K. Clearly, the least-cost location of production depends upon the production costs in various markets, together with the level of tariff and transportation costs. In the hypothetical illustration, producing the requirements of both markets in the U.S. enables unit production costs in the U.S. to be lower than otherwise, but shipments to the U.K. are saddled with tariff and transportation costs which would be eliminated if production were split between the U.S. and the U.K.

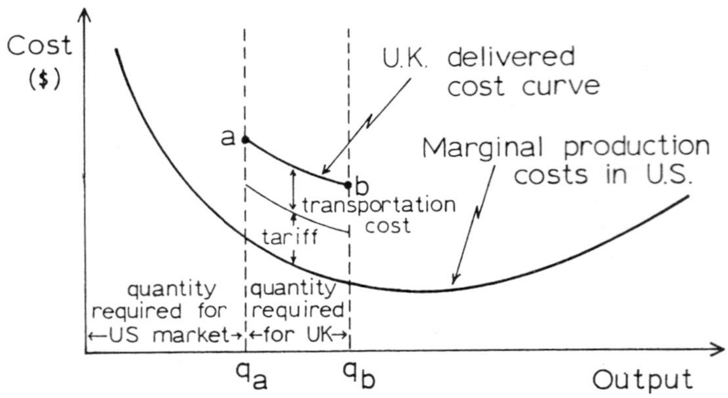

Fig. 1. Hypothetical cost curve for U.S. publisher contemplating printing in the U.S. to supply both the U.S. and U.K. markets.

The diagram makes it clear that higher tariffs tend to insulate markets, encouraging domestic production and reducing trade. On the other hand, low unit transportation costs, such as with the manufacturing of integrated circuits and many other components in the microelectronics business, tends to encourage the location of production where production costs are low, which is generally in the Far East, away from the major consuming markets.

From the point of view of the theory of foreign direct investment and multinational enterprise, the issue, however, is not the international location of production but *the national identity of the firms engaging in it.* In the example above, we wish to know if printing is performed in the U.K., whether it will be conducted in a British or U.S. owned facility.

The answer to this question is not at all obvious if one adopts the paradigm of perfect competition and applies it internationally. Perfect competition assumes that all technologies are available to all firms, and costs are the same to all firms, indigenous and foreign. If one relaxes the assumptions of perfect competition to allow nationals to have a slight cost advantage over foreigners when operating at home by virtue of familiarity with the language, customs, institutions, and traditions of their kindred, then foreign firms will have an initial handicap, and businesses will take on the nationality of the home market.

In reality, of course, foreign firms may have certain offsetting advantages. For instance, they may have certain trade secrets and proprietary information which enable them to produce goods which indigenous firms do not know how to produce. Alternatively, the possession of trade secrets and proprietary information may cause the foreign firms to have lower costs than indigenous firms. If this is the case, the foreign firms may well be able to overcome the cost disadvantage associated with doing business in markets away from home.

This advantage does not automatically imply that the profit seeking foreign firm will gain from establishing a subsidiary and competing against indigenous firms in their home market. The reason is that the foreign firm has the option of selling the service of the assets which are the source of its competitive advantage – such as trade secrets and proprietary information – to the indigenous firm for use at home. If the market for these services is efficient, the foreign firm should find this strategy at least as advantageous as setting up a foreign subsidiary. The foreign firm thereby avoids the cost penalty assumed to be associated with its alien status. Given that the relevant factor services will be more valuable if employed abroad in the hands of the indigenous firm, the foreign firm will be able to extract more from the assets which are the source of its competitive advantage (such as trade secrets and proprietary know-how) by selling them or their services to indigenous firms than through setting up a foreign subsidiary.

However, there are important circumstances where, because of contracting problems and associated transactions costs, the markets in question will not be efficient. In these instances, a profit-seeking foreign firm will discover that the best strategy may be to substitute arm's-length market transactions with internal transactions. Internal transactions – as between parent and foreign subsidiary – enable the firm to overcome certain problems associated with market transactions, thereby increasing the returns available from the firm's assets.

To summarize, a firm is likely to become multinational if (a) it has certain special assets which give it a competitive advantage over indigenous firms (the *strategic advantage* factor), (b) these assets are more economically utilized in production facilities in parts of the world beyond the firm's domestic markets (the *location* factor), and (c) the best way to obtain full value from employing the asset in foreign markets is to transfer the asset internally within the firm to another affiliated business unit (the *transactions cost* factor). *All three factors must be present to explain foreign direct investment.*[5] Put differently, *the multinational enterprise and foreign direct investment represent a response to high transactions costs by firms with unique assets/capabilities which have value when utilized in production facilities located in foreign markets.*

The locational factors identified will not concern us further. It will suffice to observe that factor costs, transportation costs, tariffs, and industrial policy often make it advantageous to locate production in the market being served rather than to attempt to supply the market via exports. The determination of a locational advantage requires comparing the unit costs associated with

[5]This tripartite taxonomy is implicit in Hymer and has been expounded and extended by Dunning (1981). However, Dunning's use of the term 'ownership' is confusing, particularly to Chicago-trained economists, as ownership is often considered to be tantamount to control. Hence I use the term strategic advantage.

production at home and exporting against the costs of producing abroad and supplying the consumers from a location within the foreign market, an exercise which is conceptually quite straightforward.

Nor is it necessary to dwell further on the role of rent-yielding special assets. Clearly, practically every firm by virtue of its history possesses some kind of unique asset which is the potential source of a stream of rents to the owner. It might be a technological, marketing, or managerial capability or a natural resource position not fully possessed by other firms. The principal reason for not focusing on locational or asset factors is that standing alone they do not have a decisive impact on the selection of the firm's boundaries. But transactional advantages are central to the understanding of the multinational firm as an organizational form. Accordingly, they are now explored in some detail.

4. Transactions costs and multinational enterprise

4.1. The transactions cost properties of the horizontally integrated multinational enterprise

Consider a firm whose production activities are purely domestic in scope. Assume, furthermore, that the firm possesses certain unique assets which are potentially the source of a stream of rents (ownership factors), and that locational factors are such that foreign production rather than the exporting of the final product is the cheapest way to access world markets. The firm in question is confronted with the choice of selling its special assets or their services (licensing in the case of know-how) to a foreign firm or in establishing a foreign affiliate.

Suppose the source of a firm's competitive advantage is its technological know-how. Consider what is needed for two firms to strike a deal over the purchase and sale of know-how. In order to work out a mutually acceptable licensing arrangement, it is necessary to discover with whom one wishes to deal, and to inform the other party that one wishes to deal, and the terms upon which this might be possible. It is also necessary to conduct negotiations leading up to the bargain, to draw up a license contract with all contingencies unambiguously specified. Once the agreement is executed, it is necessary to undertake the inspection needed to make sure that the terms of the agreement are being observed, and to enforce the contract if the terms are being violated. The ease or difficulty of performing these activities is likely to depend on many factors, including what can be called the 'regime of appropriability' within which the firm operates [Teece (1986)]. For instance, if patenting is quick and enforceable, as is sometimes the case in chemicals, and certain simple forms of mechanical devices [Levin, Klevorick, Nelson and Winter (1984)] the market for know-how might work quite satisfactorily, especially if performance guarantees can be used to avoid early disclosure of what is in the black box. In these instances, patent and know-how licensing will often enable the innovating firm to obtain the maximum possible returns from innovation. On the other hand, if the regime of appropriability permits only weak legal enforcement of rights over intellectual property, transactional problems will abound, and alternative governance modes are likely to be preferred.

Transactions cost problems will often arise because of difficulties associated with disclosing value to buyers in ways that is convincing and that does not destroy the basis for exchange. Because of information assymmetries, the less informed party (in many instances, the buyer) must be wary of exaggerated representation by the seller as to the performance properties of the technology in question. Moreover, if there is sufficient disclosure to completely assure the buyer that the information possesses value commensurate with the seller's claims, the fundamental paradox of information arises: its value to the purchaser is not known until he has the information, but then he has in effect acquired it without cost. In the absence of unassailable (legal) protection over the intellectual property in question, which will depend on the regime of appropriability that the innovator confronts, the owner of the know-how in question will be exposed to a capital loss.

Suppose that these hurdles are overcome – that opportunities for mutually profitable trade are recognized, that buyers concede value when value exists, and that complete and costless contract enforcement is possible. The transaction in question may still fail if it must take place between unaffiliated enterprises with different internal cultures, codes of communication, and capabilities. The reason is that technology transfer generally requires a symmetry in technological capability between the transferor and the transferee. Although there may be some circumstances where technology transfer can be effectuated by the simple disembodied transfer of codified information – as is sometimes the case with the formulae for a chemical compound, the code of a computer software program, or the blueprints of a mechanical device – more is frequently needed. Know-how often cannot be codified, since it has an important tacit component; even when it can, it is not always readily understood by the receiver. Such know-how is tacit and is extremely difficult to transfer without intimate personal contact, involving teaching, demonstration, and participation [Teece (1981b), Nelson and Winter (1982)].

Moreover, it will often not suffice just to transfer individuals. While a single individual may sometimes hold the key to much organizational knowledge, group support is often needed, since organizational routines [Nelson and Winter (1982, ch. 5)] may need to be transferred. In some instances, this transfer can be effected through a tightly specified one-time contract, requiring the licensor to provide a team to assist the licensee to absorb the technology. However, such contracts are often highly incomplete and the necessary adaptation difficult to accommodate. The failure to provide flexibility is likely to lead to dissatisfaction during execution. The dissatisfaction may be an unavoidable result and an irremediable characteristic of contracts.

Even when transactional difficulties are apparent, establishing a foreign subsidiary is an extreme response to the needs of a one-shot transfer. If repeated transfers of technology are contemplated, intrafirm transfer to a foreign subsidiary (which avoids the need for repeated negotiations, contract specification, monitoring and enforcement) has advantages on how contractual difficulties arise. Better disclosure, easier agreement, better enforcement, and more efficient transfer result. Here lies an incentive for developing the assets and investing in the creation of the organizational infrastructure associated with horizontal direct foreign investment.

The above arguments, while couched in the context of technological know-how, are in fact general and extend to many different kinds of difficult-to-trade assets. For instance, managerial (including organizational) know-how and goodwill (including brand loyalty) represent types of assets for which markets may falter as effective exchange mechanisms. Accordingly, the existence of high transactions costs is what lies behind horizontal direct foreign investment.[6]

Fig. 2 illustrates the argument. The diagram assumes that locational and strategic advantage factors indicate that a foreign production location will minimize costs. The issue is whether it ought to be licensed or integrated (internalized).

GC_L and GC_I (in fig. 2a) are cost schedules representing the average unit costs of employing licensing (market) channels and internal (intra firm) channels to effectuate the sale and transfer of know-how. On the horizontal axis is an index of the complexity of know-how. Contractual and transfer problems are assumed to increase with this complexity.[7] The GC schedule falls with successive transfers to the same offshore facility, as represented by the broken line GC_I in fig. 2a (n refers to the number of transfers to the same offshore facility and k is a constant number of transfers). This stems simply from spreading the overhead costs associated with setting up on the internal governance machinery associated with a foreign subsidiary.

Fig. 2b represents the differences in governance costs between licensing and internalization. Point T is that threshold level of complexity, for a given

[6]This is not to say that the existence of know-how and related assets is the sole transactional difficulty providing incentives for firms to internationalize economic activity. The evidence suggests that multinational firms also help solve other problems, such as the inability of unassisted markets to always protect product and service quality. These factors undoubtedly help explain other incentives for the horizontal expansion of the multinational firm, particularly in low-technology service-type industries [Dunning and McQueen (1981)].

[7]Complexity is probably not a good surrogate for all the factors which effect transactions costs. Appropriability regimes, for instance, do not vary monotonically with technological complexity.

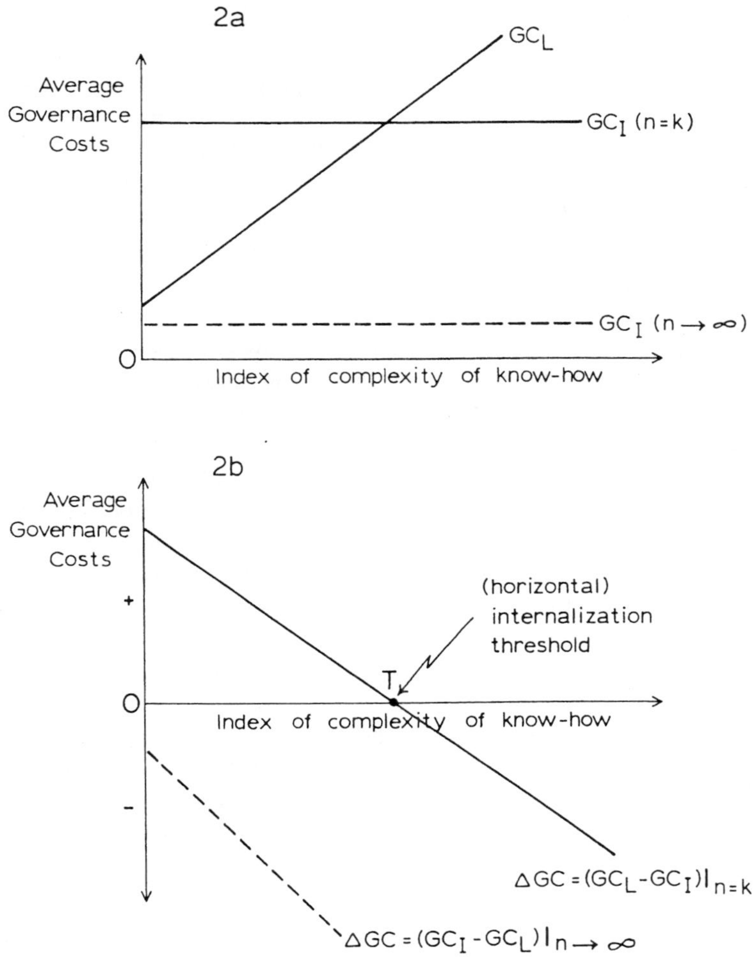

Fig. 2. Hypothetical levels of governance costs and horizontal multinational enterprise boundaries. 2a: The governance cost of licensing a (horizontal) foreign investment. 2b: The differential governance costs of licensing a (horizontal) foreign investment.

number of transfers, at which the governance costs associated with licensing become of such magnitude that internalization represents the more efficient governance mode.[8]

4.2. *The transactions cost properties of vertically integrated multinational enterprises*

A sizable portion of foreign direct investments is associated with the sourcing abroad of raw materials such as oil, copper, and alumina. The

[8]Clearly, the real world cannot be so neatly dichotomized. See Teece (1985a) for a treatment of mixed modes such as joint ventures and non-equity 'partnerships'.

vertically integrated multinational enterprise – which sources inputs in one country for use in facilities in another – internalizes a market for an intermediate product, just as the horizontal multinational enterprise internalizes markets for know-how and other difficult-to-trade assets. Vertical integration will represent the more efficient governance structure when a trading relationship requires the development of transaction specific assets. Relying on arm's-length contracts under these circumstances would involve exposure to recontracting hazards stemming from high switching costs. This principle is true whether the intermediate market in question is domestic or international in scope.

By way of example, consider a smelter located in the home country.[9] Assume that there are many different types and grades of ore and that to smelt the ore at the lowest cost the smelter needs to be designed to process a particular grade of ore. If this ore is found only in one or two locations abroad, then there may be recontracting hazards associated with relying on long-term supply contracts with another enterprise. This is because once the investment in idiosyncratic smelting facilities has been made, the enterprise engaged in smelting will be extremely vulnerable to changes made by the ore supplier in the conditions of sale. The supplier of the ore may be blatantly opportunistic. Despite contract specifications to the contrary, the supplier may simply raise prices or lower quality. Generally the opportunism is much more subtle.

Judicial redress may be weak; so to the extent that alternative suppliers are not available, the owner of the smelter will be obliged to go along with the new terms imposed by the supplier. In extreme cases of dependence, the supplier will be able to extract a pecuniary advantage from the owner, up to the value of the smelting facility. Ex ante backward vertical integration can eliminate this risk;[10] for if supplier and purchaser are one and the same, the incentive for postcontractual recontracting is attenuated, and trading relations can proceed in a fashion which is relatively smooth and efficient.[11]

Fig. 3 illustrates the argument. As with fig. 2, the diagram assumes that locational and strategic advantage factors indicate that a foreign production location will minimize costs. The issue is whether the sourcing (in the case of backward integration) or supply (in the case of forward integration) will be

[9]John Stuckey (1983) has a comprehensive treatment of the reasons for vertical integration in the aluminum industry.

[10]This treatment does not exhaust the economies associated with vertical integration. For further elaboration, see Teece (1976) and Buckley and Casson (1976). Both emphasize scheduling and coordination economies. When complete futures markets are lacking, firms can bring interdependent activities in house and adjust to changed circumstances in a sequential fashion.

[11]However, while problems in intermediate product markets can often be ameliorated by vertical integration and the establishment of a wholly or partially owned supply division, foreign governments are often well positioned to attempt recontracting with respect to the terms and conditions under which the multinational enterprise supply division can operate. These issues are addressed in section 4.3.

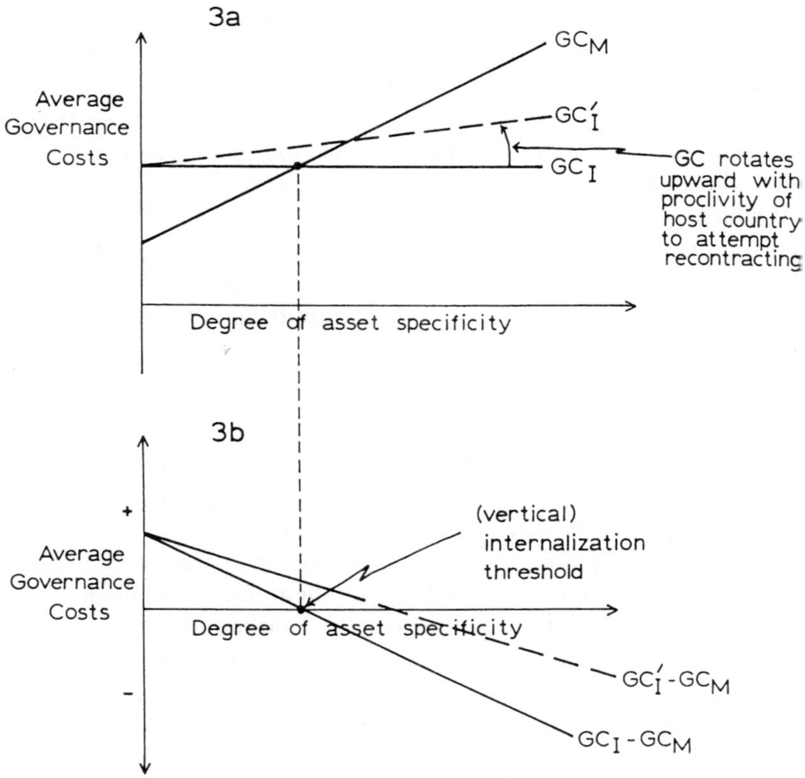

Fig. 3. Hypothetical levels of governance costs and vertical multinational enterprise boundaries. 3a: The governance costs of intermediate product markets and (vertical) foreign investment. 3b: The differentical governance costs of intermediate product markets and (vertical) foreign investment.

more efficiently governed by contracts of various durations or by internal organization. Fig. 3a depicts the governance costs associated with markets GC_M rising with the specificity of the assets needed to support the transaction at least cost. The average governance costs associated with vertical integration, while initially higher because of set up costs, do not vary with the degree of asset specificity because incentives for recontracting, and the associated costs, are assumed to be eliminated within the firm. The relationship is complicated in the international context as the governance costs associated with markets depend importantly on expropriation risks. Put differently, foreign direct investment, whether it is vertical or horizontal, internalizes intermediate contracts but creates a direct interface between the multinational enterprise and the host government. To the extent that host governments treat multinational enterprises differently from indigenous enterprises, the foreign firm may have circumvented one set of potential recontracting hazards through direct investment only to encounter another. The risk of expropriation or of host-government recontracting is such as to represent an important hazard for the multinational enterprise. It is likely to vary with the specificity of the multinational enterprise's assets deployed upstream (in the case of backward integration) or downstream (in the case of forward integration), as is discussed in section 4.3.[12] Hence, the GC_{VI} schedule will shift according to host-country recontracting proclivities, the costs of which will also vary with asset specificity. As before, T represents the internalization threshold. For asset specificity levels above this, vertical integration is indicated, though the GC_I schedule rotates upwards above a point on the vertical axis according to host-country proclivities.

4.3. Evidence

The empirical evidence supporting the transactions costs theory of the multinational enterprise is scattered and incomplete. A comprehensive bibliography and discussion can be found in Caves (1982). The treatment here draws partly on Caves as well as original sources.

Caves, Crookell and Killing (1982) have examined the various infirmities in the market for know-how, although their focus is mainly on the circumstances in which firms will enter the market as licensees, and the configurations of terms likely to occur in license agreements. Nevertheless, their findings indicate that the market is riddled with imperfections and that rent extraction by licensors is very incomplete.

Evidence on the direct investment-licensing choice tends to be much more indirect. A common strategy involves correlating the prevalence of multinational enterprises in an industry with structural traits of the industry. Thus, if the theory indicates that an attribute promotes the formation of multinational enterprises and that firms in an industry display this attribute, then multinational enterprises should be prevalent in the industry in question. Pursuing this line, researchers have analyzed the market shares of sales held by foreign subsidiaries in nations such as Canada, Australia, and the United Kingdom to determine whether high shares occur in industries marked by the traits that the theory predicts should give rise to multinational enterprises. They have also performed similar tests by examining inter-industry differences in the relative size of foreign assets held by companies based in the various manufacturing industries of different nations, comparing the activities of national companies to those of multinational enterprises. The conclusions that have emerged from these studies are summarized below.

[12]Even generalized assets are exposed if, in fact, the host government takes the expropriation route. However, for threats short of expropriation, the vulnerability of the multinational enterprise depends on the extent to which its assets are dedicated to particular activities. The multinational enterprise can sell the assets to other foreign firms or to indigenous firms if they are not transaction specific and the host government does not put restrictions on such sales.

The presence of difficult-to-trade assets as an incentive to *horizontal* investment has been confirmed in numerous studies. Since such assets are difficult to measure directly, their prevalence is proxied by the outlays that companies make for the purpose of developing them. Many researchers have seized upon the research and development and outlays undertaken by firms as the best indicators. Several studies have confirmed that the proportion of foreign subsidiary assets in the total assets of U.S. corporations is positively and significantly correlated with the importance of advertising and research and development outlays (R&D) to the industry. A partial list must include Gruber, Mehta and Vernon (1967), Horst (1972, 1974), Pugel (1978, ch. 4), Bergsten, Horst and Moran (1978, pp. 242–5), Goedde (1978, ch. 2), and Lall (1980). Goedde demonstrates that advertising exerts a more powerful influence on food and chemicals, including drugs, while R&D is more important in machinery. As Caves (1982, p. 7) remarks, 'this pattern closely matches one's sense of the apparent importance of different sorts of intangible assets in those industries'.

Various attempts have also been made to investigate the role of specialized managerial know-how as a driving force behind the international expansion of U.S. firms. Pugel (1978) and Swedenborg (1979) have shown that the influence of managerial skill levels is statistically significant.[13] So far, however, the empirical tests have failed to disaggregate down to the level of the individual transactions; so support for transactions cost considerations is only indirect.

Tests of the transaction cost theory of *vertical* integration in a purely international context are absent.[14] Most of the evidence is only tangentially relevant to the hypothesis. For instance, Pugel (1978) concluded that American manufacturing industries having greater involvement with natural resources invest larger proportions of their assets abroad, and Stuckey (1983) found the international aluminum industry to contain multinational enterprises integrated from the mining of bauxite through the fabrication of aluminum projects; but Stuckey also found a network of long-term contracts and joint ventures. These features of the industry indicate the hesitation of firms to use spot markets for bauxite (the raw ore) and alumina (the output

[13]There is also evidence to support the notion that internalization offers advantages in low-technology activities where quality control is important to market success. Dunning and McQueen (1981) have examined the involvement of multinational enterprises in the hotel industry, and their data show that multinational enterprises are focused on the high-quality end of the market, catering principally to international business travelers. These hotels carry a standardized brand name, offer an international reservation system, and train their own managers in the skills needed to efficiently provide high quality service. Thus, the internationalization of the hotel business appears to be based in part on an intangible difficult-to-trade asset (management), but there are also elements of the externality principal at work, since protecting and advancing an international brand name is a critical dimension of the business.

[14]For evidence in a domestic context, see Monteverde and Teece (1982a, b) and Masten (1984).

of the first processing stage). Stuckey also observed that switching costs are high (he actually estimated them) for the ore refiner since alumina refining facilities need to be located physically close to bauxite mines to minimize transport costs, and they are also designed to handle the properties of specific ores. Alumina smelters likewise are somewhat tied for technical and transportation cost reasons to particular sources of alumina.[15] Very similar circumstances characterize the international petroleum industry [Teece (1976)] and explain the substantial degree of integration between refining and crude production.

In recent years, however, vertical integration relationships, domestic and international, have been much affected by political considerations, particularly in aircraft, automobiles, and electronics. In order to sell abroad, many firms are discovering that foreign governments require them to provide 'offsets' to domestic producers. Subcontracting is probably the most single frequent type of offset,[16] although countertrade, coproduction, and licensed production are also important. To the extent that governments are involved in negotiating or influencing these outcomes, as is generally the case, one ought not expect observed vertical relationships to be driven entirely or even primarily by the economic considerations advanced earlier. Indeed, many companies forced to participate in farm-outs of this kind reported higher costs and other inefficiencies than would otherwise have been the case.[17]

The evidence in international banana production [Read (1983)] also appears broadly consistent with the theory, but it is not the asset specificity principal which comes through strongest. Rather, it is the coordination and scheduling of production and distribution which appears to constitute the distinct advantage of multinational banana producers, like the United Fruit Company. Finally, country specific studies of the foreign investment process also underline vertical multinational enterprises as the outcome of failed arm's-length market transactions. According to Tsurumi (1976, ch. 2), Japanese companies tended to become involved with extractive foreign investments only after the experience of having arm's-length suppliers renege on long-term contracts.

4.4. Dynamic considerations

The literature on the multinational enterprise, whether emphasizing market power or efficiency, suffers from a common deficiency: underemphasis on dynamics. The historical evidence shows American and British multinational

[15] For a brief review of Stuckey, see Teece (1983).

[16] See 'Report of a Survey on Offset and Coproduction Requirements' prepared by the U.S. Department of Treasury in cooperation with the Aerospace Industries Association and the Electronics Industries Association, May 24, 1983 (U.S. Dept. of Treasury, Washington, DC 20220).

[17] Treasury Report, ibid., Annex B, pl.

enterprises transitioning to offshore production after first establishing a sales branch abroad which in turn commonly preceded the establishment of contractual relationship with a foreign sales agent [Nicholas (1983)]. Agency contracts were unsatisfactory, containing vague and difficult-to-enforce performance clauses. Attempts to shore these up by specifying, quantitatively, budgets for traveling, advertising, engineering support, and the like, as well as certain inventory levels on the agents' premises proved unsatisfactory. The transition from agency to branch sales office was facilitated by the manufacturer's gradual accumulation of information about the foreign market, acquired through monitoring its foreign agents and by expansion in sales volumes to levels which would support a facility of minimum economic size. The establishment of a sales branch also demonstrated to customers a more solid commitment on the part of the manufacturer to support the market in question. Often triggered by the failure or termination of an agent, the establishment of a foreign sales subsidiary subsequently became the platform upon which a manufacturing investment was made [Chandler (1977, p. 369)]. However, the transition to manufacturing depended upon the relationship of production costs abroad to production costs at home plus tariffs and transportation, as well as governance considerations. Nevertheless, as Aharoni (1966) points out, the direct investment process is governed by more than just economic incentives. An initiating force, galvanizing the organization into action, is often needed. The presence of a sales office also assists information collection, thereby significantly lowering perceptions of uncertainty, and raising the probability that a firm will engage in direct foreign investment if the underlying cost conditions permit it. Clearly, transactions cost economics must be married to organizational decision theory if the dynamics of channel selection are to be better understood.

4.5. Implications for export competitiveness

The analysis of the multinational presented here indicates that foreign direct investment, at least of the horizontal kind, stems from the possession by a firm of certain unique assets and that developing and protecting the rent stream associated with these assets often requires the extension of some kind of hierarchical control structure over productive assets which are distributed internationally. Within this hierarchical structure, technology and other intangible, difficult-to-protect-and-transfer assets can often be deployed more economically and securely. The result is that when contracting problems arise, the existence of multinational firms facilitates the technology-transfer process and enables the assets in question to work in combination with foreign factors of production, according to the dictates of cost minimization.

This analysis sheds light on the difference between the export competitiveness of the U.S. economy and the export competitiveness of U.S.-based multinationals.[18] The available data indicate that in the last two decades or so, U.S.-based multinationals have overcome some of the relative decline in the competitiveness of the U.S. as a production location by shifting their production to export platforms abroad from which they could more effectively exploit their intangible, difficult-to-market assets. Between 1957 and 1977 (the last date for which comparable data is available at time of writing) exports of manufactured goods from the U.S. fell from 21.3% to 14.4% of world totals, whereas the exports of U.S. firms, including majority-owned affiliates, fell from 25.8% to 23.5% [Lipsey and Kravis (1985)]. Over the period 1966–1977, the U.S. share of world exports fell (from 16.4% to 12.3%) while exports by U.S. multinationals, inluding majority-owned affiliates, actually rose slightly.[19]

The data seem to indicate a large shift in the geographical origins of exports by U.S. firms. This reflects a relative diminution in the advantage of the U.S. as a location for manufacturing. It also suggests that the U.S. advantage in developing and exploiting difficult-to-market assets, such as know-how, has been maintained relatively well. Put differently, U.S.-based multinationals appear to have developed firm-specific comparative advantages separate from that of the U.S. economy. Hence, in fashioning U.S. policy, it is important to recognize that some policies which contribute to the export competitiveness of U.S. firms, such as certain R&D policies which assist in knowledge generation, may not contribute as strongly to the export competitiveness of the U.S. economy as to the export competitiveness of U.S. multinationals. The latter is likely to be more sensitive to policies which impact manufacturing productivity, rather than just R&D.

5. Transactions cost analysis of multinational enterprise-host country relations

Transactions cost principles can also be used to shed new light on host-country/multinational-enterprise relationships in at least two important ways. First, the widely observed transformation of the bargaining relationship between the host-country and the multinational enterprise can be analyzed by using transactions cost concepts. These concepts can also be employed to derive normative implications with respect to the governance machinery needed for both sides to extract maximum benefit from the relationship. Second, transactions cost principles can be used to analyze the implications of nationalization and the substitution of direct investment with management contracts.

[18]This definition of competitiveness refers to the share of world exports of manufactured goods provided by the entity in question. For a more robust definition of competitiveness, see Cohen, Teece, Tyson and Zysman (1985).

[19]According to Lipsey and Kravis (1985, p. 5), the difference in performance between the share of U.S. firms is probably not due to the entry of additional U.S. firms into the status of multinationals.

In recent decades, the equity position of multinationals in natural-resource industries has declined markedly, particularly in developing nations and especially in the last two decades. For instance, whereas about 95% of Mideast oil was from equity interest properties in 1970, the percentage had fallen to about 10% a decade later [Griffin and Teece (1982, p. 10)]. Similarly, between 1945 and 1970, Chile moved in on the multinational-enterprise-dominated copper industry and had nationalized it by 1970.[20] The same trend is apparent in many other industries.

In many cases, the multinational enterprise is offered a management contract after its foreign assets have been nationalized. Since in a fundamental sense the effect of these trends is to take transactions out of the firm and place them back into the market, it is important to assess the ramifications. Kindelberger (1984, p. 92) asks

> 'If Saudi Arabia chooses to take over Aramco but permits its former owners to organize a new corporation, Stemco... to handle the logistics of planning, transporting, and marketing the bulk of the oil production of Saudia Arabia, what difference does it make, assuming that the management and marketing fees of Stemco do not differ too widely from the aftertax profits of Aramco before nationalization?'

One significant difference noted by Kindleberger is that 'ownership is a continuing relationship, whereas contracts bear a terminal date' (p. 93). To the extent that renewal is not expected, then there is an incentive problem as contract expiration approaches. The Venezuelan petroleum concessions, due to expire in the early 1980s, were terminated ahead of schedule because of government allegations that the multinational enterprises were extracting the reserves faster than good oil-field practice indicated.

A second major difference relates to the ability of the nationalized firm to adapt, quickly and sequentially, to changed conditions in downstream markets. Divestment imposes an interface which may impede rapid adjustment in the face of changed circumstances. As noted elsewhere [Teece (1976)], the problems associated with using contracts in vertical supply relations are less severe when the environment is characterized by stability. Kindleberger argues persuasively that the faster response of the Danish as compared to the British farmers following the decline in wheat prices after 1880 was due to the fact that the Danes had owner-occupied farms whereas the British farmers were managed by renters. The required shift from wheat to dairying required substantial new investments in animals and equipment,

[20]However, in the case of copper in Africa, as with Saudi oil, management contracts were negotiated between the host government and the previous concessionaires. For instance, Zaire gave the Belgian multinational enterprise Union Miniere a 25-year management contract after its assets were nationalized. Zambia provided similar contracts when it nationalized various foreign companies in the 1960s. See Kindleberger (1983, p. 91).

which required an entirely new relationship between owner and renter. New sharing rules had to be devised and performance monitoring arrangements modified. The transactions costs were apparently higher in the British context than in the Danish.

The analogy obviously carries over to international business: Once direct investment has been displaced by contracts, the buyer must negotiate with the host-country, even if a management contract is involved. This will slow adjustment and add to its costs. Ultimately, the impact of nationalization depends upon the nature of the contractual relationships which are forged between the nationalized enterprises and downstream firms.

The most familiar frameworks for analyzing multinational-enterprise/host-country dynamics rely on game theoretic bargaining models, especially models of bilateral monopoly [e.g., Mikesell (1971)]. On one side, the multinational enterprise is depicted as having the skills, experience, employment and taxation opportunities, and access to finance and markets that the host country needs. On the other side, the host country is depicted as having the mineral resources, labor force (generally unskilled) and control over taxes and other factors which can often be configured to provide an attractive investment opportunity for a multinational enterprise. This sets the conditions of the bargain. As both the multinational enterprise and the host country try to increase their return from the industry, each side has threats to make and benefits to offer. The struggle centers on the sharing of the rents associated with developing the natural resources.

A transactions cost perspective of the multinational-firm/host-country relationship would appear to be more revealing. The essence of the foreign direct-investment process is that it embodies what Williamson has called 'the fundamental transformation'. Before the multinational enterprise has deployed fixed assets in the host country, it can bargain for favorable terms from the host government. Once the deal has been signed and the investment sunk, the multinational enterprise bargaining position has been transformed into one which is much weaker. Because assets have been deployed which generally cannot be withdrawn, the multinational enterprise is vulnerable to opportunistic recontracting by the host country. With the investment sunk and successful, the host country is able to bring pressure to bear for renegotiation. 'In the perception of the foreign companies, the host country can begin to cheat them' [Moran (1974, p. 160)]. This is borne out empirically: few successful concession agreements in developing countries since the end of World War II have remained unaltered. The multinational enterprise finds itself confronted by 'renegotiations', 'adjustments', 'surtaxes', 'backtaxes', 'recomputations', and the like – all manifestations of ex post recontracting. Moreover, the shifts in bargaining power are repeated before and after each new major corporate commitment. Before the investment for new productive capacity, a new smelter, or a new refinery, the relative weight of bargaining strength tilts in favor of its foreign investor; after the operation is successful, it begins to tilt back toward the host country.

The bargaining condition of the multinational enterprise is also conditioned by the host country's acquisition of skills. This not only makes the host country a more informed bargainer, but also makes the management and production skills of the multinational enterprise less indispensable. The result is that the relative bargaining position of the host country is likely to improve over time – as with the case of copper in Chile – and the multinational enterprise's threat to withdraw or not to make further investment could be expected to have less force. As Moran (1974) points out, a point-in-time (static) bargaining analysis is unlikely to capture the role of ex post recontracting and host-country skill acquisition.

Nationalization is often the ultimate consequence. At least two possibilities are likely. One is that the now independent national enterprises attempt to build supply relationships with downstream multinational enterprises, the result being that relational contracts [Williamson (1979)] – in which there is a good deal of give and take, goodwill, and mutual understanding – substitute for intra-firm exchange. Relational contracting requires that the now stand-alone producers develop and scrupulously maintain regular supply relations with downstream companies similar to what previously took place in integrated channels. Investment schedules must be adjusted to meet the needs of customers, and the temptation to be opportunistic with respect to short-run supply situations must be avoided at all cost. The challenge to the supplier is to obtain the confidence of buyers so that they do not have to spend large sums re-engineering their facilities to obtain greater flexibility with respect to sources of supply. If buyers have to put in place costly investments to avoid the possibility of holdup, they will either have to offer lower purchase prices for raw materials or have to raise prices to consumers, which may not always be possible.

History indicates, however, that there is a strong possibility that the host-country enterprises will break ties, eschewing relational contracting. This may yield some short-run gains to the host country as there will undoubtedly be some 'lock in' situations where the multinational enterprise has become dependent. The host country may well decide to take advantage of these circumstances and the associated short-run inelasticity of demand. However, once a reputation for opportunistic behavior is acquired, it is likely to be difficult to shake off. Moreover, the disadvantaged party is likely to take costly steps to avoid future dependence. These costs will eventually be borne in part by the upstream natural-resource owner.

The possibility that relational contracts will emerge from an environment of hostility and mutual suspicion is remote. Therefore, in many instances, the governance machinery of international investment must be carefully configured to reduce the likelihood that costly haggling will break out between the multinational enterprise and the host country, and also to protect transactions and contracts between independent host-country firms and downstream buyers. Several devices to assist will be discussed briefly.

One approach with interesting transactions-cost ramifications has been proposed by Albert Hirschman (1971). The idea is to anticipate host-country demands for control – and costly efforts to wrest it from the multinational firm – by establishing an option for systematic divestment with a compensation schedule agreed to ex ante. The obvious problem with this proposal is that it encourages the multinational enterprise to have a very high discount rate – treating plant and equipment and natural resources as if there were no tomorrow. To safeguard such an agreement, the host country could place a bond in the hands of an independent third party equal to the amount of the agreed-upon compensation. This needs to be done at the same time as the foreign firm deploys assets which it is unable to withdraw should the terms of the original agreement be violated.[21] With such a bond in place, of appropriate magnitude to offset the multinational enterprise's dedicated investment, the host country could be free to trigger the expropriation option at any time. The mere existence of this option is likely to defuse hostility towards the multinational enterprise and increase the probability that a mutually profitable relationship between the host government and the multinational enterprise could develop. The end-period problem could be dealt with by having independent arbiters monitor behavior as contract expiration approaches. Requirements for new investment could be met by the host country increasing the magnitude of its bond.

The analysis so far assumes that the host country itself is not putting into place specialized investment dedicated to the multinational enterprise. It might be the case, however, that the host country dedicates investments, such as roads and electric utilities, to support a particular multinational enterprise. When specialized assets are developed for the multinational enterprise, they cannot be redeployed easily should the multinational enterprise decide to pull out or to otherwise adjust its plans opportunistically. The host country has incentives to develop investment safeguards in these circumstances. These can take many forms, including some type of penalty for early withdrawal or delayed expansion, or the creation of specialized governance structures for resolving disputes between the multinational enterprise and the host country. Alternatively, the multinational enterprise could also be asked to post a bond, thereby engaging with the host country in what Williamson (1983) would refer to as an exchange of hostages. Despite safeguards of the

[21]Moran (1974, p. 255) makes a related proposal, but it is insensitive to transactions cost issues. The proposal is for the host country to place a certain portion of the incremental tax/royalty receipts in escrow. This is unsatisfactory as these are likely to be modest at first, just when the multinational enterprises irreversible investments are likely to have their highest value. The bond should offset the value of the transaction specific assets deployed.

type indicated, additional benefits might accrue from disclosure of host-country investment plans by the multinational enterprise as this might enable the host country and workers with firm-specific skills to anticipate future developments and plan accordingly. Failure to design such governance structures may result in host countries failing to provide the necessary infrastructure. It bears repeating, however, that special governance or regulatory machinery is not needed unless the host country is deploying specialized assets to support the foreign firm.

If safeguards of the kind proposed – the posting of bonds in the hands of independent third parties – cannot be devised, then there is need for an agreement akin to GATT for foreign direct investment. As Kindleberger (1984, p. 252) points out, 'the qualities of the GATT that are appealing if an attempt is made to extend it from trade to the multinational enterprise are the frank exchange of information, the discussion of principles and standards of behavior, and their concentration into a set of rules'. Such an approach is likely to be decidedly inferior to other forms of safeguards since enforcement of international agreements of this kind has been notoriously ineffective.

6. Conclusion

Transactions cost economics is more than just a new set of terms for describing the multinational enterprise. As demonstrated in this paper, it extends the internalization framework and enables managers to systematically analyze where and when to internalize. For the policy analyst, it is a framework which can assist in the analysis of the economic welfare aspects of direct foreign investment. It directs attention to the distinctive governance properties of the multinational enterprise and the consequences of restricting the employment of this particular organizational form. Commanding insights into the dynamics of multinational-enterprise/host-country relationships and export competitiveness are also provided, along with a framework which permits a degree of objective analysis into what has previously been an emotionally charged subject.

References

Aharoni, Yair, 1966, The foreign investment decision process (Division of Research, Harvard Business School, Boston, MA).

Bergsten, C.F., T. Horst and T.H. Moran, 1978, American multinationals and American interests (Brookings Institution, Washington, DC).

Buckley, P.J. and M. Casson, 1976, The future of the multinational enterprise (Holms and Meier, London).

Casson, M., ed., 1982, The growth of international business (Allen and Unwin, London .

Caves, R.E., 1971, International corporations: The industrial economics of foreign investment, Economica 38, 1–27.

Caves, R.E., 1982, Multinational enterprise and economic analysis (Cambridge University Press, Cambridge, MA).

Caves, R.E., H. Crookell and J.P. Killing, 1982, The imperfect market for technology licenses, Discussion paper no. 903 (Harvard Institute of Economic Research, Cambridge, MA).
Chandler, Alfred, 1977, The visible hand (Harvard University Press, Cambridge, MA).
Coase, R.H., 1937, The nature of the firm, Economica 4, Nov., 386–405.
Cohen, S., D. Teece, L. Tyson and J. Zysman, 1985, Competitiveness, published as vol. III, Report of the President's Commission on Industrial Competitiveness (U.S. Government Printing Office, Washington, DC).
Dunning, John, 1981, International production and the multinational enterprise (Allen and Unwin, London).
Dunning, John, 1984, Non-equity forms of foreign economic involvement and the theory of international production, in: Research in International Business and Finance 4, part A (JAI Press, Greenwich, CT) 29–61.
Dunning, John and M. McQueen, 1981, The eclectic theory of international production: A case study of the international hotel industry, Managerial and Decision Economics 2, 197–210.
Erdilek, Asim, ed., 1985, Multinationals as mutual invaders: Intraindustry direct foreign investment (Croom Helm, London).
Goedde, A.G., 1978, U.S. multinational manufacturing firms: The determinants and effects of foreign investment, Ph.D. dissertation (Duke University, Durham, NC).
Griffin, J. and D. Teece, 1982, OPEC behavior and world oil prices (Allen and Unwin, London).
Gruber, W., D. Mehta and R. Vernon, 1967, The R&D factor in international trade and international investment of U.S. industries, Journal of Political Economy 75, 20–37.
Hennart, Jean-Francois, 1982, A theory of multinational enterprise (University of Michigan Press, Ann Arbor, MI).
Hirschman, Albert, 1971, How to divest in Latin America, and why, in: A bias for hope: Essays in development and Latin America (Yale University Press, New Haven, CT) 225–252.
Horst, T., 1972, The industrial composition of U.S. exports and subsidiary sales to the Canadian market, American Economic Review 62, 37–45.
Hymer, Stephen, 1970, The efficiency (contradictions) of multinational corporations, American Economic Review 60, 441–448.
Hymer, Stephen, 1976, The international operations of national firms: A study of direct foreign investment (MIT Press, Cambridge, MA).
Kindleberger, C.P., 1969, American business abroad (Yale University Press, New Haven, CT).
Kindleberger, C.P., 1984, Multinational excursions (MIT Press, Cambridge, MA).
Klein, B., R. Crawford and A. Alchian, 1978, Vertical integration, appropriable rents, and the competitive contracting process, Journal of Law and Economics 21, 297–326.
Lall, S., 1980, Monopolistic advantages and foreign involvement by U.S. manufacturing industry, Oxford Economic Papers 32, 102–122.
Levin, R., A. Klevorick, R. Nelson and S. Winter, 1984, Survey research on R&D Appropriability and technological opportunity, Unpublished manuscript (Yale University, New Haven, CT).
Lipsey, R.E. and I.B. Kravis, 1985, The competitive position of U.S. manufacturing firms, Unpublished manuscript (National Bureau of Economic Research, New York).
Masten, S., 1984, The organization of production: Evidence from the aerospace industry, Journal of Law and Economics 27(2), 403–417.
Mikesell, R., ed., 1971, Foreign investment in the petroleum and mineral industries: Case studies in investor–host country relations (Johns Hopkins University, Baltimore, MD).
Monteverde, K. and D.J. Teece, 1982, Supplier switching costs and vertical integration in the automobile industry, Bell Journal of Economics 13(1), 206–213.
Monteverde, K. and D.J. Teece, 1982, Appropriable rents and quasi-vertical integration, Journal of Law and Economics 25(2), 321–328.
Moran, T.H., 1974, Multinational corporations and the politics of dependence: Copper in Chile (Princeton University Press, Princeton, NJ).
Nelson, R.R. and S.G. Winter, 1982, An evolutionary theory of economic change (Harvard University Press, Cambridge, MA).

Nicholas, Steven, 1983, Agency contracts, institutional modes, and the transition to foreign direct investment by British manufacturing multinationals before 1939, Journal of Economic History 43(3), 675–686.
Pugel, T.A., 1978, International market linkages and U.S. manufacturing: Prices, profits, and patterns (Ballinger, Cambridge, MA).
Read, Robert, 1983, The growth and structure of multinationals in the banana export trade, in: Mark Casson, ed., The growth of international business (Allen and Unwin, London) 180–213.
Richardson, G.B., 1960, Information and investment (Oxford University Press, Oxford).
Rugman, Alan M., 1981, Inside the multinationals: The economics of internal markets (Croom Helm, London).
Rugman, Alan M., ed., 1982, New theories of multinational enterprises (St. Martin Press, New York).
Stuckey, John, 1983, Vertical integration and joint ventures in the aluminum industry (Harvard University Press, Cambridge, MA).
Swedenborg, B., 1979, The multinational operations of Swedish firms: An analysis of determinants and effects (Industrial Institute for Economic and Social Research, Stockholm).
Teece, David, 1976, Vertical integration and vertical divestiture in the U.S. petroleum industry (Institute for Energy Studies, Stanford, CA).
Teece, David, 1981a, Multinational enterprise: Market failure and market power considerations, Sloan Management Review 22(3), 3–17.
Teece, David, 1981b, The market for know-how and the efficient international transfer of technology, Annals of the American Academy of Political and Social Science 458, 81–96.
Teece, David, 1983, Technological and organizational factors in the theory of multinational enterprise, in: Mark Casson ed., Growth of international business (Allen and Unwin, London) 51–62.
Teece, David, 1984, Review of vertical integration and joint ventures in the international aluminum industry, Journal of Economic Literature 22, 1151–1153.
Teece, David, 1985, Multinational enterprise, internal governance, and industrial organization, American Economic Review 75, 233–238.
Teece, David, 1986, Firm boundaries, technological innovation, and strategic management, in: Lacy Thomas, ed., The economics of strategic planning, forthcoming.
Tsurumi, Y., 1976, The Japanese are coming: A multinational spread of Japanese firms (Ballinger, Cambridge, MA).
United Nations, 1978, United Nations Transnational Corporation in world development: A reexamination (New York).
Williamson, O.E., 1975, Markets and hierarchies (Free Press, New York).
Williamson, O.E., 1979, Transactions cost economics: The governance of conctractual relations, Journal of Law and Economics 22, 233–261.
Williamson, O.E., 1981, The modern corporation: Origins, evolution, attributes, Journal of Economic Literature 19(4), 1537–1568.
Williamson, O.E., 1983, Credible commitment: Using hostages to support exchange, American Economic Review 73, 519–540.
Williamson, O.E., 1984, Corporate governance, The Yale Law Journal 93, 1197.

mir vol. 47, 2007/2, pp. 175–192

mir
Management International Review

Mie Augier/David J. Teece

Dynamic Capabilities and Multinational Enterprise: Penrosean Insights and Omissions

Abstract and Key Results

- Penrose's legacy is a curious one. Much cited, but little read, her work is recognized as one of the main intellectual foundations for modern resource based theories of business strategy and theories of organizational routines and capabilities.
- However, Penrose did not aim to contribute to the field of strategy; her goal was to advance understanding of the nature of the firm and its growth. Nevertheless, there are important insights in Penrose's work that have implications for international business and for strategy.
- We discuss some of the implications of Penrose's work as well as its limitations. We also briefly discuss the usefulness of adopting a "Penrosean" capability perspective in multinational enterprise (MNE) strategy analysis.
- The dynamic capabilities framework puts entrepreneurial management into the theory of multinational enterprise, a task Penrose left untouched.

Key Words

Penrose, Multinational Enterprise, Strategy Analysis, Strategic Management, Entrepreneurship, Dynamic Capabilities

Authors

Mie Augier, Post-doctoral Fellow, Graduate School of Business, Stanford University, Stanford, California, USA.
David J. Teece, Mitsubishi Bank Professor of International Business and Finance, Institute of Management, Innovation and Organization, Haas School of Business, University of California, Berkeley, Berkeley, California, USA.

Manuscript received December 2004, revised June 2005, final version received January 2006.

Introduction

Edith Penrose's many and varied contributions to business studies deserve recognition. In her later years, she focused on the oil industry and on multinational enterprises (MNEs).[1] In this paper, we note some of her earlier contributions which helped initiate important streams of research, including the resource based theory of the firm. Her influence has also extended to new streams of research on dynamic capabilities and entrepreneurship. We discuss some implications for MNEs of the dynamic capabilities framework.

In her most important scholarly journey, Edith Penrose set out to develop a theory of the growth of the firm. Indeed, this was the title of her now well-known 1959 treatise.[2] Along the way she made several other astute observations about firms that turned out to be provocative to scholars interested in the theory of the firm and business strategy. It is these observations – particularly the notion that the firm is best thought of as a bundle of resources – which now constitute her better-known legacy.

The Resource Based Theory of the Firm

Penrose defined the internal resources of the firm as "the productive services available to a firm from its own resources, particularly the productive services available from management with experience within the firm" (p. 5). She presents the firm as an "autonomous administrative planning unit, the activities of which are interrelated and are coordinated" by management (pp. 15 et seq.). "A firm is more than an administrative unit; it is a collection of productive resources the disposal of which between uses and over time is determined by administrative decision – the physical resources of the firm consist of tangible things – there are also human resources available in a firm – strictly speaking, it is never resources themselves that are the 'inputs' in the productive process, but only the services that they render" (pp. 24 et seq.).

Put succinctly, Edith Penrose saw the firm as a "pool of resources the utilization of which is organized in an administrative framework. In a sense, the final products being produced by a firm at any given time merely represent one of several ways in which the firm could be using its resources" (pp. 149 et seq.).

As with the dynamic capabilities approach (which we shall discuss later), Penrose was enlightened enough to see a role in economic theory not only for managers but for entrepreneurs. "A theory of the growth of firms is essentially an examination of the changing productive opportunities of firms..." (pp. 31 et seq.). Penrose furthermore saw the business environment as an «image» in the entrepreneur's mind. This is an important insight about entrepreneurship as well as leadership (and the importance of having an entrepreneurial element in leadership). Innovation is very much about the ability of the entrepreneur to look at markets, technologies and business models and to interpret them "differently". Being able to see market and technological opportunities through different lenses (and in new ways) is an important entrepreneurial capability. It enables one to see opportunities that others might miss.

Penrose also recognized that as managers embrace growth, they are forced to decentralize, thereby shifting responsibility down the hierarchy. "New men are brought in and the existing personnel of the firm all gain further experience" (p. 52).[3] Critically, "many of the productive services created through an increase in knowledge that occurs as a result of experience gained in the operation of the firm as time passes will remain unused if the firm fails to expand" (p. 54). These unused resources aren't manifested in the form of idleness, but "in the concealed form of unused abilities" (p. 54). Penrose therefore saw the capacities of management – not exhaustion of technologically based economies of scale – as setting the limit to which a firm could grow. In her view, there was always a limit to the amount of expansion any firm, no matter how large, could undertake in a given period.[4]

It was the unused capacities of management, coupled with the tangibility of certain resources, which also enabled diversification in the Penrosian firm. Industrial R&D could assist by drawing firms into entirely new areas, particularly if the firm focused on more generic R&D activities. Sales and marketing relationships could also be leveraged to support the roll out of new products (pp. 116 et seq).

Edith Penrose's ideas influenced the work of Teece (1980, 1982) on diversification. In particular, Teece (1982) built on Penrose's observation that "[o]f all outstanding characteristics of business firms, perhaps the most inadequately treated in economic analysis is the diversification of their activities" (Penrose 1959, p. 104) in outlining a theory of the multi-product firm. This in turn alerted the strategy field to her work on resources, impacting Wernerfelt (1984) and others. But it wasn't so much her claim that managers learn and develop unused capacities that has received the most attention in recent years.[5] Rather, it was her representation of the firm as a pool of resources that has caught the imagination of scholars in the field of business strategy.

However, what Penrose precisely meant by resources remains rather vague.[6] Moreover, the Penrosian view that growth is fueled primarily by underutilized managerial capabilities can be challenged.[7] In particular, enterprise growth can be attributed to market and technological factors as well as to the strong financial rewards that both managers and shareholders receive as the business enterprise grows. Growth also flows from investment in R&D, as pointed out by several business historians and economists.[8]

From the perspective of modern (strategic) management, a missing dimension in Penrose is an understanding of the basis for competitive advantage. Penrose implicitly adopts a profit-seeking framework; but other than a very general discussion of the competitive strength of small and large firms, she does not address the question of how firms develop competitive advantage. While she does recognize the importance of managerial skills, she underplays the role of intangible assets, though they are mentioned.[9] In this sense, she is not "modern"; but she was ahead of her time in many ways, not least of which is that she did recognize the importance of the entrepreneurial activities of management. However, this was only mentioned in passing, and the importance of managerial action in sensing and seizing emerging opportunities and managing threats.

The importance of knowledge assets is also underplayed. This ought not be surprising since the world Penrose was observing was one in which there were still significant barriers to trade and investment, and in such environments know-how is less critical as a factor in determining competitive advantage (Teece 2000, Chapter 1). Outsourcing and off shoring debates were not center stage in the early post war economy which was her laboratory.

Nevertheless, the Penrosian conceptualization of the firm remains relevant. Her insights remain good starting points for developing a theory of the firm, and for understanding the role of the manager. Her perspective is compatible with the recent emphasis on the importance of routines and processes. Routines and processes can be thought of as providing underutilized capacity that management can leverage for growth.

Penrose and the Theory of Dynamic Capabilities

As noted, and with the benefit of hindsight, Penrose appears to have underplayed growth driven by the entrepreneurial elements of management. She seems to recognize that know-how can be used to convert physical assets to different uses.[10] The firm, she said, was "both an administrative organization and a collection of productive resources, both human and material" (p. 320). The services rendered by these resources are the primary inputs into a firm's production processes and are firm specific in the sense that they are a function of the knowledge and experience that the firm has acquired over time. This is in essence a recognition of the path-dependent nature of organizational processes and routines and their roles in carrying knowledge (later emphasized by Cyert and March (1963) and Nelson and Winter (1982).[11]

When services that are currently going unused are applied to new lines of business, these services can also function as a growth engine for the firm through diversification (Teece 1980, 1982). Learning likewise enables the organization to use its resources more efficiently. As a result, even firms that have weak balance sheets may nevertheless be able to grow as managerial capacity is freed up for new uses as a result of managerial and organizational learning.[12] Penrose appears to be articulating a weak form of what is now referred to as the dynamic capabilities approach.

The dynamic capabilities approach seeks to provide a coherent (and evolutionary) framework for how firms develop competitive advantage, and maintain it over time. In essence, dynamic capabilities are about identifying the foundations that undergird long run enterprise growth and prosperity. First outlined in working papers (Teece/Pisano/Shuen 1990), and then published in Teece and Pisano (1994) and in Teece, Pisano and Shuen (1997),[13] the dynamic capabilities approach builds upon the theoretical foundations provided by Schumpeter (1934), Williamson (1975, 1985), Cyert and March (1963), Rumelt (1984), Nelson and Winter (1982), Teece (1982) and Teece and Pisano (1994). As discussed above, it is consistent with certain elements of Penrose's framework too. If one can explain the foundations of long run profitability, one is quite some distance down the road to a theory of the growth of the enterprise. This was of course Penrose's ambition.

Dynamic capabilities refer to the (inimitable) capacity firms have to shape, reshape, configure and reconfigure the firm's asset base so as to respond to changing technologies and markets. Dynamic capabilities relate to the firm's ability to proactively adapt in order to generate and exploit internal and external *firm specific competences,* and to address the firm's *changing environment* (Teece/Pisano/Shuen 1997). As Collis (1994) and Winter (2003) note, one element of dynamic capabilities is that they govern the rate of change of ordinary capabilities.[14] If a firm possesses resources/competences but lacks dynamic capabilities, it has a chance to make a competitive return for a short period, but superior returns cannot be sustained. It may earn Ricardian (quasi) rents, but such quasi rents will be competed away, often rather quickly. It cannot earn Schumpeterian rents because it hasn't built the capacity to be continually innovative. Nor is it likely to be able to earn monopoly (Porterian) rents since these require market power coupled with exclusive behavior or strategic manipulation (Teece/Pisano/Shuen 1997). Dynamic capabilities thus not only include an organization's (non-imitable) ability to sense changing customer needs, technological opportunities, and competitive developments; but also its ability to adapt to – and possibly even to shape – the business environment in a timely and efficient manner. A significant element of intentionality is involved.

The development and astute management of intangible assets/intellectual capital is now central to sustained enterprise competitiveness, requiring new conceptual frameworks for business and economic analysis. As former U.S. Federal Reserve Chairman Alan Greenspan remarked, "we must begin the important work of developing a framework capable of analyzing the growth of an economy increasingly dominated by conceptual products".[15] Dynamic capability theory is a framework that is well equipped to meet this challenge (Teece 2006b).

Penrose's framework is consistent with elements of the dynamic capabilities framework. Her emphasis on the fungible nature of resources obviously provides scope for the notion that a firm's competencies can be reshaped. But as noted, her framework was bereft of considerations of competitive advantage.[16] The whole inimitability story is missing.[17] Nor did she emphasize the role of the changing environment and the constant need to improve and renew capabilities. She saw learning as an opportunity, not a necessity. She also underplayed the resource allocation role of management. She recognized the importance of entrepreneurship but did not develop this concept much nor did she show how entrepreneurship could be important to the erection of new markets.

Other Growth Issues and the Penrose Effect

We have emphasized Penrose's contribution to the resource-based theory of the firm. Some of her ideas are consistent with the dynamic capabilities framework; yet until two decades ago when strategy scholars picked up on this work (Teece 1982), Penrose's emphasis on fungible resources had not received much attention in either the economics or the strategic management literature. Rather, it was her work on constraints on firm level growth and on the role of learning that received attention. While she recognized how the fungible nature of a firm's resources could create the foundation for lateral enterprise expansion, it was her emphasis on the administrative and managerial constraints on growth that captured the attention of scholars.

Penrose argued that the human resources required for firm growth and the management of change are firm specific. As a corollary, at any moment in time these resources are constrained by their internal availability. Put differently, managerial capacity cannot be expanded indefinitely and at will. Rather, expansion requires the recruitment and development of additional high-level human resources.[18] Accordingly, the level of current efficiency will, beyond a point, diminish with the rate of change in size.

The above constraints on firm growth became known as the "Penrose effect". Both microeconomic and macroeconomic scholars recognized the Penrose effect in the 1960s. These scholars incorporated Penrosean thinking into their work (e.g. Marris 1964, Uzawa 1969). However, as noted above, we think the more enduring legacy will be Penrose's conceptualization of the firm as a bundle of (quasi fungible) resources.

Interestingly, the Economic Journal (1961) predicted that the "Theory of the Growth of the Firm" would be an influential book; however, that influence has been far greater in the field of strategy than in the field of economics. Economists in the main are resistant to her teachings, as they imply the total inadequacy of the neoclassical theory of the firm.[19]

Entrepreneurship, Enterprise Design, and the Role of Markets

As implicitly recognized by Penrose, firms need to be viewed as human organizations, not computer controlled machines. As such, firms must confront challenges in the realm of organizational design including imperfect incentive alignment, imperfect governance, and bureaucratic decision-making. Organizations facilitate decisions because they constrain the set of alternatives as well as the relevant parameters to be considered. Organizations can be rendered more effective and efficient by improving the ways in which those limits are defined and imposed (Simon 1947, March/Simon 1958).

While Penrose may have recognized the human element in organizations, she did not really explore issues of organizational *design*. As noted by Herbert Simon, "[d]esign calls for initiative, focus of attention on major problems, search for alternatives. One cannot choose the best, one cannot even satisfice, until one has alternatives to choose from." Nowhere is this clearer than in the *entrepreneurial* activities of organizations. As Simon has observed: "Especially in the case of new or expanding firms, the entrepreneur does not face an abstract capital market. He or she exerts much effort to induce potential investors to share the company's views (often optimistic) about its prospects. This executive is much closer to Schumpeter's entrepreneur [and to the Penrosian manager] than to the entrepreneur of current neoclassical theory. Whether the firm expands or contracts is determined not just by how its customers respond to it, but by how insightful, sanguine and energetic its owners and managers are about its opportunities – by how much they possess of the "animal spirits" that Keynes was obliged to introduce into his account of the trade cycle (Nelson/Winter 1982)." (1991, pp. 35 et seq.) These factors go beyond the managerial elements highlighted by Penrose. Arguably, they are more important.

One example of the importance of design is in the development of the "architecture" of a business firm. This element of design is embedded in part in management's choice of (or creation of) a business model. A business model defines the manner in which a business enterprise delivers value to customers, entices customers to pay for value, and converts those payments to profit. It reflects the firm-specific assumptions about what customers want and how an enterprise can be profitable as a result of the value delivered. The business model determines: (1) how the revenue and cost structure of business is to be "designed" and then possibly "redesigned" to meet customer needs; (2) the ways in which the resources are to be assembled and the relevant market segments can be identified; (3) the mechanisms through which value can be created and captured. The purpose of a business model is to "articulate" the value proposition, identify targeted market segments, define the structure of the value chain, and estimate the cost structure and profit potential (Chesbrough/Rosenbloom 2002, pp. 533 et seq.). In short, a business model is a plan for the financial and organizational "architecture" of a business that makes valid assumptions about costs, scale, and customer and competitor behavior. It outlines the contours of the solution required to win in the market place. Getting the business model right is critical to the success of a new business; adjusting and/or improving the model is likely to be critical for continued success. However, the importance of "business models" has been largely neglected in the management and economics literature, at least until recently. A firm's capacity to create, adjust, hone and replace business models is a critical building block of this firm's dynamic capabilities.

Design issues are also important when considering the *changing nature and dynamics* of international business. In recent decades, increased globalization, and in particular outsourcing and off shoring, appear to have gained momentum. However, while globalization has expanded, it is by no means "complete". Precisely because cross-border integration is incomplete (i.e., the world is characterized by semi-globalization) the study of international business and multinational enterprise remains an important scholarly activity.[20] Otherwise, mainstream strategy and management content would suffice for international business too. Because of incomplete integration and differences in business environments, locational factors and institutional differences must be taken into account. Such differences do not merely indicate the presence of barriers to the internationalization of business; they can also be beneficial to MNEs.

In recent decades, the MNE has been shaped by three key developments critical to its nature and scope: (i) the simultaneous increase in both the outsourcing and the off-shoring of production, (ii) the emergence of a distributed and open innovation model, i.e. not only production and manufacturing are being outsourced, but so is innovation, and, (iii) the development of low cost information and computer technology, which enables small firms to perform transactions, and adopt business models, previously only available to large enterprises. For instance, internet-based companies such as Amazon, eBay, Google, and Yahoo make it possible for small businesses to reach global markets that were previously inaccessible, except at considerable cost. This has led to the emergence of what might be thought of as "mini"-multinationals, sometimes employing only a handful of workers, and using internet-based technology to anchor the coordination of their global activities. In short, information and computer technology has enabled efficient global operations for very small as well as small, medium, and large enterprises. Small enterprises in particular may be launched from multiple jurisdictions – rendering the home/host country dichotomy irrelevant from the time of organizational founding. Also, these mini-multinationals are often founded by individuals collaborating across boundaries, and they exhibit MNE characteristics from their birth. Designing and orchestrating the business model and organizational structure of such firms has become increasingly complex. In the realm of the external environment, markets for such expanding firms must be "seized" and, sometimes, created.

Although Penrose did recognize the importance of creating markets as a result of entrepreneurship, she did not address the *simultaneous* role of entrepreneurs in creating markets and designing organizations.[21] An essential characteristic of organizations/firms is that they embody knowledge, which can't be easily bought and sold. Sometimes, the only way to capitalize on knowledge is to start a firm and build the necessary complementary assets (Teece 1986).[22] Profit flows from innovation, buttressed by the development of complementary technologies, and the astute deployment of complementary assets.

Penrose's work differs from Coase's (1937), in terms of the rationale for the firm's existence and expansion. She does not assume that "in the beginning there were markets". Her perspective is more in keeping with Simon's (1991) perspective that "in the beginning there were firms"; entrepreneurs create new markets by starting entrepreneurial organizations because the relevant external capabilities simply aren't there (cf. Langlois 1992).[23]

Penrose emphasized entrepreneurial imagination and the non-market nature of entrepreneurial knowledge. She noted that it is: "evident that such management [entrepreneurship] cannot be hired in the market place" (p. 45). A few decades earlier Frank Knight (1921) perceptively linked the existence of firms to entrepreneurs seizing opportunities for profit in the face of uncertainty: "It is ... true uncertainty which ... gives the characteristic form of 'enterprise' to economic organization as a whole and accounts for the peculiar income of the entrepreneur" (1921, p. 232).[24]

Her vision of entrepreneurship is very close to that of Frank Knight. She emphasized entrepreneurship as explicitly an organizational phenomenon: "The productive activities of ... a firm are governed by what we shall call its 'productive opportunity', which comprises all of the productive possibilities that its 'entrepreneurs' can see and take advantage of" (p. 31). Entrepreneurs have "intimate knowledge of the resources, structure, history, operations and personnel of the firm" (p. 54).

There are other ideas in strategic management that were not directly anticipated by Penrose, perhaps in part because she did not self-consciously endeavor to provide normative frameworks for managers. Hence, her neglect of certain issues now considered important to the field of management today should not be construed as a criticism, but merely as an observation. One such example is the idea that markets need to be developed. It is common in economics to assume that markets exist. As Arrow (1974) observed: "Although we are not usually explicit about it, we really postulate that when a market could be created, it would be." If it is not, this reflects market failure, and such failure can in turn be attributed to "transaction costs" or "adverse selection". The absence of certain insurance markets is a typical example. As a general rule, economics suggests that markets fail because inputs or outputs are not priced properly. For example, gasoline that pollutes is consumed "too much" because the costs of using it are not fully internalized. Arrow (1956) and Arrow and Debreu (1959) do discuss the absence of fully developed contingent claims markets, but in the main such lacunae are explained by the absence of demand, or just simply transaction costs.

Moreover, in commercializing new technologies, pioneering entrepreneurs often find that formal market research and expert forecasts, however sophisticated from a methodological perspective, fail to predict which new markets will come into existence, and where and when these markets will actually materialize. Christensen (1997), Mintzberg (1994) and others have documented a wide variety of cases that illustrate this unpredictability in business. Human history also attests to this unpredictability in other areas – such as Columbus' discovery of the New World or the fall of the Berlin Wall.

There is little in economics to suggest that markets can be shaped by the purposeful decisions of managers, i.e. by firms. Penrose chose not to develop that point either. For her theory of the growth of the firm, markets were not specifically treated. However, firm behavior shapes markets just as markets shape firm behavior and firm growth. Consistent with this view, Herbert Simon argued that perhaps we should not assume an explanation is needed of why *firms* actually exist. Simon uses the illustration of a visitor from Mars approaching earth observing economic exchanges, with organizations appearing as green areas, and market transactions as red lines in between. What does the visitor see? Organizations, green areas, would be dominant. We live in an "organization economy" rather than a "market economy", and organizations are more ubiquitous than market transactions; even more so if we go back in history (Simon 1991). Simon suggests that the more natural question to ask than the Coasian one, is "why do markets exist"? Instead of focusing on theories to explain the existence of firms, Simon raises questions such as, why do particular organizational forms (such as professional services firms) exist, and how should they be structured? How do these organizational forms relate to broader processes of (technological, cultural, etc.) change in the modern economy? What motivates people in real organizations (authority, rewards, loyalty, identification, coordination)? These are issues to be addressed by strategic management in the future; not all of these are Penrosian themes sensu stricto.

Dynamic Capabilities, Resources and Competitive Advantage: Implications for MNEs

While Penrose may not have fully developed the capability concept, the subsequent development of the (dynamic) capabilities approach can be usefully applied to MNEs. Somewhat under-researched in mainstream MNE theory (at least as far as internalization theory is concerned), has been consideration of the importance and the particulars of the firm's managerial and organizational capabilities, although this is now being addressed.[25] To the extent that notions of organizational capability have been around for decades, and have received much attention recently, more efforts to embed the capability concept into MNE theory would appear useful so as to align more closely academic research on the MNE and strategic management theory.

As discussed above, Edith Penrose had provided elements of a resource-based/capabilities perspective. She viewed the firm as an administrative organization, and as a pool of production resources: "At all times there exist, within every firm, pools of unused productive resources and these together with the changing knowledge of management, create a productive opportunity which is unique to each firm. Unused productive services are, for the enterprising firm, at the same time a challenge to innovate, an incentive to expand, and a source of competitive advantage". (Penrose 1960, p. 2). As Pitelis (2000) notes, unused resources are critical to Penrose's theory of internal or "organic"/endogenous enterprise expansion.

Penrose certainly did not overplay, from a theoretical perspective, the international aspects of large corporations, believing that the differences do not, in fact, require theoretical distinction (1987, p. 56). However, she did note that: "the managerial, technological, or financial contribution from the parent may be considerable and generally make new real resources available to the local economy", (1968, p. 43).

The general framework advanced by dynamic capability theory sees difficulty-to-imitate and globally exercised dynamic capabilities (and resources) as foundational to the competitive advantage of MNEs.[26] The greater the diversity and rate of change in business environments, the more critical dynamic capabilities become for the MNE's financial performance.

Some observers have identified a modality of competition, referred to as hypercompetition. It is a modality "characterized by intense and rapid competitive moves, in which competitors must move quickly to build [new] advantages and erode the advantages of their rivals" (D'Aveni/Gunther 1994, pp. 217 et seq.). Hypercompetition appears to be the result of rapid innovation, globalization, and deregulation. Dynamic capabilities are likely to be essential to the survival of MNE in industries and environments characterized as hypercompetitive.

As noted above, it is necessary that the MNE build capabilities that are "sustainable" i.e. inimitable. Inimitability is more likely to occur in the presence of "isolating mechanisms" and "tight appropriability regimes" (Rumelt 1987; Teece 1986, 2000).[27] When the appropriability regime is "tight", differential performance can be more readily sustained, at least for some length of time.[28]

The dynamic capabilities perspective on the MNE addresses more than simply the need for rapid innovation, adaptation, and flexibility. It also identifies the importance of proactive entrepreneurial behavior shaping the MNE's footprint. In the presence of significant gaps between the cost structures and growth rates of national economies, the MNE's ability to respond to – and shape – the changing kaleidoscope of opportunities at home and abroad is critical to success. Outsourcing and off-shoring activities to foreign subsidiaries and alliance partners, involves establishing quality control and product/service evaluation protocols on a global basis.

Indeed, dynamic capabilities are resident in a firm's processes and routines as well as within the firm's top management team. Maintaining dynamic capabilities within the MNE requires continuous entrepreneurial activity on a global scale. Entrepreneurial activity is different from – but related to – managerial activity. It is about understanding opportunities, getting things started, and finding new and better ways of putting things together. It is about coordinating on a global basis the assembly of disparate and usually co-specialized resources, getting "approvals" for non-routine activities, sensing business opportunities, and finding ways to deploy capabilities globally as well as locally. We have come to associate the entrepreneur with the individual who starts a new business providing a new or improved product or service. Such action is clearly entrepreneurial; but the entrepreneurial function required in the MNE context should not be thought of as confined to new enterprise startup activities.

The replication of capabilities involves transferring or redeploying competences (technological or organizational) from one concrete economic setting to another. Since productive knowledge is usually embodied, the transfer of skill cannot be accomplished by simply transmitting information. Only in those instances where all relevant knowledge is fully codified and understood can replication be collapsed into a simple problem of information transfer. Too often, the contextual dependence of original performance in the home market is poorly appreciated, so unless the MNE has already replicated its systems of productive knowledge in other markets, the act of replication is likely to be difficult (Teece 1976). Indeed, replication and transfer are often impossible absent the transfer of people, though this can be minimized if investments are made to convert tacit knowledge to codified knowledge. Often, however, this is simply not possible.

In short, competences and capabilities, and the routines upon which they rest, are usually rather difficult to replicate. Even understanding what all the relevant routines are that support a particular competence may not be transparent. Indeed, Lippman and Rumelt (1982) have argued that some sources of competitive advantage are so complex that the firm itself, let alone its competitors, does not understand them.[29] As Nelson and Winter (1982) and Teece (1981, 1982) have explained, many organizational routines are quite tacit in nature. Imitation can also be hindered by the fact that few routines are 'stand-alone'; coherence may require that a change in one set of routines in one part of the firm (e.g. production) be accompanied by changes in some other part (e.g. R&D).

Some routines and competences seem to be attributable to local or regional forces that shape firms' capabilities at early states in their lives. Porter (1990), for example, shows that differences in local product markets, local factor markets, and institutions play an important role in shaping competitive capabilities. Differences also exist within populations of firms from the same country. Various studies of the automobile industry, for example, show that not all Japanese automobile companies are top performers in terms of quality, productivity, or product development (see, for example, Clark/Fujimoto 1991). The role of firm-specific history has been highlighted as a critical factor explaining such firm-level (as opposed to regional or national-level) differences (Nelson/Winter 1982).[30] Replication in a different context may thus be rather difficult.[31]

At least two types of strategic value flow from replication. One is the ability to support geographic expansion, and has been emphasized here. The other is the ability to support product line expansion. To the extent that the capabilities in question are relevant to customer needs elsewhere, replication can confer value.[32] Another is that the ability to replicate also indicates that the enterprise has the foundations in place for learning and improvement. Considerable empirical evidence supports the notion that the understanding of processes, both in production and in management, is the key to process improvement. In short, an organization cannot improve that which it does not understand.

Factors that make replication difficult also make imitation difficult. Thus, when the MNE's productive knowledge is more tacit, it becomes harder for the MNE itself to replicate it, and for competitors to imitate it. When the tacit component is high, imitation may well be impossible, absent the hiring away of key individuals and the transfer of key organizational processes.

In conclusion, the concept of dynamic capabilities, when applied to the MNE, highlights organizational and managerial competences, critical to achieve superior performance. Key ingredients are difficult-to-replicate routinized processes, the basic manner in which a business is designed, as well as the decision frames, heuristics and protocols that enable MNEs to avoid poor investment choices and embrace astute ones. Once assets are within management's orbit, their effective utilization and continuous orchestration on a global basis becomes essential. Indeed, orchestration directed at achieving new combinations and new asset co-alignments is central to the dynamic capabilities framework. Preventing imitation and internal rent dissipation are key elements too.

Lying at the heart of dynamic capabilities are several fundamental management/organizational skills including: (1) learning and innovation processes; (2) business "design" competence (what business model to employ); (3) investment allocation decision heuristics; (4) asset orchestration, bargaining and transactional competence, and (5) efficient governance and incentive alignment (Teece 2006). Buttressing these is an understanding of the processes of imitation and the strategies and processes that can be used to protect intellectual property. Widely diffused managerial and organizational competence cannot be core elements of an MNE's dynamic capabilities.

Note that dynamic capabilities flow from more than just learning and technological accumulation. This is not meant to downplay the importance of technological accumulation. Technological innovation and learning remain important mechanisms by which firms build from specific (technological) capabilities. However, in a world where the global outsourcing of R&D is common (Teece/Pisano/Shan 1988, Chesbrough 2003) it becomes problematic to rely too much on in house R&D as the sole foundation of competitive advantage. Orchestrating a global portfolio of technological assets inside and outside the enterprise is now essential.

The dynamic capabilities framework relegates an MNE's administrative competence to secondary importance, unless such competence is embedded in distinct and difficult to replicate business processes. Stable administrative functions can typically be outsourced to multiple vendors. Of course, there may well be circumstances where administration is complex, novel, and difficult to imitate in which case it can be the source of competitive advantage.

The distinct skills, which constitute an MNE's dynamic capabilities cannot generally be bought or "outsourced"; they must be built, or at least assembled. Once co-specialized assets are assembled, they must be skillfully orchestrated on a global basis. Such orchestration skills require astute decision-making on a global basis and an entrepreneurial capacity built into the management team. These skills and processes are instrumental to long-run enterprise performance and cannot be outsourced without loss of competitive advantage. They lie at the core of the MNE's capabilities. MNEs possessing dynamic capabilities are able to quickly respond to – and shape – evolving technologies and marketplaces. Accordingly, such firms should exhibit superior enterprise performance over multiple product life cycles.

While Penrose did not anticipate most and certainly not all elements critical to successful international expansion, she did play an important role by being an important inspiration to dynamic capabilities. Her search for a theory of the growth of the firm is in some measure answered by the dynamic capabilities framework.

Conclusion

Within the field of strategic management, Penrose's work has often been extensively cited while also being mis-characterized. What is needed is careful scholarship, initiated by a careful reading of her work, especially the Theory of the Growth of the Firm. In this paper, we have discussed some of her insights relevant to strategic management, entrepreneurship, international business, and industrial organization. We have also indicated certain limitations to her framework and analyses. A critical reading of her writings can nevertheless provide fresh insights to economics and management.

Acknowledgements

We are grateful to Giovanni Dosi, Richard Nelson, Christos Pitelis, Alain Verbeke and two anonymous referees for comments on earlier drafts; to the Kauffmann Foundation, the Sloan Foundation and the Lester Center for Entrepreneurship & Innovation for support and to Frances Darnley and Patricia Lonergan for skillful assistance.

Endnotes

1 For an extensive coverage of Penrose's overall contribution, see Penrose and Pitelis (1999).
2 Less well known, but also elaborating the theme of the growth of the firm, is her case study of the Hercules Powder Company, published in 1960. It was originally intended for inclusion in the Theory of the Growth of the Firm but was omitted to keep down the size of the book though the case study was designed to illustrate the theory outlined in the book.
3 This has subsequently come to be known as the "Penrose Effect".
4 In her own words, describing the limits of growth as being generated by the same dynamics underlying the growth process itself: "[B]ecause the very nature of a firm as an administrative and planning organization requires that the existing responsible officials of the firm at least know and approve, even if they do not in detail control, all aspects of the plans and operations of the firm ... the capacities of the existing managerial personnel of the firm necessarily set a limit to the expansion of that firm in any given period of time, for it is self-evident that such management cannot be hired in the market place" (p. 45). Note the emphasis on the missing markets for management – we shall return to that issue later in this paper in sections 4 and 5.
5 This is discussed in more detail in section 4 below.
6 Teece, Pisano, and Shuen (1997) tried to tighten this by defining resources as firm specific assets that are difficult if not impossible to imitate. Trade secrets and certain specialized production facilities are examples. These assets are difficult to transfer because of transfer and transaction costs, amplified in the presence of tacit knowledge.
7 The flip side of this is of course that a firm's growth is limited by the capabilities of its incumbent management (the "Penrose Effect" discussed later).
8 Moreover the use of "excess resources" may involve positive costs, see Pitelis (2002).
9 At least industrial R&D is discussed along with customer relationships.
10 As Penrose writes: "For physical resources the range of services inherent in any given resource depends on the physical characteristics of the resource, and it is probably safe to assume that at any given time the known productive services inherent in a resource do not exhaust the full potential of the resource... The possibilities of using services change with changes in knowledge... there is a close connection between the type of knowledge possessed by the personnel in the firm and the services obtainable from its material resources" (1959, p. 76).
11 The links between Penrose and Cyert & March are discussed in Pitelis (2006).
12 Teece's paper on the multiproduct firm (Teece 1982) was the first to apply Penrose's ideas to strategic management issues. This paper focused on developing further Penrose's idea that human capital in firms is usually not entirely 'specialized' and can therefore be (re)deployed to allow the firm's diversification into new products and services. He also extended the Penrosian notion that firms' possess excess resources which can be used for diversification. Later, Wernerfelt (1984) cites Penrose for "the idea of looking at firms as a broader set of resources ... [and] the optimal growth of the firm involves a balance between exploitation of existing resources and development of new ones".
13 This explains why references to dynamic capabilities began before the publication of this paper. In the early to mid 90's, the working paper versions were quoted. See for instance Mahoney and Pandian (1992).
14 For the particulars on the specific nature of different types of dynamic capabilities, see Teece (2006b).
15 Chairman Alan Greenspan also noted recently, "over the past half century, the increase in the value of raw materials has accounted for only a fraction of the overall growth of U.S. gross domestic product (GDP). The rest of that growth reflects the embodiment of ideas in products and services that consumers value. This shift of emphasis from physical materials to ideas as the core of value creation appears to have accelerated in recent decades" (Remarks of Alan Greenspan, Stanford Institute for Economic Policy Research 2004).
16 See also Rugman and Verbeke (2002).
17 Except perhaps for her discussion on "impregnable bases" see Pitelis (2004).

18 As an example, consider Google's expansion of online network advertising into new markets around the world. According to Google's CEO Eric Schmidt, this is limited only by the speed at which the company can hire local staff, "set up bank accounts and collect the money". "Google sees no limit to global drive", Financial Times, February 3, 2005, p. 17.

19 See also Penrose and Pitelis (1999).

20 See also Rugman and Verbeke (2004).

21 The dynamic process of market creation is illustrated in Penrose's study of the Hercules Powder Company where she talks about "the creation of consumer demand as a consequence of entrepreneurial desire to find a use for available productive resources" (Penrose 1960, p. 9).

22 That was essentially also what Frank Knight had in mind: "The receipt of profit in a particular case may be argued to be the result of superior judgment. But it is judgment of judgment, especially one's own judgment, and in an individual case there is no way of telling good judgment from good luck, and a succession of cases sufficient to evaluate the judgment or determine its probable value transforms the profit into a wage. ... If ... capacities were known, the compensation for exercising them can be competitively imputed and is a wage; only, in so far as they are unknown or known only to the possessor himself, do they give rise to a profit" (1921, p. 311). For a discussion of Knight's theory of the firm, see Langlois and Cosgel (1993).

23 She did also recognize the role of intentionality, an important part of entrepreneurship, as reflected in her early contributions to the debate in the American Economic Review about biological analogies.

24 His full argument is as follows: "With uncertainty entirely absent, every individual being in possession of perfect knowledge, there would be no occasion for anything of the nature of responsible management or control of productive activities. ... its [business firm's] existence in the world is a direct result of the fact of uncertainty" (p. 271).

25 See, for example, the various contributions by Rugman and Verbeke (2001, 2003 and 2005). In addition, others have emphasized management expertise in the theory of the MNE e.g. Hood and Young (1979, p. 56) in discussing firm-specific factors, reference management expertise. Indeed, they state clearly (p. 92) that "large corporations do possess, and lay much store by, acquired managerial experience through which profit opportunities are diagnosed. Such experience is an important dimension of an MNE's comparative advantage". The framework developed here endeavors to specify what particular management expertise is likely to be critical.

26 For applications of Penrose's ideas to the MNE, see Pitelis (2000, 2004), Dunning (2003), Rugman and Verbeke (2002, 2004) and various contributions in this special issue.

27 In addition to the importance of intellectual property rights protection, the tacit nature of know-how, and the inherent difficulty of technology transfer, another factor is the importance of the unique coalignment of specific assets. Specific assets may not simply be ubiquitously available.

28 Competitive advantages are continuously eroded by actions of other players that lead again to higher levels of competition and the need to react faster. In the end, these dynamic interactions between firm learning and adaptation, on the one hand, and higher levels of competition and selection, on the other hand, can cancel each other out. This is often dubbed an 'arms race' or 'the Red Queen effect' (Kaufman 1995) after the comment to Alice in Wonderland: "it takes all the running you can do to keep in the same place" (Carroll 1946). Companies adapt faster and faster, but as a consequence of the resulting increase in competition, they do not make any progress. When isolating mechanisms are operative and appropriability regimes are tight, Red Queen effects can be overcome.

29 If so, the firm's advantage is likely to fade, as luck does run out.

30 See also Bartlett and Ghoshal (1989), Rugman and Verbeke (2005).

31 See for example, Rugman and Verbeke (2004).

32 Needless to say, there are many examples of firms replicating their capabilities inappropriately by applying extant routines to circumstances where they may not be applicable e.g. Nestle's transfer of developed-country marketing methods for infant formula to the Third World (Hartley 1989). A key strategic need is for firms to screen capabilities for their applicability to new environments.

References

Alchian, A., Uncertainty, Evolution, and Economic Theory, *Journal of Political Economy*, 58, 3, 1950, pp. 211–222.

Alchian, A., Biological Analogies in the Theory of the Firm: Comment, *American Economic Review*, 43, 4, 1953, pp. 600–603.

Bartlett, C./Ghoshal, S., *Managing Across Borders: The Transnational Solution*, Boston, MA: Harvard Business School Press 1989.

Chesbrough, H./Rosenbloom, R. S., The Role of the Business Model in Capturing Value from Innovation: Evidence from Xerox Corporation's Technology, *Industrial and Corporate Change*, 11, 3, 2002, pp. 529–555.

Chesbrough, H., *Open Innovation: The New Imperative for Creating and Profiting from Technology*, Boston: Harvard Business School Press 2003.

Christensen, C. M., *The Innovator's Dilemma: When New Technologies Cause Great Firms to Fail*, Boston, MA: Harvard Business School Press 1997.

Collis, D. J., Research Note: How Valuable are Organisational Capabilities?, *Strategic Management Journal*, 15, Winter Special Issue, 1994, pp. 143–152.

Cyert, R./March, J.G., *A Behavioral Theory of the Firm*, Englewood Cliffs: Prentice Hall 1963.

D'Aveni, R. A./Gunther, R., *Hypercompetition: Managing the Dynamics of Strategic Maneuvering*, New York: The Free Press 1994.

Dunning, J. H., The Contribution of Edith Penrose to International Business Scholarship, *Management International Review*, 43, 1, 2003, pp. 3–19.

Hood, N./Young, S., *The Economics of Multinational Enterprise*, London: Longman 1979.

Jacobides, M., The Architecture of Organizational Capabilities, *Industrial and Corporate Change*, forthcoming.

Knight, F., *Risk, Uncertainty and Profit*, Boston, MA: Houghton Mifflin 1921.

Langlois, R., Transactions-cost Economics in Real Time, *Industrial and Corporate Change*, 1, 1, 1992, pp. 99–127.

Langlois, R./Cosgel, M., Frank Knight on Risk, Uncertainty, and the Firm, *Economic Inquiry*, 31, 3, 1993, pp. 456–465.

Loasby, B., *Choice, Complexity and Ignorance*, Oxford: Oxford University Press 1976.

Loasby, B., *The Mind and Method of the Economist*, Cheltenham: Edward Elgar 1989.

March, J.G./Simon, H.A., *Organizations*, New York: Wiley 1958.

Marshall, A., *Principles of Economics*, London: McMillan 1925.

Marris, R.L., *The Economic Theory of Managerial Capitalism*. New York: Free Press of Glencoe 1964.

Mintzberg, H., *The Rise and Fall of Strategic Planning*, New York: Free Press 1994.

Nelson, R./Winter, S.G., *An Evolutionary Theory of Economic Change*, Cambridge: Belknap Press 1982.

Penrose, E., Biological Analogies in the Theory of the Firm, *American Economic Review*, 42, 6, 1952, pp. 804–819.

Penrose, E., Biological Analogies in the Theory of the Firm: Rejoinder, *American Economic Review*, 43, 4, 1953, pp. 603–609.

Penrose, E. T., *The Theory of the Growth of the Firm*, 3rd edition, Oxford: Oxford University Press 1959/1995.

Penrose, E., The Growth of the Firm: A Case Study: The Hercules Powder Company, *Business History Review* 34, 1, 1960, pp. 1–23.

Penrose, P./Pitelis, C. N., Edith Elura Tilton Penrose: Life, Contribution and Influence, *Contributions to Political Economy*, 18, 1, 1999, pp. 3–22.

Pitelis, C., A Theory of the (Growth of the) Transnational Firm: A Penrosian Perspective, *Contributions to Political Economy*. 19, 2000, p. 71–79.

Pitelis, C. N. (ed.), *The Growth of the Firm: The Legacy of Edith Penrose*, Oxford: Oxford University Press 2002.

Pitelis, C. N., Edith Penrose and the Resource-based View of (International) Business Strategy, *International Business Review*, 13, 4, 2004, pp. 523–532.
Pitelis, C., A Note on Cyert and March (1963) and Penrose (1959): A Case for Synergy, Organization Science Special Issue, 'A Behavioral Theory of the Firm? 40 Years and Counting', Submission 2006.
Rugman, A./Verbeke, A., A Final Word on Edith Penrose, *Journal of Management Studies*, 41, 1, 2004, pp. 205–217.
Rugman, A. M./Verbeke, A., A Note on the Transnational Solution and the Transaction Cost Theory of Multinational Strategic Management, *Journal of International Business Studies*, 23, 4, 1992, pp.761–71.
Rugman, A. M./Verbeke, A., Subsidiary-specific Advantages in Multinational Enterprises, *Strategic Management Journal*, 22, 3, 2001, pp. 237–250.
Rugman, A. M./Verbeke, A., Edith Penrose's Contribution to the Resource-based View of Strategic Management, *Strategic Management Journal*, 23, 2002, pp. 769–780.
Rugman, A. M./Verbeke, A., Extending the Theory of the Multinational Enterprise: Internalization and Strategic Management Perspectives, *Journal of International Business Studies*, 34, 2, 2003, pp. 125–137.
Rugman, A. M./Verbeke, A., A Perspective on Regional and Global Strategies of Multinational Enterprises, *Journal of International Business Studies*, 35, 1, 2004, pp. 3–18.
Rumelt, R., Towards a Strategic Theory of the Firm, in Lamb, R. B. (ed.), *Competitive Strategic Management*, Englewood Cliffs, NJ: Prentice Hall 1984.
Rumelt, R./Schendel, D./Teece, D., Introduction, in Rumelt, R./Schendel, D./Teece, D. (eds.), *Fundamental Issues in Strategy*, Boston: Harvard Business School Press 1994.
Schumpeter, J., *The Theory of Economic Development*, Cambridge: Harvard University Press 1934.
Simon, H. A., Organizations and Markets, *Journal of Economic Perspectives*, 5, 2, 1991, pp. 25–44.
Simon, H.A., Strategy and Organizational Evolution, *Strategic Management Journal*, 14, Winter Special Issue, 1993, pp.131–142.
Teece, D., The Market for Know-how and the Efficient International Transfer of Technology, *The Annals of the Academy of Political and Social Science*, 458, 1981, pp. 81–96.
Teece, D., Towards an Economic Theory of the Multiproduct Firm, *Journal of Economic Behavior and Organization*, 3, 1, 1982, pp. 39–63.
Teece, D., *Managing Intellectual Capital*, Oxford: Oxford University Press 2000.
Teece, D., Reflections on the Hymer Thesis, *International Business Review*, 15, 2, 2006, pp. 124–139.
Teece, D., *Explicating Dynamic Capabilities*, Seminar held at the Judge Business School, Cambridge May 2006.
Teece, D./Pisano, G., The Dynamic Capabilities of Firms: An Introduction, *Industrial and Corporate Change*, 3, 3, 1994, pp. 537–556.
Teece, D./Pisano, G./Shuen, A., Dynamic Capabilities and Strategic Management, *Strategic Management Journal*, 18, 7, 1997, pp. 537–533.
Uzawa, H., Time Preference and the Penrose Effect in a Two-Class Model of Economic Growth, *Journal of Public Economy*, 77, 4, 1969, pp. 628–652.
Wernerfelt, B., A Resource-based View of the Firm, *Strategic Management Journal*, 5, 2, 1984, pp. 171–180.
Williamson, O. E., *Markets and Hierarchies: Analysis and Antitrust Implications*, New York: Free Press 1975.
Williamson, O. E., *The Economic Institutions of Capitalism*, New York: Free Press 1985.
Winter, S., Understanding Dynamic Capabilities, *Strategic Management Journal*, 24, 10, 2003, pp. 991–995.